APPLYING
AutoCAD® 2000
A STEP-BY-STEP APPROACH

Terry T. Wohlers

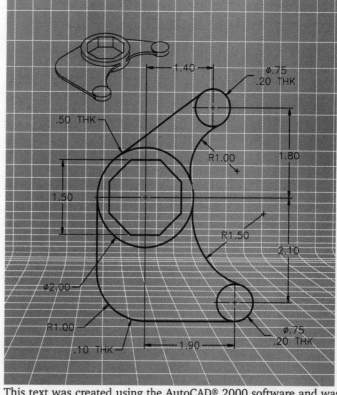

This text was created using the AutoCAD® 2000 software and was fully tested with Wi... ...ws NT 4.0.

D1065985

Glencoe
McGraw-Hill

New York, New York   Columbus, Ohio   Woodland Hills, California   Peoria, Illinois

Contributors:

Don Grout, Ed.D.
Professor, Champlain College,
Burlington Vermont

Gary J. Hordemann
Professor and Chair, Gonzaga
Engineering Department, Gonzaga
University, Spokane, Washington

Donald Sanborn
Consultant, Unique Solutions, Inc.,
Colorado Springs, Colorado

Stuart Soman, Ed.D.
Drafting and Design Instructor,
West Hempstead Public Schools,
West Hempstead, New York

Thanks also to **Cynde Hargrave, Jim
Quanci, and the members of the
AutoCAD 2000 beta team at
Autodesk, Inc.**, for answering
questions and providing assistance.

Except where otherwise credited, all CAD drawings in this book were developed with AutoCAD. Cover design was provided courtesy of Pudik Graphics, Inc.; images on the front cover were provided courtesy of Gary J. Hordemann, Gonzaga University.

Applying AutoCAD 2000: A Step-by-Step Approach is a textbook for those who wish to learn how to use the AutoCAD software. AutoCAD is a computer-aided drafting and design package produced by Autodesk, Inc. For information on how to obtain the AutoCAD software, contact Autodesk.

Applying AutoCAD 2000: A Step-by-Step Approach is not an Autodesk product and is not warranted by Autodesk. Autodesk, AutoCAD, AutoLISP, Visual LISP, 3D Studio, DXF, Kinetix, ObjectARX, and *WHIP!* are trademarks of Autodesk, Inc.

Internet listings throughout this book provide a source for extended information related to this text. We have made every effort to recommend sites that are informative and accurate. However, these sites are not under the control of Glencoe/McGraw-Hill, and, therefore, Glencoe/McGraw-Hill makes no representation concerning the content of these sites. Many sites may eventually contain "hot links" to other sites that could lead to exposure to inappropriate material. Internet sites are sometimes "under construction" and may not always be available. Sites may also move or have been discontinued completely by the time you attempt to access them.

Glencoe/McGraw-Hill

A Division of The **McGraw·Hill** Companies

Printed in the United States of America

Send all inquires to:
Glencoe/McGraw-Hill
3008 W. Willow Knolls Drive
Peoria, IL 61615-1083

ISBN 0-02-668589-2 Student Edition

ISBN 0-07-821231-6 QuickRef (Package of 15)

ISBN 0-02-668591-4 Instructor's Resource Guide

4 5 6 7 8 9 10   009   04 03 02 01

About the Author

Since 1977, Terry Wohlers has focused his education, research, and practice on design and manufacturing. In 1983, he taught one of the nation's first university credit courses on AutoCAD. Since then, he has taught many courses and given countless lectures on subjects related to computer-aided design, prototyping, and manufacturing. Presently, Terry serves as president of Wohlers Associates, Inc., an independent consulting firm he founded in 1986 after leaving his faculty and research position at Colorado State University.

He has published more than 200 books, articles, reports, and technical papers on engineering and manufacturing automation. His work has appeared in many popular magazines and journals, including *Manufacturing Engineering, Computer-Aided Engineering, PC Magazine, InfoWorld, CADENCE, CADalyst,* and *T.H.E. Journal.* Terry has been a contributing editor for *Computer Graphics World* since 1987, and his "Wohlers Talk" column appears regularly in *Prototyping Technology International.*

Terry earned a master's degree in industrial sciences and a bachelor's degree in industrial technology from Colorado State University and the University of Nebraska at Kearney, respectively.

Dedication

Applying AutoCAD 2000: A Step-by-Step Approach is dedicated to my wife, Diane, and kids, Chad and Heather, for their constant support and encouragement.

Contents in Brief

Contents in Brief — continued

Autodesk's Vision

In March of 1999, Autodesk officially unveiled AutoCAD® 2000, a landmark release of the most popular design software. AutoCAD 2000 software is designed to aid users of AutoCAD in a wide array of disciplines. The features of AutoCAD 2000 are grouped in the following five categories:

❏ **Improved Access and Usability** – Features and options are placed where you intuitively expect them to be.

❏ **Streamlined Output** – A new, high-performance hard-copy output system and user interface increase user productivity.

❏ **Heads-up Design™** – AutoCAD's new interactive interfaces for design content creation minimize time spent looking away from the design or model.

❏ **Expanded Reach** – New features connect AutoCAD to the world via the Internet.

❏ **Improved Extensibility and Customization** – AutoCAD provides a Visual LISP™ programming environment, VBA, and a complete object-oriented kernel in ObjectARX™ for quick but flexible customization.

AutoCAD 2000 software delivers 400 useful features within a powerful technology framework. Visit the Autodesk Web site at www.autodesk.com/autocad for more detailed information.

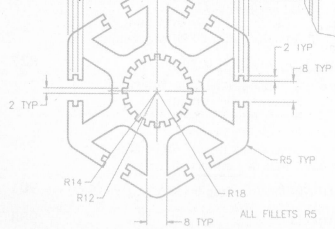

A New Century

Our Response: *Applying AutoCAD 2000*

Glencoe/McGraw-Hill has taken this opportunity to respond to your requests for more real-world applications, more problem solving, and more challenging projects. *Applying AutoCAD 2000: A Step-by-Step Approach* will help you master the AutoCAD 2000 software. This text presents complex concepts using terms that are easy to grasp and retain. You will be presented with:

- ❑ *Examples* of how AutoCAD is being applied in industry.

- ❑ *Techniques* that conform to industry standard practices.

- ❑ *Exercises* in the context of new product design and production drafting.

- ❑ *Problem solving* that extends beyond the basic use of AutoCAD.

- ❑ *Three-dimensional modeling methods* that reflect industry trends.

- ❑ *Shortcut methods* that speed your work with AutoCAD.

- ❑ *Review questions and real-world problems* that reinforce and expand your knowledge and skills.

- ❑ *Drawings and illustrations* that support the key points in each chapter.

- ❑ *Buttons in the margins* that serve as a visual aid to help you find the right toolbar and button to use.

- ❑ *Careers Using AutoCAD* pages that provide examples of how AutoCAD is used in manufacturing, architecture, civil engineering, mapping, and other disciplines.

- ❑ *Additional problems* that further challenge and motivate beginning, intermediate, and advanced users.

- ❑ *Appendices* filled with important reference information.

> *Applying AutoCAD 2000: A Step-by-Step Approach* is based on the AutoCAD 2000 software. This textbook takes you from the beginning to the advanced levels of computer-aided drafting and design.

A Practical Approach

Applying AutoCAD 2000: A Step-by-Step Approach presents each feature of the AutoCAD® 2000 software in a logical, sequential format. You will build beginning to advanced skills as you read about and apply techniques, solve problems, and practice computer-aided drafting and design.

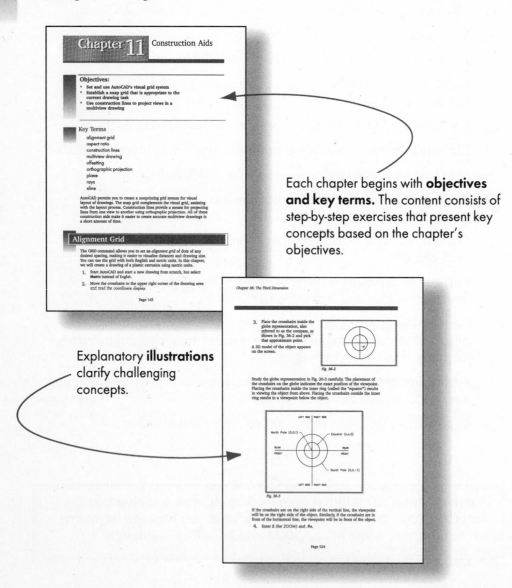

Each chapter begins with **objectives and key terms.** The content consists of step-by-step exercises that present key concepts based on the chapter's objectives.

Explanatory **illustrations** clarify challenging concepts.

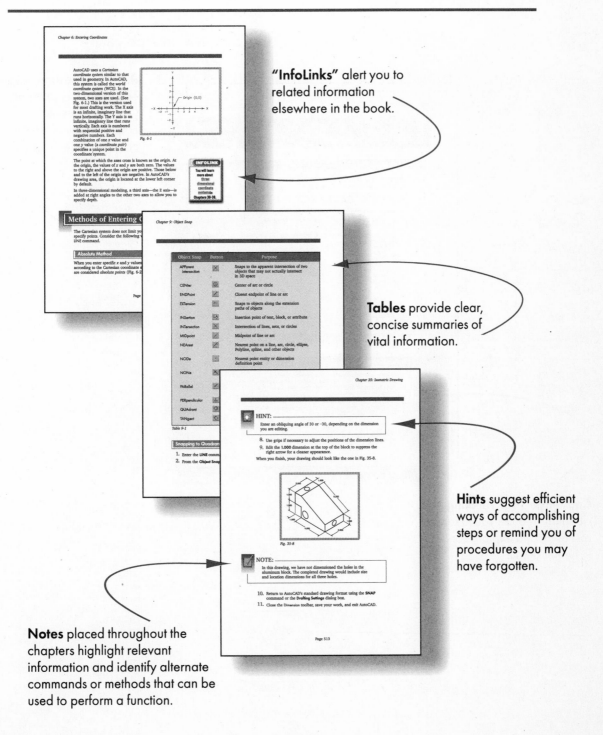

"InfoLinks" alert you to related information elsewhere in the book.

Tables provide clear, concise summaries of vital information.

Hints suggest efficient ways of accomplishing steps or remind you of procedures you may have forgotten.

Notes placed throughout the chapters highlight relevant information and identify alternate commands or methods that can be used to perform a function.

Using the Student Edition (Continued)

Chapter Review & Activities

Each chapter concludes with activities that review, reinforce, and expand learning.

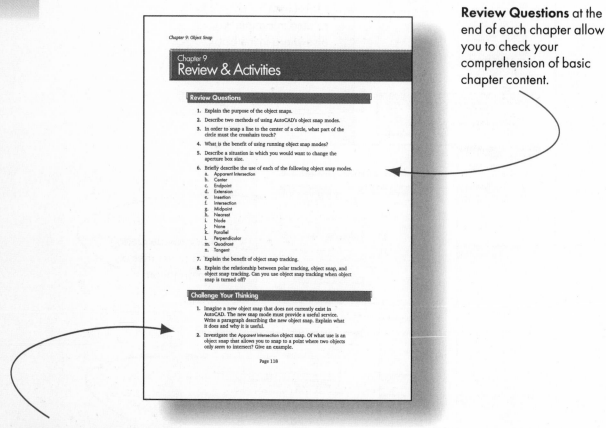

Review Questions at the end of each chapter allow you to check your comprehension of basic chapter content.

Challenge Your Thinking questions require you to reason, research, and explore concepts in further detail.

Using the Student Edition (Continued)

Applying AutoCAD Skills

Problems at the end of each chapter help you gauge your
understanding of the skills taught in the chapter.

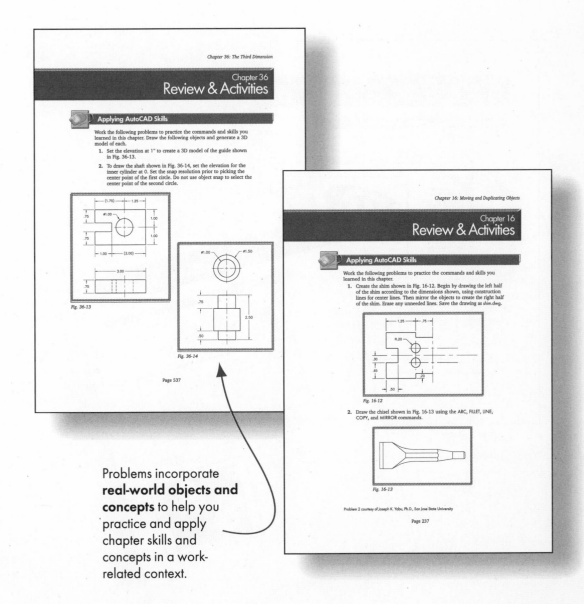

Problems incorporate
**real-world objects and
concepts** to help you
practice and apply
chapter skills and
concepts in a work-
related context.

Using the Student Edition (Continued)

Using Problem-Solving Skills

New problem-solving activities incorporate representative tasks that you might encounter in commerce or industry. These problems require you to synthesize the AutoCAD skills presented throughout the text to arrive at practical solutions.

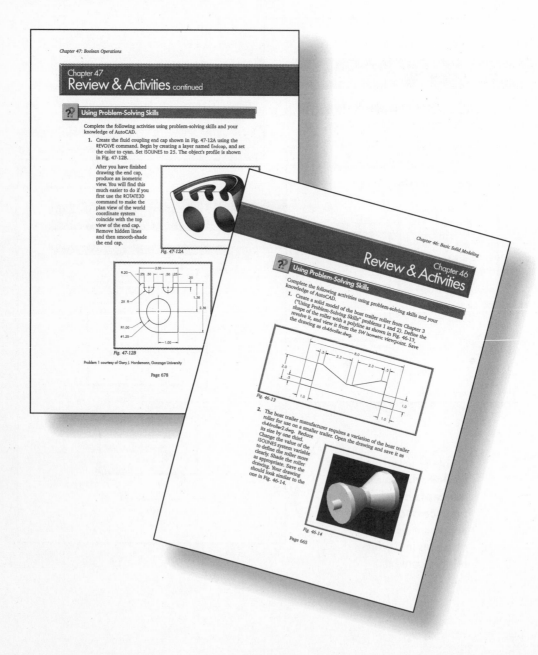

Using the Student Edition (Continued)

Career Leads

Careers Using AutoCAD pages help you explore the different types of careers open to people with AutoCAD knowledge and skills.

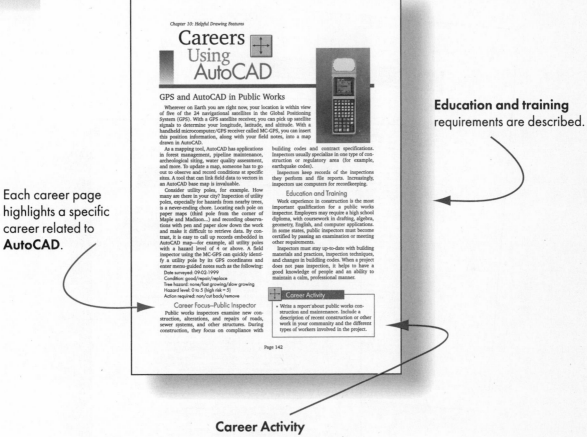

Each career page highlights a specific career related to **AutoCAD**.

Education and training requirements are described.

Career Activity encourages you to explore the career in further detail.

Using the Student Edition (Continued)

Part Projects

A real-world project is located at the end of each of the twelve parts in this book. These projects tie together key concepts and skills and help you apply the skills and techniques you have learned in a realistic manner.

Using the Student Edition (Continued)

Additional Problems

Additional Problems provide further opportunities to put your AutoCAD skills to work. Each problem is rated according to the skill level required and is accompanied by specific instructions.

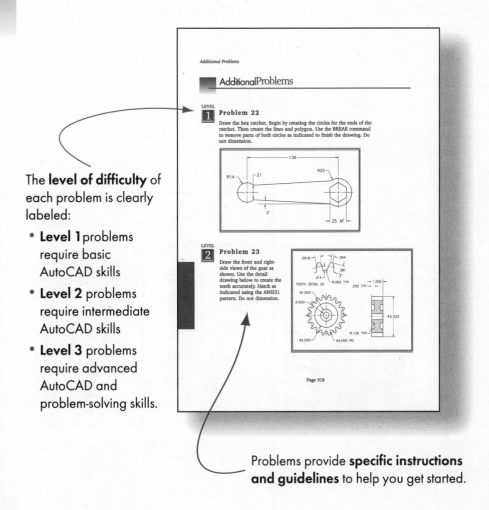

The **level of difficulty** of each problem is clearly labeled:

- **Level 1** problems require basic AutoCAD skills
- **Level 2** problems require intermediate AutoCAD skills
- **Level 3** problems require advanced AutoCAD and problem-solving skills.

Problems provide **specific instructions and guidelines** to help you get started.

Using the Student Edition (Continued)

Advanced Projects

Each project presents a problem that you might encounter on the job. Little instruction is given—you are required to plan a solution and then carry it out. This is an opportunity for you to show initiative and creativity on a long-term or group project.

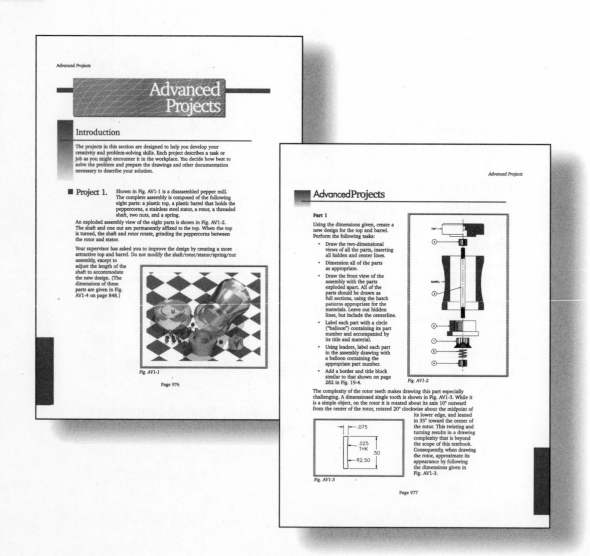

Using the Student Edition (Continued)

Appendices

The appendices serve as important reference information on various topics:

- Customizing AutoCAD by setting preferences and using individual user profiles
- Managing an AutoCAD installation
- Table of drawing area guidelines
- Using dimensioning and tolerancing symbols
- Sample of fonts that are packaged with AutoCAD
- Visual reference for hatch patterns packaged with AutoCAD
- Toolbar reference
- Complete command reference
- Complete list and explanation of AutoCAD system variables
- Glossary of terms

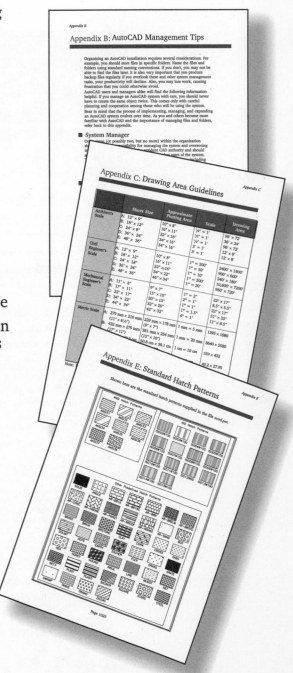

Applying AutoCAD 2000 QuickRef

QuickRef provides quick and convenient access to key information.
Your QuickRef can be found inside the back cover of this text.
QuickRefs are also available in sets of 15.

Functions of the buttons
on the status bar

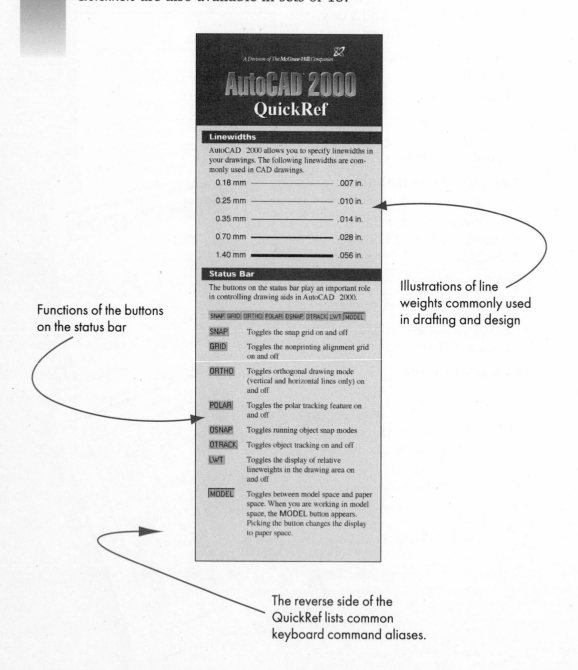

Illustrations of line
weights commonly used
in drafting and design

The reverse side of the
QuickRef lists common
keyboard command aliases.

Table of Contents

Table of Contents — continued

Part 2 – Drawing Aids and Controls

Table of Contents — continued

Part 3 – Drawing and Editing

Table of Contents — continued

Part 4 – Preparing and Printing a Drawing

Table of Contents — continued

Part 5– Dimensioning and Tolerancing

Table of Contents — continued

Part 8 – Surface Modeling and Rendering

Table of Contents — continued

Part 9 – Solid Modeling

Table of Contents — continued

Part 10 – Menus

Table of Contents — continued

Table of Contents — continued

Careers Using AutoCAD

Table of Contents — continued

Additional Problems

Table of Contents — continued

Additional Problems (Continued)

Advanced Projects

Part 1

Groundwork

Chapter 1 Tour of AutoCAD

Objectives:

- **Start AutoCAD**
- **Preview drawing files**
- **Open drawing files**
- **View details on an AutoCAD drawing**
- **Shade a complex model**
- **Exit AutoCAD**

Key Terms

buttons
Command prompt
dialog box
double-clicking
drawing area
resolution
toolbar

AutoCAD provides countless methods and tools for producing, viewing, and editing two-dimensional drawings and three-dimensional models. The software permits designers, drafters, engineers, and others to create, revise, model, and document industrial parts and assemblies for prototyping, mold-making, and manufacturing. Around the world, organizations also use AutoCAD for the design of maps, buildings, bridges, factories, and about every product imaginable, ranging from car parts and stereo equipment to snow skis and cellular phones.

Starting AutoCAD

The first step in using AutoCAD is to learn how to start the software and open a drawing file.

1. Start AutoCAD by double-clicking the **AutoCAD 2000** icon found on the Windows desktop.

HINT:

Double-clicking means to position the pointer on the item and press the left button on the pointing device twice very quickly. If the AutoCAD 2000 icon is not present, pick the Windows Start button, move the pointer to Programs, move it to AutoCAD 2000, and pick the AutoCAD 2000 option.

The AutoCAD window appears as shown in Figure 1-1. The exact configuration may vary depending on how it was left by the last person to use the software. The display hardware can cause the appearance of screen elements to vary. For example, at lower resolutions, the buttons at the top may extend across the entire window. The *resolution* is the number of pixels per inch on a display system; this determines the amount of detail that you can see on the screen.

Buttons in AutoCAD are the small areas that contain pictures, called icons, normally found along the top and left side of the AutoCAD drawing area. Pressing a button enters its associated command or function. The *drawing area* is the main portion of the window where the drawing appears.

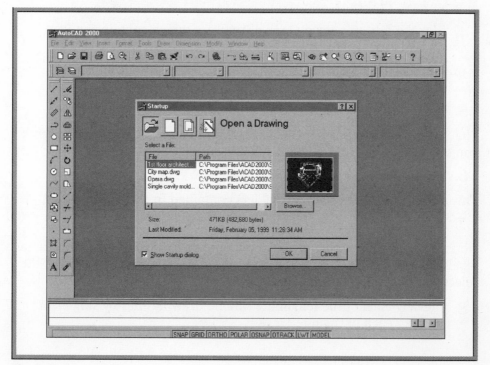

Fig. 1-1

The Startup dialog box appears in the middle of the screen. A *dialog box* provides information and permits you to make selections and enter information. In this case, the Startup dialog box enables you to open a drawing file, as well as select other options. Notice the four buttons located in the upper left area of the dialog box. The last button used will be the one currently depressed.

You will learn more about the options in the Startup dialog box in Chapter 3.

2. Position the pointer on top of each of the four buttons, allowing it to rest a couple of seconds on each one.

AutoCAD displays the name of the button when the pointer rests on it.

3. Pick each of the four buttons and notice how the contents of the dialog box change.

HINT:

Position the pointer over each button and press the left button on the pointing device.

Previewing Drawing Files

4. Pick the button named **Open a Drawing**, which contains an icon of a folder.

If drawing files were opened prior to the current session, AutoCAD provides a list of these files as shown in the dialog box in Fig. 1-1.

5. From the dialog box, pick the **Browse...** button.

The Select File dialog box appears as shown in Figure 1-2. The folders that you see are those stored in the main AutoCAD 2000 folder.

6. Double-click the folder named **Sample**.

This displays all of the files and folders contained in the Sample folder.

7. Single-click any one of the files in the list.

Notice that a small picture of the drawing appears at the right in the area labeled Preview. This gives you a quick view of the graphic content of the selected file.

8. Preview each of the remaining sample drawing files by single-clicking each of them.

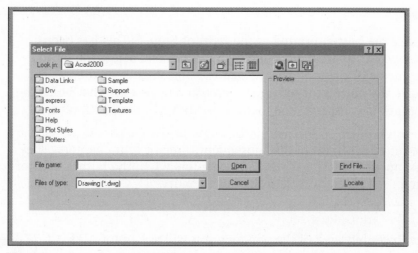

Fig. 1-2

Opening a Drawing File

9. Double-click the file named **R300-20.dwg** or single-click it and pick the **Open** button.

A three-dimensional (3D) drawing of an air cylinder assembly appears. It may take a few seconds to load. Notice the tabs at the bottom of the drawing area. The tab named ANSI Plot of exploded view is in the foreground. This shows that you are viewing a layout of the drawing as it would appear on a printed or plotted page.

10. Pick the **Model** tab located to the left of the **ANSI Plot of exploded view** tab.

This displays the model view of the drawing. This is the view used for most drafting and design work in AutoCAD.

INFOLINK

See Chapter 24 for more about model space, paper space, and layouts.

Viewing Details

AutoCAD offers powerful tools for viewing details on a drawing. For example, the zooming function allows you to take a closer look at very small sections of the drawing. Shading makes surfaces of three-dimensional models appear solid so that they are easier to view.

1. Near the top of the window, pick the **Zoom Window** button. (See the following hint.)

Standard

HINT:

Notice the button printed at the right of Step 1. This has been provided as a visual aid to help you locate and select the button. The name of the toolbar in which the button is located appears above the button for reference. A *toolbar* is the strip that holds the buttons. If you need more help, refer to Appendix H, "Toolbars."

Buttons appear in the margin throughout the chapter in which they are introduced and in the subsequent three or four chapters. They also appear occasionally in later chapters when the button has not been used recently, to refresh your memory.

In the lower left area of the window, you will see AutoCAD's *Command prompt,* which allows you to key in commands and information. AutoCAD also displays important information in this area. Notice that it is now asking you to specify the first corner. Imagine a small window surrounding the lower left area of the drawing.

INFOLINK

You will learn more about zooming in Chapter 12.

2. Using the pointing device, pick a point (using the left button) to specify one of the four corners of the imaginary window. (Fig. 1-3.)

3. Move the pointing device to the opposite corner to create the window and pick another point.

4. AutoCAD zooms in to fill the drawing area with the area represented by the zoom window.

Shading the Model

Complex drawings such as the air cylinder assembly can be hard to see clearly. AutoCAD's shading feature can help you distinguish among parts of the assembly.

1. At the **Command** prompt, type the command **SHADE** using upper- or lowercase letters and press the **ENTER** key.

AutoCAD shades the assembly in several colors.

INFOLINK

You will learn more about shading in Chapters 42 and 43.

2. With the pointing device, click on the yellow coil spring in the assembly.

AutoCAD highlights the spring, showing that it has been selected.

3. Press the **ESC** key twice to remove the selection.

You will learn much more about object selection in Chapter 5.

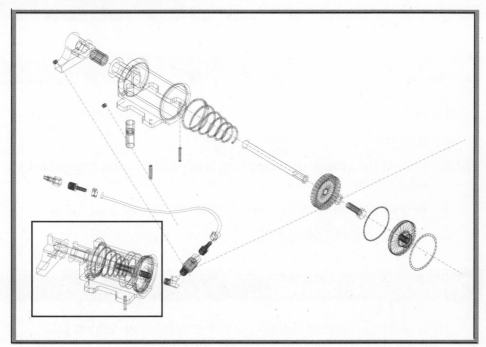

Fig. 1-3

Exiting AutoCAD

1. At the Command prompt, type the **EXIT** command using upper- or lowercase letters and press **ENTER**.

AutoCAD asks whether you want to save your changes.

2. Pick the **No** button.

This exits AutoCAD.

 NOTE: ————————————————————————

It is very important that you exit AutoCAD at the end of each chapter. The appearance of some of the buttons may change as you work, depending on your selections. Exiting AutoCAD resets the buttons in the toolbars to their default settings.

Chapter 1
Review & Activities

Review Questions

1. Name five examples of products or industries that use AutoCAD.

2. How do you start AutoCAD?

3. Explain the process of previewing AutoCAD drawing files prior to opening one.

4. Why might you want to shade a complex model?

5. How do you exit AutoCAD?

Challenge Your Thinking

1. Explore the concept behind the screen resolution. Why is this resolution important, and how do you change it?

2. AutoCAD offers four buttons in the Startup dialog box. In this chapter, you used the Open a Drawing button, but not the other three. What are their purposes, and when would you use one instead of another?

 ## Applying AutoCAD Skills

Work the following problems to practice the commands and skills you learned in this chapter.

1. Start AutoCAD.

2. Preview the drawing files located in AutoCAD's Template folder.

3. Open the drawing file named watch.dwg located in AutoCAD's Sample folder. Pick the Model tab and shade the watch.

4. Open the drawing file named campus.dwg in the Samples folder. Pick the Zoom Window button on the docked Standard toolbar and make a window to take a closer look at the entrance, steps, and covered walkway in the foreground. Shade the drawing.

5. Exit AutoCAD. Do not save your changes.

Chapter 1
Review & Activities

 Using Problem-Solving Skills

Complete the following activities using problem-solving skills and your knowledge of AutoCAD.

1. Start AutoCAD. Open Single cavity mold.dwg, which is located in AutoCAD's Sample folder. Notice the tabs at the bottom of the screen. Pick each layout tab, noticing the name of the tab and the contents that are displayed. Then view the file in model space. Zoom in on different areas of the drawing. Why did the drafter create four layout views? Explain the advantage of AutoCAD's multiple layout capabilities. Exit AutoCAD without saving.

2. Start AutoCAD. Using AutoCAD's preview feature, find and open each of the following files in AutoCAD's Sample folder:

    Lineweights.dwg
    Plot Shading and Fill Patterns.dwg
    Tablet 2000.dwg
    Truetype.dwg

    Explore the drawings, including the model and layout tabs used. What is the purpose of these files? Explain how knowing these files exist can help you as you use AutoCAD in the future. Exit AutoCAD without saving.

Chapter 2 — User Interface

Objectives:

- Identify the parts that make up the AutoCAD window and describe their function
- Navigate AutoCAD's pull-down menus
- Display and reorganize docked and floating toolbars.
- Describe the function of the **Command** window, status bar, and scrollbars

Key Terms

cascading menus

crosshairs

docked toolbar

floating toolbar

pull-down menus

status bar

AutoCAD's user interface includes many parts that are important to the efficient operation of the software. For example, the proper use of docked and floating toolbars improves the speed with which you create, edit, and dimension drawings. Productive users of AutoCAD use a combination of toolbars, pull-down menus, dialog boxes, and the keyboard to enter commands and interact with AutoCAD. All of these elements are a part of the AutoCAD window.

Overview of the AutoCAD Window

Figure 2-1 describes the parts that make up the AutoCAD window.

1. To display the AutoCAD window, start AutoCAD by double-clicking the icon on the Windows desktop.

2. Pick **Cancel** in the **Startup** dialog box to finish loading AutoCAD.

The AutoCAD window appears. Note the locations of the parts labeled pull-down menus, docked and floating toolbars, command window, status bar, and scrollbars. Most of the rest of the AutoCAD window is the drawing area.

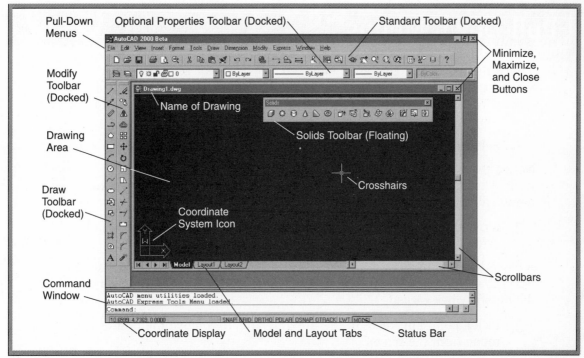

Fig. 2-1

As you move the pointing device in the drawing area, notice that the pointer is a special cursor called the *crosshairs*. You will use the crosshairs to pick points and select objects in the drawing area. When you move outside the drawing area, the normal Windows pointer appears. You will learn more about the features in the drawing area in future chapters.

NOTE:

As explained in Chapter 1, the exact configuration of the AutoCAD window may vary depending on the screen resolution and on how it was left by the last person to use the software. If the Standard and Object Properties toolbars are not present, that's okay. You will learn how to display and position them later in this chapter, in the section titled "Docked Toolbars."

Pull-Down Menus

Notice the words File, Edit, View, Insert, etc., that appear in the upper left area of the AutoCAD window. These words are the names of *pull-down menus*. These menus allow you to perform many drawing, editing, and file manipulation tasks.

1. Select the **View** pull-down menu.

Notice that several menu items contain a small arrow pointing to the right. This means that when you point to them, they will display another menu called a *cascading menu*.

2. Move the pointer to any one of the menu items containing a small arrow and allow it to rest for a moment.

A cascading menu appears, displaying another set of menu options. Menu items that contain ellipsis points (...), as in the Toolbars... item, display a dialog box when picked.

3. Pick a point anywhere in the drawing area to remove the pull-down menu from the screen.

Toolbars

Toolbars contain collections of buttons that can save you time when working with AutoCAD. Under the menu bar (File, Edit, View, etc.) you should see the Standard toolbar docked horizontally across the top of the screen. Below it, you should see the Object Properties toolbar, also docked horizontally. A *docked toolbar* is one that appears to be a part of the top, bottom, left, or right border of the drawing area. When a toolbar is docked, the name of the toolbar does not appear.

Floating Toolbars

Let's focus on displaying and positioning toolbars. Toolbars that are not attached are called *floating toolbars*. The Solids toolbar in the previous illustration is an example.

1. If any floating toolbars are present, close them by picking the **x** (close button) in the upper right corner of the toolbar.

2. Select the **View** pull-down menu.

3. Pick the **Toolbars...** item.

This displays a dialog box with a list of toolbars that AutoCAD makes available to you, as shown in Fig. 2-2.

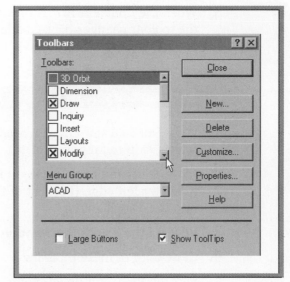

4. Scroll down the list until **Solids** appears and click the check box beside the word **Solids**.

The Solids toolbar appears in the drawing area.

5. Pick the **Close** button in the dialog box.

6. Move the pointer to any one of the docked or floating toolbars and right-click.

Fig. 2-2

This is a convenient way of displaying the same list of toolbars. Those that are checked are the ones that are present on the screen.

7. Using this method, display the **Shade** toolbar.

8. Close the **Shade** toolbar.

It is easy to move toolbars to a new location. As you work with AutoCAD, you should move floating toolbars to a location that is convenient, yet does not interfere with the drawing.

9. Move the **Solids** toolbar by clicking the bar located at the top of the toolbar and dragging it. (This bar contains the word Solids.)

Changing a Toolbar's Shape

AutoCAD makes it easy to change the shape of toolbars. You might want to change the shape of a toolbar, for example, to make it fit more readily in the area of the drawing in which you are working.

1. Move the pointer to the bottom edge of the **Solids** toolbar, slowly and carefully positioning it until a double arrow appears.

2. When the double arrow is present, click and drag downward until the toolbar changes to a vertical shape and then release the button.

3. Move the pointer to the right edge of the **Solids** toolbar, positioning it until a double arrow appears.

4. When the double arrow is present, click and drag to the right until the toolbar changes back to a horizontal shape.

5. Close the **Solids** toolbar.

Docked Toolbars

1. If the Standard and Object Properties toolbars are not present on the screen, skip to Step 5.

2. Move the pointer to the **Object Properties** toolbar and click and drag the outer left edge (border) of the toolbar, moving the toolbar down into the drawing area.

3. Repeat Step 2 with the **Standard** toolbar.

4. Close both of these toolbars.

5. Pick the **View** pull-down menu, select the **Toolbars...** item, and check **Object Properties** and **Standard**.

6. Pick the **Close** button in the dialog box.

7. Click and drag the **Standard** toolbar and carefully dock it under the menu bar (File, Edit, View, etc.).

The toolbar locks into place.

8. Click and drag the **Object Properties** toolbar and carefully dock it under the **Standard** toolbar.

9. If the **Draw** and **Modify** toolbars are not present and locked into place, position them now as shown on page 3.

You should now see four docked toolbars on the screen.

 NOTE: ————————————————

In future chapters, *Applying AutoCAD* assumes that these four toolbars are present on the screen in a configuration similar to the one on page 3. This book has been written to work with AutoCAD straight from the installation. It assumes that the screen color, crosshairs, and other elements of the AutoCAD window have not been customized.

10. If you did not complete Steps 2 through 4, do so now.

Other Features

The AutoCAD window offers other parts that you will use many times throughout this book. The Command window and status bar are key parts of AutoCAD, so pay close attention to them as you work with the software.

Command Window

Notice the Command window near the bottom of the screen. Keep your eye on this area because this is where you receive important messages from AutoCAD. Also, as you learned in Chapter 1, you can enter AutoCAD commands at the Command prompt.

Status Bar

At the bottom of the screen is the status bar. The *status bar* tells you the coordinates of the screen crosshairs and the status of various AutoCAD modes and settings. The more you work with AutoCAD, the more you will appreciate the availability of the buttons labeled SNAP, GRID, ORTHO, POLAR, OSNAP, OTRACK, LWT, and MODEL. They will save you time and effort as you create drawings.

Scrollbars

Scrollbars appear along the bottom and right side of the drawing area. The scrollbars are used to move a zoomed area of the drawing up and down and back and forth.

1. Exit AutoCAD by picking the **x** button located in the upper right corner of the AutoCAD window.

Chapter 2
Review & Activities

Review Questions

1. What is a cascading menu?

2. Describe two ways of displaying a toolbar that is not on the screen.

3. How do you move a toolbar?

4. How do you make a floating toolbar disappear?

5. On a separate sheet of paper, write the names of items A through P in Fig. 2-3.

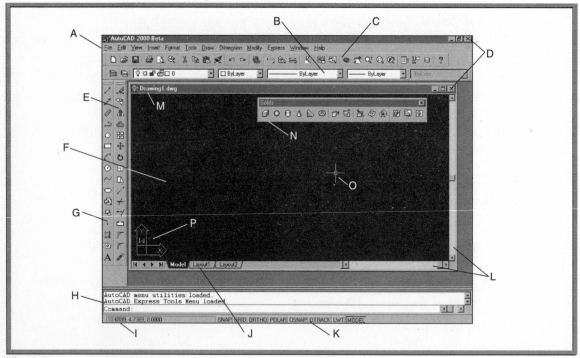

Fig. 2-3

Chapter 2
Review & Activities

Challenge Your Thinking

1. AutoCAD provides ways to change the appearance and colors of the AutoCAD window. For example, if you prefer to work on a gray surface, you can change the drawing area from black to gray. Find out how to customize AutoCAD's appearance and colors. Then write a short paragraph explaining how to customize the graphics screen and why it might sometimes be necessary to do so.

2. The AutoCAD Window includes two sets of Windows-standard maximize, minimize, and close buttons. Explain why there are two sets. What is the function of each?

Applying AutoCAD Skills

Work the following problems to practice the commands and skills you learned in this chapter.

1–3. Rearrange the toolbars so that they are similar to the ones in Figs. 2-4 through 2-6.

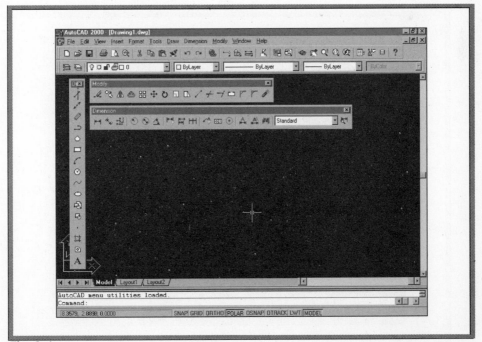

Fig. 2-4

Chapter 2
Review & Activities continued

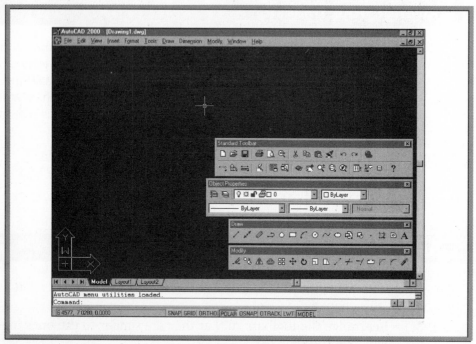

Fig. 2-5

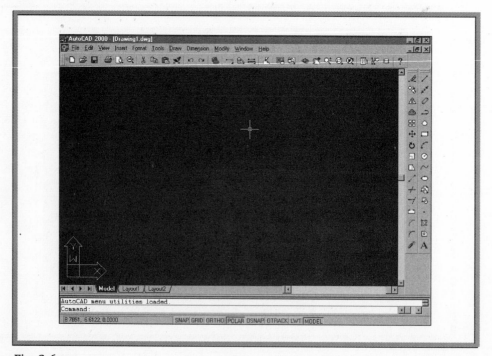

Fig. 2-6

Chapter 2
Review & Activities

4. From the View pull-down menu, pick Toolbars, and uncheck all toolbars. Then close the dialog box. Notice that the drawing area has increased, but you have lost the convenience of toolbar command selection. Reopen the Toolbars dialog box and recheck the toolbars that appear by default in AutoCAD: Object Properties, Standard Toolbar, Draw, and Modify. Close the dialog box. If some of the toolbars are docked on top of others, move the toolbars so that they are docked in their default positions.

5. Change the positions of the toolbars back to their standard appearance as shown on page 3.

 ## Using Problem-Solving Skills

Complete the following activities using problem-solving skills and your knowledge of AutoCAD.

1. Open the City base map.dwg file in AutoCAD's Sample folder. Use the Zoom Window button to zoom in on a small area in the center of the drawing. Then use the horizontal and vertical scrollbars to navigate the drawing. Pick and drag the movable box within the scrollbars to move quickly around the drawing. Experiment to find other ways to navigate. What other methods does AutoCAD provide?

2. Using the Toolbars dialog box, open all of the toolbars provided in AutoCAD. Dock them along the sides, top, and bottom of the screen. Using the tooltips, investigate the buttons that make up the toolbars. What two problems might you encounter if you left all the toolbars open on the screen? When you have finished, close all the toolbars except the four that are docked by default in AutoCAD (Standard, Object Properties, Draw, and Modify).

Careers
Using
AutoCAD

Clothing Design and Layout

"You'll need 15 yards," Beverly Knox told the bride-to-be. Knox had designed a wedding dress for the bride in the traditional way, using pencil and paper to plan how the dress pieces would be laid out on standard-width fabric. The fabric was to be purchased in India by a friend of the bride, who happened to be traveling there at just the right time.

The friend came back with 15 yards of beautiful satin taffeta and twice-embossed lace. But the fabric was too narrow. The pattern pieces for the dress would have to be arranged into a new layout on the narrower fabric. There was about a square yard less fabric to work with.

Knox decided that this was the time to try CAD design techniques. The alternative was to lay out pattern pieces on the 15 yards of fabric by trial and error, using a worktable no bigger than an average-size door.

She chose to use AutoCAD, along with CADTERNS—a custom patternmaking program. With the client's measurements as input, CADTERNS generates a garment template, or sloper. The sloper is then styled in AutoCAD into a custom pattern. Plotted pattern pieces are the output.

This system allowed Knox to modify the dress design on-screen and try out different arrangements of pattern pieces for a layout that would work. "The narrower fabric left literally no margin for error," recalls Knox.

Wedding bells rang, and the bride was beautiful in her dress with the 5-foot train, redesigned by computer. Knox has used this system in her business ever since.

Career Focus—Custom Designer

Beverly Knox is a custom garment designer. Custom designers create designs for women's, men's, and children's clothing and accessories.

To do their jobs, they analyze fashion trends and predictions and select appropriate fabrics. Then they use their knowledge of design to create new designs for clothing and accessories. They sketch rough and detailed drawings of apparel and write specifications for factors such as color scheme, construction, and type of material to be used.

Designers also supervise piece workers who carry out their designs. Those who run their own businesses, such as Beverly Knox, devote a considerable amount of time to business and administrative tasks.

Education and Training

Formal training for designers (in general) is available in 2- and 3-year professional schools. Of course, clothing designers also need to take courses in computer-aided design.

 ### Career Activity

- Research clothing design on the Internet.
- Write a report detailing what types of companies employ clothing designers. What range of income could they expect to earn?

Chapter 3 — Entering Commands

Objectives:

- Create, save, and open an AutoCAD drawing file
- Enter commands using the keyboard and toolbars
- Apply shortcut methods to enter commands
- Reenter commands

Key Terms

command alias

context-sensitive

rubber-band effect

shortcut menu

tooltips

wizard

AutoCAD offers several methods of entering and reentering commands. When entering them frequently, as you will practice in this chapter, you will want to use methods that minimize the time it takes. Command aliases and other shortcuts permit you to enter commands with as little effort as possible. If you take advantage of these time-saving features of AutoCAD, you may require less time to complete a drawing, compared to someone who ignores them. Consequently, you become more attractive to prospective companies looking for highly productive individuals.

In the steps that follow, you will practice entering commands by doing some basic operations such as creating a file and drawing lines. You will apply the concepts you learn in this chapter nearly every time you use AutoCAD.

Creating and Saving a Drawing File

1. Start AutoCAD.

It is important that you start (or restart) AutoCAD at the beginning of each chapter, unless instructed otherwise. This ensures that the buttons in the toolbars are set to their default settings.

2. From the **Startup** dialog box, pick the **Use a Wizard** button.

A *wizard* is a series of dialog boxes that steps you through a sequence; in this case, a setup sequence.

3. In the middle of the dialog box, single-click **Quick Setup**.

4. Read **Wizard Description**.

Note that many of the settings are based on the acad.dwt template file. Template files are drawing files that contain predefined settings that you can use to reduce duplication of effort. If you were to pick the Start from Scratch button instead of the Use a Wizard button, all of the settings would come from the acad.dwt template file. In many future chapters, you will use the Start from Scratch button.

5. Pick the **OK** button.

AutoCAD displays another dialog box that focuses on the unit of measurement. Decimal is the default selection. Notice the sample in the right area of the box.

6. Pick the other units of measurement and watch how the sample changes.

7. Pick **Decimal** and then pick the **Next** button.

This dialog box concentrates on the drawing area. As you can see, the default drawing area is 12 × 9 units. These units can represent inches, feet, meters, kilometers, or any length you wish.

8. Pick the **Finish** button.

9. Pick either **Save** or **Saveas...** from the **File** pull-down menu.

The Save Drawing As dialog box appears as shown in Fig. 3-1. If you use other Windows software, the dialog box should look familiar to you.

10. Pick the **Create New Folder** button (as shown in Fig. 3-1), type your first and last name, and press **ENTER**.

This creates a new folder with your name.

11. Double-click this new folder to make it current.

12. In the **File name** text box, highlight the name of the drawing file (e.g., **Drawing1.dwg**) by double-clicking it, and enter **stuff**.

13. Pick the **Save** button to create a file named **stuff.dwg**.

INFOLINK

You will learn more about template files in Chapter 21.

Standard

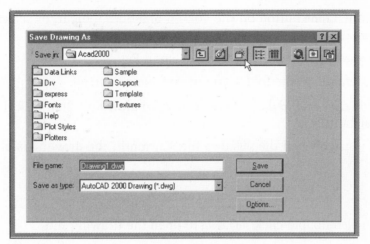

Fig. 3-1

 NOTE: ───────────────────────

If you do not see the file extension (".dwg") after the names of drawing files, this feature may be turned off on your computer. Your work with AutoCAD will be much easier if you turn the extension display on. To turn it on, pick the Start button on the Windows desktop, and then pick Settings, and Folder Options.... In the Folder Options dialog box, pick the View tab and *uncheck* the check box next to Hide file extensions for known file types.

Notice that the drawing file name appears in the upper left corner of the drawing area.

14. From the **File** pull-down menu, pick **Close** to close the stuff.dwg file.

The stuff.dwg file disappears, and the drawing area becomes gray because no drawing file is open. Notice also that many of the pull-down menus and all of the toolbars disappear except for an abbreviated version of the Standard toolbar.

Opening a Drawing File

As you may have noticed from previous illustrations, the Startup dialog box allows you to open drawing files that have previously been saved. This is the method you would use when you have just started AutoCAD. If you are already working in AutoCAD, however, you can open a file more easily using another method.

1. Pick the **Open** button on the docked **Standard** toolbar, or select **Open...** from the **File** pull-down menu.

Standard

The Select File dialog box appears as shown in Fig. 3-2.

AutoCAD defaults to the folder you named after yourself.

2. Single-click **stuff.dwg** in the selection window and then pick the **Open** button.

You have just reopened the stuff.dwg file. Currently, the drawing is blank.

HINT:

You can also double-click stuff.dwg to open the drawing file.

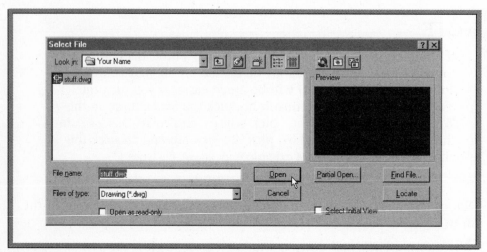

Fig. 3-2

Entering Commands

For the purpose of learning to enter commands, we will use the LINE command. You will learn more about using various AutoCAD commands in the chapters that follow.

Keyboard Entry

The original way to enter a command in AutoCAD was to enter the command name on the Command line. This method is still available, although other methods have been added in recent years to give you a choice.

1. Type **LINE** using upper- or lowercase letters and press the **ENTER** key.

In the Command window, notice that AutoCAD asks you to specify the first point.

2. Pick a point anywhere in the drawing area.

AutoCAD requests the next point.

3. Pick a second point anywhere on the screen.

AutoCAD requests another point.

4. Move the crosshairs away from the last point and notice the stretching effect of the line segment.

This is known as the *rubber-band effect.*

5. Produce second and third line segments to create a triangle and then press **ENTER** to terminate the LINE command.

 NOTE: ──────────────────────────────

Depending on the current AutoCAD settings, you may notice that temporary dotted lines and yellow symbols appear on the screen as you complete the triangle in Step 5. These are produced by the AutoCAD features known as object snap and AutoTracking. You will use these features extensively in the chapters that follow. For now, you can ignore them.

Picking Buttons on a Toolbar

It can be faster to pick a button than to enter a command at the keyboard or from a pull-down menu.

Draw

1. Rest the pointer on the **Line** button located at the top of the docked **Draw** toolbar, but do not pick it.

As discussed briefly in Chapter 1, one or more words appear after about one second. These words, called *tooltips,* help you understand the purpose of the button. Notice also the status bar at the bottom of the AutoCAD window. When a tooltip appears, the status bar changes to provide more detailed information about the button.

2. Slowly position the pointer on top of other buttons and read the information that AutoCAD displays.

Draw

3. Pick the **Line** button.

You have just entered the LINE command. At the command line, you can see that AutoCAD is again asking where you want the line to begin.

4. Pick a point anywhere in the drawing area; then pick a second point to form a line.

5. Press the **ENTER** key to terminate the LINE command.

6. Pick another button (any one of them) from the **Draw** toolbar.

7. Press the **ESC** key to cancel the last entry at the Command line.

HINT:

As you work with AutoCAD, you will be entering commands by picking buttons, selecting them from menus, or typing them at the keyboard. Occasionally you might accidentally select the wrong command or make a typing error. It's easy to correct such mistakes.

If you catch a typing error *before* you press ENTER...	use the backspace key to delete the incorrect character(s). Then continue typing.
If you select the wrong icon or pull-down menu item...	press the ESC key to clear the command line.
If you type the wrong command...	press the ESC key to clear the command line.
If you accidentally pick a point or object on the screen...	press the ESC key twice.

Command Aliases

AutoCAD permits you to issue commands by entering the first few characters of the command. Command abbreviations such as these are called *command aliases*.

1. Type **L** in upper or lower case and press **ENTER**.

This enters the LINE command. You can also enter other commands using aliases. Some of the more common command aliases are listed in Table 3-1.

2. Press the **ESC** key to cancel the LINE command.

3. Enter each of the commands listed previously using the alias method. For now, press **ESC** to cancel each command.

You will learn more about these commands in future chapters.

4. Reenter the **LINE** command using the command alias method and create another line.

NOTE:

Command aliases are defined in the AutoCAD file acad.pgp, which is located in the \Program Files\AutoCAD 2000\Support folder. You can view this file or create additional command aliases by modifying it using a text editor.

Command	Entry	Command	Entry
ARC	A	LINE	L
CIRCLE	C	MOVE	M
COPY	CO	PAN	P
DONUT	DO	POLYGON	POL
ELLIPSE	EL	TOOLBAR	TO
ERASE	E	ZOOM	Z

Table 3-1

Reentering the Last Command

After you have entered a command, you have several choices for repeating it.

1. Press the spacebar or **ENTER**.

This enters the last-entered command, in this case the LINE command. You will find that this saves time over alternative methods of entering the command.

2. Create a line segment and then press **ENTER** to end the LINE command.

3. Press the right button on the pointing device.

Right-clicking in AutoCAD displays a context-sensitive *shortcut menu*. The top item on the menu, when selected, repeats the last command you entered.

4. Pick **Repeat LINE**.

5. Draw two of the polygons shown in Fig. 3-3.

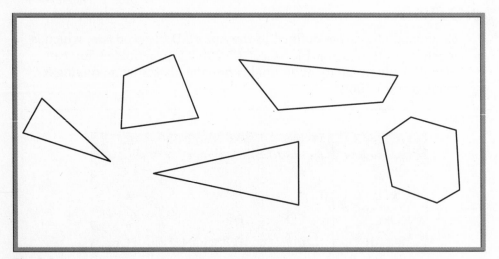

Fig. 3-3

The Repeat option is *context-sensitive*. This means that the software remembers the last command you used.

6. Draw the remaining polygons shown in Fig. 3-3 using the **LINE** command. Be sure to use one of the shortcut methods of reentering the last command.

HINT:

After you have entered at least two line segments, the prompt changes to include the Close option. This option allows you to close a polygon automatically and terminate the LINE command. To choose the Close option, type the letter C and press ENTER.

Draw

7. Enter the **LINE** command and draw a couple of connecting line segments, but do not terminate the command.

8. In reply to Specify next point or [Undo], enter **U** for **Undo**.

AutoCAD backs up one segment, undoing it so that you can recreate it.

9. Practice the **Undo** option with additional polygons.

Standard

10. Pick the **Save** button from the docked **Standard** toolbar to save your work.

As you work with the different methods of entering commands, you may prefer one method over another. Picking the Line button from the toolbar may or may not be faster than entering L at the keyboard. Keep in mind that there is no right or wrong method of entering commands. Experienced users of AutoCAD use a combination of methods.

11. Exit AutoCAD.

Chapter 3
Review & Activities

Review Questions

1. Describe the purpose of the Open button located on the docked Standard toolbar and the Open... item on the File pull-down menu.

2. When a file has not yet been saved for the first time, what appears when you pick either Save or Saveas... from the File pull-down menu?

3. What appears when resting the pointer on a button contained in a toolbar?

4. What is the fastest and simplest method of reentering the previously entered command?

5. How can you enter commands such as LINE, CIRCLE, and ERASE quickly at the keyboard?

6. What is the fast method of closing a polygon when using the LINE command?

7. Explain the use of the LINE Undo option.

Challenge Your Thinking

1. Discuss the advantages and disadvantages of having more than one way to enter a command. Which method of entering a command is most efficient? Collect opinions from other AutoCAD users and consider ergonomic factors.

2. Describe a way to open a drawing file in AutoCAD when AutoCAD is not currently running on the computer.

Chapter 3
Review & Activities

Applying AutoCAD Skills

Work the following problems to practice the commands and skills you learned in this chapter.

1. Create a new drawing file named prb3-1.dwg and store it in the folder named after yourself. Create the bookcase shown in Fig. 3-4. For practice, enter the LINE command in various ways: by entering the command at the keyboard, picking a button, and using the various shortcut methods. Save the drawing.

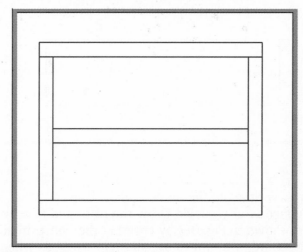

Fig. 3-4

2. Create a new drawing file named prb3-4.dwg and store it in the folder named after yourself. Draw the concrete block shown in Fig. 3-5 using the most efficient method(s) of entering the LINE command.

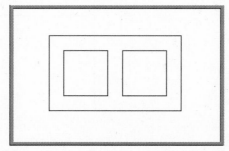

Fig. 3-5

Chapter 3°
Review & Activities continued

3. Create a new drawing file named prb3-2.dwg and store it in the folder named after yourself. Draw the simple house elevation drawing shown in Fig. 3–6 by picking the Line button from the docked Draw toolbar. When reentering the command, use a shortcut method.

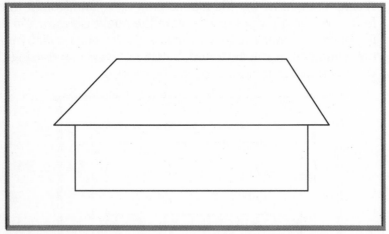

Fig. 3-6

4. Create a new drawing file named prb3-3.dwg and store it in the folder named after yourself. Draw the front and side views of the sawhorse shown in Fig. 3-7 by entering the LINE command's alias at the keyboard. Be sure to use shortcut methods when reentering the command.

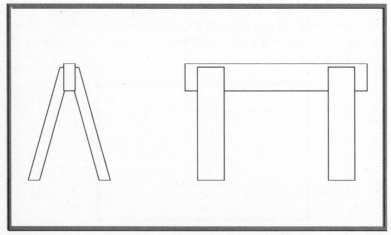

Fig. 3-7

Problem 4 courtesy of Joseph K. Yabu, Ph.D., San Jose State University

Chapter 3
Review & Activities

 Using Problem-Solving Skills

Complete the following problems using problem-solving skills and your knowledge of AutoCAD.

1. A boat trailer manufacturer needs a new design for the rollers that facilitate raising a boat onto the trailer. The engineering division of your company, which designs the rollers, has proposed the design shown in Fig. 3-8. Create a drawing to be submitted to the trailer manufacturer for approval. Notice that the drawing is dimensioned. Experiment with the buttons on the status bar and see if you can figure out a way to draw the roller approximately to size. Do not dimension. Save the drawing as roller1.dwg.

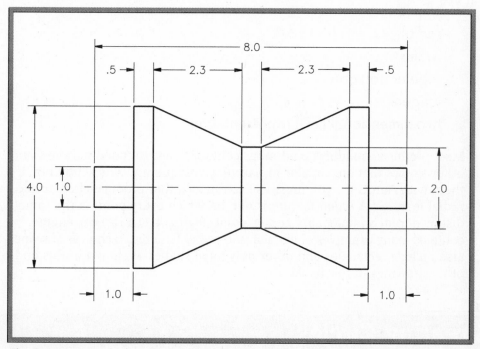

Fig. 3-8

2. The trailer manufacturer accepted the roller design with the following modification: the 4-inch overall diameter is to be increased to 5 inches. All other dimensions are to remain the same. Open roller1.dwg and make the change. Save the drawing as roller2.dwg.

Objectives:

- **Draw two-dimensional views of holes, cylinders, and rounded and polygonal features**
- **Create curved objects such as circles, arcs, ellipses, and donuts**
- **Create rectangles and regular polygons**

Key Terms

concentric

diameter

donuts

ellipse

radial

radius

regular polygon

tangent

two-dimensional (2D) representation

Most products, buildings, and maps contain holes, rounded features, and other shapes that are circular in nature. Consequently, when drawing them with AutoCAD, you must use commands that produce circular and radial features. A *radial* feature is one on which every point is the same distance from an imaginary center point. Figure 4-1, a racecar engine, contains many examples of circular and radial features. Technical drawings also include rectangles and other polygonal shapes as shown in this and other drawings in this book.

Creating Circles

AutoCAD makes it easy to draw *two-dimensional (2D) representations* of holes, cylinders, and other round shapes. A 2D representation is a single profile view of an object seen typically from the top, front, or side.

1. Start AutoCAD and pick the **Start from Scratch** button, **English**, and **OK** in the **Startup** dialog box.

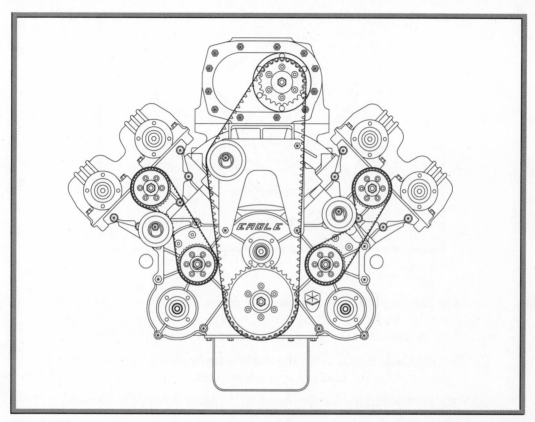

Fig. 4-1   Courtesy of Joe Schubeck, Eagle Engine Manufacturing

 NOTE:

As a reminder, be sure to exit and restart AutoCAD if it was loaded prior to Step 1.

Standard

2. Pick the **Save** button from the **Standard** toolbar, or select **Save As...** from the **File** pull-down menu.

This causes the Save Drawing As dialog box to appear.

3. Double-click the folder with your name and double-click **Drawing1.dwg**.

4. Type **engine** in upper- or lowercase letters and pick **Save** or press **ENTER**.

This creates and stores a new drawing file named engine.dwg in your named folder.

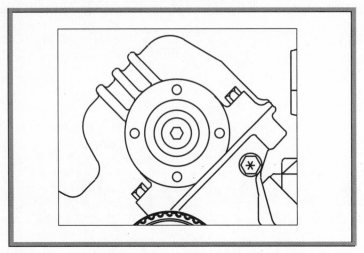

Fig. 4-2

The detail drawing in Fig. 4-2 shows one of the engine's cylinder heads. Let's create the center portion of the cylinder head. (For now, disregard the bolt at the center of the head. You will add it later in this chapter.)

Draw

5. Pick the **Circle** button from the docked **Draw** toolbar, or enter **C** (the command alias for **CIRCLE**) at the keyboard.

Notice the instruction on the Command line: Specify center point for circle.

6. Use the pointing device to pick a center point near the center of the drawing area.

7. Move the crosshairs and notice that you can "drag" the radius of the circle.

 NOTE: _____

> If the radius of the circle does not change as you move the crosshairs, an AutoCAD feature called DRAGMODE may be turned off on your system. To turn it back on, enter DRAGMODE at the Command line and then enter A (for Auto).

Notice that AutoCAD is requesting the radius or diameter of the circle on the command line. The *radius* of a circle is the length of a line extending from its center to one side of the circle. The *diameter* of a circle is the length of a line that extends from one side of a circle to the other and passes through its center.

8. Pick a point a short distance from the center as shown in Fig. 4-3.

This completes the command and forms the innermost circular feature of the cylinder head.

Now let's draw a concentric circle to represent the next feature. *Concentric* circles are ones that share a common center.

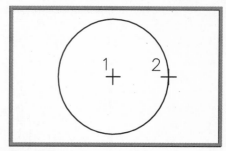

Fig. 4-3

Concentric Circles

Draw

9. Reenter the **CIRCLE** command.

HINT:

Use one of the shortcuts to reenter the command.

10. Move the crosshairs to the center of the circle you created.

A small yellow circle appears at the center of the circle. This is the Center object snap. Object snaps are like magnetic points that permit you to lock onto a specific point on an object easily and accurately.

11. While the small yellow circle is present, pick a point.

AutoCAD snaps to the center point of the existing circle.

12. Drag the radius into place to create a second circle (Fig. 4-4). Review the detail of the cylinder head to approximate the size of the circle.

INFOLINK

You will learn more about object snaps in Chapter 9.

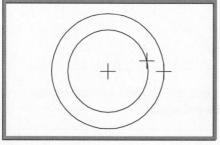

Fig. 4-4

13. Repeat Steps 9 through 12 to place the remaining concentric circles as shown in the detail of the cylinder head.

14. Reenter the **CIRCLE** command and create the four small circles to represent bolt holes. Approximate their size and location.

Draw

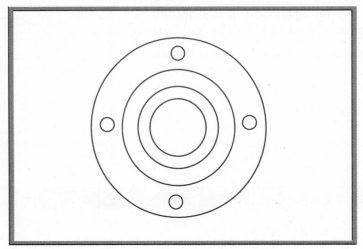

Fig. 4-5

Your drawing should now look like the one in Fig. 4-5.

15. Save and close the drawing file.

Tangent Circles

Tangent circles are circles that meet at a single point. The Ttr option allows you to create a circle with a specified radius that is tangent to two other circles or arcs.

1. Create a new drawing from scratch; select **English** and pick **OK**.

2. Create two circles and position them relatively close to one another.

Draw

3. Reenter the **CIRCLE** command, and enter **T** to select the Ttr (tan, tan, radius) option.

4. Read the instruction on the Command line and then move the crosshairs over one of the two circles.

A small yellow symbol appears, showing that the center of the crosshairs is tangent to the circle.

NOTE: ─────────────────────

If the yellow symbol does not appear, pick the OSNAP button on the status bar to depress it, and then try Step 4 again.

5. Pick a point on the first circle.

6. Move the crosshairs to another circle and pick a point on the second circle.

7. For the radius of the new circle, enter **2**.

AutoCAD creates a new circle that is tangent to the two circles. It's okay if the circle extends off the screen.

8. Save your work.

Creating Arcs

The person who created the engine drawing shown in Fig. 4-1 used arcs to produce the radial features on some of the individual parts. In AutoCAD, the ARC command is often used to produce arcs.

1. Pick the **Arc** button from the **Draw** toolbar, or enter **A** at the keyboard.

Draw

This enters the ARC command.

2. Pick three consecutive points anywhere on the screen.

3. Reenter the **ARC** command and produce two additional arcs using the same method.

Options for Creating Arcs

AutoCAD offers several options for creating arcs. The option you choose depends on the information you know about the arc. Suppose you want to create an arc for an engine part and you know the start point, the endpoint, and the radius of the arc.

1. Enter the **ARC** command and pick a start point.

Draw

2. Enter **E** for End and pick an endpoint a short distance from the start point.

3. Move the crosshairs and watch the arc form dynamically on the screen.

4. Enter **R** for Radius and enter **1**.

An arc forms on the screen.

Now suppose you don't know the radius of the arc or where the endpoint should be. However, you know where the center point of the arc should be, and you know how long the arc must be.

Draw

5. Reenter the **ARC** command and pick a point.

6. Enter **C** for Center and pick a point.

7. Enter **A** for Angle and enter a number (positive or negative) up to 360. (The number specifies the angle in degrees.)

An arc forms on the screen.

8. Experiment with the other options on your own.

As you can see, AutoCAD's options allow you to create an arc accurately using the information that is available to you. Understanding these options can save time when you need to produce accurate production drawings that contain circular or radial features.

Series of Tangent Arcs

In some cases, you may need to draw a series of arcs that are connected to one another. The Continue option of the ARC command provides an efficient way to draw such a series.

The line in Fig. 4-6 is really a series of tangent arcs. This line was developed for a map. It represents a stretch of highway 34 through the Rocky Mountain National Park, west of Estes Park, Colorado. Let's reproduce the map line to practice creating tangent arcs.

Draw

1. Enter the **ARC** command and create an arc using any of the options.

2. Reenter the **ARC** command by pressing the spacebar or **ENTER**; press the spacebar or **ENTER** a second time.

Notice how another arc segment develops as you move the crosshairs.

3. Create the next arc segment.

Notice that it is tangent to the first. (The point of tangency is the point at which the two arcs join.)

4. Repeat Steps 2 and 3 until you have finished the road.

5. Create the concentric circles to show Estes Park, but do not include the text.

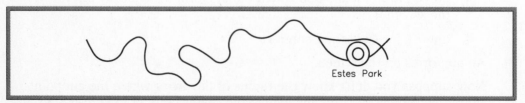

Fig. 4-6

Drawing Ellipses

The ELLIPSE command enables you to create mathematically correct ellipses. An *ellipse* is a regular oval shape that has two centers of equal radius. Ellipses are used to construct shapes that become a part of the engineering drawings. For example, the race car engine includes a partial ellipse, as shown in Fig. 4-7.

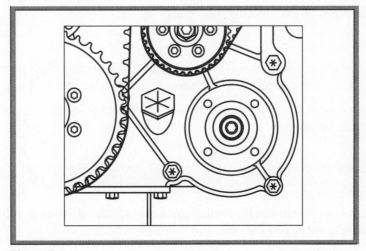

Fig. 4-7

Draw

1. From the **Draw** toolbar, pick the **Ellipse** button to enter the ELLIPSE command.

2. Pick a point for **Axis endpoint 1** as shown in Fig. 4-8.

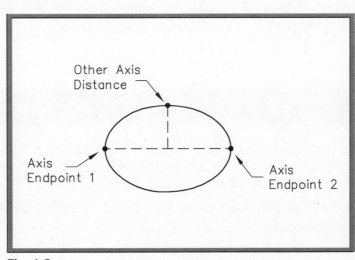

Fig. 4-8

3. Pick a second point for **Axis endpoint 2** directly to the right of the first point.

HINT:

As noted in the previous chapter, temporary dotted lines appear when you move the crosshairs over the top of existing objects on the screen. This feature allows you to align and create new points from the existing objects. Take advantage of the dotted horizontal line that appears when creating axis endpoint 2.

4. Move the crosshairs and watch the ellipse develop. Pick a point, or enter a numeric value such as **.3**, and the ellipse will appear.

5. Experiment with the **Center** option on your own. How might this option be useful?

Elliptical Arcs

Many parts that include elliptical shapes do not use the entire ellipse. To draw these shapes, you can use the ELLIPSE Arc option.

Draw

1. Pick the **Ellipse** button and enter **A** for Arc.

2. Pick three points similar to those shown in Fig. 4-8.

A temporary ellipse forms.

3. In reply to Specify start angle, pick a point anywhere on the ellipse.

4. As you slowly move the crosshairs counterclockwise, notice that an elliptical arc forms.

5. Pick a point to form an arc.

Donuts

The DONUT command allows you to create thick-walled or solid circles, known in AutoCAD as *donuts*. Drafters commonly use donuts to represent features on a machine part or architectural drawing.

1. From the **Draw** pull-down menu, pick **Donut**, or enter **DONUT** at the Command prompt.

2. Specify an inside diameter of **.25**...

3. ...and an outside diameter of **.5**.

The outline of a donut locks onto the crosshairs and is ready to be dragged and positioned by its center.

4. Place the donut anywhere in the drawing by picking a point.

5. Move the crosshairs away from the new solid-filled donut and notice the prompt line at the bottom of the screen.

6. Place several additional donuts in the drawing.

7. Press **ENTER** to terminate the command.

That's all there is to the DONUT command.

Rectangles

One of the most basic shapes used by drafters is the rectangle. You can create rectangles using the LINE command, but doing so has some disadvantages. For example, you would have to take the time to make sure that the corner angles are exactly 90°. Also, each line segment would be a separate object. Therefore, AutoCAD provides the RECTANG command, which allows you to create rectangles with perfect corners. Clicking anywhere on a rectangle created with the RECTANG command selects the entire rectangle.

Draw

1. From the **Draw** toolbar, pick the **Rectangle** button.

As you can see at the Command prompt, AutoCAD is asking for the first corner of the rectangle. Other options appear in brackets.

2. In reply to Specify first corner point, pick a point at any location.

3. Move the pointing device in any direction and notice that a rectangle begins to form.

4. Pick a second point at any location to create the rectangle.

5. Create a second rectangle. Since the RECTANG command was just entered, reenter it by pressing the spacebar or **ENTER**, or right-click and pick **Repeat Rectangle** from the shortcut menu.

6. Create a third rectangle.

7. Close the drawing without saving it.

Polygons

The POLYGON command enables you to create regular polygons with 3 to 1024 sides. A *regular polygon* is one with sides of equal length. Using the POLYGON command, let's insert the bolt head into the engine drawing you started earlier in this chapter.

1. Open the **engine.dwg** file.

2. From the **Draw** toolbar, pick the **Polygon** button, or enter the **POL** alias.

Draw

Notice the <4> at the end of the Command line. This is the default value, meaning that if you were to press ENTER now, AutoCAD would enter 4 in reply to Enter number of sides. To represent the bolt head, you will need a hexagon (six-sided polygon).

3. Enter **6**.

AutoCAD now needs to know if you want to define an edge of the polygon or select a center point. Let's specify a center.

4. Move the pointing device over any of the larger circles in the drawing. When the small yellow symbol appears at the center of the cylinder head, pick a point to select the center of the circle as the center point of the hexagon.

AutoCAD allows you to create a polygon by inscribing it in a circle of a specified diameter or by circumscribing it around the specified circle. Figure 4-9 shows the difference.

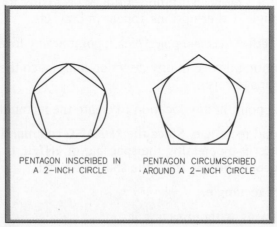

Fig. 4-9

5. Press **ENTER** to select the I (Inscribed) default value.

AutoCAD now wants to know the radius of the circle within which the polygon will appear.

6. With the pointing device, move the crosshairs from the center of the polygon and notice that a hexagon begins to form.

7. Pick a point to create the hexagonal bolt head at an appropriate size relative to the cylinder head. Refer to the detail drawing on page 41 if necessary.

 NOTE: ———————————————————————————

> To create the polygon with a more accurate size, you could have entered a specific numeric value, such as .5, at the keyboard. (Entering a 0 before the decimal point is optional.)

The cylinder head is now complete. Your drawing should look similar to the one in Fig. 4-10.

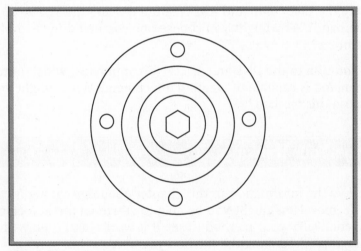

Fig. 4-10

8. Save your work and exit AutoCAD.

Standard

 HINT: ———————————————————————————

> Back up your files regularly. The few seconds required to make a backup copy may save you hours of work. Experienced users back up faithfully because they know the consequences of not doing so.

Chapter 4
Review & Activities

Review Questions

1. Briefly describe the following methods of producing circles.
   a. 2 points
   b. 3 points
   c. tan tan radius

2. In which AutoCAD toolbar are the ARC and CIRCLE commands found?

3. What function does the ARC Continue feature serve?

4. Explain the purpose of the DONUT command.

5. When creating a polygon using the Inscribed in circle option, does the polygon appear inside or outside the imaginary circle?

6. What are the practical differences between creating a rectangle using the LINE command and creating it using the RECTANG command? Why might you choose one command instead of the other?

7. In addition to the LINE and RECTANG commands, which other command is capable of creating a rectangle? When might you choose this method?

Challenge Your Thinking

1. Review the information in this chapter about specifying an angle in degrees. How might you be able to create an arc in a clockwise direction? Try your method to see if it works, and then write a paragraph describing the method you used.

2. Experiment further with the ARC command. Is it possible to create a single arc that has a noncircular curve using the options available for this command? (A *noncircular curve* is one in which not all the points are exactly the same distance from a common center point.) Explain your answer.

Chapter 4
Review & Activities

Applying AutoCAD Skills

Work the following problems to practice the commands and skills you learned in this chapter.

1. Create a new drawing named pumpkin.dwg. Create a pattern for a hand-carved pumpkin face. Use the commands in this chapter to create the features. Define the face using two arcs: one for the top and another for the lower part of the face. Use Fig. 4-11 for guidance, but be as creative as you wish. Save the drawing.

Fig. 4-11

2. Using the commands you just learned, complete the drawings shown in Figs. 4-12, 4-13, and 4-14. Don't worry about text matter or exact shapes, sizes, or locations, but do try to make your drawings look similar to the ones in the figures.

Fig. 4-12

Chapter 4
Review & Activities continued

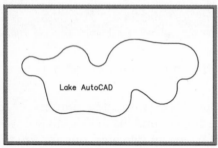

Fig. 4-13

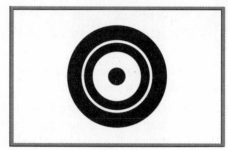

Fig. 4-14

3. Use the LINE, POLYGON, CIRCLE, and ARC commands to create the hex bolt shown in Fig. 4-15.

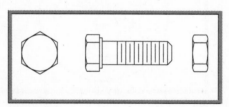

Fig. 4-15

4. Create the object shown in Fig. 4-16 by first drawing a six-sided polygon with the POLYGON command. Then draw the six-pointed star using the LINE command.

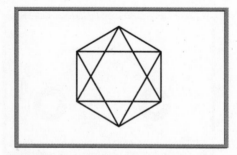

Fig. 4-16

Problem 3 courtesy of Joseph K. Yabu, Ph.D., San Jose State University
Problem 4 courtesy of Gary J. Hordemann, Gonzaga University

Chapter 4
Review & Activities

5. Create the same object again, but draw the star using two three-sided polygons.

6. How would you draw a five-pointed star? Draw it.

7. Draw a block with a rectangular cavity like the one shown in Fig. 4-17 using the RECTANG and LINE commands.

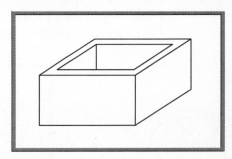

Fig. 4-17

8. The two objects shown in Fig. 4-18 are composed entirely of equal-sided and equal-sized polygons with common edges surrounding a central polygon. Can this be done with equal-sided and equal-sized polygons of any number of sides? Answer this question by trying to draw such objects using the POLYGON command with polygons of five, six, seven, and eight sides.

9. Draw the screw heads shown in Fig. 4-19.

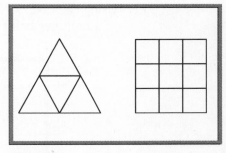

Fig. 4-18

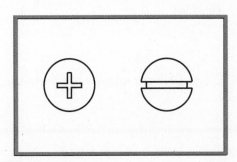

Fig. 4-19

Problems 5 through 8 courtesy of Gary J. Hordemann, Gonzaga University

Chapter 4
Review & Activities continued

10. Draw the eyebolt in Fig. 4-20 using all of the commands you have learned.

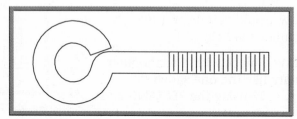

Fig. 4-20

11. Using the LINE, ARC, and CIRCLE commands, draw the two views of a hammer head as shown in Fig. 4-21.

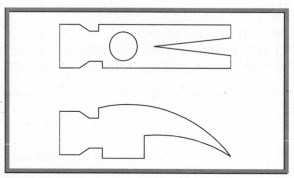

Fig. 4-21

12. Use the ARC and LINE commands to create the screwdriver shown in Fig. 4-22.

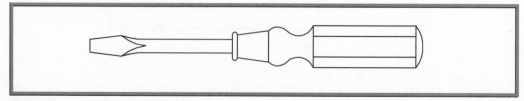

Fig. 4-22

Problems 10 through 12 courtesy of Joseph K. Yabu, Ph.D., San Jose State University

Chapter 4
Review & Activities

Using Problem-Solving Skills

Complete the following activities using problem-solving skills and your knowledge of AutoCAD.

1. Draw the ski lift rocker arm for a design study by the gondola manufacturer (Fig. 4-23). Do not include dimensions. Save your drawing as rockerarm.dwg.

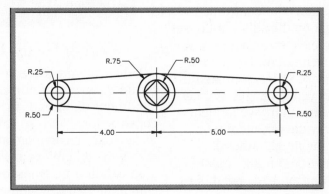

Fig. 4-23

2. The traditional mantle clock body drawing shown in Fig. 4-24 is necessary for the reconstruction of antique and collectible clocks. Draw the clock body. Set the appropriate units and area in the Startup or Create New Drawing dialog box. Do not include dimensions. Save your drawing as mantle.dwg.

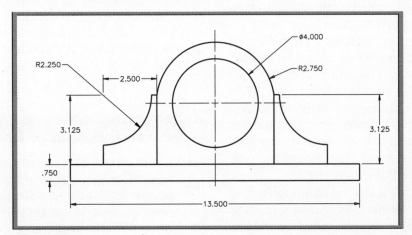

Fig. 4-24

Careers
Using
AutoCAD

Concerned Neighbors Visualize a Problem

Members of the Windsor Park Community in Calgary were concerned about plans for a three-story apartment complex. How would the new building and the influx of new residents affect traffic? Would the building block sunlight on the adjacent houses?

These questions were easy to answer using computer visualization techniques. Dr. Richard M. Levy at the University of Calgary created a multimedia presentation using images from a computer model. His presentation gave residents a preview of how this new development would impact the character of the neighborhood.

Dr. Levy created in AutoCAD a 3D model of several blocks of the Windsor Park neighborhood complete with an accurate representation of the proposed development. The final model was rendered in 3D Studio MAX. Digitized photographs taken from critical locations in the neighborhood were merged with images of the model. Dr. Levy then created a sequence of images showing the impact of the proposed apartment on shade. The model also served as a tool in discussions on neighborhood character, traffic, and crime prevention.

In Dr. Levy's courses, students learn principles of urban design using 3D computer modeling. With site plans and information from 2D CAD files, they create 3D models to investigate likely results from specific changes in urban policy. For example, how would zoning changes impact the construction of new commercial development in the area? How would the new construction affect traffic circulation?

The ability to develop walkthroughs is a valuable tool for planners. By allowing them to recognize potential problems, costly problems can be avoided.

Career Focus—Urban Planner

Urban planners create plans that shape the future direction of our cities. They study local conditions and trends in housing, employment, construction, and social services. They also make policy recommendations to the mayor and members of city council. Planners may also plan studies in transportation, tourism, and community and economic development.

Education and Training

Many schools offer a two-year master's degree in urban planning. Students usually have an undergraduate degree in one of several areas: urban studies, geography, economics, urban sociology, architecture, or engineering. Graduate students in urban planning spend time in seminars, studios, and workshops learning to analyze real-world problems. They also work in government offices and planning firms as part of their education.

 Career Activity

- Write a report on a recent case of planning and development in your community—for example, a recently constructed shopping center or park. What agencies were involved? What involvement of the public and private sectors were required?

Chapter 5 Object Selection

Objectives:

- **Identify AutoCAD objects**
- **Select objects to create a selection set**
- **Add and remove objects from the selection set**
- **Erase and restore objects in AutoCAD**

Key Terms

entity

grips

noun/verb selection

object

pickbox

selection set

verb/noun selection

When using AutoCAD, you will find yourself selecting objects so that you can move, copy, erase, or perform some other operation. Because drawings can become very dense with lines, dimensions, and text, it is important to use the most efficient methods of selecting these objects. Usually, the quickest way to move or copy an object is to click and drag. This is true whether you are using an inexpensive shareware program on a low-cost computer or sophisticated CAD software on an expensive computer. This chapter covers this and alternative methods of selecting objects and performing basic editing operations, such as erasing.

Objects

An *object,* also called an *entity,* is an individual predefined element in AutoCAD. The smallest element that you can add to or erase from a drawing is an object.

The following list gives examples of object types in AutoCAD.

3Dface	line	shape
3Dsolid	mline	solid
arc	mtext	spline
attrib	oleframe	text
body	point	tolerance
circle	polyline	trace
dimension	ray	viewport
ellipse	region	xline
leader		

You'll learn more about these objects and how to use them in future chapters.

Selecting Objects

Basic object selection in AutoCAD is straightforward—you select the object by picking it with the left button on the pointing device. However, AutoCAD also offers more advanced object selection techniques. This chapter illustrates some of these techniques.

1. Start AutoCAD and start a new drawing from scratch.

 NOTE: ───────────────────────

In this work-text, all new drawings use English units unless otherwise specified.

Draw

2. Using the **LINE** and **CIRCLE** commands, draw the triangle and circle shown in Fig. 5-1.

Draw

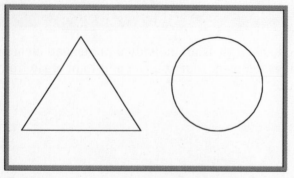

Fig. 5-1

Modify

Window Selection

3. Enter the ERASE command by picking the **Erase** button from the **Modify** toolbar or by entering **E** at the keyboard.

The Command prompt changes to Select objects. Notice that the crosshairs have changed to a small box called the *pickbox*. The pickbox is used to pick objects.

4. Type **W** for Window and press **ENTER**.

The crosshairs return and AutoCAD asks you to specify the first corner.

5. Pick a point anywhere to the left of the triangle.

6. Move the cursor and notice that a rubber-band box forms.

This box defines the selection window.

7. Move the cursor so that the triangle fits completely in the selection window and pick a point.

Notice what happens. The objects to be erased are highlighted with broken lines as shown in Fig. 5-2. The highlighted objects make up the *selection set.*

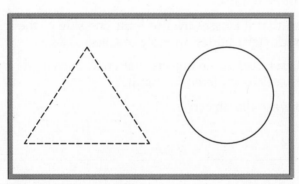

Fig. 5-2

8. Press **ENTER** to make the triangle disappear. (This also terminates the ERASE command.)

HINT:

You can also select objects using a window without formally entering the Window option. When the Select objects prompt appears, you can produce a window by picking two corner points. Be sure to specify the *left* side of the selection window first.

Restoring Erased Objects

What if you have erased an object by mistake and you want to restore it? Let's restore the triangle.

9. Pick the **Undo** button on the docked **Standard** toolbar.

Standard

The triangle reappears. The Undo button conforms to standard Windows function, so you may be familiar with it from other Windows applications. Picking this button repeatedly undoes your previous actions, one at a time, beginning with the most recent.

Using a Crossing Window

The Crossing object selection option is similar to the Window option, but it selects all objects that cross the window boundary as well as those that lie completely within it.

1. Enter the **ERASE** command.

Modify

2. Type **C**, press **ENTER**, and pick a point to the right of the triangle.

3. Move the crosshairs to the left to form a box and notice that the box is made up of broken lines.

4. Form the box so that it crosses over at least one side of the triangle and press the pick (left) button in reply to Other corner.

Notice that AutoCAD selected those objects that cross over the box as well as any objects that lie completely within it.

5. Press **ENTER** to erase the object(s).

 HINT:

> You can also select objects using a crossing window without formally entering the Crossing option. When the Select objects prompt appears, simply pick two points to form a window, being sure to pick the *right-most* point first.

Polygonal Selection Options

The WPolygon and CPolygon object selection options are similar to the Window and Crossing options. However, they offer more flexibility because they are not restricted to the rectangular window shape used by the Window and Crossing options.

1. Fill the screen with several objects, such as lines, circles, arcs, rectangles, polygons, ellipses, and donuts.

2. Enter the **ERASE** command.

Modify

3. In reply to Select objects, enter **WP** for WPolygon. (W is for Window.)

4. Pick a series of points that form a polygon of any shape around one or more objects and press **ENTER**.

NOTE: ─────────────────────────────────

When forming the polygon, AutoCAD automatically connects the last point to the first point.

AutoCAD selects those objects that lie entirely within the polygon.

5. Press **ENTER** to erase the objects.

The CPolygon option is similar to WPolygon.

Modify

6. Enter the **ERASE** command or press the spacebar.

7. In reply to Select objects, enter **CP** for CPolygon. (C is for Crossing.)

8. Pick a series of points to form a polygon. Make part of the polygon cross over at least one object and press **ENTER**. (As you create the polygon, notice that it is made up of broken lines, indicating that the Crossing option is in effect.)

AutoCAD selects those objects that cross the polygon, as well as those that lie completely within it. As you can see, CPolygon is similar to the Crossing option, whereas WPolygon is similar to the Window option.

9. Press **ENTER** to complete the erasure.

Using a Fence

The Fence option is similar to the CPolygon option except that you do not close a fence as you do a polygon. When you select objects using the Fence option, AutoCAD looks for objects that touch the fence. Objects "inside" the fence are not selected unless they actually touch the fence.

Modify

1. Enter **ERASE**.

2. Enter **F** for Fence.

3. Draw a line that crosses over one or more objects and press **ENTER**.

4. Press **ENTER** to complete the erasure.

Selecting Single Objects

What if you want to select single objects, such as individual line segments?

1. Recreate the triangle and circle you drew previously using the **LINE** and **CIRCLE** commands.

2. Pick the **Erase** button or enter **E** at the keyboard.

The prompt changes to Select objects.

Modify

3. Using the pickbox, pick two of the line segments in the triangle.

4. Press **ENTER** to complete the command and to make the highlighted objects disappear.

HINT:

As you may recall, if you are using a pointing device that has more than one button, pressing one of them, usually the one on the right, is the same as pressing the **ENTER** key. This is usually faster than pressing **ENTER** at the keyboard.

Selecting the Last Object Drawn

The Last object selection option automatically selects the last object you drew. This option works whenever the Select objects prompt is present at the Command line.

Standard

1. Pick the **Undo** button to restore the line segments.

Notice that because you erased both line segments in a single ERASE operation, you only have to pick the Undo icon once to make both lines reappear.

Modify

2. Enter the **ERASE** command.

3. Type **L** (for Last) and press **ENTER**.

AutoCAD highlights the last object you drew.

NOTE:

The last object you drew may be different from the last object you selected in the drawing. If you want to reselect the last object you selected, enter P (for Previous) instead of Last. The Previous option is especially useful when you need to reselect an entire group of objects.

4. Press **ENTER**.

If you continue to enter ERASE and Last, you will erase objects in the reverse order from which you created them. However, it is usually faster to use the Undo button for this purpose.

5. Pick the **Undo** button to restore the object you deleted using the Last option.

Standard

Selecting the Entire Drawing

The All object selection option permits you to select all the objects in the drawing file quickly.

1. Enter the **ERASE** command.

2. In reply to Select objects, enter **All** and press **ENTER**.

AutoCAD selects all objects in the drawing.

4. Press **ENTER** again to erase all the objects.

5. Pick **Undo** to restore the objects.

Modify

Standard

Editing Selection Sets

AutoCAD allows you to add and remove objects from a selection set while you are in the process of creating it. For example, if you accidentally pick an object that you did not mean to select, you can remove it without affecting the rest of the selected objects.

Removing Objects from a Selection Set

To remove an unwanted object from a selection set, you must enter the Remove objects mode.

1. Enter the **ERASE** command.

2. Use a selection window to select the triangle and the circle.

Modify

Both the triangle and the circle should now be highlighted. Let's say you have decided not to erase the circle.

3. Type **R** (for Remove) and press **ENTER**.

Notice that the Command line changes to Remove objects. You can now remove one or more lines from the selection set. The line(s) you remove from the selection set will not be erased.

4. Remove the circle from the selection set by picking the line that makes up the circle.

Note that the line you picked is no longer highlighted.

5. Press **ENTER** to complete the ERASE operation.

6. Pick the **Undo** button to restore the triangle.

Standard

Adding to the Selection Set

After you have used the Remove option to remove objects from a selection set, you can select additional objects to include in the set.

7. Repeat Steps 1 through 4 in the previous section ("Removing Objects from a Selection Set").

8. Instead of pressing ENTER, type **A** for Add and press **ENTER**.

Notice that the prompt line changes back to Select objects.

9. Select the circle.

10. Press **ENTER** to complete the ERASE operation.

So you see, you can add and remove objects as you wish until you are ready to perform the operation. The objects selected are indicated by broken lines. These selection procedures work not only with the ERASE command, but also with other commands that require object selection, such as MOVE, COPY, MIRROR, ARRAY, and many others.

Standard

11. Pick the **Undo** button to restore the objects to the screen.

Using Grips

Another method of selecting objects is to pick them using the pointing device. When you pick an object without first entering a command, small boxes called *grips* appear at key points on the object. Using these grips, you can perform several basic operations such as moving, copying, or changing the shape or size of an object.

Draw

1. Enter the **RECTANG** command and draw a rectangle of any size and at any location.

2. Select any point on the rectangle. (No command should be entered at the Command prompt.)

Notice that grips appear at each corner of the rectangle.

Moving Objects with Grips

3. Pick any one of the four grips. (As you pick the grip, notice how the pickbox locks onto the selected grip.)

The grip turns red, showing that you've selected it. Notice also that the command line has changed, indicating that AutoCAD is in Stretch mode. This mode allows you to change an object by moving one or more grips or by moving or copying the entire object.

4. Press the right button on the pointing device (right-click) to display a shortcut menu.

5. Pick **Move**, drag the rectangle to a new location, and pick a point.

6. Select one of the other editing options on the shortcut menu, such as **Rotate**, and complete the operation.

7. Press **ESC** twice to remove the grips.

Copying Objects with Grips

You can also copy objects using grips.

1. Draw a circle of any size and at any location.

2. Select the circle and pick one of the four grips that lie on it.

3. Move the pointing device and notice that you can adjust the circle's radius.

4. Pick a point to change the circle's radius.

5. Pick the circle's center grip, right-click, and pick **Copy**.

6. Copy the circle three times by picking three points anywhere on the screen.

7. Press **ENTER** to terminate the copying.

Modifying the Grips Feature

For most AutoCAD users, the standard grips characteristics, such as color and size of the grip boxes, are adequate. In some companies or circumstances, however, it may be necessary to adjust the grips so that you can work with them more easily. For example, if your company has customized the background of the drawing area to dark gray, you may want to change the color of unselected grips boxes from blue to yellow so that you can see them more easily. You may need to adjust the size of

grip boxes depending on your display resolution. Grip boxes viewed at
1600 × 1200 resolution are twice as small as those viewed at 800 × 600.
AutoCAD allows you to make changes such as these using the Options
dialog box.

1. Select the **Tools** pull-down menu and select the **Options...** item.

The Options dialog box includes nine tabs with important information and
settings in each one.

2. Pick the **Selection** tab.

Information and settings related to the grips feature appear in the right
half of the dialog box, as shown in Fig. 5-3.

3. In the area labeled Grip Size, move the slider bar and watch how it
   changes the size of the box to the left of it.

4. Increase the size of the grips box and pick the **OK** button.

5. Select one of the objects and notice the increased size of the grips.

6. Pick one of the grips and perform an editing operation.

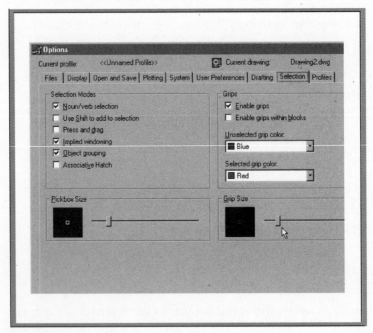

Fig. 5-3

7. Display the **Options** dialog box and adjust the size of the grips box, but do not close the dialog box.

The Selection tab also permits you to enable and disable the grips feature, as well as control the assignment of grips within blocks. You can also make changes to the color of selected and unselected grips. For now, do not change these settings.

8. Pick the **OK** button to close the Options dialog box.

Using Grips with Commands

The traditional AutoCAD method of entering a command and then selecting the object to be edited is sometimes called *verb/noun selection*. First you tell the software what to do (using a verb such as *erase*), and then you select the object to be acted upon (such as a *line*). When you use grips with editing commands, you are using *noun/verb selection*. First you select the object, and then you tell the software what to do to the object. You may find the grips (noun/verb) method convenient to use, especially if you are familiar with other graphics programs that use this technique.

1. Select one of the circles.

2. Pick the **Erase** button or enter **E** for ERASE.

Modify

As you can see, AutoCAD erased the circle without first asking you to select objects. You have just used the noun/verb technique.

3. Pick another circle and press the **Delete** key on the keyboard.

As you can see, pressing the Delete key is the same as entering the ERASE command.

4. Pick the **Undo** icon from the docked **Standard** toolbar.

5. Save your work and exit AutoCAD.

Standard

Chapter 5
Review & Activities

Review Questions

1. In AutoCAD, what is an object? Give three examples.

2. After you enter the ERASE command, what does AutoCAD ask you to do?

3. Experiment with and describe each of these object selection options.
   a. Last
   b. Previous
   c. Window
   d. Crossing
   e. Add
   f. Remove
   g. Undo
   h. WPolygon
   i. CPolygon
   j. Fence
   k. All

4. How do you place a window around a figure during object selection?

5. If you erase an object by mistake, how can you restore it?

6. How can you retain part of what has been selected for erasure while remaining in the ERASE command?

7. What is the fastest way of erasing the last object you drew?

8. When you use a graphics program, such as AutoCAD, what usually is the quickest way to edit an object?

9. Describe how you would remove one or more objects from a selection set.

10. Explain the primary benefit of using the grips feature.

11. What editing functions become available to you when you use grips?

12. Explain how to copy an object using grips.

13. Explain the difference between noun/verb and verb/noun selection techniques.

Chapter 5
Review & Activities

Challenge Your Thinking

1. As you have seen, AutoCAD's grips feature is both easy and convenient to use. With that in mind, discuss possible reasons AutoCAD also includes specific commands that you must enter to perform some of the same functions you can do easily with grips.

2. Describe a situation in which the Remove option can be useful for more than just removing an object that was added to a selection set by mistake.

Applying AutoCAD Skills

1. Create a new drawing from scratch. Draw a circle and a rectangle as illustrated in Fig. 5-4. Use the POLYGON command to create the rectangle. Enter the ERASE command and try to erase one line of the rectangle. At another location in the drawing, create another rectangle using the LINE command. Try to erase one line of the second rectangle. What is the difference? Save the drawing as prb5-1.dwg.

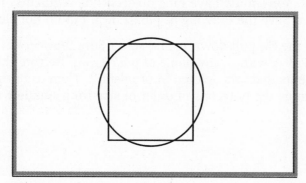

Fig. 5-4

Chapter 5
Review & Activities continued

2. Create a new drawing. Use the LINE command to draw polygons *a* through *e* as shown in Fig. 5-5.

   Enter the ERASE command and use the various object selection options to accomplish the following without pressing ENTER.

   - Place a window around polygon *a*.
   - Pick two of polygon *b*'s lines for erasure.
   - Use the Crossing option to select polygons *c* and *d*.
   - Remove one line selection from polygon *c* and one from polygon *d*.
   - Pick two lines from polygon *e* for erasure, but then remove one of the lines so it won't be erased.

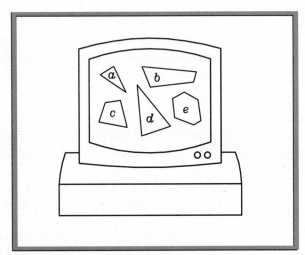

Fig. 5-5

   After you have completed the five operations listed above, press ENTER. You should have nine objects (line segments) left on the screen. Save the drawing as prb5-2.dwg, but do not close the file.

3. From the File pull-down menu, select Save Drawing As... and save prb5-2.dwg with a new name of prb5-3.dwg. Restore the line segments that you deleted in problem 2. Then use grips to rearrange the polygons in order of size from smallest to largest.

Chapter 5
Review & Activities

Using Problem-Solving Skills

Complete the following activities using problem-solving skills and your knowledge of AutoCAD.

1. The computer sketch in Fig. 5-6 was sent to the project engineer for comments and has now been routed to your desk with the attached note. Use the ELLIPSE, ARC, and LINE commands to create the sketch of the bushing. Because this is only a rough sketch, you may approximate the dimensions. Make the change requested by the engineering department. Save the drawing as ch5-bushing.dwg.

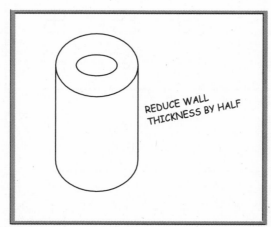

REDUCE WALL THICKNESS BY HALF

Fig. 5-6

Chapter 5
Review & Activities continued

2. Your architectural firm's customer wants the gable-end roof changed to a hip roof (Fig. 5-7). Draw the front elevation with both roof lines. Since this is only a roof line change, approximate the size and location of the windows and door. Save the drawing as Ch5-roof.dwg.

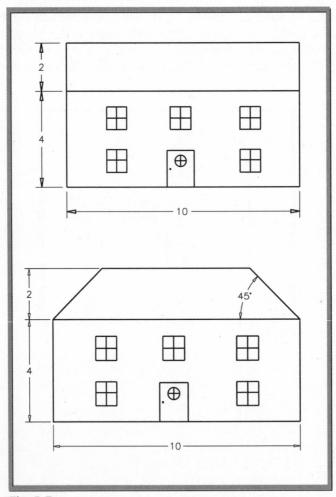

Fig. 5-7

Objectives:

- Describe several methods of entering *x,y* coordinates
- Locate points and draw objects using the Cartesian coordinate system
- Apply the absolute, relative, polar, polar tracking, and direct distance methods of entering coordinates

Key Terms

absolute points

alignment paths

Cartesian coordinate system

coordinates

coordinate pair

direct distance method

origin

polar method

polar tracking method

relative method

world coordinate system

Coordinates are sets of numbers used to describe specific locations. In AutoCAD, coordinates are used to specify the location of objects in a drawing. In previous chapters, when you used the pointing device to pick the endpoints of lines, etc., AutoCAD recorded the coordinates of the points that you picked on the screen.

You can also use the keyboard to enter coordinates. The keyboard method allows you to specify points and draw lines of any specific length and angle. This method also applies to creating arcs, circles, and other object types. By specifying the coordinates of objects, you can achieve the accuracy needed for engineering drawings.

AutoCAD uses a *Cartesian coordinate system* similar to that used in geometry. In AutoCAD, this system is called the *world coordinate system (WCS)*. In the two-dimensional version of this system, two axes are used. (See Fig. 6-1.) This is the version used for most drafting work. The X axis is an infinite, imaginary line that runs horizontally. The Y axis is an infinite, imaginary line that runs vertically. Each axis is numbered with sequential positive and negative numbers. Each combination of one *x* value and one *y* value (a *coordinate pair*) specifies a unique point in the coordinate system.

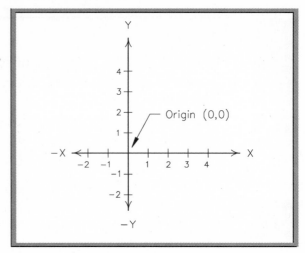

Fig. 6-1

The point at which the axes cross is known as the *origin*. At the origin, the values of *x* and *y* are both zero. The values to the right and above the origin are positive. Those below and to the left of the origin are negative. In AutoCAD's drawing area, the origin is located at the lower left corner by default.

In three-dimensional modeling, a third axis—the Z axis—is added at right angles to the other two axes to allow you to specify depth.

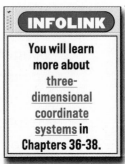

INFOLINK

You will learn more about three-dimensional coordinate systems in Chapters 36-38.

Methods of Entering Coordinates

The Cartesian system does not limit you to entering coordinate pairs to specify points. Consider the following ways to specify points using the LINE command.

Absolute Method

When you enter specific *x* and *y* values, AutoCAD places the points according to the Cartesian coordinate system. Points entered in this way are considered *absolute points* (Fig. 6-2).

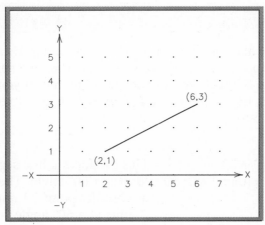

Fig. 6-2

Example:

LINE Specify first point: 2,1
Specify next point or [Undo]: 6,3

This begins the line at absolute point 2,1 and ends it at absolute point 6,3

Relative Method

The *relative method* allows you to enter points based on the position of a point that has already been defined. AutoCAD determines the position of the new point relative to a previous point (Fig. 6-3). Notice the use of the @ symbol in the following example. Use of this symbol tells AutoCAD that the following coordinate pair should be read relative to the previous point.

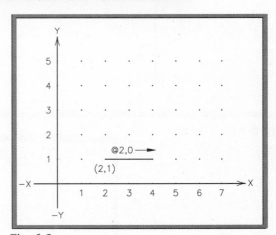

Fig. 6-3

Example:

LINE Specify first point: 2,1
Specify next point or [Undo]: @2,0

This draws a line 2 units in the positive X direction and 0 units in the Y direction from absolute point 2,1. In other words, the distances 2,0 are relative to the location of the first point.

Polar Method

The *polar method* is another form of relative positioning. Using this method, you can produce lines at precise angles. The area surrounding a point is divided into 360 degrees, as shown in Fig. 6-4. If you specify an angle of 90 degrees, for example, the line extends upward vertically from the last point (Fig. 6-5).

Example:

LINE Specify first point: 2,1
Specify next point or [Undo]: @3<60

This produces a line segment 3 units long at a 60° angle. The line begins at absolute point 2,1 and extends for 3 units at a 60° angle.

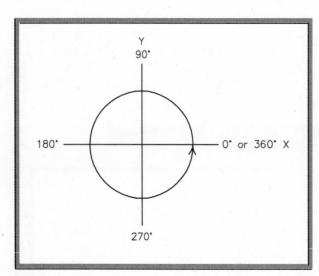

Fig. 6-4

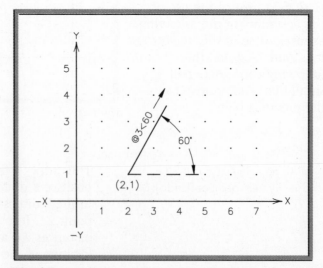

Fig. 6-5

Polar Tracking Method

The *polar tracking method* is similar to the polar method, except it's faster and easier. With this method, you can produce lines at precise angles and lengths with minimal keyboard entry (Fig. 6-6).

Polar tracking causes AutoCAD to display temporary *alignment paths* at prespecified angles to help you create objects at precise positions and angles. Polar tracking is on when the POLAR button on the status bar is depressed.

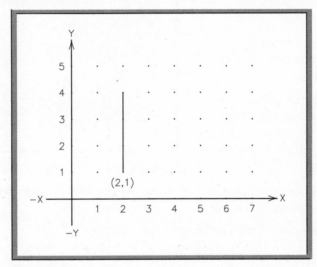

Fig. 6-6

Example:

LINE Specify first point: 2,1
Specify next point or [Undo]: Use the 90° alignment path that appears automatically on the screen and enter 3.

This produces a line segment 3 units long at a 90° angle. The line begins at absolute point 2,1 and extends 3 units upward vertically.

The alignment paths appear at 90° angles by default. You can change the angle of the alignment paths to meet your needs for specific drawing tasks. The following steps guide you through this process.

1. Create a new drawing from scratch.

2. From the **Tools** pull-down menu, select the **Drafting Settings...** item and pick the **Polar Tracking** tab.

The Drafting Settings dialog box appears as shown in Fig. 6-7.

3. In the Polar Angle Settings area of the dialog box, pick the down arrow and change the increment angle to **30.0**.

4. Pick the **OK** button to close the dialog box.

5. Enter the **LINE** command and pick a point anywhere on the screen.

6. Move the crosshairs away from and around the point.

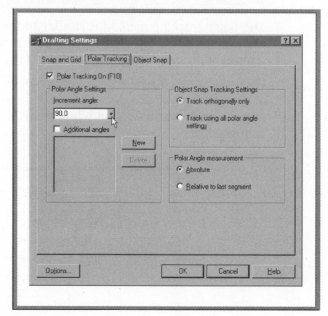

At 30° increments around the point, AutoCAD displays an alignment path and a small box showing the distance and angle from the point.

Fig. 6-7

7. Position the crosshairs so that they are at a 60° (1 o'clock) position from the point and enter **2**.

AutoCAD produces a line 2 units long at a 60° angle.

8. Use the polar tracking feature to produce additional lines.

9. Erase everything in the drawing file.

10. Display the **Drafting Settings** dialog box, set the polar tracking increment angle to **90**, and pick the **OK** button.

Direct Distance Method

AutoCAD also offers a *direct distance method* of entering points. This method is similar to polar tracking. After you specify the first point, you can "point" at any angle using the pointing device (without using alignment paths) and enter a number at the keyboard to specify the distance. This method is fast, but it is not as accurate as polar tracking because you cannot specify an exact angle.

Example:

LINE Specify first point: 2,1
Specify next point or [Undo]: Use the pointing device to point toward the lower left corner of the screen and enter 2.

This produces a line segment 2 units long at a 90° angle. The line begins at absolute point 2,1 and extends 3 units upward vertically.

Applying Methods of Coordinate Entry

Some methods of coordinate entry work better than others in a given situation. The best way to become familiar with the various methods of coordinate entry, and which to use when, is to practice using them.

1. Enter the **LINE** command.

2. Using the keyboard, create the drawing of a gasket shown in Fig. 6-8. Don't worry about the exact locations of the holes, and don't dimension the drawing. However, do make the drawing exactly this size. The top and bottom edges of the gasket are perfectly horizontal.

HINT:

As you draw the holes in the gasket, keep in mind that the ø (*theta*) symbol indicates that the dimensions given are diameters.

NOTE:

You can enter negative values for line lengths and angles.

    LINE Specify first point: 5,5
    Specify next point or [Undo]: @2<-90

This polar point specification produces a line segment 2 units long downward vertically from absolute point 5,5. This is the same as entering @2<270. Try it and then erase the line.

3. Save your work in a file named gasket.dwg and exit AutoCAD.

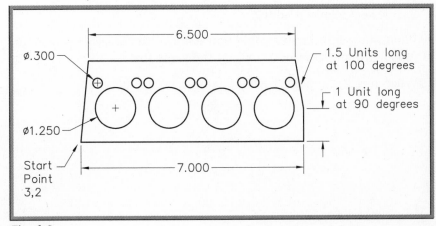

Fig. 6-8

Chapter 6
Review & Activities

Review Questions

1. What is the relationship between AutoCAD's world coordinate system and the Cartesian coordinate system?

2. Briefly describe the differences between the absolute, relative, polar, and polar tracking methods of point specification.

3. Explain the advantage of specifying endpoints from the keyboard rather than with the pointing device.

4. What is the advantage of specifying endpoints with the pointing device rather than the keyboard?

5. What happens if you enter the @ character at the Specify first point prompt after entering the LINE command?

6. What is the effect on the direction of the new line segment when you specify a negative number for polar coordinate entry?

7. What is the difference between direct distance entry and the polar tracking method of entering coordinates?

Challenge Your Thinking

1. Why may entering absolute points be impractical much of the time when completing drawings?

2. Consider the following command sequence:

   LINE Specify first point: 4,3
   Specify next point or [Undo]: 5,1

   What polar coordinates could you enter at the Specify next point or [Undo] prompt to achieve exactly the same line?

3. When might the polar tracking feature be useful?

Chapter 6
Review & Activities

Applying AutoCAD Skills

Work the following problems to practice the commands and skills you learned in this chapter.

1. Figure 6-9 shows the end view of a structural member for an architectural drawing. Create the rectangle using the LINE command and the following keyboard entries.

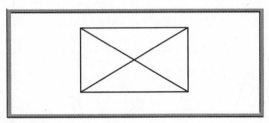

Fig. 6-9

Specify first point: 1,1
Specify next point or [Undo]: 10.5,1
Specify next point or [Undo]: @0,5.5
Specify next point or [Close/Undo]: @9.5<180
Specify next point or [Close/Undo]: c

Draw a line from the intersection at the lower left corner of the rectangle to the upper right corner of the rectangle. Use coordinate pairs to place the endpoints of the diagonal line exactly at the corners of the triangle. The new line defines the rectangle's diagonal distance, or the hypotenuse of each of the two triangles you have created. Then draw a line from the lower right corner to the upper left corner of the rectangle to create the other diagonal. Save the drawing as prb6-1.dwg.

Chapter 6
Review & Activities continued

2-3. List on a separate sheet of paper exactly what you would enter when using the LINE command to produce the drawings of sheet metal parts in Figs. 6-10 and 6-11. Then step through the sequence in AutoCAD. Try to use four methods—absolute, relative, polar, and polar tracking—to enter the points. (Note that the horizontal lines in the drawings are perfectly horizontal.)

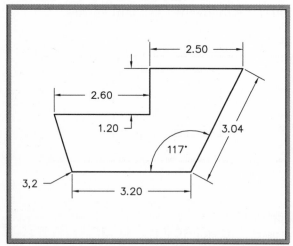

Fig. 6-10

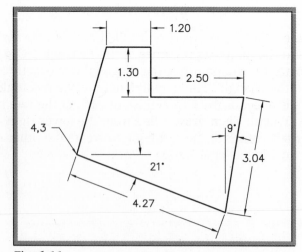

Fig. 6-11

4. Using the polar tracking option, create the drawing of a 5¼″ disk shown in Fig. 6-12. The slotted hole is optional.

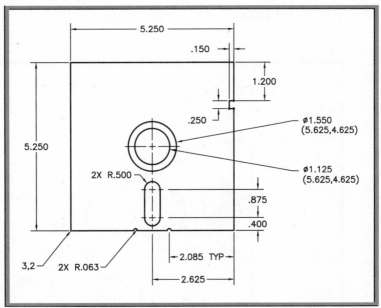

Fig. 6-12

5. Draw the 3½″disk shown in Fig. 6-13.

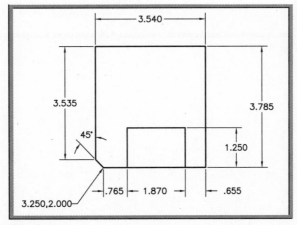

Fig. 6-13

Chapter 6
Review & Activities continued

6. Draw the front view of the spacer plate shown in Fig. 6-14. Start the drawing at the lower left corner using absolute coordinates of 3,3. Determine and use the best methods of coordinate entry to draw the spacer plate.

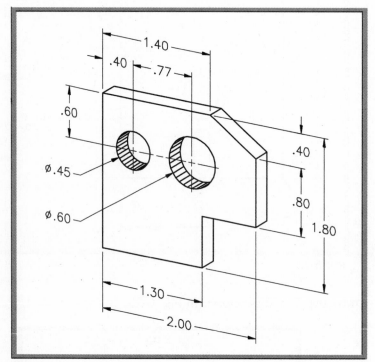

Fig. 6-14

7. Draw the front view of the locking end cap shown in Fig. 6-15. Start the drawing at the center of the object using absolute coordinates of 5,5. Use absolute coordinates to locate the centers of the four holes, and use polar tracking with the RECTANG command to draw the rectangular piece.

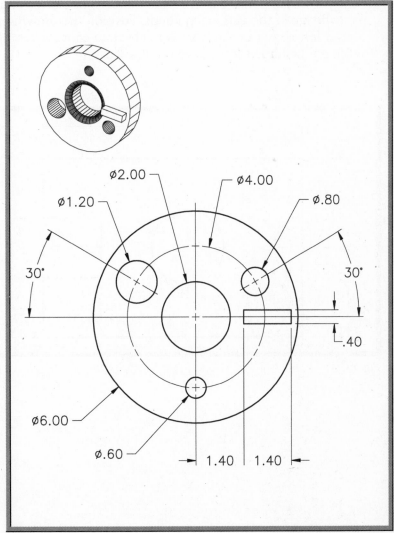

Fig. 6-15

Chapter 6
Review & Activities continued

Using Problem-Solving Skills

Complete the following activities using problem-solving skills and your knowledge of AutoCAD.

1. Draw the shim (Fig. 6-16). Decide on the best method of coordinate entry for each point, based on the dimensions shown and your knowledge of the various methods. Position the drawing so that the lower left corner of the shim is at absolute coordinates 1,1. Do not include the dimensions. Save the drawing.

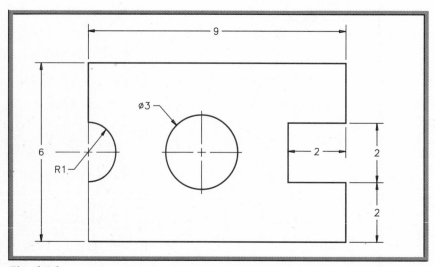

Fig. 6-16

2. Draw the front view of the slider block shown in Fig. 6-17. Assume the spacing between dots to be .125. Start the drawing at the lower left corner using absolute coordinates of 1,3. Use the RECTANG command to draw the rectangular slot. Use relative polar coordinates to locate the endpoints of the lines, and absolute rectangular coordinates to locate the centers of the three holes.

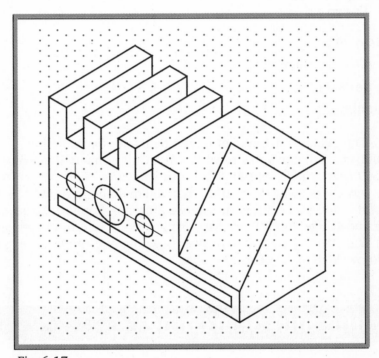

Fig. 6-17

Chapter 7 Securing Help

Objectives:

- **Browse a group of files**
- **Obtain help on AutoCAD commands and topics**
- **Use AutoCAD's context-sensitive help features**

Key Terms

browsing

context-sensitive help

hypertext link

thumbnails

AutoCAD provides a rich mix of online help and file searching capabilities. Its digital versions of the *AutoCAD Command Reference, AutoCAD User's Guide,* and *AutoCAD Customization Guide* are no more than a few clicks away no matter what you're doing in AutoCAD. The powerful AutoCAD Learning Assistance program, Support Assistance online knowledgebase, What's New explanations, and resources on the World Wide Web allow you to answer puzzling questions and expand your knowledge of AutoCAD.

Browsing and Searching

When you open a drawing, AutoCAD provides visual help, making the file selection process faster.

1. Start AutoCAD.

2. Pick the **Open a Drawing** button in the **Startup** dialog box.

3. Locate and click once on the **engine.dwg** file. (Engine.dwg is the file you created in Chapter 4.)

As you have seen in previous chapters, the drawing appears in a preview box in the right area of the dialog box. This allows you to find more easily drawings that you or others have created.

Browsing Files

AutoCAD also provides more advanced options for browsing files. In AutoCAD, *browsing* means looking quickly through *thumbnails,* or small representations, of a large number of files to find the one you want to open.

4. Pick the **Browse...** button in the **Startup** dialog box to display the Select File dialog box.

5. Find and double-click the **Sample** folder, which is located in the **Acad2000** folder.

6. Pick the **Find File...** button in the Select File dialog box.

The Browse/Search dialog box appears. Near the top, you should see two tabs labeled Browse and Search. AutoCAD displays either the browse or search information, depending on how it was left by the last user.

7. If the browse information is not present, pick the **Browse** tab.

AutoCAD displays a thumbnail of each drawing file in the Sample folder, making it easier to find a particular file.

8. In the **Directories** list box, double-click **Acad2000**.

9. In the same list box, double-click **SUPPORT**.

This displays the contents of the SUPPORT directory.

10. In the lower right area of the dialog box, change Size to **Medium** using the down arrow. (It may already be set to Medium.)

This displays medium-size images of the drawings.

11. Use the scrollbar to browse the files.

Double-clicking any of these images will open the file. Do not open a file at this time.

12. Change Size to **Small**.

Performing Searches

The Search tab allows you to search for files according to their type and date of creation.

1. Pick the **Search** tab, causing the search information to appear in front of the browse information.

2. Change the text in the Search Pattern edit box to ***.dwg.** (It may already be set to this.)

3. In the Date edit box, change the date to **1/1/97**.

4. In the Search Location area, pick the **Path** radio button and enter **\Program Files\Acad2000** in the Path text box, assuming that this is the path to your AutoCAD folder.

5. Pick the **Search** button.

AutoCAD searches for all DWG files in the Acad2000 folder, as well as in all subfolders in Acad2000, that were created after January 1, 1997.

6. Use the scrollbar to browse the files.

7. Find the file named **WATCH.DWG** and double-click it.

HINT: ───

You may need to use both the vertical and the horizontal scrollbars to find the drawing.

This opens and displays the watch.dwg file.

Obtaining Help

In addition to the visual help AutoCAD provides through its thumbnail browsing feature, the software offers many other forms of help. Most of these can be obtained by entering the HELP command.

1. Find and open the **engine.dwg** file. (You may need to use the **Up One Level** button in the **Select File** dialog box.)

Notice that the watch.dwg file remains open behind engine.dwg. The cylinder head you created in Chapter 4 should now be visible on the screen.

2. Pick the **Help** button from the **Standard** toolbar, or enter **HELP** or **?** at the keyboard.

Standard

HINT: ───

You can also obtain help by pressing the F1 function key.

Fig. 7-1

The AutoCAD Help dialog box appears as shown in Fig. 7-1.

Searching by Content Area

At this point, you can obtain instructions on how to use a particular command, or you can obtain an alphabetical list of AutoCAD topics.

3. Pick the **Contents** tab.

4. Double-click **Command Reference**.

5. Under Command Reference, double-click **Commands**.

This displays a window with an alphabetical listing of AutoCAD commands. Each of the commands is underlined, indicating that it is linked to additional information using hypertext. A *hypertext link,* when clicked, takes you to related information.

NOTE: ——————————————————————

The commands with the word NEW beside them are new to AutoCAD 2000.

Notice the horizontal row of buttons near the top of the window. These alphabetical buttons allow you to jump quickly to topics that start with each letter of the alphabet.

6. Click the **L** button located above the list of commands.

This displays all commands that begin with the letter L.

7. Find the command named **LWEIGHT** and click it.

This displays detailed information on the command. Notice the additional words that are underlined. These are also hypertext links. Below the information on the LWEIGHT command, a SEE ALSO section refers you to other areas of AutoCAD help that contain related information.

8. Click on the **LAYER** link in the SEE ALSO section.

This provides information and additional links related to the LAYER command. So you see, much of AutoCAD's help information is linked using hypertext.

9. In the upper left area of the window, pick the **Back** button.

This returns you to the previous screen.

10. Pick the **Back** button again.

11. Pick the **Glossary** button.

This produces a glossary of AutoCAD terms. Notice that the glossary also contains hyperlinks to relevant AutoCAD topics.

12. Pick the **Help Topics** button.

The original AutoCAD Help dialog box reappears.

13. Close both help windows by clicking the **x** in the upper right corner of the windows.

Using the Help Index

Another way to search for help in AutoCAD is to use the help index. Using this method, you can find entries in all of the online help documents by entering just the first few letters of the command or topic for which you need help.

Standard

1. Reopen the **AutoCAD Help** dialog box.

2. Pick the **Index** tab.

The dialog box appears as shown in Fig. 7-2.

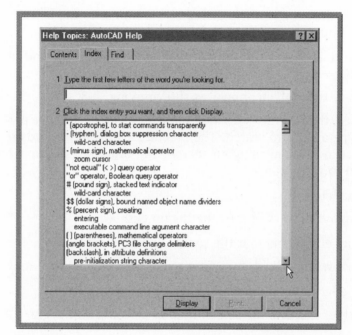

Fig. 7-2

3. Use the scrollbar to scroll down through the list of topics.

Notice a blinking cursor inside the empty entry box.

4. Type **P** in upper or lower case, but do not press ENTER.

AutoCAD very quickly finds the first topic in the list box that begins with the letter P.

5. After the P, type **L** and then **O**, but do not press ENTER.

AutoCAD finds the first topic that begins with PLO. In this case, it finds the PLOT command.

6. Pick the **Display** button.

Information on plotting appears from the online *AutoCAD User's Guide*.

7. Read the information.

8. Close the help window when you are finished.

Finding Help Topics

The Find tab in the AutoCAD Help dialog box is similar to the Index tab, although it produces many more options.

Standard

1. Pick the **Help** button from the docked **Standard** toolbar and pick the **Find** tab.

The contents of the Find tab appear as shown in Fig. 7-3.

As you can see, the Find tab is divided into three parts. The first part is where you type a word to search. The second part displays words that you can select to help narrow your search. The third part displays the topics related to your search.

2. Type the letter **M**.

AutoCAD displays a list of entries that begin with M.

3. Type the letter **O**, so that the search line now reads MO.

AutoCAD narrows the search by listing words that begin with MO.

4. Type the letter **V**, so that the search line reads MOV.

AutoCAD further narrows the search by listing words that begin with MOV.

5. Select **MOVE** (the entry that appears in all capital letters).

In the third section, AutoCAD lists all topics related to MOVE.

6. Scroll down the list and click on **MOVE Command [ACR – AutoCAD Command Reference]**.

7. Pick the **Display** button to display information on the MOVE command from the *AutoCAD Command Reference* and review this information.

8. In the upper left corner of the window, pick the **Help Topics** button.

This displays the AutoCAD Help dialog box.

9. Experiment further with the Find capability on your own.

10. Close any help windows that might be open.

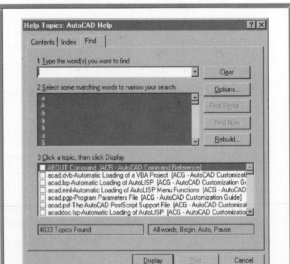

Fig. 7-3

Other Sources of Help

The Help pull-down menu provides several sources of help. Selecting the first item on the menu, AutoCAD Help, is the same as entering the HELP command. The remaining items on the pull-down menu offer other types of assistance, as shown below.

What's New	Overview of what's new in AutoCAD 2000
Learning Assistance	Self-instructional learning aid with graphics, motion, and sound
Support Assistance	Solutions index categorized by topic
Autodesk on the Web	Websites for AutoCAD users
Connect to AutoCAD Website	AutoCAD Website

Each option is easy to follow. Note that you will need Internet access for the last two options. You will need the AutoCAD 2000 AutoCAD Learning Assistance CD-ROM to use AutoCAD Learning Assistance.

1. Experiment with each option on your own.

2. Close any help-related windows before proceeding to the next section.

Context-Sensitive Help

AutoCAD allows you to obtain help in the middle of a command, when you're most likely to need it. If another command is active when you enter the HELP command, AutoCAD automatically displays help for the active command. This is known as *context-sensitive help*. When you close the help window, the command resumes.

1. Enter the **GRID** command at the keyboard.

2. Pick the **Help** button.

Standard

AutoCAD displays help information for the GRID command.

3. After closing the help window, press the **ESC** key to cancel the GRID command.

4. Open the **AutoCAD Help** dialog box, pick the **Find** tab, and experiment with it on your own.

5. Close any help windows that might be open and exit AutoCAD without saving.

Chapter 7
Review & Activities

Review Questions

1. Suppose you want to open a specific file in the folder you named after yourself, but you've forgotten the name of it. Explain how to browse through the files in that folder to find the file you want to open.

2. When you pick the HELP button, what does AutoCAD display?

3. How do you obtain a listing of all AutoCAD topics?

4. Suppose you have entered the MIRROR command. At this point, what is the fastest way of obtaining help on the command?

5. Why are hypertext links useful?

6. Name and briefly describe each of the options available in the Help pull-down menu.

7. What is context-sensitive help? Explain how to use it in AutoCAD.

Challenge Your Thinking

1. For some commands, the help information may be long and complicated. Discuss ways to make a permanent hard copy of commonly used help information to keep for reference.

2. Explain the benefit of using the Find capability in the AutoCAD Help dialog box. When would you use it instead of the Index?

Applying AutoCAD Skills

Work the following problems to practice the commands and skills you learned in this chapter.

1. You know you have an elevation drawing of a public building somewhere on your hard drive, but you don't remember the full name of the drawing. You recall that it starts with the letters "oc," and because it's a drawing file, you know that it has a DWG extension. Use the Browse/Search dialog box to find the drawing. What is the name of the file, and what does it contain?

Chapter 7
Review & Activities

2. Obtain AutoCAD help on the following commands and state the basic purpose of each.
   a. EXPORT
   b. SNAP
   c. TIME
   d. SCALE
   e. TRIM

3. Locate each of the above commands in the online *AutoCAD Command Reference* and read what it says about each of them.

 Using Problem-Solving Skills

1. Your company has purchased a new plotter for the AutoCAD network. You have been asked to configure it, but since you don't do that many configurations, you need some help. Find the instructions in the AutoCAD help system for configuring a network nonsystem plotter, and write down the path for future reference.

2. You need to draw two concentric circles (circles that share the same center point). You have already drawn the inner circle. The outer circle needs to be .5″ larger. Enter the OFFSET command from the Draw toolbar and access context-sensitive help to offset the diameter of the first circle by .5″.

Careers Using AutoCAD

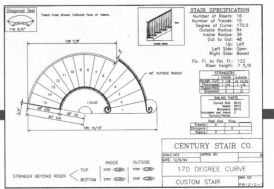

Adapted from an article by Wayne Turner in CADENCE magazine. © 1990, Miller Freeman, Inc. Revised 1999.

Staircase Manufacturing

During a busy season, the Century Stair Co. (Haymarket, VA) cuts a thousand risers a day for straight staircases, which is the staircase type most customers want. A curved staircase, though elegant, takes much longer to build and costs more.

On a curved staircase, every step must be cut to its own specifications. According to Production Manager Bob Snitzer, a machine operator might need half a day to cut out risers, treads, and stringer boards for a typical curved staircase. However, using a system called TRIAD, Century Stair Co. can deliver a curved staircase five times faster.

TRIAD is an integration of "adjustable template" software and robotic machining to make the most of AutoCAD. Here is how it works: When an order for a curved staircase comes in, Century Stair Co. makes an AutoCAD drawing. The drawing is generated quickly with the aid of Synthesis, a parameter-control program that works like an adjustable template. The user specifies certain dimensions as inputs; Synthesis calculates the remaining dimensions.

On the manufacturing side, the AutoCAD drawing serves as the data source for robotic machining. A program called NC-AUTO-CODE implants numeric code into the AutoCAD drawing to guide a CNC (computer numeric control) machine. Just one two-head router can turn out ten exquisitely curved staircases per week.

Career Focus—CNC Operator

A CNC machine combines two major systems: a computer and a machine tool. Operators are responsible for setting up and monitoring both systems. The operator loads job-specific infor-

mation into the computer and ensures that the workpiece is positioned and secured correctly.

The operator examines the first workpiece closely to make sure that cutting angles, hole depths, and other dimensions are within specification. As the job continues, the operator periodically checks measurements, machine stock, and lubrication. The operator may be responsible for two or more machines running at once.

Education and Training

Employers generally require training in a 2-year technical program for operators of CNC machines. A typical curriculum includes classes in machining as well as in blueprint interpretation, technical mathematics, and technical reading/writing.

Programming languages vary by machine, so operators often receive additional training as new equipment or software is introduced. Operators who get sufficient training in CNC programming may advance to the position of CNC programmer.

Career Activity

- Interview someone who can tell you about work as a machine operator. What is the typical pay? What is the most difficult part of the job?

Chapter 8 — File Maintenance

Objectives:

- **Copy, rename, move, and delete drawing files**
- **Create and delete folders**
- **Audit and recover damaged drawing files**

Key Terms

audits

file attributes

When working with AutoCAD, there will be times when you want to review the contents of a folder or delete, rename, copy, or move files. This chapter gives you practice in using these functions, as well as commands that attempt to repair drawing files containing errors.

File and Folder Navigation

AutoCAD's Select File dialog box can help you organize and manage AutoCAD-related files.

1. Start AutoCAD.

2. Pick the **Open a Drawing** button in the **Startup** dialog box.

3. Pick the **Browse...** button.

AutoCAD displays the Select File dialog box (Fig. 8-1).

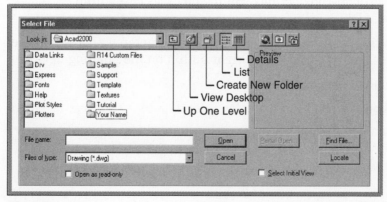

Fig. 8-1

Notice the buttons along the top of the dialog box.

4. Double-click the folder with your name—the one you created in Chapter 3.

Viewing File Details

5. Pick the **Details** button.

This provides a listing of each file, along with its size, type, and the date and time it was last modified. These items are known as *file attributes*. In folders that contain many drawing files, it is sometimes easier to find files if they are listed in a certain order. By default, files are listed in alphabetical order by name. You can change the order in which the files are listed by clicking the column names, which are actually buttons.

6. Pick the top of the **Name** column.

This reverses the order of the files. They are now in reverse alphabetical order (from Z to A).

7. Pick the top of the **Name** column again.

The files are once again listed in alphabetical order.

8. Pick the top of the **Size** column.

This places the files in order of size from smallest to largest. Picking Size again would reorder the files from largest to smallest. Picking the top of the Type and Modified columns organizes the files according to type and date, respectively.

9. Pick **Modified**; pick it again.

10. Pick the **List** button to produce a basic listing of the files.

Copying Files

The Select File dialog box permits you to copy files easily from within AutoCAD.

1. Right-click and drag the **stuff.dwg** file to an empty area within the list box area and release the right button on the pointing device.

A shortcut menu appears.

2. Pick **Copy Here**.

A new file named Copy of stuff.dwg appears.

Renaming Files

You can also rename files using the Select File dialog box.

1. Single-click the new file named **Copy of stuff.dwg**.

2. Single-click the file again.

You should now see a blinking cursor at the end of the file name with a box around the file name.

3. Type the name **junk.dwg** and press **ENTER**.

NOTE: ————————————————————————

> Instead of entering an entirely new name, you can also edit the name of the file by picking a new cursor location and typing new text. This method of renaming also permits you to use the backspace key and the spacebar.

The new file named junk.dwg contains the same contents as stuff.dwg.

Creating New Folders

It's also fast and easy to create new folders.

1. Pick the **Create New Folder** button.

A new folder named New Folder appears. Notice that New Folder is highlighted.

2. To assign a new name to the folder, type **Junk Files** while the **New Folder** name is highlighted and press **ENTER**.

Moving Files

Let's move a file to the new folder.

1. Right-click **junk.dwg** and drag and drop it into the **Junk Files** folder.

A shortcut menu appears.

2. Pick **Move Here**.

The junk.dwg file is now located in the Junk Files folder.

Deleting Files and Folders

AutoCAD also allows you to delete unwanted files and folders using the Select File dialog box.

1. Double-click the **Junk Files** folder.

2. Select **junk.dwg** and press the **Delete** key on the keyboard.

3. Pick the **Yes** button in the **Confirm File Delete** dialog box if you are sure you have selected junk.dwg.

4. Pick the **Up One Level** button.

5. Select the **Junk Files** folder and press the **Delete** key.

6. If you are sure that you selected the Junk Files folder, pick the **Yes** button in the **Confirm Folder Delete** dialog box to send this folder to the Windows Recycle Bin.

Diagnosing and Repairing Files

Occasionally, AutoCAD drawing files may become corrupted. AutoCAD supplies two commands to help fix corrupted files.

Auditing a Drawing File

The AUDIT command is available as a diagnostic tool. It *audits,* or examines the validity of, the current drawing files. It can also correct some of the errors it finds. The AUDIT command automatically creates an audit report file containing an ADT file extension when the AUDITCTL system variable is set to 1 (On). The default setting is 0.

Probable causes of damaged files include:

• An AutoCAD system crash

• A power surge or disk error while AutoCAD is writing the file to disk

When a file becomes damaged, AutoCAD may refuse to edit or plot the drawing. In some cases, a damaged file may cause an AutoCAD internal error or fatal error.

1. Open the file named **stuff.dwg**.

2. From the **File** pull-down menu, select **Drawing Utilities** and **Audit**.

This enters the AUDIT command.

3. In reply to Fix any errors detected? press **ENTER** to accept the No default.

4. Press the **F2** function key to open the AutoCAD Text Window and read all of the message.

Information similar to that shown in Fig. 8-2 appears.

```
6 Blocks audited
Pass 1 54   objects audited
Pass 2 54   objects audited
Total errors found  0   fixed  0
```

Fig. 8-2

If you had entered Yes in reply to Fix any errors detected? the report would have been the same if this file has no errors.

 NOTE:

AutoCAD offers a command similar to AUDIT called RECOVER. This command attempts to open and repair damaged drawing files. The "recover" process has been embedded in the OPEN command, so when you try to open a damaged file, AutoCAD automatically executes the RECOVER command and attempts to repair it.

Permanently Damaged Files

A drawing file may be damaged beyond repair. If so, the drawing recovery process will not be successful.

Each time you save an AutoCAD file, AutoCAD saves the changes to the current DWG file. AutoCAD also creates a second file with a BAK file extension. This file contains the previous version of the file—the version prior to saving. If the DWG file becomes damaged beyond repair, you can rename the BAK file to a DWG file and use it instead. To prevent the loss of data, save often and produce a backup copy of your DWG files frequently.

5. Exit AutoCAD without saving unless the file was damaged and recovered. In that case, save the changes to the file.

Chapter 8
Review & Activities

Review Questions

1. Explain the purpose of each of the five buttons located in the top middle area of the Select File dialog box.

2. How do you copy a file in the Select File dialog box?

3. In the Select File dialog box, what information can you display by picking the Details button?

4. Explain how to create a folder from within AutoCAD.

5. What is the primary purpose of the AUDIT command?

Challenge Your Thinking

1. Is it possible to move and copy files from the Select File dialog box to folders on the Windows desktop? Is it possible to move and copy files from folders on the Windows desktop to the Select File dialog box? Explain why this might be useful.

2. AutoCAD gives you the option of creating a report file when you run the AUDIT command. Explain why such a report might be useful.

Applying AutoCAD Skills

Using the features described in this unit, complete the following activities.

1. Display a list of drawing files from one of AutoCAD's folders.

2. Display a list of template files found in one of AutoCAD's folders.

3. Rename one of your drawing files. Then change it back to its original name.

4. Make a copy of the file Truck model.dwg in the AutoCAD Sample folder. Rename it TM.dwg, and move it to your named folder.

5. Delete TM.dwg from the folder with your name.

6. Diagnose one of your drawing files and fix any errors detected. Display the entire text of the AutoCAD message. How many passes did the diagnostic tool run? How many objects were included?

Chapter 8
Review & Activities

 Using Problem-Solving Skills

Complete the following activities using problem-solving skills and your knowledge of AutoCAD.

1. It's time to build the bookcase you designed in Chapter 3. Make a copy of the bookcase you saved in the folder with your name as prb3-1.dwg. Modify the drawing as shown in Fig. 8-3, and save the modification as Bookcase.dwg. Delete the copy of prb3-1.dwg.

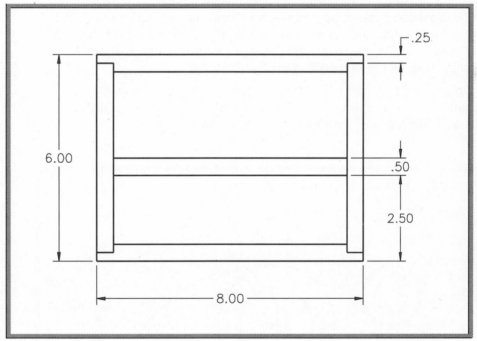

Fig. 8-3

2. To keep track of projects, you need to organize them into folders. Make a new folder called Build within your named folder. Move bookcase.dwg into the new folder as shelves.dwg. Check the file for corruption. You are now ready to build the bookcase/shelves!

Part 1 Project

Rocker Arm

Mechanical drafting refers to the creation of highly accurate drawings using drafting instruments or, more commonly today, CAD software. Mechanical drawings are used, among other things, to describe a part so that the manufacturer can build it correctly.

Description

The rocker arm shown in Fig. P1-1 was submitted to your manufacturing company as a hand-drafted drawing. The design has been approved, and now the marketing department needs an electronic version of the drawing for use in a new product brochure. The drawing must reflect the actual dimensions of the rocker arm, although for this use, it is not necessary to show the dimensions on the drawing.

Your task is to draw the rocker arm accurately using the AutoCAD commands and procedures you learned in Part 1. Be sure to read the "Hints and Suggestions" for this project before you begin.

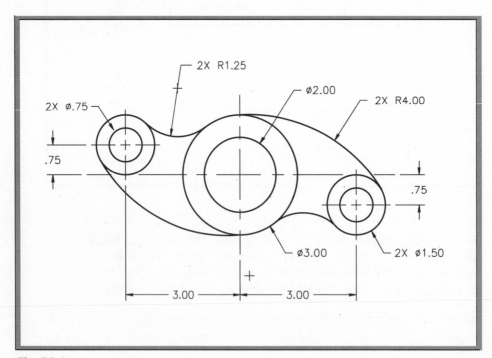

Fig. P1-1

Part 1 Project

Hints and Suggestions

1. Examine the drawing to determine an appropriate drawing area and the type of units you should specify. Then use AutoCAD's Quick Setup wizard to set up the drawing.

2. The lines made up of short and long dashes are center lines (lines that intersect at the exact center of a feature). Use these lines and your knowledge of coordinates to draw the rocker arm accurately.

3. Study the drawing to determine the best order in which to draw the parts of the rocker arm. Keep in mind that it is *not* necessary to construct the part linearly, so that each line connects to the next. For example, you may choose to create all the holes before you draw the actual arm.

4. If you have trouble visualizing how to draw the part in AutoCAD, try sketching it first on graph paper.

Summary Questions/Self-Evaluation

Your instructor may direct you to answer these questions orally or in writing, or simply to think about your responses as a means of self-evaluation.

1. Pinpoint any difficulties you had in drawing the rocker arm and explain how you solved them.

2. Describe the order in which you drew the parts to make up the rocker arm. Why did you choose this order?

3. Study the drawing in P1-1 once more. Does this drawing provide everything needed to manufacture the actual part? If not, what is missing?

4. In this project, you have created an electronic drawing file to be used by the marketing department because marketing needed the file in electronic format. However, there are other advantages to creating the file electronically. How else can this file be used? How would it need to be modified to be useful to other departments? Explain your answer.

Careers
Using
AutoCAD

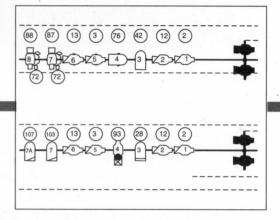

Theater Lighting

The actor "struts and frets his hour upon the stage," as Shakespeare said. Then the crew dismantles everything, making way for the next show. Temporariness is just a part of the challenge for the stage lighting director. Where residential/commercial lighting is standard, permanent, and governed by codes, stage lighting gets reconfigured with every change of the marquee.

Yet a lighting director relies on schematic drawings like any other electrical technician. The paperwork can get complicated because, in addition to the multiple rows of spotlights (attached to a grid of pipes raised up near the ceiling), there are boom-mounted lights and footlights to keep track of. The versatility of stage lighting makes it possible to set a misty mood for the double-double-toil-and-trouble scene in Act I and then crank everything up for the blaze of battle in Act V.

Enter Light "Works," a program that runs with AutoCAD to help lighting directors manage the versatility of stage lighting. With a built-in library of lighting symbols, Light "Works" makes it easy to draw a schematic and even easier to modify a schematic to suit an upcoming show. With only six commands (plus eight AutoCAD commands), the user can make yesterday's floods become twinks tonight.

Light "Works" is the brainchild of Mark Weaver, who developed the software while a graduate student at the Yale School of Drama's Technical Design and Production Department. Now a staff lighting director for television's ABC News, Mark has also done lighting for the network's sports and entertainment divisions.

Career Focus—Lighting Director

A theater lighting director plans and implements stage lighting to fit the requirements of a particular performance, whether it is a drama, a concert, or any other theatrical performance. For a play, the lighting director consults the play director about how lighting can contribute to the performance. Then the lighting director draws up a plan that shows, scene by scene, how the stage will be lit.

During a performance, the lighting director or a technician operates stage lighting from a control panel, typically situated above and behind the audience. Before a production, the lighting director supervises the work of stagehands in installing and repositioning different types of lights.

Education and Training

The curriculum in theater arts programs offered at two- and four-year colleges usually emphasizes literature, history, and cultural aspects of theater as much or more than performance and production skills. At a few schools, such as the Yale School of Drama, postgraduate programs focus on performance and production, including methods and technology. Schools with professional-level programs tend to be clustered in or near cities with plenty of theatrical activity, such as New York.

Career Activity

- Write a report about opportunities for training and employment in the entertainment industry, including stage productions and broadcasting.

Part 2

Drawing Aids and Controls

Chapter 9 Object Snap

Objectives:

- Set and use running object snap modes
- Apply object snaps that are not currently turned on
- Change object snap settings to increase productivity

Key Terms

aperture box

flyout toolbars

object snap

object snap tracking

perpendicular

quadrant

quadrant point

running object snap modes

running object snaps

Object snaps are like magnets that permit you to pick endpoints, midpoints, center points, points of tangency, and other specific points easily and accurately. *Object snap tracking,* which is used in conjunction with object snap, provides alignment paths that help you produce objects at precise positions and angles. Combined with polar tracking, these features speed the precision drawing process.

Running Object Snaps

You can use object snaps in different ways to make a drawing task easier. When you know that you will be need one or more specific object snaps several times, you can preset them to "run" in the background as you are working. Object snaps that have been preset in this way are known as *running object snaps.*

Specifying Object Snap Modes

1. Start AutoCAD.

2. In the **Startup** dialog box, pick the **Use a Wizard** button, **Quick Setup**, and **OK**.

3. Choose **Decimal** units and pick the **Next** button.

4. For Width, enter **95**, and for Length, enter **70**.

5. Pick the **Finish** button and save your work in a file named **bike.dwg**.

The drawing area now represents 95 × 70 units. These units can represent millimeters, kilometers, feet, miles, or whatever we want them to be. We will use them as inches.

6. Enter **Z** for ZOOM and **A** for All.

This zooms the drawing area to the full 95 × 70 inches.

7. Use the pick button on the pointing device to depress the **POLAR**, **OSNAP**, and **OTRACK** buttons on the status bar, unless they are already depressed.

The MODEL button should also be depressed, but the SNAP, GRID, and ORTHO buttons should not be.

8. Enter the **OSNAP** command.

Entering the OSNAP command is equivalent to selecting Drafting Settings... from the Tools menu and selecting the Object Snap tab in the dialog box (Fig. 9-1).

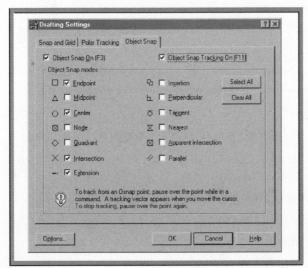

Fig. 9-1

9. Review the object snap modes.

The ones that are checked are called *running object snap modes*. These object snaps are "on" and function automatically as you work with drawing and editing commands in AutoCAD.

10. Check **Endpoint**, **Center**, **Intersection**, and **Extension**, and uncheck all the others.

Notice that Object Snap On and Object Snap Track On are checked. This is because the OSNAP and OTRACK buttons on the status bar are on (depressed). A third way to turn them on and off is to press the F3 and F11 function keys, as indicated in the dialog box.

11. Pick the **OK** button.

Using Preset Object Snaps

Let's use AutoCAD's object snaps to create a basic drawing of a new mountain bike frame design (Fig. 9-2). Be sure to save your work every few minutes as you work through this chapter.

1. In the lower left area of the screen, draw a 22″-diameter circle to represent the front wheel as shown in Fig. 9-2.

Fig. 9-2

The Center object snap selects the exact center of a circle or arc. Let's use it to start a 24″ line at the center of the bike's front wheel. This line will begin the outline of the bike frame.

2. Enter the **LINE** command.

Draw

3. At the Specify first point prompt, move the crosshairs so that it touches the circle or move the crosshairs to the circle's center.

A small yellow marker appears at the circle's center. The yellow marker indicates that AutoCAD has selected the point at the center of the circle.

4. While the yellow marker is present, pick a point.

AutoCAD snaps to the circle's center.

5. For the next endpoint, use polar coordinates to extend the line **24"** up and to the right at a **70°** angle.

HINT:

Enter @24<70.

6. Press **ENTER** to terminate the LINE command.

Object Snap Tracking

Next, we are going to draw the horizontal line using AutoCAD's object snap tracking feature.

1. Reenter the **LINE** command and move the crosshairs to the second endpoint of the line until the Endpoint object snap marker appears.

The purpose of doing this is to "acquire" the Endpoint object snap. When you acquire an object snap, a small yellow + (plus sign) appears at the snap point. An acquired object snap also displays a small white x near the crosshairs. To remove an acquired object snap, pass over the snap point a second time.

2. Slowly move the crosshairs around the endpoint and notice the temporary alignment paths that appear.

As you may recall, alignment paths are temporary lines that display at specific angles from an acquired object snap. Object snap tracking is active when the alignment paths appear from one or more acquired object snaps. This feature is a part of AutoCAD's AutoTracking. You can toggle Auto-Tracking on and off with the POLAR and OTRACK buttons on the status bar. (OTRACK stands for "object snap tracking.") Because object snap tracking works in conjunction with object snaps, one or more object snaps must be set in advance before you can track from an object's snap point.

3. Move the crosshairs along the line so that a small x appears on the line.

4. Enter **3**.

This starts the new line 3″ from the endpoint of the first line.

5. Draw a horizontal line segment **24″** to the right.

HINT: ───────────────────────────────

> When creating the second point, use polar tracking to force the line horizontal and enter 24.

6. In reply to Specify next point, enter **@21<250** to create the next line segment.

7. In reply to Specify next point, move the crosshairs to the intersection of the first and second line segments.

NOTE: ───────────────────────────────

> The intersection of the two lines is also an endpoint of the second line. Therefore, either the Intersection or Endpoint object snap marker will appear. A yellow x-shaped marker depicts the Intersection object snap, and a square marker depicts the Endpoint object snap.

8. Snap to this point and press **ENTER** to terminate the LINE command.

9. Using polar tracking, create the second horizontal line. Make it **20″** long.

10. Create the final segment of the bike frame, as shown in Fig. 9-2.

11. Create the second wheel for the bike. Use the **Center** object snap to place the center of the wheel at the right endpoint of the lower horizontal line.

The drawing of the new bike design is now complete.

12. Save your work.

Specifying Object Snaps Individually

Another way to use object snaps is to specify them individually as you need them. You can do this either from the keyboard or by displaying and using the Object Snap toolbar. Using this toolbar streamlines the selection of object snaps.

1. Right-click on any of the four docked toolbars and select **Object Snap** from the shortcut menu.

The Object Snap toolbar appears as shown in Fig. 9-3.

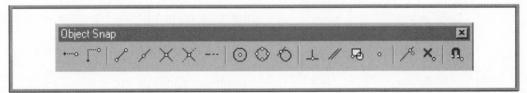

Fig. 9-3

2. Rest the pointer on each of the buttons and read the tooltips.

Table 9-1 on page 112 contains a list of object snap modes and their functions. To enter an object snap mode at the keyboard, enter only the capitalized letters.

You can also access the Object Snap toolbar from the Standard toolbar.

3. Press and hold the **Temporary Tracking Point** button found in the **Standard** toolbar.

Standard

NOTE:

Buttons containing a small black triangle in their lower right corner display a toolbar when you click and hold down the pick button on the pointing device. These are called *flyout toolbars*.

4. Release the button without selecting any of the object snap modes.

Object Snap	Button	Purpose
APParent intersection		Snaps to the apparent intersection of two objects that may not actually intersect in 3D space
CENter		Center of arc or circle
ENDPoint		Closest endpoint of line or arc
EXTension		Snaps to objects along the extension paths of objects
INSertion		Insertion point of text, block, or attribute
INTersection		Intersection of lines, arcs, or circles
MIDpoint		Midpoint of line or arc
NEArest		Nearest point on a line, arc, circle, ellipse, Polyline, spline, and other objects
NODe		Nearest point entity or dimension definition point
NONe		Temporarily cancels all running object snaps (for one operation only)
PARallel		Snaps to a point on an alignment path that is parallel to the selected object
PERpendicular		Perpendicular to line, arc, or circle
QUAdrant		Quadrant point of arc or circle
TANgent		Tangent to arc or circle

Table 9-1

Snapping to Quadrant Points

1. Enter the **LINE** command.
2. From the **Object Snap** toolbar, pick the **Snap to Quadrant** button.

Object Snap

Quadrant points on a circle are the points at 0°, 90°, 180°, and 270°. If you were to connect these points with horizontal and vertical lines, as shown in Fig. 9-4, you would create four equal parts. Each part represents a *quadrant* of the circle.

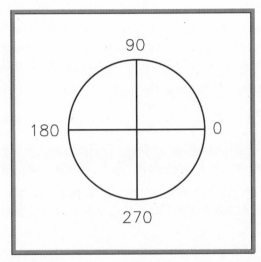

Fig. 9-4

3. Move the crosshairs over one of the two wheels until a small yellow diamond marker appears, along with the tooltip Quadrant.

4. Move to each of the four quadrant points and pick one of them.

5. Enter **ESC** to cancel the LINE command.

Snapping Perpendicularly

The Perpendicular object snap allows you to snap to a point perpendicular to an object such as a line, multiline, polyline, ray, spline, construction line, or even a point on a circle or arc. (A line is *perpendicular* to another line when the two form a right angle.)

1. Enter the **LINE** command and pick a point anywhere above the bike.

Object Snap

2. Pick the **Snap to Perpendicular** button from the **Object Snap** toolbar.

3. Move the crosshairs along the top horizontal segment of the frame until a right angle marker appears, along with the tooltip Perpendicular, and pick a point.

AutoCAD snaps the line perfectly perpendicular to the horizontal segment of the frame.

Standard

4. Pick the **Undo** button twice.

Snapping Parallel Lines

You can use the Parallel object snap to produce parallel line segments.

1. Pick another point above the bike and pick the **Parallel** object snap from the toolbar.

2. Touch the crosshairs on the fork of the frame (the first line you drew) and then move the crosshairs back until a parallel marker appears, as well as a temporary line that is parallel to the fork.

3. Pick a point anywhere to form the parallel line.

4. Pick **Undo** twice.

Snapping to Points of Tangency

The Tangent object snap permits you to snap to a point on an arc or circle that, when connected to the last point, forms a line tangent to that object.

1. Pick the **Snap to Tangent** button from the **Object Snap** toolbar.

Object Snap

2. Move the crosshairs along the bottom half of the front wheel until a Snap to Tangent marker appears, along with the tooltip Deferred Tangent.

AutoCAD defers the point of tangency until after you've picked both endpoints of the line segment.

3. Pick a point anywhere along the bottom half of the front wheel.

4. Pick the **Snap to Tangent** button again.

Object Snap

5. Move the crosshairs along the bottom half of the rear wheel until the **Snap to Tangent** marker appears and pick a point.

AutoCAD draws a line tangent to both wheels.

6. Using polar tracking, extend the line **12"** to the right and press **ENTER** to end the LINE command.

Snapping to the Nearest Point

The Nearest Point object snap mode snaps to a location on the object that is closest to the crosshairs. This is particularly useful when you want to make certain that the point you pick lies precisely on the object and not a short distance away from it.

Object Snap

1. Pick the **Object Snap Settings** button from the **Object Snap** toolbar.

This also displays the Object Snap tab of the Drafting Settings dialog box. (You entered the OSNAP command to display it the last time.)

2. Pick the **Select All** button and pick **OK**.

All object snaps are now on.

3. Enter the **LINE** command and slowly move the crosshairs over the snap points of the objects you've drawn.

A yellow hourglass-shaped marker depicts the Nearest object snap.

4. Snap to a Nearest point on one of the objects.

Cycling Through Snap Modes

Pressing the TAB key cycles through all of the running snap modes associated with the object. This is especially useful when an area is densely covered with lines and other objects, making it difficult to snap to a particular point.

1. Rest the crosshairs on the rear wheel.

2. Press the **TAB** key.

3. Press **TAB** again; and again; and several more times, stopping for a moment between each.

As you can see, AutoCAD moves from one object snap to the next.

4. Press **ESC** to cancel the LINE command.

Object Snap

5. Pick the **Object Snap Settings** button from the **Object Snap** toolbar.

6. Pick the **Clear All** button, check **Endpoint, Center, Intersection,** and **Extension,** and pick **OK**.

Object Snap Settings

Other settings allow you to tailor the behavior of object snaps to meet your individual needs.

Changing the Settings

Object Snap

1. Pick the **Object Snap Settings** button from the **Object Snap** toolbar.

2. Pick the **Options...** button from the dialog box.

Marker, Magnet, and Display AutoSnap tooltip should be checked. When Marker is checked, AutoCAD geometric markers appear at each of the snap points. (These are the yellow markers you have used throughout this chapter.) When Magnet is checked, AutoCAD locks the crosshairs onto the snap target. When Display AutoSnap tooltip is checked, AutoCAD displays a tooltip that describes the name of the snap location.

3. Check the **Display AutoSnap aperture box** check box.

This causes an aperture box to appear at the center of the crosshairs when you snap to objects. The *aperture box* defines an area around the center of the crosshairs within which an object or point will be selected when you press the pick button.

4. Under **AutoSnap marker color**, pick the down arrow to display a list of colors and select green.

5. In the **AutoSnap Marker Size** area, move the slider bar slightly to the right, increasing the size of the marker.

In the AutoTracking Settings area of the dialog box, all three check boxes should be checked. When Display polar tracking vector is checked, AutoCAD displays an alignment path for polar tracking. When Display full-screen tracking vector is checked, AutoCAD displays alignment vectors as infinite lines. When Display AutoTracking tooltip is checked, AutoCAD displays AutoTracking tooltips.

The Alignment Point Acquisition area controls the method of displaying alignment vectors in a drawing. When Automatic is checked, AutoCAD displays tracking vectors automatically when the aperture moves over an object snap. When Shift to acquire is checked, AutoCAD displays the tracking vectors only if you press the SHIFT key while you move the aperture over an object snap.

6. In the **Aperture size** area, slightly increase the size of the aperture using the slider bar.

7. Pick the **Apply** and **OK** buttons.

8. Pick the **OK** button in the Drafting Settings dialog box.

Applying the New Settings

Now let's examine how the changes made in the dialog box affect the drawing process.

1. Enter the **ARC** command and move the crosshairs over the drawing.

Draw

Notice that the object snap markers do not appear until the aperture box contacts the object.

2. Press **ESC** to cancel the **ARC** command.

The AutoSnap settings, including the display of the aperture box and the color and size of the markers, are stored with AutoCAD, not with the drawing. This means that AutoCAD remembers the changes. Likewise, running object snap modes are stored with AutoCAD.

Object Snap

3. Pick the **Object Snap Settings** button from the **Object Snap** toolbar and pick the **Options...** button.

4. Uncheck **Display AutoSnap aperture box**, change the marker color to yellow, and reduce the size of the marker and aperture box.

5. Pick **Apply** and **OK**; pick **OK** again.

As you work with polar tracking, object snap, and object snap tracking, keep in mind that you can toggle them on and off at any time using the POLAR, OSNAP, and OTRACK buttons on the status bar. Most of the time, you will want to use these features, but there will be times when you will want to disable them.

6. Save your work.

7. Practice using the remaining object snaps on your own.

8. Close the Object Snap toolbar and exit AutoCAD.

Chapter 9
Review & Activities

Review Questions

1. Explain the purpose of the object snaps.

2. Describe two methods of using AutoCAD's object snap modes.

3. In order to snap a line to the center of a circle, what part of the circle must the crosshairs touch?

4. What is the benefit of using running object snap modes?

5. Describe a situation in which you would want to change the aperture box size.

6. Briefly describe the use of each of the following object snap modes.
   a. Apparent Intersection
   b. Center
   c. Endpoint
   d. Extension
   e. Insertion
   f. Intersection
   g. Midpoint
   h. Nearest
   i. Node
   j. None
   k. Parallel
   l. Perpendicular
   m. Quadrant
   n. Tangent

7. Explain the benefit of object snap tracking.

8. Explain the relationship between polar tracking, object snap, and object snap tracking. Can you use object snap tracking when object snap is turned off?

Challenge Your Thinking

1. Imagine a new object snap that does not currently exist in AutoCAD. The new snap mode must provide a useful service. Write a paragraph describing the new object snap. Explain what it does and why it is useful.

2. Investigate the Apparent Intersection object snap. Of what use is an object snap that allows you to snap to a point where two objects only *seem* to intersect? Give an example.

Chapter 9
Review & Activities

Applying AutoCAD Skills

1. Draw the square on the left in Fig. 9-5. Then use object snap to make the additions shown on the right to produce a top view of a jeweler's new design of a cut gemstone.

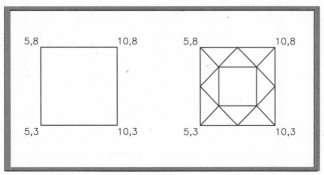

Fig. 9-5

2. Create a new drawing from scratch and set running snap modes for Center and Endpoint. Then create the shaft bearing in shown Fig. 9-6 using the dimensions shown. Start the lower left corner of the drawing at absolute coordinates 1,1, and work counterclockwise. Save the file as shaftbearing.dwg.

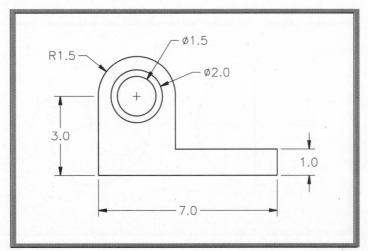

Fig. 9-6

Chapter 9
Review & Activities continued

3. Draw the top view of the video game part shown in Fig. 9-7.

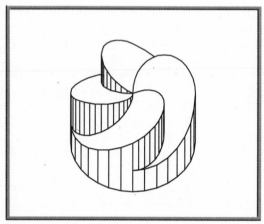

Fig. 9-7

Start by drawing a circle of any size at any location. Then use the Quadrant and Center object snaps to draw the four lines to divide the circle into quadrants, as shown on the left side of Fig. 9-8. (You must draw four lines, not two.) Use the ARC command with the Start, Center, End option and Quadrant, Midpoint, and Center snap modes to draw the arcs. Erase the quadrant lines.

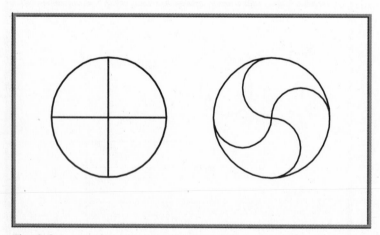

Fig. 9-8

Problem 3 courtesy of Gary J. Hordemann, Gonzaga University

Using Problem-Solving Skills

1. Your company provides special machine parts for an injection molding company. Draw the new design for a pivot arm to swing the mold halves apart (Fig. 9-9). Use the Quick Setup and select the appropriate drawing unit of measurement and drawing area. Set the appropriate object snap modes. Do not include dimensions. Save the drawing as injection.dwg.

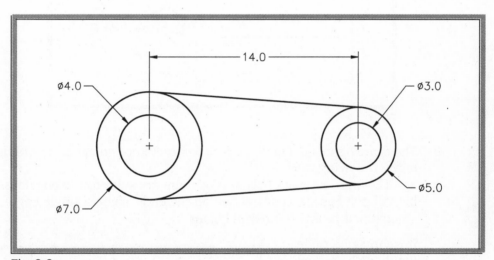

Fig. 9-9

Chapter 9
Review & Activities continued

2. The drawing in Fig. 9-10 represents a variation of a magnet for a new motor design. Draw the magnet. Select the appropriate drawing unit of measurement, drawing area, and object snap modes. Do not include dimensions, but save the drawing.

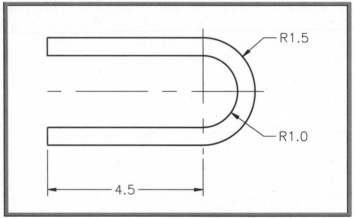

Fig. 9-10

3. The three figures in Fig. 9-11 are constructed of polygons according to the following rules:
   a. There are as many polygons as there are sides to the polygons.
   b. All polygons, except the two on the ends, share a single edge with each of two other polygons.

If we imagine these figures to be groups of blocks sitting on a horizontal surface, then all three groups are stable (assuming a fair amount of friction between the surfaces). Using the

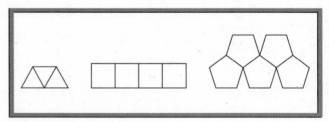

Fig. 9-11

Edge option of the POLYGON command with either the Intersection or the Endpoint object snap, draw similar groupings of hexagons and heptagons. Draw them as though they are sitting on a horizontal surface, and be sure to follow the two rules stated above. Is either grouping unstable? Can you formulate a general rule about the stability of such groupings that would apply to polygons with any number of sides?

Problem 3 courtesy of Gary J. Hordemann, Gonzaga University

Chapter 10 | Helpful Drawing Features

Objectives:

- Use and change the display of coordinate information
- Restrict lines to vertical or horizontal orientations
- Track time spent on a drawing file
- Use the **AutoCAD Text** and **AutoCAD — Command Line** windows
- Undo and redo multiple operations

Key Terms

ortho

orthogonal

Several features in AutoCAD provide information to help users become more productive. The coordinate display, for example, gives size information on the drawing area. The ortho feature helps you draw horizontal and vertical lines quickly and accurately. The TIME command tracks the time you spend working with AutoCAD. The AutoCAD Text and AutoCAD — Command Line windows provide historical information on your interactions with the software. With AutoCAD's undo and redo features, you don't have to worry about making a wrong move.

Coordinate Display

Coordinate information, in digital form, is located in the lower left corner of the AutoCAD window and is part of the status bar. The coordinate display tracks the current position (or coordinate position) of the crosshairs as you move the pointing device. It also gives the length and angle of line segments as you draw.

1. Start AutoCAD and start a new drawing from scratch.

2. Move the crosshairs in the drawing area by moving the pointing device.

Note how the coordinate display changes with the movement of the crosshairs.

3. Enter the **LINE** command and draw the first line segment of a polygon. Note the coordinate display as you draw the line segment.

NOTE:

Most keyboards have function keys called F1, F2, F3, and so on. AutoCAD has assigned specific functions, such as the coordinate display and the ortho feature, to many of these keys.

4. Press the **F6** function key once, draw another line segment, and notice the coordinate display.

The coordinate display should now be off.

5. Press **F6** again, draw another line segment or two, and watch the coordinate display.

As you can see, it displays polar coordinate information.

6. While in the **LINE** command, click the coordinate display and note the change in the coordinate display as you move the crosshairs.

7. Create additional line segments, clicking the coordinate display between each.

As you can see, clicking the coordinate display serves the same function as pressing the F6 key.

The Ortho Mode

Now let's focus on an AutoCAD feature called ortho. *Ortho,* short for "orthogonal," allows you to draw horizontal or vertical lines quickly and easily. *Orthogonal,* in this context, means to draw at right angles.

Ortho is on when the ORTHO button on the status bar is depressed. You can toggle ortho on and off by clicking the ORTHO button.

1. Turn off object snap.

2. Click the **ORTHO** button in the status bar or press the **F8** function key. (Both actions perform the same function.)

3. Experiment by drawing lines with ortho turned on and then with ortho off. Note the difference.

 HINT: ————————————————

Like the coordinate display feature, ortho can be toggled on and off at any time, even while you're in the middle of a command.

4. Attempt to draw an angular line with ortho on.

Can it be done?

5. Clear the screen if necessary and draw the plug shown in Fig. 10-1, first with ortho off and then with ortho on. Don't worry about exact sizes and locations.

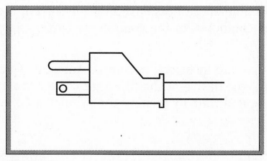

Fig. 10-1

Was it faster with ortho on?

 NOTE: ————————————————

In many situations, polar tracking is faster than ortho and easier to use. However, ortho can be very useful when polar is set to angles of other than 90° or when polar is turned off.

Tracking Time

AutoCAD keeps track of time while you work. With the TIME command, you can review this information.

1. Enter **TIME** at the keyboard or select the **Tools** pull-down menu and pick **Inquiry** and **Time**.

Text appears on the screen, providing information similar to that shown in Fig. 10-2. Dates and times will of course be different.

Here's what this information means.

Current time:	Current date and time
Created:	Date and time drawing was created
Last updated:	Date and time drawing was last updated
Total editing time:	Total time spent editing the current drawing
Elapsed timer:	A timer you can reset or turn on or off
Next automatic save in:	Time before the next automatic save occurs

If the current date and time are not displayed on the screen, they were not set correctly in the computer, or the computer's battery may be weak or dead.

2. If the date and time are incorrect, minimize AutoCAD (pick the minimize button—the dash—at the top right corner of the screen), and reset the date and time by picking the Windows **Control Panel** and **Date/Time**.

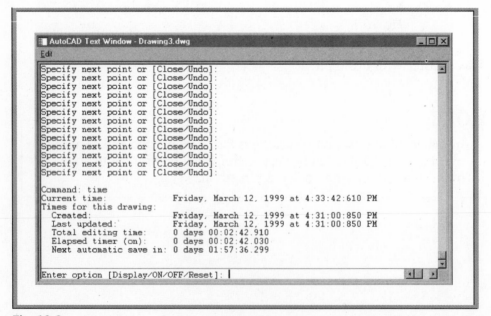

Fig. 10-2

3. If you minimized AutoCAD, maximize it now by clicking the **AutoCAD** button on the Windows taskbar and reenter the **TIME** command.

What information is different? Check the current time, total editing time, and elapsed time.

4. With AutoCAD's **TIME** command entered, type **D** for Display, and notice what displays on the screen.

This provides updated time information.

5. Enter the **OFF** option, and display the time information again.

The elapsed timer should now be off.

 NOTE: ─────────────────────────────

An example of when you might specify OFF is when you want to leave the computer to take a break. When you return, you turn the time back ON. This keeps an accurate record of the actual time (elapsed time) you spend working on the project.

6. Reset the timer by entering **R**, and display the time information once again.

The elapsed timer should show 0 days 00:00:00.000, with the exception of a second or two that might have passed.

7. Last, turn on the timer and display the time information.

Notice that the elapsed timer has kept track of the time only while the timer was turned on.

Why is all of this time information important? In a work environment, the TIME command can track the amount of time spent on each project or job, making it easier to charge time to clients.

8. Press **ENTER** to terminate the TIME command, but do not close the AutoCAD Text Window.

AutoCAD Text Window

As you are aware, AutoCAD displays the time information in the AutoCAD Text Window. AutoCAD uses this window for various purposes, as you will see in future chapters.

Displaying the Window

1. Pick a point in the drawing area but outside the AutoCAD Text Window.

This causes the graphics screen to appear in front of the AutoCAD Text Window, hiding it from view.

2. Press the **F2** function key to make the AutoCAD Text Window come to the front.

3. In the upper right corner of the AutoCAD Text Window, pick the **Minimize** button (the dash).

4. Press **F2** or pick the **AutoCAD Text Window** button on the Windows taskbar to make it reappear.

NOTE:

You can press F2 at any time to display the AutoCAD Text Window. F2 serves as a toggle switch, permitting you to toggle the display of the window.

5. Using the vertical scrollbar, scroll up the list of text in the AutoCAD Text Window.

Observe that AutoCAD maintains a complete history of your activity. Also, notice that the AutoCAD Text Window offers an Edit pull-down menu.

Copying and Pasting

1. Select the **Edit** pull-down menu from the AutoCAD Text Window.

Paste to CmdLine pastes highlighted text to the Command line. This option is grayed out and not available because text has not been highlighted. The Copy item permits you to copy a selected portion of the text to the Windows Clipboard. The Clipboard is a memory space in the computer that temporarily stores information (text and graphics). After you have copied information to the Clipboard, it is easy to paste it into another software application.

Copy History copies the entire contents of the AutoCAD Text Window to the Windows Clipboard. Paste enables you to paste the contents of the Clipboard to the Command line. The Options... menu item permits you to display the Options dialog box, which allows you to change various settings to meet individual needs.

2. Using the pointer, highlight the most recent time information, as shown in Fig. 10-3.

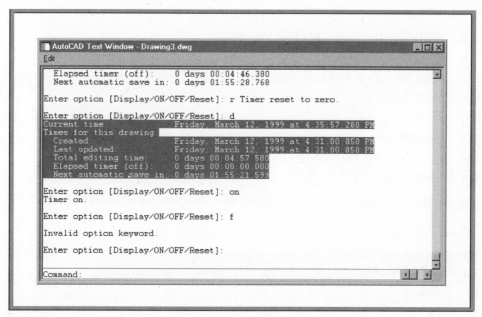

Fig. 10-3

3. Pick the **Copy** item from the **Edit** pull-down menu in the AutoCAD Text Window.

4. Close the AutoCAD Text Window by clicking the **Close** button (the **x**) in the upper right corner of the window.

5. Minimize AutoCAD by picking the **Minimize** button (the dash) located in the upper right corner.

6. Launch a text editor such as **Notepad.** (Notepad is a standard Windows program and is part of the Accessories group of Windows utilities.

7. In the text editor, select **Paste** from the **Edit** menu.

This pastes the time information from the Windows Clipboard. At this point, you could name and save this file and/or print this information.

8. Exit the text editor without saving the file.

9. Click the **AutoCAD** button on the Windows taskbar to make the AutoCAD window reappear.

Command Line Window

Now focus your attention on the AutoCAD – Command Line window (the window that contains the Command prompt).

1. Redisplay the **AutoCAD Text Window**.

2. Position the pointer so that its tip is touching any part of the AutoCAD – Command Line window's border. (Using the left or right edge works well.)

3. With the standard Windows pointer, click and drag the window upward into the drawing area until a floating window forms.

You can move the window anywhere on the screen, and you can resize its width and height.

4. Drag the AutoCAD – Command Line window downward and dock it into its original position.

Copying and Pasting

It is possible to copy and paste text in the AutoCAD – Command Line window. Later in this book, you may find this feature useful as you work with AutoLISP, an AutoCAD programming language.

1. At the right side of the AutoCAD – Command Line window, scroll up using the up arrow.

2. Identify a command (any command) and highlight it by dragging the cursor across it while pressing the pick (left) button on the pointing device.

3. With the pointer inside the AutoCAD – Command Line window, press the right button on the pointing device, causing a pop-up menu to appear.

4. Pick **Paste to CmdLine**.

AutoCAD pastes the highlighted text at the Command line.

5. With the pointer inside the AutoCAD – Command Line window, right-click again and highlight **Recent Commands**.

6. Pick one of the commands that is listed.

AutoCAD enters the command.

7. Press the **ESC** key.

Editing Command Entries

AutoCAD allows you to correct misspellings at the Command line.

8. Type **TME**, a misspelled version of the TIME command, but do not press ENTER.

9. Using the left arrow key, back up and stop between the T and M.

10. Type the letter **I**.

 NOTE:

The Insert key serves as a toggle to turn on and off the insert mode. If the Insert function is not active, pick the Insert key to turn it on.

11. Assuming that TIME is now spelled correctly, press ENTER.

12. Press **ESC** to cancel.

13. Close the AutoCAD Text Window.

14. On your own, experiment with the Command line navigation options shown in Table 10-1.

Key	Action
LEFT ARROW	Moves cursor back (to the left)
RIGHT ARROW	Moves cursor forward (to the right)
UP ARROW	Displays the previous line in the command history
DOWN ARROW	Displays the next line in the command history
HOME	Places the cursor at the beginning of the line
END	Places the cursor at the end of the line
INSERT	Turns on and off the insertion mode
DEL	Deletes the character to the right of the cursor
BACKSPACE	Deletes the character to the left of the cursor
CTRL+V	Pastes text from the Clipboard

Table 10-1

Undoing Your Work

As you may recall from earlier chapters, you can undo or redo one or several actions by picking the Windows-standard Undo and Redo buttons in the docked Standard toolbar. AutoCAD also offers an AutoCAD-specific UNDO command that provides further options. Let's try a few.

Reversing Multiple Operations

The basic UNDO command is similar to the Undo button on the Standard toolbar. It undoes, or reverses, one or more operations, beginning with the most recent.

1. Draw the objects shown in Fig. 10-4 at any size and location.

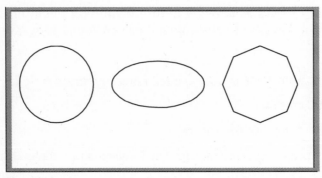

Fig. 10-4

2. Enter the **UNDO** command at the keyboard.

The following options appear on the Command line.

3. Review the list of command options, and then enter **2** for the number of operations.

By entering 2, you told AutoCAD to back up two steps.

NOTE: ───────────────────────────────

AutoCAD also includes an older command, the U command, which allows you to back up one step at a time. The action of the U command can be reversed by the REDO command.

Returning to a Previous Drawing Status

Suppose you want to proceed with drawing and editing, but you would like the option of returning to this point in the session at a later time. The UNDO command's Mark and Back options allow you to do this efficiently.

Draw

1. Draw a small rectangle.

2. Enter the **UNDO** command.

3. Enter the Mark option by typing **M** and pressing **ENTER**.

AutoCAD has (internally) marked this point in the session.

4. Perform several operations, such as drawing and erasing objects and changing the buttons on the status bar.

Now suppose you decide to back up to the point where you drew the rectangle in Step 1.

5. Enter the **UNDO** command and then the **Back** option.

You have just practiced two of the most common uses of the UNDO command. Other UNDO options exist, such as BEgin and End. Refer to AutoCAD's on-line help for more information on these options.

Controlling the Undo Feature

The undo feature can use a large amount of disk space and can cause a "disk full" situation when only a small amount of disk space is available. You may want to disable the UNDO command partially or entirely by using the UNDO Control option.

Let's experiment with UNDO Control.

1. Enter the **UNDO** command and then the **Control** option.

2. Enter the **One** option.

3. Draw two arcs anywhere in the drawing area.

4. Enter the **UNDO** command.

UNDO now permits you to undo only your last operation.

5. Press **ENTER**.

6. Enter **UNDO** and **All**.

The undo feature is once again fully enabled.

7. Enter **UNDO** and note the complete list of options.

8. Enter **ESC** to cancel the command.

Chapter 10
Review & Activities

Review Questions

1. Which function key turns on the coordinate display feature?

2. Of what value is the coordinate display?

3. What is the name of the feature that forces all lines to be drawn only vertically or horizontally? What function key controls this feature?

4. Of what value is the TIME command?

5. On a separate sheet of paper, briefly explain each of the following components of the TIME command.
   a. Current time
   b. Created
   c. Last updated
   d. Total editing time
   e. Elapsed timer
   f. Next automatic save in

6. Which function key toggles the display of the AutoCAD Text Window on and off?

7. Explain how you would copy a few lines from the AutoCAD Text Window to a text editor such as Notepad.

8. Describe two methods to back up or undo your last five operations quickly.

9. Explain the use of the UNDO Mark and Back options.

Challenge Your Thinking

1. Experiment with ortho in combination with various object snap modes using the LINE command. What happens when you try to snap to a point that is not exactly horizontal or vertical to the previous point?

2. Why might you want to copy parts of the AutoCAD Text Window or AutoCAD – Command Line window to the Windows Clipboard?

Chapter 10
Review & Activities

Applying AutoCAD Skills

1. Set the elapsed timer to ON. Complete the design shown in Fig. 10-5 for an over-the-hood storage area for a garage. Use ortho to keep the lines perfectly horizontal and vertical. When you have completed the drawing, set the elapsed timer OFF and write down the total time it took you to complete the drawing.

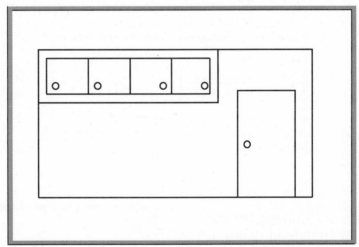

Fig. 10-5

2–4. Practice using ortho by drawing the objects shown in Figs. 10-6 through 10-8. Create a new drawing to hold them, and use ortho when appropriate. Turn on the coordinate display, and note the display as you construct each of the objects.

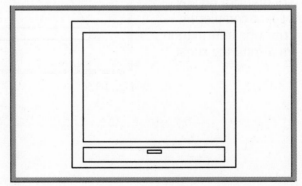

Fig. 10-6

Chapter 10
Review & Activities continued

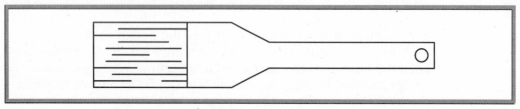

Fig. 10-7

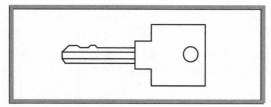

Fig. 10-8

5. Create a new drawing file and draw the lamp head shown in Fig. 10-9. Copy everything entered in the AutoCAD – Command Line window to the Clipboard and paste it into a text editor such as Notepad. Then print it. (Read problem 6 before completing this problem.)

6. Use AutoCAD's timer to review the time you spent completing problem 5. Print this information or record it on a separate sheet of paper.

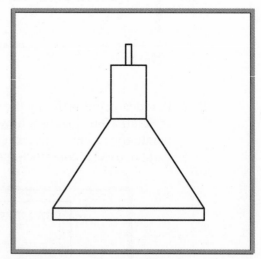

Fig. 10-9

7. Draw the simple house elevation shown in Fig. 10-10A at any size with ortho on. Prior to drawing the roof, use UNDO to mark the current location in the drawing. Then draw the roof.

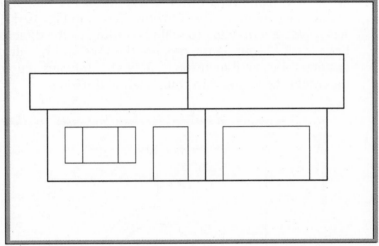

Fig. 10-10A

With UNDO Back, return to the point prior to drawing the roof. Draw the roof shown in Fig. 10-10B in place of the old roof. Use the UNDO command as necessary as you complete the drawing.

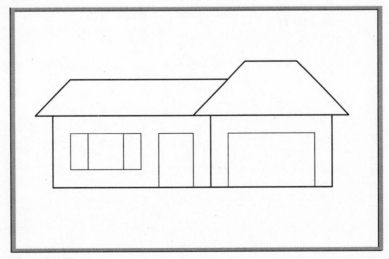

Fig. 10-10B

Chapter 10
Review & Activities continued

? Using Problem-Solving Skills

1. As the packaging engineer for a major manufacturer, you have been asked to design a support for a new product to hold it in place within the container. The drawing, shown in Fig. 10-11, is not to scale. Create a drawing to scale according to the dimensions given. Keep track of your hours, because the time will be billed to the customer. Do not dimension it. Save the drawing and write down the total time required to complete the drawing.

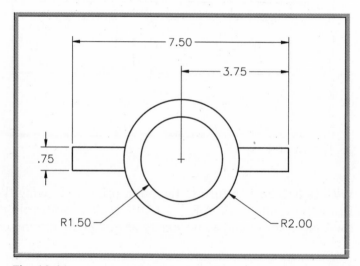

Fig. 10-11

Chapter 10
Review & Activities

2. The core of an electromagnet is not made up of one solid piece of metal. Rather, it is made up of a series of thin "slices," or laminations. These laminations fit together to form the core of the electromagnet. The laminations concentrate the electromagnetic field, adding strength. Draw the lamination shown in Fig. 10-12. Keep track of your time with the elapsed timer. Do not dimension, and save the drawing.

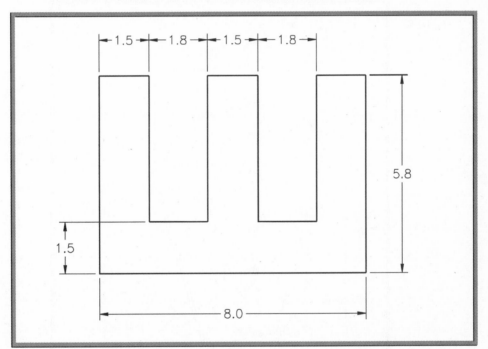

Fig. 10-12

Chapter 10
Review & Activities continued

3. Draw the front view of the tube bulkhead shown in Fig. 10-13. Before drawing the holes, use the UNDO command to mark the current location in the drawing. After completing the bulkhead, use UNDO to replace the two holes with two new ones. Draw both of the new holes with a diameter of 1.00. Change the center-to-center distance from 1.40 to 1.60.

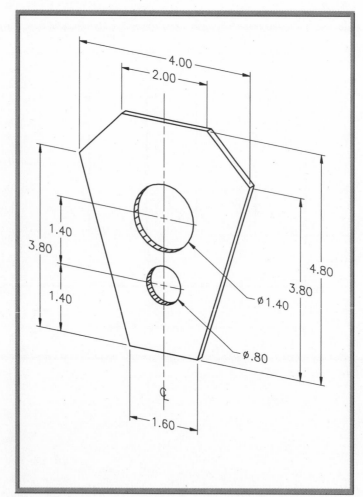

Fig. 10-13

Problem 3 courtesy of Gary J. Hordemann, Gonzaga University

4. The two identical hubs shown in Fig. 10-14 are meant to be mounted back-to-back in an assembly. Draw the front view of one hub. Before drawing the four small holes, use the UNDO command to mark the current location in the drawing. After completing the hub, use UNDO to replace the four holes with eight new ones, equally spaced circumferentially. Use a new hole diameter of .284 and a new center line diameter of 2.50.

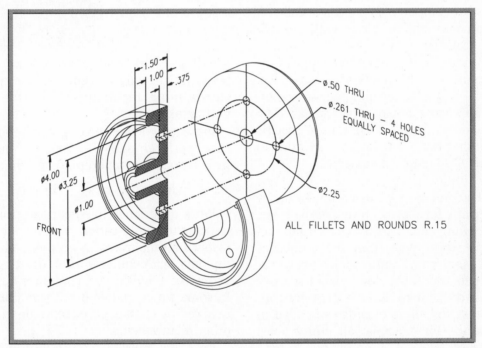

Fig. 10-14

Careers  Using AutoCAD

GPS and AutoCAD in Public Works

Wherever on Earth you are right now, your location is within view of five of the 24 navigational satellites in the Global Positioning System (GPS). With a GPS satellite receiver, you can pick up satellite signals to determine your longitude, latitude, and altitude. With a handheld microcomputer/GPS receiver called MC-GPS, you can insert this position information, along with your field notes, into a map drawn in AutoCAD.

As a mapping tool, AutoCAD has applications in forest management, pipeline maintenance, archeological siting, water quality assessment, and more. To update a map, someone has to go out to observe and record conditions at specific sites. A tool that can link field data to vectors in an AutoCAD base map is invaluable.

Consider utility poles, for example. How many are there in your city? Inspection of utility poles, especially for hazards from nearby trees, is a never-ending chore. Locating each pole on paper maps (third pole from the corner of Maple and Madison...) and recording observations with pen and paper slow down the work and make it difficult to retrieve data. By contrast, it is easy to call up records embedded in AutoCAD map—for example, all utility poles with a hazard level of 4 or above. A field inspector using the MC-GPS can quickly identify a utility pole by its GPS coordinates and enter menu-guided notes such as the following:

Date surveyed: 09-02-1999
Condition: good/repair/replace
Tree hazard: none/fast growing/slow growing
Hazard level: 0 to 5 (high risk = 5)
Action required: non/cut back/remove

Career Focus—Public Inspector

Public works inspectors examine new construction, alterations, and repairs of roads, sewer systems, and other structures. During construction, they focus on compliance with building codes and contract specifications. Inspectors usually specialize in one type of construction or regulatory area (for example, earthquake codes).

Inspectors keep records of the inspections they perform and file reports. Increasingly, inspectors use computers for recordkeeping.

Education and Training

Work experience in construction is the most important qualification for a public works inspector. Employers may require a high school diploma, with coursework in drafting, algebra, geometry, English, and computer applications. In some states, public inspectors must become certified by passing an examination or meeting other requirements.

Inspectors must stay up-to-date with building materials and practices, inspection techniques, and changes in buildling codes. When a project does not pass inspection, it helps to have a good knowledge of people and an ability to maintain a calm, professional manner.

 Career Activity

- Write a report about public works construction and maintenance. Include a description of recent construction or other work in your community and the different types of workers involved in the project.

Chapter 11 Construction Aids

Objectives:

- **Set and use AutoCAD's visual grid system**
- **Establish a snap grid that is appropriate to the current drawing task**
- **Use construction lines to project views in a multiview drawing**

Key Terms

alignment grid

aspect ratio

construction lines

multiview drawing

offsetting

orthographic projection

plane

rays

xline

AutoCAD permits you to create a nonprinting grid system for visual layout of drawings. The snap grid complements the visual grid, assisting with the layout process. Construction lines provide a means for projecting lines from one view to another using orthographic projection. All of these construction aids make it easier to create accurate multiview drawings in a short amount of time.

Alignment Grid

The GRID command allows you to set an *alignment grid* of dots of any desired spacing, making it easier to visualize distances and drawing size. You can use the grid with both English and metric units. In this chapter, we will create a drawing of a plastic extrusion using metric units.

1. Start AutoCAD and start a new drawing from scratch, but select **Metric** instead of English.

2. Move the crosshairs to the upper right corner of the drawing area and read the coordinate display.

Since you are now working in the metric system, the unit of measure is millimeters. Instead of a drawing area of 12 × 9, it is now approximately 420 × 297 millimeters, depending on the size of the window representing the drawing area on your computer.

3. Pick the **GRID** button located in the status bar.

A grid of dots appears on the screen.

4. Enter the **GRID** command at the keyboard.

Notice that the default value for the grid spacing is 10. This means that the dots are currently spaced 10 millimeters apart.

5. Change the spacing to 20 millimeters by entering **20**.

6. Reenter **GRID** and change the spacing back to **10** millimeters.

7. Turn off the grid by picking the **GRID** button.

8. Make the grid visible again.

NOTE: ────────────────────────────────

You can also turn the grid on and off by pressing CTRL G or the F7 function key.

Snap Grid

The snap grid is similar to the visual grid, but it is an invisible one. You cannot see the snap feature, but you can see the effects of it as you move the crosshairs. It is like a set of invisible magnetic points. The crosshairs jump from point to point as you move the pointing device. This allows you to lay out drawings quickly, yet you have the freedom to toggle snap off at any time.

1. Pick the **SNAP** button in the status bar to turn on the snap grid.

NOTE: ────────────────────────────────

The snap grid is different from object snap, which permits you to snap to points on objects.

You can also turn the snap grid on and off by pressing CTRL B or the F9 function key.

When you turn on the snap grid, it appears at first as though nothing has happened.

2. Slowly move the pointing device and watch closely the movement of the crosshairs.

The crosshairs jump (snap) from point to point.

3. Enter the **SNAP** command at the keyboard.

4. Enter **5** to specify the snap spacing.

5. Move the pointing device and note the crosshairs movement.

HINT:

If the grid is not on, turn it on to see better the movement of the crosshairs. Notice the spacing relationship between the snap resolution and the grid. They are independent of one another.

6. Draw a line.

Changing the Aspect Ratio

The *aspect ratio* is the ratio of height to width of a rectangular region. You can change this ratio so that the crosshairs snap one distance vertically and a different distance horizontally. This can be useful when you are laying out a drawing.

1. Enter the **SNAP** command.

Several options appear at the Command line.

2. Enter the Aspect option by entering **A**.

AutoCAD asks for the horizontal spacing.

3. Enter **5**.

Next AutoCAD asks for the vertical spacing.

4. For now, enter **10**.

5. Move the crosshairs up and down and back and forth. Note the difference between the amount of vertical movement and horizontal movement.

6. Change the vertical snap back to **5**.

Rotating the Grid

By default, the grid appears at a 0° angle, which aligns the points in vertical and horizontal rows. However, AutoCAD allows you to change the angle of the grid. This can be useful when you want to create a drawing at an angle, while taking advantage of the snap grid.

1. Enter the **SNAP** command again, and this time enter the **Rotate** option.

2. Leave the base point at 0,0 by giving a null response. (Simply press the space bar or **ENTER**).

3. Enter a rotation angle of **30** (degrees).

The snap, grid, and crosshairs rotate 30° counterclockwise.

4. Draw a small object at this rotation angle near the top of the screen.

5. Return to the original snap rotation of **0** and erase all the objects.

Applying the Snap and Grid Features

1. Set snap at **5** millimeters both horizontally and vertically.

2. Create the drawing of a plastic extrusion shown in Fig. 11-1. Use the pointing device (not the keyboard) to specify all points, and do not include the dimensions.

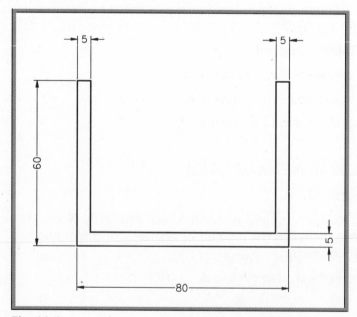

Fig. 11-1

HINT:

Use the snap grid and coordinate display when specifying endpoints. You may need to deselect OSNAP on the status bar to create the extrusion accurately.

Changing Grid and Snap Settings

The Drafting Settings dialog box permits you to review and make changes to the snap and grid settings.

1. From the **Tools** pull-down menu, select the **Drafting Settings...** item.

The Drafting Settings dialog box appears.

2. If the **Snap and Grid** tab is not in the foreground, select it.

As you can see, this dialog box enables you to change the *x* and *y* spacing of both the snap and grid. AutoCAD saves these settings with the drawing, so if you or someone else starts a new drawing, the old settings return.

3. Close the dialog box.

4. Save your work in a file named **snap.dwg**, but do not close the drawing or exit AutoCAD.

Construction Lines

Construction lines, also called *xlines,* are lines of infinite length. They can be used to construct objects and lay out new drawings.

Standard

1. Start a new drawing from scratch and select **Metric**.

2. Set the grid to **10** millimeters and the snap to **5** millimeters.

Draw

3. From the **Draw** toolbar, pick the **Construction Line** button.

This enters the XLINE command.

4. Pick a point anywhere on the screen.

5. Move the crosshairs and notice what happens.

A line of infinite length passes through the first point and crosshairs.

6. Pick a second point anywhere on the screen.

A line freezes into place. As you move the crosshairs, notice that a second line forms.

7. Pick another point so that the intersecting lines approximately form a right angle.

8. Press **ENTER** to terminate the XLINE command.

Unlike the snap and grid construction aids, xlines are AutoCAD objects. You can select them for use with commands, and you can snap to them using object snap.

9. Enter the **LINE** command and use the **Intersection** object snap to snap to the intersection of the two lines.

Object Snap

 HINT: ───────────────────────

> If Intersection is not currently set as a running object snap, use the object snap flyout located on the docked Standard toolbar to select it.

10. Press **ESC** to terminate the LINE command.

11. Erase the xlines.

Constructing Horizontal and Vertical Xlines

The XLINE command provides several options that allow you to create specific types of construction lines.

1. Enter the **XLINE** command and enter **H** for Horizontal.

2. Pick a point anywhere on the screen; then pick a second point, and a third.

Because the Horizontal option is in effect, all the construction lines are horizontal.

Draw

3. Press **ENTER** to terminate the XLINE command and press **ENTER** or the space bar to reenter it.

4. Enter **V** for Vertical and pick three consecutive points anywhere on the screen.

The Vertical option forces all the lines to be exactly vertical.

Constructing Xlines at Other Angles

To create construction lines at angles other than 0 or 90 degrees, you can use the Angle option.

1. Press **ENTER** to terminate the XLINE command and reenter it.

2. Enter **A** for Angle and enter **45**.

3. Place three construction lines.

4. Terminate XLINE.

5. Enter **ERASE** and **All** to clear the screen.

Offsetting Construction Lines

You may find many occasions to use the Offset option of the XLINE command. *Offsetting* a line means creating the line at a specific distance from another line. For example, suppose you have created a line that represents the outer edge of a sidewalk on an architectural site drawing. If the sidewalk is to be 3 feet wide, you can create a construction line at an offset distance of 3 feet to show the width of the sidewalk.

1. Terminate **XLINE**, reenter it, and enter **O** for Offset.

2. Enter **7** (for 7 millimeters), select the construction line, and pick a point on either side of it.

INFOLINK

You will learn more about offsetting ordinary lines and circles in Chapter 15.

Creating Perpendicular Construction Lines

1. Terminate XLINE, reenter it, and select or enter the **Perpendicular** object snap.

2. Pick the last construction line you created and pick another point elsewhere on the screen.

The new construction line is perpendicular to the one you selected.

3. Terminate XLINE and erase all the objects from the screen.

Rays

Rays are a different kind of construction line. They extend from a single point into infinity in a radial fashion. Rays are useful when you are laying out objects radially.

1. From the **Draw** pull-down menu, pick **Ray**.

This enters the RAY command.

2. Pick a point in the center of the screen.

3. Move the pointing device in any direction and then pick several points around the first point.

4. Terminate the RAY command.

5. Close the current drawing file and do not save your work.

Orthographic Projection

Construction lines, rays, and the grid and snap features are often used to create multiview drawings. A *multiview drawing* is one that describes a three-dimensional object completely using two or more two-dimensional views. These views commonly include a front, top, and side view, although others are often necessary. The AutoCAD features discussed in this chapter, coupled with object snap tracking, help you to produce multiview drawings using a technique known as orthographic projection. As mentioned earlier in this book, *ortho* means "at right angles." *Orthographic projection,* therefore, is the projection of views at right angles to one another. Each adjacent view is projected at a right angle onto a plane, resulting in a two-dimensional view. A *plane* is an imaginary flat surface used to construct the two-dimensional view.

Figure 11-2 shows an example of a simple multiview drawing created using orthographic projection. Notice that the top view is located above the front view. The right view is located to the right of the front view. This positioning is standard. It is important to use the standard positions because doing so allows other people to understand the position of these views at a glance. The thin lines are temporary construction lines.

1. Using grips, move the drawing of the plastic extrusion to the lower left area of the screen.

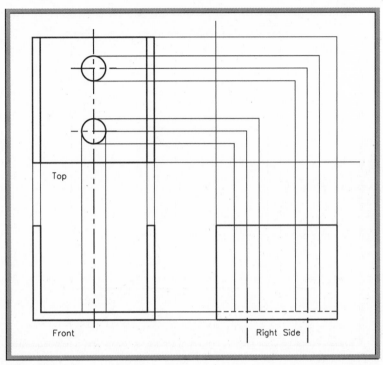

Fig. 11-2

2. Using object snap tracking, snap, and grid, create the top view as shown in the previous illustration. Use the following information when constructing the top view.

   • From Fig. 11-3 (page 152), obtain the length dimension of the extrusion and the location and size of the two holes.

   • The broken lines through the holes are center lines. For now, create these lines using a series of line segments. Later in this book, you will learn an easier way to draw dashed lines to create center and hidden lines.

   • You may need to turn off object snap when picking some of the points. However, object snap must be on for object snap tracking to work.

   • Using the **XLINE** command, consider drawing a vertical construction line snapped to the midpoint of one of the horizontal lines to represent the center of the holes. Then erase the xline.

3. Save your work.

INFOLINK

You will learn more about alternative linetypes in Chapter 22.

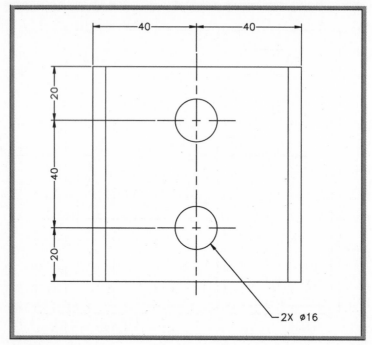

Fig. 11-3

4. After creating the top view, project the holes down to the front view. Consider the following suggestions.

   • Set **Quadrant** as a running object snap mode.

   • Use object snap tracking to acquire the diameter of the holes.

   • Draw hidden lines to show the holes in the front view. (*Hidden lines* are the short dashed lines in the multiview drawing that are used to indicate features that are hidden from view.) For now, create them using a series of short line segments.

5. Save your work.

6. Create the right-side view by projecting lines from the top and front views. Consider the following suggestions.

   • Create temporary lines as shown in Fig. 11-4. The vertical line is at an arbitrary distance from the top view.

   • Draw the angled line at a 45° angle. (Use the Angle option of the XLINE command.) Use it to project the dimensions from the top view to the right-side view.

   • Project lines from the top view to the 45° line. Then project vertical lines down from the intersections of the horizontally projected lines and the 45° lines to locate the right-side view.

   • Use the height dimensions from the front view (Fig. 11-3) to finish the right-side view.

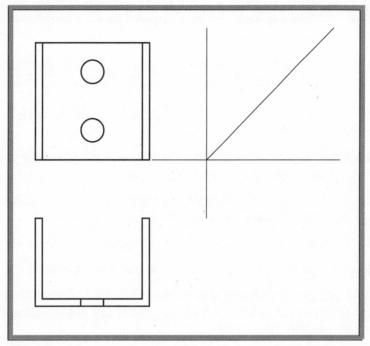

Fig. 11-4

 NOTE: ——————————————————————————

In practice, the views in a multiview drawing are laid out carefully to ensure that the distance between the front and top views is approximately equal to the space between the front and side views. Keep this in mind as you create your construction lines in Step 6.

7. Erase all temporary construction lines.

Your multiview drawing should now look similar to the multiview drawing shown previously, except that your drawing will not include hidden, center, or construction lines. You will learn how to create hidden and center lines later in this book.

8. Save your work and exit AutoCAD.

INFOLINK

You will learn more about setting up multiview drawings in Chapters 24 and 50.

Chapter 11
Review & Activities

Review Questions

1. What is the purpose of the grid? How can the grid be toggled on and off quickly?

2. When would you use the snap feature? When would you toggle off the snap feature?

3. How can you set the snap feature so that the crosshairs move a different distance horizontally than vertically?

4. Explain how to rotate the grid, snap, and crosshairs 45°.

5. What is the purpose of the XLINE and RAY commands?

6. Explain the purpose of construction lines and rays. Describe a drawing situation in which you could use construction lines or rays to simplify your work.

7. What is orthographic projection? How is it used in drafting?

8. What is the purpose of a multiview drawing?

Challenge Your Thinking

1. AutoCAD provides several different grid and snap systems. Write a short paper describing their differences and similarities.

2. It is possible to accomplish the same tasks using either construction lines or AutoCAD's AutoTrack feature. Describe a situation in which you would prefer to use one instead of the other.

Chapter 11
Review & Activities

Applying AutoCAD Skills

Work the following problems to practice the commands and skills you learned in this chapter. Use English units, not metric, for all of the problems in this section.

1–3. Draw the objects in Figs. 11-5 through 11-7 using the grid and snap settings provided beside each object.

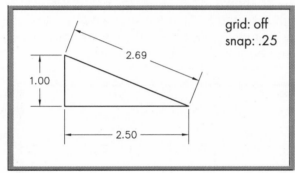

grid: off
snap: .25

2.69

1.00

2.50

Fig. 11-5

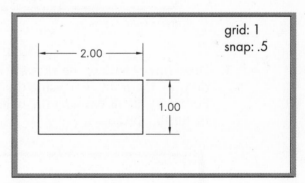

grid: 1
snap: .5

2.00

1.00

Fig. 11-6

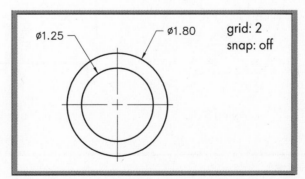

ø1.25   ø1.80   grid: 2
snap: off

Fig. 11-7

Chapter 11
Review & Activities continued

4. Draw the two views of the rubberized hand grip for closing the safety bar on a roller coaster car (Fig. 11-8B). Begin by drawing two construction lines as shown in Fig. 11-8A.

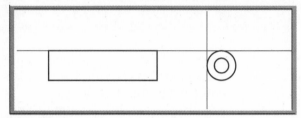

Fig. 11-8A

Draw the front view as shown. Then draw the ø40 circle using the tangent-tangent-radius method. Then locate the ø20 circle to be concentric with the first circle, and erase the construction lines. Do not dimension it, and do not draw the center lines.

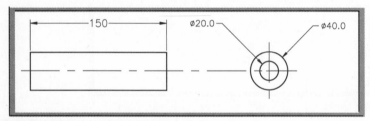

Fig. 11-8B

5. Draw three views of the slotted block shown in Fig. 11-9A. Start by drawing the front view using the LINE and CIRCLE commands. Do not worry about drawing the object exactly. Make the task easier by turning ortho on.

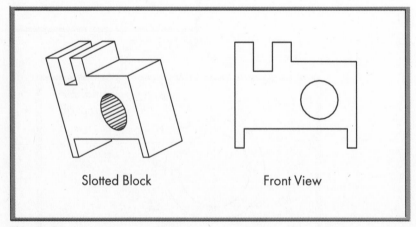

Fig. 11-9A

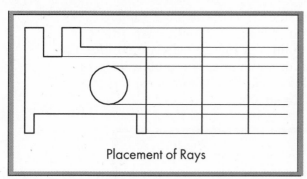

Placement of Rays

Fig. 11-9B

Use the RAY command to extend the edges in the front view into what will become the right-side view. Use the appropriate object snaps to begin the rays at the edges of the front view, as shown in Fig. 11-9B.

To begin drawing the right-side view, add two vertical lines (estimate the space between them), as shown in Fig. 11-9B. Use the construction lines (rays) to draw the visible lines. Draw the hidden lines as a series of very short lines.

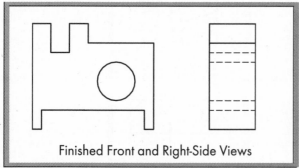

Finished Front and Right-Side Views

Fig. 11-9C

When you are finished, erase the construction lines. The drawing should look like the one in Fig. 11-9C.

Repeat the process to draw the top view. The finished drawing should look like the one in Fig. 11-9D.

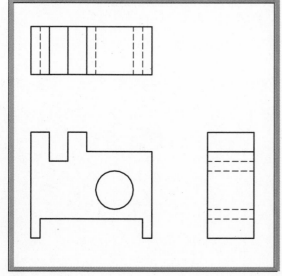

Fig. 11-9D

Problem 5 courtesy of Gary J. Hordemann, Gonzaga University

Chapter 11
Review & Activities continued

6. Draw the top, front, and right-side views of the block support shown in Fig. 11-10. Use a grid spacing of .2 and a snap setting of .1. Consider using construction lines or rays to help you position the three views. For now, draw all of the lines as continuous lines, or draw the hidden and center lines using very short lines. You may wish to change the snap spacing to .05.

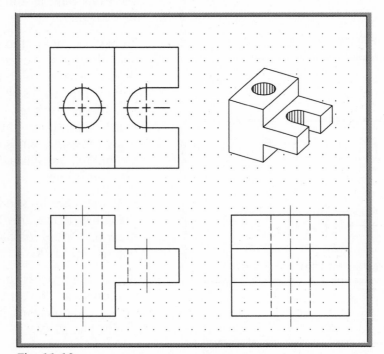

Fig. 11-10

Problem 6 courtesy of Gary J. Hordemann, Gonzaga University

7. Draw the three orthographic views of the bushing holder shown in Fig. 11-11. Use a grid spacing of .2 and a snap setting of .1. Consider using construction lines or rays to help you position the views. The front view is shown as a full section view; *i.e.,* it is shown as if the object has been cut in half. The diagonal lines indicate cut material; you may want to rotate the crosshairs using the SNAP command to create these lines. AutoCAD provides a way to insert such hatch patterns easily. You will learn how to do this later.

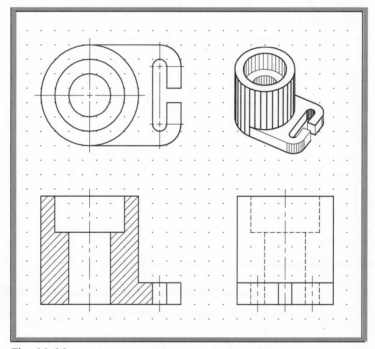

Fig. 11-11

Problem 7 courtesy of Gary J. Hordemann, Gonzaga University

Chapter 11
Review & Activities continued

 Using Problem-Solving Skills

Complete the following activities using problem-solving skills and your knowledge of AutoCAD.

1–2. When the drafting department converted to CAD from board drafting, many vellum drawings had to be copied into the computer. The two drawings shown in Figs. 11–12 and 11–13 were not copied completely. Use the skills you have acquired to complete the missing views. Determine the dimensions from the .5 grid spacing. Save both drawings.

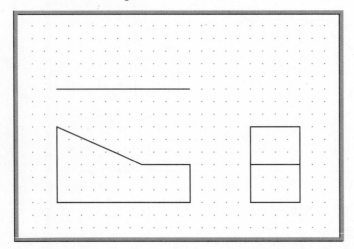

Fig. 11-12

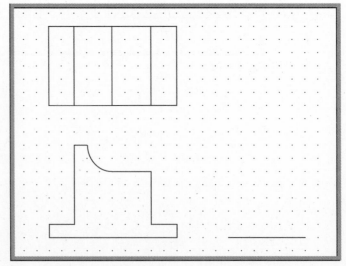

Fig. 11-13

Chapter 12 | AutoCAD's Magnifying Glass

Objectives:

- Zoom in on portions of a drawing to view or add details
- Use the **ZOOM** command while another command is entered
- Explain the relationship between screen regenerations and view resolution
- Control view resolution

Key Terms

floating-point format
realtime zooming
screen regeneration
transparent zooms
view resolution

Large drawings with detail can be difficult, and at times impossible, to view legibly on even the largest computer monitors. AutoCAD offers a zooming capability that magnifies areas of the drawing. Remarkably, this capability can magnify objects as much as ten trillion times. While this may seem impractical for most work, scientists and astronomers produce drawings and models of incredible scale. Consider a solar system, in which distances are measured in millions or billions of miles.

Zooming

Let's apply the ZOOM command.

1. Start AutoCAD, pick the **Use a Wizard** button, and pick **Quick Setup** and **OK**.

2. Pick the **Architectural** radio button and **Next**.

3. Enter **20'** for the Width and **15'** for the Length of the drawing area and pick **Finish**.

HINT:

Be sure to include an apostrophe (') for the foot mark. If you don't, AutoCAD will assume that you mean inches, which would be much too small for the following office drawing. Note that when you work in inches, adding a double quote (") after the number is optional.

Zoom

4. Enter **Z** for ZOOM and **A** for All.

This zooms the drawing to the 20' × 15' area.

5. Enter **GRID** and enter **1'** or **12** (inches).

6. Set the snap grid to **2"**.

7. Draw the room shown in Fig. 12-1, including the table and chair.

8. Save your work in a file named **zoom.dwg**.

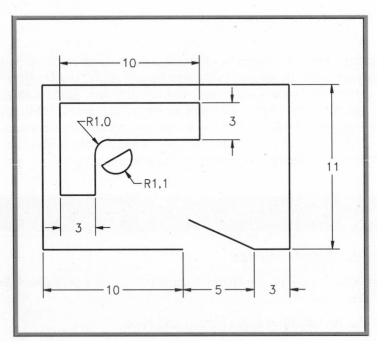

Fig. 12-1

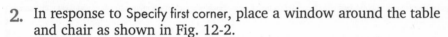

Methods of Zooming

AutoCAD provides several options for zooming in or out on a drawing. You can enter these options by typing Z (for ZOOM) at the keyboard and then entering the first letter of the option, or you can use the many zooming buttons. These buttons are located on the flyout on the Zoom Window button in the Standard toolbar and in the Zoom toolbar.

Standard

1. Pick the **Zoom Window** button from the **Standard** toolbar.

This enters the ZOOM command and displays several options at the Command line.

2. In response to Specify first corner, place a window around the table and chair as shown in Fig. 12-2.

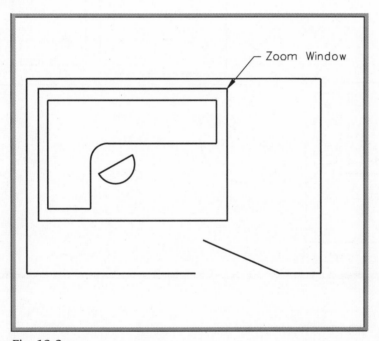

Zoom Window

Fig. 12-2

The table and chair should magnify to fill most of the screen.

3. Next, draw basic representations of several components that make up a CAD system, as shown in Fig. 12-3. Approximate their sizes, and omit the text.

 HINT:

Set the snap grid to a smaller increment, such as .2, to make this task easier.

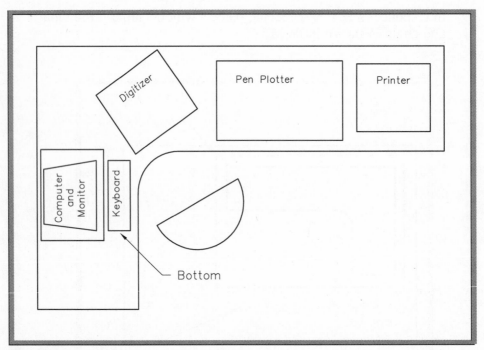

Fig. 12-3

4. Save your work.

Now let's zoom in on the keyboard.

5. Pick the **Zoom Window** button again, and this time place a window around the lower half of the keyboard.

Let's zoom in further on the keyboard, this time using the Zoom In option.

6. From the **Zoom** toolbar or the **Zoom** flyout on the **Standard** toolbar, pick the **Zoom In** button.

Standard

Zoom

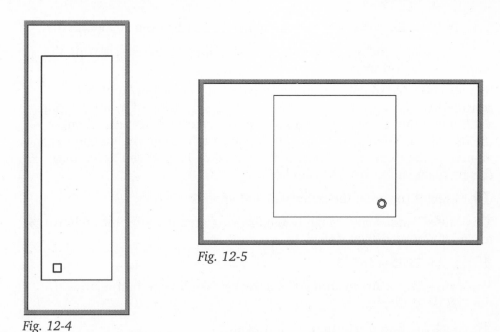

Fig. 12-5

Fig. 12-4

7. If you cannot see the bottom of the keyboard, zoom again (using the **Zoom In** or **Zoom Window** button) so that the bottom of the keyboard fills most of the screen.

8. In the lower left corner of the keyboard, draw a small square to represent a key, as shown in Fig. 12-4. (You may need to turn snap off first.)

9. Zoom in on the key, this time using the **Zoom Center** button from the **Zoom** flyout. Pick a center point by picking the center of the key, and then specify the height by picking points below and above the key.

10. Using the **POLYGON** command, draw a small trademark on the key as shown in Fig. 12-5.

11. Next, pick the **Zoom Previous** button located on the Standard toolbar and watch what happens.

You should now be at the previous zoom factor. The trademark should look smaller.

12. Pick the **Zoom All** button, or enter **Z** and **A** at the keyboard.

AutoCAD zooms the drawing to its original size.

Standard

Zoom

13. Pick the **Zoom Extents** button or enter **Z** and **E** at the keyboard.

Zoom

AutoCAD zooms the drawing as large as possible while still showing the entire drawing on the screen.

It is also possible to enter a specific magnification factor. AutoCAD considers the entire drawing area to be a magnification of 1. By entering a different number with the ZOOM command entered, you can change the magnification relative to this. If you need to change the magnification relative to the current magnification (instead of 1), you can accomplish this by typing an x after the number.

14. Enter **Z** (or press the spacebar) and enter **2**.

This causes AutoCAD to magnify the drawing area by two times relative to the ZOOM All display (All = 1).

15. Enter **ZOOM** and **.5**.

This causes AutoCAD to shrink the drawing area by one half relative to the ZOOM All display.

Zoom

16. Pick the **Zoom In** button; pick it again.

17. Pick the **Zoom Out** button.

Zoom

18. Pick the **Zoom Extents** button again.

19. Enter **ZOOM** and **.9x**.

Zoom

Because you placed an x after the number, the zoom magnification decreases to nine tenths of its current size, not nine tenths of the ZOOM All display area.

20. Continue to practice using the **ZOOM** command by zooming in on the different CAD components of the drawing and including detail on each.

21. Save your work.

Realtime Zooming

AutoCAD's *realtime zooming* allows you to change the level of magnification by moving the cursor. This is one of the fastest ways to zoom.

Standard

1. From the **Standard** toolbar, pick the **Zoom Realtime** button.

This places you into realtime zoom mode. Notice that the crosshairs has changed.

2. Using the pick button, click and drag upward to zoom in on the drawing.

3. Click and drag downward to zoom out on the drawing.

4. Press **ESC** or **ENTER** to cancel realtime zooming.

NOTE:

AutoCAD offers a time-saving shortcut for using realtime zooming. At the keyboard, enter Z (for ZOOM) and then press ENTER or the spacebar. Realtime zooming becomes available.

Zooming Transparently

With AutoCAD, you can perform *transparent zooms*. This means you can use ZOOM while another command is in progress. To do this, enter 'ZOOM (notice the apostrophe) at any prompt that is not asking for a text string.

1. Enter the **LINE** command and pick a point on the screen.

2. At the prompt, enter **'Z**.

AutoCAD responds by listing two lines of options at the Command line. Notice that the first line is preceded by ≫. This reminds you that you're in transparent mode.

3. Zoom in on any portion of the screen.

Notice the message Resuming LINE command in the Command window.

4. Pick an endpoint for the line and terminate the command.

NOTE:

Picking any of the zoom buttons also permits you to perform a transparent zoom.

Transparent zooms give you great zoom flexibility while you are using other commands. For instance, if the line endpoints require greater accuracy than the present display will allow, the transparent zoom provides you with a solution.

Zoom

5. Erase the line and enter **ZOOM All**.

6. Save your work, but do not close the drawing file.

Screen Regenerations

A *screen regeneration* is the recalculation of each vector (line segment) in the drawing. All AutoCAD objects (including arcs, text, and other curved shapes) are made up of vectors stored in a high-precision, floating-point format. (*Floating-point format* is a mathematical format in which the decimal point can be manipulated.) As a result, the regeneration of large, complex drawings can take considerable time.

1. Open the drawing file named **EXPO98 base.dwg** located in AutoCAD's **Sample** folder.

2. Toggle off the snap mode.

Standard

3. Zoom in on a small area repeatedly until you force a screen regeneration.

AutoCAD displays a warning: About to regen – proceed?

4. Pick the **OK** button.

Regenerating model appears at the Command window. The time it takes to regenerate depends on the size of the drawing and the speed of the computer.

You can also cause screen regenerations by using other AutoCAD commands, including the REGEN command.

5. Enter **REGEN**.

As you work with AutoCAD commands such as QTEXT and FILL, you will use the REGEN command.

Zoom

6. Enter **ZOOM All** and close the file. Do not save changes.

Controlling View Resolution

View resolution refers to the accuracy with which curved lines appear on the screen. It is inversely related to regeneration speed. The lower you set the view resolution, the faster the screen will regenerate. Setting the view resolution to a higher value increases the time required for regeneration. The VIEWRES command allows you to set the view resolution for objects in the current drawing.

1. In the drawing named zoom.dwg, enter the **VIEWRES** command.

2. In reply to Do you want fast zooms? press **ENTER**.

NOTE:

The Fast Zoom mode is no longer functional. This option has been retained only for compatibility with AutoCAD scripts that were written using previous releases of AutoCAD.

Notice that the circle zoom percent (screen resolution) is 100 by default. AutoCAD allows you to enter any number from 1 to 20,000 in answer to this prompt.

3. Enter **20** for the circle zoom percent.

4. Construct an arc, circle, or ellipse and notice its "coarse" appearance.

As you can see, it contains fewer vectors and therefore generates more quickly on the screen.

5. Enter **VIEWRES** again; press **ENTER** and then enter **150**.

Notice the smooth appearance of the objects on the screen.

6. Erase the arc, circle, or ellipse.

7. Save your work and exit AutoCAD.

Chapter 12
Review & Activities

Review Questions

1. Explain why the ZOOM command is useful.

2. Cite one example of when it would be necessary to use the ZOOM command to complete a technical drawing and explain why.

3. Describe each of the following ZOOM options.
   a. All
   b. Center
   c. Extents
   d. Previous
   e. Window

4. What is a transparent zoom? When might a transparent zoom be useful?

5. What is a screen regeneration?

6. Why might you often choose to use the realtime zoom option instead of other zoom methods?

7. Explain the purpose of the VIEWRES command.

Challenge Your Thinking

1. You read in this chapter that AutoCAD drawings are stored in a vector format. Many other graphics programs store their drawings in a raster format. Find out the differences between the raster and vector formats. Can you convert a drawing in vector format to a raster format? Can you convert a raster drawing into a vector format? Explain.

2. Investigate the ZOOM Center option. What is its purpose? When might you want to use it? Explain.

Chapter 12
Review & Activities

Applying AutoCAD Skills

Work the following problems to practice the commands and skills you learned in this chapter.

1. As a general rule, the outside diameter of the washer face on a hex bolt head is the same as the distance across the hex flats. This dimension is approximately 1½ times the diameter of the bolt shank. Draw the head and washer face of the ¼" hex bolt shown in Fig. 12-6. Use the following guidelines:

   • Create a new drawing using Quick Setup, fractional units, and a drawing area of 20,15.

   • Create the outer edge of the washer face using a diameter of ⁷⁄₁₆. Place it near the center of the screen.

   • Zoom a window so that the circle almost fills the drawing area. Do a REGEN if necessary to smooth the circle.

   • Draw a hexagon (6-sided polygon) with the same center as the circle and a distance across the flats of ⁷⁄₁₆. In other words, circumscribe the hexagon around a ⁷⁄₁₆" circle.

   • Draw a second circle with a diameter of ¼" to represent the shank of the bolt.

   • Enter ZOOM All. Notice how small the bolt head is. Even if you had set the drawing area for the default of 12,9, it would be impossible to draw the bolt head accurately without using the ZOOM command.

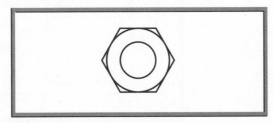

Fig. 12-6

Chapter 12
Review & Activities continued

2. Create a drawing such as an elevation plan of a building, a site plan of a land development, or a view of a mechanical part. Using the ZOOM command, zoom in on the drawing and include details. Zoom in and out on the drawing as necessary, using the different ZOOM options and transparent zooms.

3. Draw the kitchen floor plan shown in Fig. 12-7A. Then zoom in on the kitchen sink. Edit the sink to include the details shown in Fig. 12-7B.

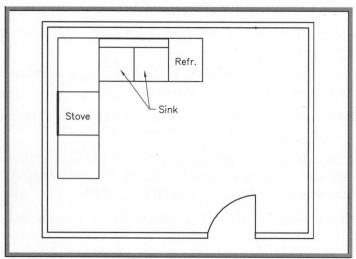

Fig. 12-7A

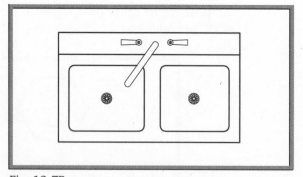

Fig. 12-7B

4. Refer to the file named land.dwg and instructions contained on the Instructor's Resource CD-ROM.

Problem 3 courtesy of Kathleen P. King, Fordham University, Lincoln Center.

Chapter 12
Review & Activities

Using Problem-Solving Skills

Complete the following activities using problem-solving skills and your knowledge of AutoCAD.

1. Use the ZOOM command to draw the door with a fan window shown in Fig. 12-8. Approximate all dimensions.

2. Draw the gasket shown in Fig. 12-9. Do not dimension it. Use construction lines for center lines and erase them when the drawing is complete. Are you sure the 1.50 radius is in fact a radius? Use the ZOOM command to find out. One way to create this drawing is to draw two R.75 circles and one R1.50 circle, and locate the four lines with the Tangent object snap. You can then erase the circles and use the ARC command to connect the lines. When you have completed the drawing, use the ZOOM command to ensure that all the intersections between lines and arcs are made correctly.

Fig. 12-8

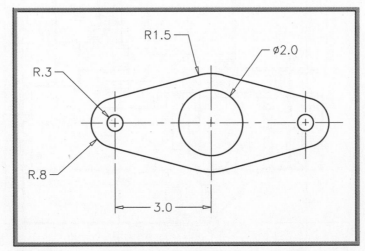

Fig. 12-9

Objectives:

- **Pan from one zoomed area to another**
- **Produce and use named views**
- **Use Aerial View to navigate in a complex drawing**

Key Terms

panning

view box

Zooming in on a drawing solves part of the problem of working with fine detail on a drawing. A method called panning solves the other part. Panning permits you to move the zoom window to another part of the drawing while keeping the same zoom magnification. This helps keep you from constantly zooming in and out to examine detail.

Note the degree of detail in the architectural floor plan shown in Fig. 13-1.

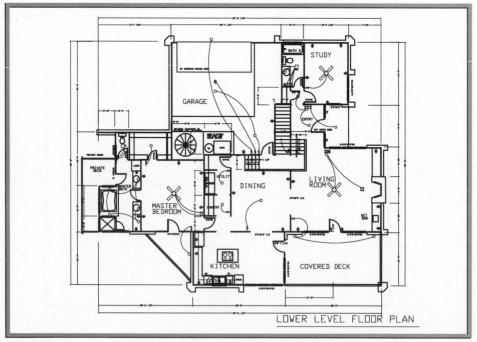

LOWER LEVEL FLOOR PLAN

Fig. 13-1

The drafter who completed the CAD drawing in Fig. 13-1 zoomed in on portions of the floor plan in order to include detail. For example, the drafter zoomed in on the kitchen to place cabinets and appliances, as shown in Fig. 13-2.

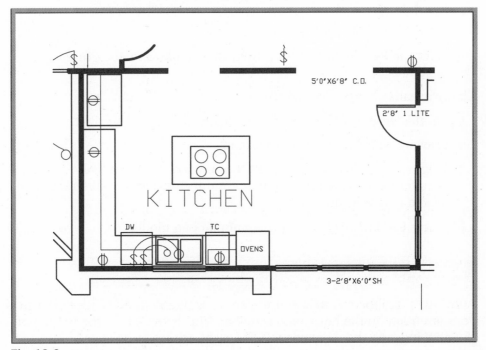

Fig. 13-2

Suppose the drafter wants to include detail in an adjacent room but wants to maintain the present zoom magnification. In other words, the drafter wants to simply "move over" to the adjacent room. This operation can be accomplished by panning.

Panning

Panning means moving the viewing window from one area to another in a drawing that has been zoomed to a high magnification. There are several ways to pan in AutoCAD. You can use the PAN command, various panning icons, or the scrollbars to move around in a drawing.

1. Open the drawing named **wilhome.dwg** located in AutoCAD's **Sample** folder.

2. From the **File** pull-down menu, pick **Save As...**, double-click the folder with your name, and pick the **Save** button.

3. Pick the **Model** tab in the lower left area of the drawing area.

4. Zoom in on the right one-third of the drawing.

Standard

Realtime Panning

Let's pan to the left side of the drawing.

Standard

5. Pick the **Pan Realtime** button from the **Standard** toolbar, or enter **P** for PAN.

The crosshairs changes to a hand.

6. In the left portion of the screen, click and drag to the right.

The drawing window moves to the right.

7. Experiment further with the **PAN** command until you feel comfortable with it. Pan in different directions and at different zoom magnifications.

8. Press **ESC** or **ENTER** to exit the realtime pan mode.

Scrollbar Panning

AutoCAD's scrollbars enable you to pan horizontally and vertically. Focus your attention on the horizontal scrollbar. This is the bar at the bottom of the drawing area that contains left and right arrows at its ends.

1. Pick the left arrow once; pick it again.

This moves the viewing window to the left.

2. In the horizontal scrollbar, click and drag the movable scroll box a short distance to the right.

This moves the viewing window to the right. You can also move to the left or right by picking a point on either side of the movable scroll box.

Now focus on the vertical scrollbar. This is the bar along the right side of the drawing area. It also contains arrows at its ends.

3. Pick the up arrow a couple of times.

This moves the viewing window in the upward direction.

4. Pick the down arrow; then drag the scroll box down a short distance.

As you can see, the scrollbars offer a convenient way of panning.

Transparent Panning

You can use the PAN command transparently. This is particularly useful for reaching points that are not currently visible.

1. Enter the **LINE** command and pick a point anywhere on the screen.

2. Pick the **Pan Realtime** button or enter **P** at the keyboard.

3. Pan to a new location, and press **ESC** or **ENTER** to exit the realtime panning mode.

4. Pick a second point to create a line segment and terminate the LINE command.

5. Erase the line.

Standard

Alternating Realtime Pans and Zooms

AutoCAD provides a fast method of alternating between realtime pans and zooms.

1. Pick the **Zoom Realtime** button from the **Standard** toolbar.

2. Zoom in on the drawing.

3. Right-click the pointing device.

This displays a shortcut menu.

4. Select **Pan** from the menu and pan to a new location.

5. Right-click the pointing device to display the menu.

6. Pick one of the **Zoom** options.

7. When you are finished, pick **Exit** from the shortcut menu or press **ESC** or **ENTER** to exit the realtime zoom and pan mode.

Standard

Capturing Views

Imagine that you are working on an architectural floor plan. You've zoomed in on the kitchen to include details such as the appliances, and now you're ready to pan over to the master bedroom. Before leaving the kitchen, you foresee a need to return to the kitchen for final touches. However, by the time you're ready to do this final work on the kitchen, you may be at a different zoom magnification or at the other end of the drawing. The VIEW command solves the problem.

1. Find and zoom in on the three-car garage.

2. Pick the **Named Views** button from the **Standard** toolbar and pick the **Named Views** tab if it is not in the foreground.

Standard

This enters the VIEW command and displays the View dialog box (Fig. 13-3).

3. Pick the **New...** button.

4. For View name, enter **garage** and pick **OK**.

AutoCAD adds the named view to the list.

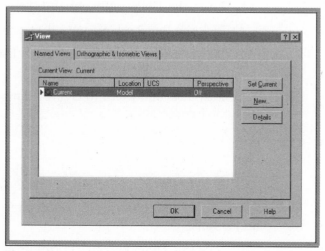

5. Pick the **OK** button.

6. Pan to a new location on the drawing.

7. Pick the **Named Views** button, pick the **garage** view from the list, pick the **Set Current** button in the dialog box, and pick **OK**.

AutoCAD restores the named view.

Fig. 13-3

8. Enter **VIEW** again and pick the **Details** button.

This displays information on the named view.

9. Read the information and then pick **OK**.

NOTE: ────────────

The Orthographic & Isometric Views tab does not apply to two-dimensional drawings. It enables you to restore orthographic or isometric views of three-dimensional drawings.

10. Pick **OK to exit the View dialog box.**

Aerial View

Aerial View is a fast and easy-to-use navigational tool for zooming and panning.

1. From the **View** pull-down menu, pick **Aerial View**.

Fig. 13-4

This displays the Aerial View window with wilhome.dwg inside, as shown in Fig. 13-4.

The buttons in the Aerial View window should look familiar. The white rectangle represents the zoomed area.

2. Move the crosshairs into the Aerial View window and click once.

New crosshairs appear.

3. Move the window (called a *view box*) around and pick a point.

AutoCAD zooms in on the area defined by the view box. Meanwhile, Aerial View maintains a view of the entire drawing. The view box, depicted by a white rectangle, shows the size and location of the current view in the graphics screen.

4. Move the view box around the screen.

5. Right-click to freeze the view box.

Aerial View shows the location of the current zoom window.

6. Press the pick button to activate the view box.

7. Press the pick button again to change the size of the view box.

8. Move the view box to a new location and right-click.

9. On your own, experiment with the buttons and menu items in the Aerial View window.

10. Close the Aerial View window.

11. **ZOOM All**, save your work, and exit AutoCAD.

Chapter 13
Review & Activities

Review Questions

1. Explain why panning is useful.

2. Is it possible to pan diagonally using the scrollbars? Explain.

3. Explain why named views are useful.

4. Explain the method of alternating between realtime pans and zooms.

5. What is the primary benefit of using Aerial View?

Challenge Your Thinking

1. Many zoom and pan variations were introduced in the last two chapters. Describe how and when you might use a combination of them, but explain also why it may be impractical to use all of them. Which methods of zoom and pan are the fastest, and which seem to be the most practical for most applications?

2. Is Aerial View an alternative to using the PAN command? Explain when you would use each most effectively.

Applying AutoCAD Skills

Work the following problems to practice the commands and skills you learned in this chapter.

1. Open the campus.dwg file in AutoCAD's Sample folder. Pick the Model tab at the bottom of the drawing area, and use the scrollbars to pan to the far right side of the drawing, where you will see an entrance with three arches. Use the Realtime Zoom and Realtime Pan buttons to zoom in on the steps leading up to the arched entrance. Count and record the number of steps at the entrance.

2. Draw each of the views shown in Fig. 13-5 using the dimensions shown. However, do not include the dimensions on your drawing. Zoom in on one of the views and store it as a named view. Then pan to each of the other views and store each as a named view. Restore each named view and alter each shape. Be as creative as you wish, but be certain that all three of the views are consistent with one another.

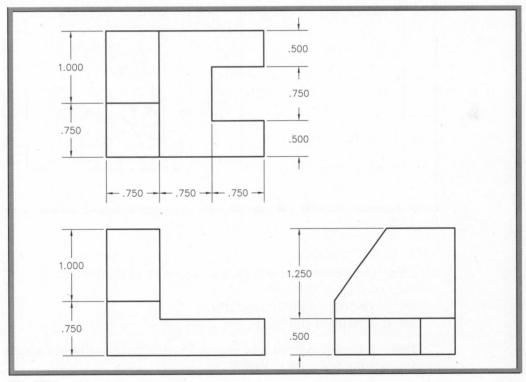

Fig. 13-5

3. Zoom in on one of the three views from the drawing in problem 2. Use the scrollbars to pan to the adjacent views.

4. Perform dynamic pans and zooms on the drawing in problem 2.

Chapter 13
Review & Activities continued

5. Open the kitchen drawing you created in problem 3 of Chapter 12. Add the details for the refrigerator and the stove as shown in Fig. 13-6. Open Aerial View and use it to position the zoom window as you work.

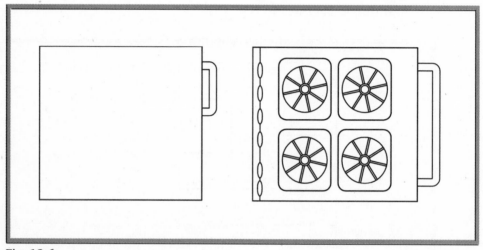

Fig. 13-6

6. Using the VIEW command, save these views of the kitchen.
   Sink: closeup of the sink
   Fridge: closeup of the refrigerator
   Stove: closeup of the stove

7. Refer to the file named land.dwg and instructions contained on the Instructor's Resource CD-ROM.

Problem 5 courtesy of Dr. Kathleen P. King, Fordham University, Lincoln Center

Chapter 13
Review & Activities

 Using Problem-Solving Skills

Complete the following activities using problem-solving skills and your knowledge of AutoCAD.

1. As the person responsible for computer allocation, you need to know how many offices are without computers. Open the AutoCAD file db_samp.dwg located in AutoCAD's Sample folder (Fig. 13-7). Using dynamic zoom and pan as necessary, make a list (by office number) of offices that do not have computers. Do not save any changes you may have made to the drawing.

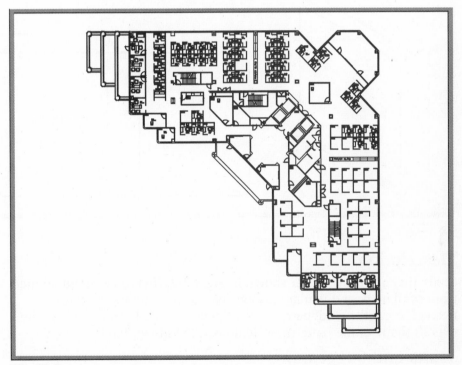

Fig. 13-7

2. Reopen db_samp.dwg and use Aerial View to make a list of rooms with designations of COPY, I.S., STOR., or COFFEE. Which of these rooms should have computers? Do you think office number 6001 should have a computer? Why or why not? Do not save your work.

Part 2 Project

Patterns and Developments

Many manufactured products begin as flat pieces of metal, plastic, or other material. Examples include cereal boxes, heating and cooling ducts, and even items of clothing. Manufacturers use flat *patterns* to cut material to the proper size and shape. The construction of these patterns is known as a *development*. Fig. P2-1 shows a box constructed for a company that distributes breakfast cereals, as well as an example of a pattern.

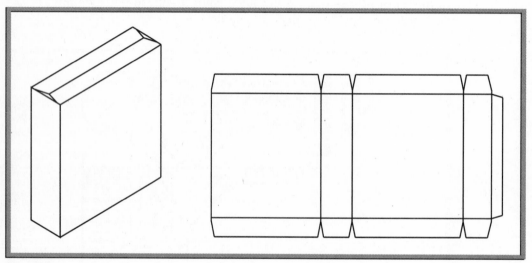

Fig. P2-1

Description

Study the magazine holder shown in Fig. P2-2. It is to be manufactured from cardboard, and you are responsible for developing the pattern needed to cut the cardboard to size. Show the fold lines as well as the overall shape of the pattern, as demonstrated in Fig. P2-1.

Hints and Suggestions

1. Draw the orthographic views of the magazine holder.

2. Referring to the orthographic views as necessary, create a stretchout line for the magazine holder. A *stretchout line* is a line that represents the total length of the finished pattern and shows the relative lengths of each part. An example of a stretchout line for the cereal box is shown at the bottom of Fig. P2-2.

3. Using the orthographic views and the stretchout line, create the development. Use object snaps and other construction aids as necessary, and use ZOOM and PAN to simplify your drawing task.

Part 2 Project

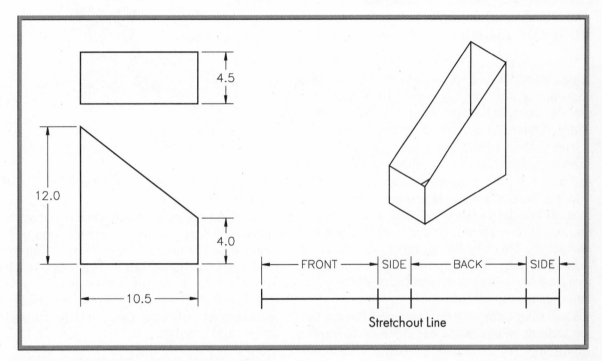

Fig. P2-2

Summary Questions/Self-Evaluation

Your instructor may direct you to answer these questions orally or in writing, or simply to think about your responses as a means of self-reflection.

1. Pinpoint any difficulties you encountered with this project and explain how you solved them.

2. Which construction aids did you use to create the orthographic views of the magazine holder? The development? Explain how these aids helped.

3. How would you create a pattern for a soup can? In what ways would the development be different from that for the magazine holder? Explain.

Careers
Using
AutoCAD

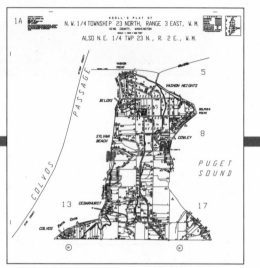

Vashon Island 911

Vashon Island, only a short ferry ride from Seattle, Washington, has about 9,000 residents. This small community has a highly capable emergency-response service, comparable to systems in big cities. Vashon Island devised its system using off-the-shelf software.

The Vashon Island 911 system integrates AutoCAD drawings, Field Notes™, and Microsoft® Access. As a call comes in, the dispatcher types in the phone number. Instantly, a map appears beside a form that shows data associated with the location: name, address, access information, 911 history, and so on. The map has zoom control to show exact locations. The map and data are relayed to a mobile, battery-operated fax machine in an emergency-response vehicle.

Kroll Map Company implemented the system for Vashon Island. Kroll has done cartography (mapping) for the Pacific Northwest for more than 85 years. This family-run company, currently under the leadership of John Locker, has been passed down through three generations. Kroll creates its base mapping from historical survey information, consultant maps, and engineering and construction drawings. The company also receives weekly updates from the county seat.

The mapping system's usefulness goes beyond 911 service because information related to other government services is attached to the Vashon Island maps. This information has applications in property tax listings, real estate, public utilities, and transportation.

Career Focus—Land Surveyor

Land surveyors establish official land and water boundaries; write descriptions of land for deeds, leases, and other legal documents; and measure construction and mineral sites. They are assisted by survey technicians, who

Adapted from a story by David White in CADENCE, January, © 1995 Miller Freeman, Inc. Revised 1999.

operate surveying instruments and collect information.

Land surveyors plan the work of survey technicians, select survey reference points, and determine the precise location of all important features in a survey area. They research legal records and look for evidence of previous boundaries. They record results, verify the accuracy of collected data, and prepare plats, maps, and reports.

Education and Training

Surveyors who establish official boundaries must be licensed by the state in which they work. Standards vary, but some states require a bachelor's degree in survey technology or in a related field such as civil engineering. Higher standards for licensure are the trend, as computer and satellite technology change operating practices in the profession. Community colleges and vocational schools offer two-year programs. On-the-job experience as a survey technician is generally required.

Career Activity

- Write a report about survey-related jobs.
- How are global positioning systems (GPS) changing the kind of work surveyors do and the level of pay they receive?

Part 3

Drawing and Editing

Chapter 14 Solid and Curved Objects

Objectives:

- Produce solid-filled objects
- Control the appearance of solid-filled objects
- Create and edit polylines
- Create and edit splines

Key Terms

B-splines

control points

polyline

splines

system variable

traces

vertex

Many technical drawings contain thick lines and solid objects. Drawings for printed circuit boards, for instance, include relatively thick lines called *traces*. The trace lines represent the metal conductor material that is etched onto the PC board. Drawings and models that use curved lines often take advantage of polylines and spline curves. Examples are consumer products and business machines ranging from phones, pagers, and cosmetic products to printers, copiers, and fax machines. Many of the plastic injection-molded parts on these products are designed using polylines and spline curves. You will learn more about polylines and spline curves later in this chapter.

Note the heavy lines in the drawing shown in Fig. 14-1. They were used to make the drawing descriptive and visually appealing.

Drawing Solid Shapes

The SOLID command produces solid-filled objects.

1. Start AutoCAD and start a new drawing from scratch using English units.

2. From the **Draw** pull-down menu, pick **Surfaces** and **2D Solid**.

This enters the SOLID command.

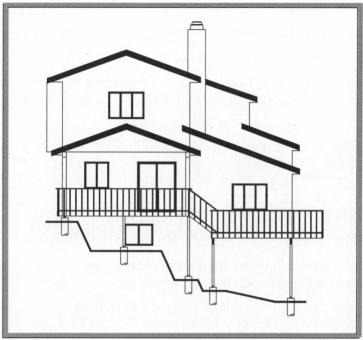

Fig. 14-1

3. Produce a solid-filled object similar to the one in Fig. 14-2. Pick the points in the exact order shown, and press ENTER to terminate the command.

 HINT: ─────────────────────────────

Consider using snap and grid.

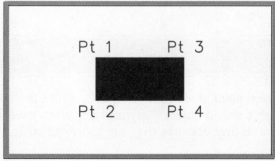

Fig. 14-2

4. Repeat Step 3, but do not press ENTER to terminate the command.

5. Pick a fifth and sixth point.

AutoCAD adds a new piece to the solid object.

6. Experiment with the **SOLID** command. (If you pick the points in a different order, AutoCAD creates an hourglass-shaped object.)

7. Press **ENTER** to terminate the SOLID command.

Keep the objects on the screen because you will need them for the following section.

Controlling Object Fill

With the FILL command, you can control the appearance of solids, wide polylines, multilines, and hatches.

FILL is either on or off. When FILL is off, only the outline of a solid is represented on the screen. This saves time when the screen is regenerated.

INFOLINK

You will learn more about multilines in Chapter 15 and hatches in Chapter 20.

1. Enter the **FILL** command.

2. Enter **OFF**.

After you turn FILL on or off, you must regenerate the screen before the change will take place.

3. Enter the **REGEN** command to force a screen regeneration.

The objects are no longer solid-filled.

4. Reenter the **FILL** command and turn it **ON**.

5. Enter **REGEN** to force another screen regeneration.

6. Erase all objects on the screen.

Polylines

A *polyline* is a connected sequence of line and arc segments that is treated by AutoCAD as a single object. Polylines are often used in lieu of conventional lines and arcs because they are more versatile.

Drawing a Thick Polyline

Let's use the PLINE command to create the electronic symbol of a resistor, as shown in Fig. 14-3.

Fig. 14-3

1. Set snap at **.5**.

2. Pick the **Polyline** button from the **Draw** toolbar, or enter **PL** (the PLINE command alias).

Draw

This enters the PLINE command.

3. Pick a point in the left portion of the screen.

The PLINE options appear at the Command line.

4. Enter **W** (for Width) and enter a starting and ending width of **.15** unit. (Notice that the ending width value defaults to the starting width value.)

5. Draw the object by approximating the location of the endpoints. If you make a mistake, undo the segment. Press **ENTER** when you have finished the object.

6. Save your work in a file named **poly.dwg**.

7. Move the polyline a short distance.

Notice that the entire polyline is treated as a single object.

Editing a Polyline

AutoCAD supplies a special editing command called PEDIT to edit polylines. PEDIT allows you to perform many advanced functions, such as joining polylines, fitting them to curve algorithms, and changing their width. Let's edit the polyline using the PEDIT command.

1. Display the **Modify II** toolbar.

2. Pick the **Edit Polyline** button from the **Modify II** toolbar.

This enters the PEDIT command.

3. Pick the polyline.

The polyline does not highlight as other objects do during object selection, but the PEDIT options appear at the Command line.

4. Enter **W** and specify a new width of **.2** unit.

The width of the entire polyline changes. As you can see, using polylines can be an advantage when there is a possibility that you will need to change the width of several connected line segments at a later time.

Let's close the polyline, as shown in Fig. 14-4.

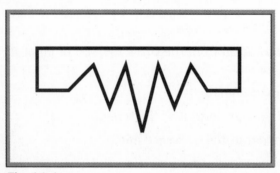

Fig. 14-4

5. Enter **C** for Close.

The PEDIT Fit option creates a smooth curve based on the locations of the vertices of the polyline. In this context, a *vertex* (plural *vertices*) is any endpoint of an individual line or arc segment in a polyline. Let's do a simple curve fitting operation.

6. Enter **F** for Fit.

Notice how the drawing changed. The new curve passes through all the vertices of the polyline.

7. Enter **D** (for Decurve) to return it to its previous form.

Next, let's move one of the vertices as shown in Fig. 14-4.

8. Enter **E** for Edit Vertex.

Notice that a new set of choices becomes available at the Command line. Also notice the x at one of the vertices of the polyline.

9. Move the x to the vertex you want to change by pressing **ENTER** several times.

10. Enter **M** for Move and pick a new point for the vertex.

11. To exit the PEDIT command, enter **X** (for eXit) twice.

12. Save your work.

PEDIT contains many more editing features. Experiment further with them on your own.

Exploding Polylines

The EXPLODE command gives you the ability to break up a polyline into individual line and arc segments.

1. Enter the **UNDO** command and select the **Mark** option. (We will return to this point at a later time.)

2. Pick the **EXPLODE** button from the **Modify** toolbar.

3. Pick the polyline and press **ENTER**.

Modify

Notice the message: Exploding this polyline has lost width information. The UNDO command will restore it.

This is the result of applying EXPLODE to a polyline that contains a width other than the default. You now have an object that contains many line objects for easier editing, but you have lost the line width.

4. To illustrate that the object is now made up of several objects, pick one of them.

Drawing Curved Polylines

In some drafting applications, there is a need to draw a series of continuous arcs to represent, for example, a river on a map. If the line requires thickness, the ARC Continue option will not work, but the PLINE Arc option can handle this task.

1. Enter the **PLINE** command and pick a point anywhere on the screen.

2. Enter **A** for Arc.

Draw

A list of arc-related options appears at the Command line.

3. Move the crosshairs and notice that an arc begins to develop.

4. Enter the **Width** option and enter a starting and ending width of **.1** unit.

5. Pick a point a short distance from the first point.

6. Pick a second point, and a third.

7. Press **ENTER** when you're finished.

8. Enter **UNDO** and **Back**.

Spline Curves

Splines are curves that use sampling points to approximate mathematical functions that define a complex curve. AutoCAD allows you to create splines, also referred to as *B-splines,* using two different methods. You can use the PEDIT command to transform a polyline into a spline curve, or you can use the SPLINE command to generate the spline directly.

Transforming a Polyline into a Spline

The PEDIT Spline option uses the vertices of the selected polyline as the control points of the curve. *Control points* are points that exert a "pull" on a curve, influencing its shape. The curve passes through the first and last control points and is pulled toward the other points, but it does not necessarily pass through them. The more control points you specify in an area, the more pull they exert on the curve.

Modify II

1. Pick the **Edit Polyline** button from the **Modify II** toolbar and select the polyline.

2. Enter **S** for Spline.

Do you see the difference between the Spline and Fit options? The Spline option uses the vertices of the selected polyline as control points, whereas the Fit option produces a curve that passes through all vertices of the polyline.

3. Enter **D** to decurve the object.

4. Enter **X** to exit the PEDIT command.

The type of spline created by the Spline option depends on the setting of the SPLFRAME system variable. (A *system variable* in AutoCAD is similar to a command, except that it holds temporary settings and values instead of performing a specific function.) The note on page 195 provides more information about the types of splines AutoCAD provides through the PEDIT command.

Producing Splines Directly

With the SPLINE command, you can create spline curves using a single command. Spline curves are used extensively in industry to produce free-form shapes for automobile body panels and the exteriors of aircraft and consumer products. Designers fit three-dimensional surfaces through two or more spline curves to model the design. The spline curve itself is the building block for producing interesting and sometimes extremely complex designs.

NOTE:

In connection with PEDIT, AutoCAD offers two spline options: quadratic B-splines and cubic B-splines. An example of each is shown below.

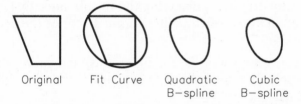

Original   Fit Curve   Quadratic   Cubic
                         B-spline   B-spline

The system variable SPLINETYPE controls the type of spline curve to be generated. Set the value of SPLINETYPE at 5 to generate quadratic B-splines. Set its value at 6 to generate cubic B-splines.

The PEDIT Decurve option enables you to turn a spline back into its frame, as illustrated in the previous Step 3. You can view the spline curve and its frame simultaneously by setting the SPLFRAME system variable to 1.

1. From the **File** pull-down menu, pick **Save As...**, and enter **nurbs** for the file name.

2. Erase all the objects in the drawing.

3. From the **Draw** toolbar, pick the **Spline** button.

This enters the SPLINE command.

Draw

4. Pick a point and then a second point.

5. Slowly move the end of the spline back and forth and notice how the spline behaves.

6. Continue to pick a series of points.

7. Press **ENTER** three times to complete the spline and terminate the command.

8. Enter **SPLINE** again and pick a series of points.

Draw

9. Enter **C** for Close.

10. Press **ENTER** to terminate the command.

NOTE:

The Fit Tolerance option changes the tolerance for fitting the spline curve to the control points. A setting of 0 causes the spline to pass through the control points. By changing the fit tolerance to 1, you allow the spline to pass through points within 1 unit of the actual control points.

You can easily convert a spline-fit polyline into spline curve. To do so, enter the SPLINE command, enter O for Object, and pick the spline-fit polyline.

Editing Splines

Using grips, you can interactively edit a spline entity. The SPLINEDIT command provides many additional editing options.

1. Without anything entered at the Command prompt, select one of the spline curves.

Grip boxes appear at the control points.

2. Lock onto one of the grips and move it to a new location.

This is the fastest way to edit a spline.

3. Press **ESC** twice.

4. From the **Modify II** toolbar, pick the **Edit Spline** button.

This enters the SPLINEDIT command.

5. Select a spline.

The editing options appear at the Command line. The Close option closes the spline curve. The Move vertex option enables you to edit a spline in a manner similar to the grips editing method you used above. The Refine option permits you to enhance the spline data.

6. Enter **R** for Refine.

The refinement options appear at the Command line.

7. Enter **A** for Add control point.

8. Pick several points on the spline to add new control points, and then press **ENTER**. (You may want to turn off object snap before adding new control points.)

9. Enter **X**.

The rEverse option reverses the order of the control points in AutoCAD's database. The Undo option undoes the last edit operation. The Fit Data option permits you to edit fit data using several options.

10. Create a new spline of any shape.

11. Pick the **Edit Spline** button and pick the new spline.

12. Enter **F** for Fit Data and pick the **Help** button to read about each of the fit options.

13. After exiting the help screen, experiment with the **Fit Data** options on your own.

14. Close the Modify II toolbar and exit AutoCAD.

Chapter 14
Review & Activities

Review Questions

1. How would you draw a solid-filled triangle using the SOLID command?

2. What is the purpose of FILL, and how is it used?

3. What object types does FILL affect?

4. What is a polyline?

5. In what situations might you decide to use the PLINE command instead of the LINE and ARC commands?

6. Briefly describe each of the following PLINE options.
   a. Arc
   b. Close
   c. Halfwidth
   d. Length
   e. Undo
   f. Width

7. Briefly describe each of the following PEDIT command options.
   a. Close
   b. Join
   c. Width
   d. Edit vertex
   e. Fit
   f. Spline
   g. Decurve
   h. Ltype gen
   i. Undo

8. Explain a situation in which you might need to explode a polyline.

9. Describe the SPLINETYPE system variable.

10. When you create a spline curve using the SPLINE command, what does a fit tolerance setting of 0 indicate?

11. What is the fastest method of editing a spline?

12. What editing function can you perform by entering SPLINEDIT, Refine, and Add control point?

Chapter 14
Review & Activities

Challenge Your Thinking

1. Discuss possible uses for the SOLID and FILL commands. Under what circumstances might you use them? Give specific examples.

2. Discuss everyday applications of NURBS curves.

Applying AutoCAD Skills

Work the following problems to practice the commands and skills you learned in this chapter.

1. Construct the building in Fig. 14-5 using the PLINE and SOLID commands. Specify a line width of .05 unit. Don't worry about the exact size and shape of the roof.

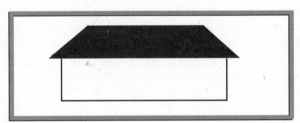

Fig. 14-5

2. After you have completed problem 1, place the solid shapes as indicated in Fig. 14-6. Don't worry about their exact sizes and locations.

Fig. 14-6

Chapter 14
Review & Activities continued

3. Use the PLINE command to draw the road sign indicating a sharp left corner (Fig. 14-7).

Fig. 14-7

4. The SI symbol shown in Fig. 14-8 is used on metric drawings that conform to the SI system and use what is called *third-angle projection*. This projection system, which places the top view of a multiview drawing above the front view, is the normal way of showing views in the United States. Draw the icon, including the outlines of the letters S and I; then use the SOLID command to fill in the letters.

Fig. 14-8

Problem 4 courtesy of Gary J. Hordemann, Gonzaga University

5. Create the approximate shape of the racetrack shown in Fig. 14-9 using PLINE. Specify .4 unit for both the starting and ending widths. Select the Arc option to draw the figure.

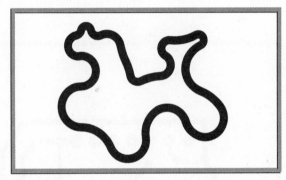

Fig. 14-9

6. Draw the symbols in Fig. 14-10 using the PLINE and PEDIT commands.

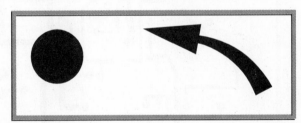

Fig. 14-10

7. Draw the car in Fig. 14-11 using the PLINE and PEDIT commands.

Fig. 14-11

Chapter 14
Review & Activities continued

8. The artwork for one side of a small printed circuit board is shown in Fig. 14-12. Reproduce the drawing using donuts and wide polylines. Use the grid to estimate the widths of the polylines and the sizes of the donuts.

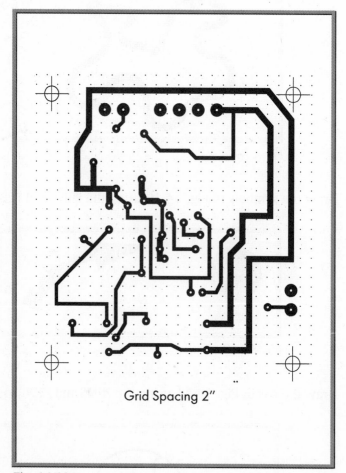

Grid Spacing 2″

Fig. 14-12

Problem 8 courtesy of Gary J. Hordemann, Gonzaga University

Chapter 14
Review & Activities

Using Problem-Solving Skills

Complete the following activities using problem-solving skills and your knowledge of AutoCAD.

1. Use the SPLINE command to approximate the route of Interstate 98 as it passes through Chittenden County, Vermont (Fig. 14-13). Include the Lake Champlain shoreline, but do not include the text. Save the drawing as 98.dwg.

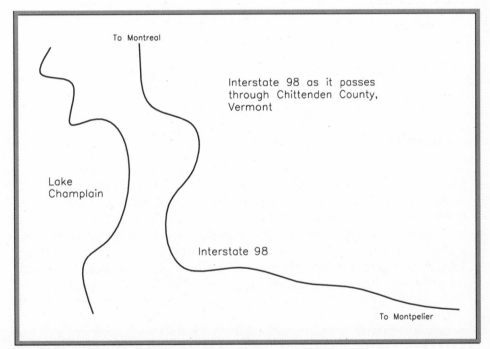

To Montreal

Interstate 98 as it passes through Chittenden County, Vermont

Lake Champlain

Interstate 98

To Montpelier

Fig. 14-13

2. Model railroads can be built to many different scales. One of the more popular scales is HO, in which $1/8'' = 1'$. With the PLINE command, design an HO scale railroad track layout in a figure-eight pattern. Make the center intersection at ninety degrees. At a convenient point in your layout, add a spur to terminate in a railroad yard with three parallel tracks for storing the rolling stock. Refer to books in your local library or search the Internet for any additional information you may need to complete this problem.

3. Construct the drawing shown in Fig. 14-1.

Chapter 15 — Adding and Altering Objects

Objectives:

- **Create chamfered corners**
- **Break pieces out of lines, circles, and arcs**
- **Produce fillets and rounds**
- **Offset lines and circles**
- **Create and edit multilines**

Key Terms

chamfers
fillets
image tiles
mline
multiline
rounds

Most consumer products contain rounded or beveled edges for improved aesthetics and functionality. AutoCAD offers quick ways to create rounded inside corners, called *fillets,* and outside corners, called *rounds,* as well as beveled edges called *chamfers*. Other useful tools include breaking lines and offsetting them to create parallel lines. AutoCAD also gives you the flexibility to produce multiple lines at one time. This is especially useful for architectural drawing, as are the other capabilities mentioned here.

Creating Chamfers

The CHAMFER command enables you to place a chamfer at the corner formed by two lines.

1. Start AutoCAD and open the drawing named **gasket.dwg**.

2. Pick the **Chamfer** button from the **Modify** toolbar.

Modify

3. Enter **D** for Distance.

The distance you specify is the distance from the intersection of the two lines (the corner) to the start of the bevel, or chamfer. You can set the chamfer distance for the two lines independently.

4. Specify a chamfer distance of **.25** unit for both the first and second distances.

5. Enter the **CHAMFER** command again and place a chamfer at each of the corners of the gasket by picking the two lines that make up each corner.

Modify

When you're finished, the drawing should look similar to the one in Fig. 15-1, with the possible exception of the holes.

6. Obtain AutoCAD on-line help to learn about the other options offered by the CHAMFER command.

HINT:

Enter the CHAMFER command and then pick the Help button or press F1.

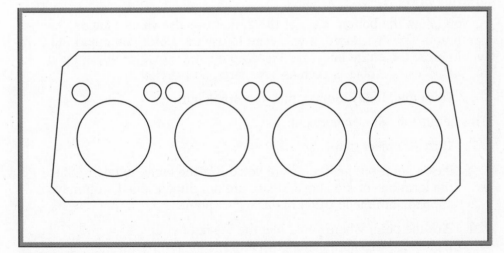

Fig. 15-1

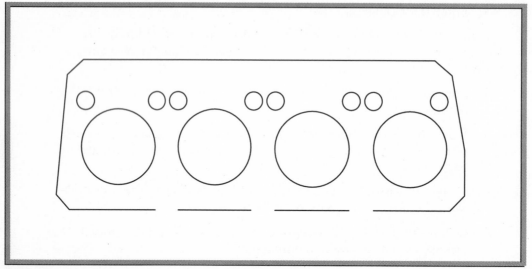

Fig. 15-2

Breaking Objects

Let's remove (break out) sections of the gasket so that it looks like the drawing in Fig. 15-2.

As you know, the bottom edge of the gasket was drawn as a single, continuous line. Therefore, if you were to use the ERASE command, it would erase the entire line. The BREAK command, however, allows you to "break" certain objects such as lines, arcs, and circles.

1. From the **Modify** toolbar, pick the **Break** button.

This enters the BREAK command.

Modify

2. Turn off object snap.

3. Pick a point on the line where you'd like the break to begin. Since the locations of the above breaks are not dimensioned, you may approximate the location of each start point.

4. Pick the point where you'd like the break to end.

The piece of the line between the first and second points disappears. Let's break out two more sections of approximately equal size, as shown in Fig. 15-2.

5. Repeat Steps 1 through 3 to create each break.

6. Experiment with the other BREAK options on your own. Refer to AutoCAD's on-line help for more information about these options.

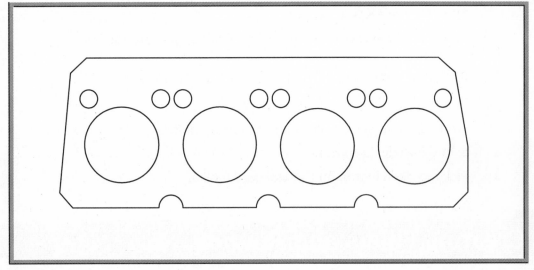

Fig. 15-3

7. Insert arcs along the broken edge of the gasket, as shown in Fig. 15-3.

8. Save your work.

Let's break out a section of one of the holes in the gasket, as shown in Fig. 15-4.

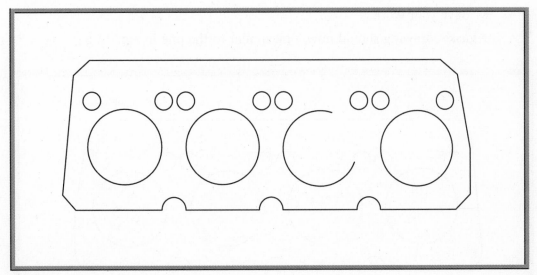

Fig. 15-4

9. Enter the **BREAK** command.

10. Pick a point anywhere on the circle. (You may need to turn off object snap.)

11. Instead of picking the second point, enter **F** for first point.

12. Pick the first (lowest) point on the circle.

13. Working counterclockwise, pick the second point.

A piece of the circle disappears.

14. Pick the **Undo** button on the **Standard** toolbar.

Modify

Standard

Creating Fillets and Rounds

In AutoCAD, the FILLET command creates both fillets and rounds on any combination of two lines, arcs, or circles. Let's change the chamfered corners on the gasket to rounded corners.

1. Erase each of the four chamfered corners.

2. From the **Modify** toolbar, pick the **Fillet** button, and enter a new radius of **.3**.

3. Reenter the **FILLET** command and produce fillets at each of the four corners of the gasket by picking each pair of lines.

4. Save your work.

Modify

The gasket drawing should now look similar to the one in Fig. 15-5.

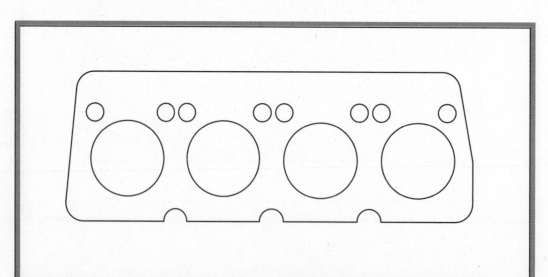

Fig. 15-4

Extending to Form a Corner

Let's move away from the gasket and try something new.

1. Above the gasket, draw lines similar to the ones in Fig. 15-6. Omit the numbers.

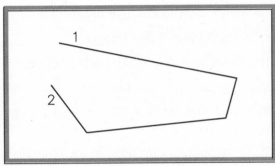

Fig. 15-6

Modify

2. Set the fillet radius at **0**.

3. Reenter the **FILLET** command and select lines 1 and 2.

This technique works with the CHAMFER command, too.

4. Experiment with the remaining FILLET options on your own. If necessary, obtain on-line help.

Offsetting Objects

Offsetting a line results in another line at a specific distance, or offset, from the original line. Offsetting a circle or arc results in another circle or arc that is concentric (shares the same center point) with the original one.

Offsetting Circles

Modify

1. From the **Modify** toolbar, pick the **Offset** button.

This enters the OFFSET command.

2. For the offset distance, enter **.2**.

3. Select one of the holes in the gasket drawing.

4. Pick a point inside the hole in reply to Specify point on side to offset.

5. Select another circle and pick a side to offset.

6. Press **ENTER** to terminate the command.

Offsetting Through a Specified Point

1. Using the **LINE** command, draw a triangle of any size.

2. Pick the **Offset** button and enter **T** (for Through).

Modify

3. Pick any one of the three lines that make up the triangle.

4. Pick a point a short distance from the line and outside the triangle.

The offset line appears. Notice that the line runs through the point you picked.

5. Do the same with the remaining two lines in the triangle so that you have an object similar to the one in Fig. 15-7. Press **ENTER** when you're finished.

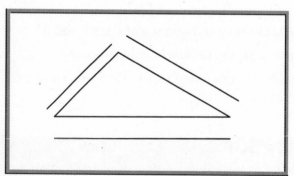

Fig. 15-7

6. Enter **CHAMFER** and set the first and second chamfer distances at **0**.

Modify

7. Enter **CHAMFER** again and pick two of the new offset lines.

8. Do this again at the remaining two corners to complete the second triangle.

9. Clean up the drawing so that only the gasket remains and save your work.

Drawing Multilines

The OFFSET command is useful for adding offset lines to existing lines, arcs and circles. If you want to produce up to 16 parallel lines, you can do so easily—and simultaneously—with the MLINE command. This command produces parallel lines that exist as a single object in AutoCAD. The object is called a *multiline* or *mline* object.

1. Begin a new drawing from scratch.

2. From the **Draw** toolbar, pick the **Multiline** button.

This enters the MLINE command.

Draw

3. Create a large triangle by picking three points and entering **C** to close the triangle.

Notice that the corners of the triangle meet perfectly. The triangle is a single object—an mline object.

4. Erase the triangle.

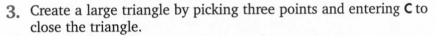
Setting Multiline Styles

The Multiline Styles dialog box permits you to change the appearance of multilines and save custom multiline styles. Let's use this dialog box to produce the drawing of a room, as shown in Fig. 15-8.

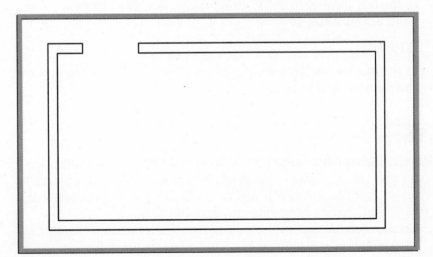

Fig. 15-8

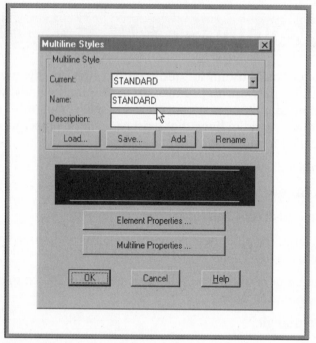

Fig. 15-9

1. Set snap at **.25** unit and turn on either polar tracking or ortho.

2. From the **Format** pull-down menu, pick the **Multiline Style...** item.

The Multiline Styles dialog box appears (Fig. 15-9).

3. In the box located to the right of Name, double-click the word **STANDARD**.

4. Type **S1**.

5. In the box located to the right of Description, pick a point and type **This style has end caps.**

NOTE:

AutoCAD does not require you to enter a description for multiline styles you create. However, it is good practice to document your styles so that you can see at a glance which style is appropriate if you need to change or update the drawing later.

6. Pick the **Save...** button.

7. In the **Save Multiline Style** dialog box, pick the **Save** button to store the S1 style.

Applying a New Style

Notice that the current style is still STANDARD. Before you can use a new style, you must load it.

8. Pick the **Load...** button. The Load Multiline Styles dialog box appears.

9. Select **S1** and pick **OK**; pick **OK** again.

10. Redisplay the **Multiline Styles** dialog box (press the spacebar) and pick the **Multiline Properties...** button.

11. Pick the **Start** and **End** check boxes located at the right of Line and pick the **OK** button.

This causes the example multiline in the Multiline Styles dialog box to change. Notice that both ends of the multiline are now "capped" with short connecting lines.

12. Pick the **OK** button.

When you create new multilines, the S1 style will apply.

Scaling Multilines

You can set the distance between lines in a multiline using the Scale option of the MLINE command.

Draw

1. Pick the **Multiline** button or enter **MLINE** at the keyboard.

2. Enter **S** for Scale and enter **.25**.

3. With **MLINE** entered, create the drawing shown in Fig. 15-8 (page 211), approximating its size.

 HINT: ————————————————

Use the Undo button if you need to back up one line segment.

4. Press **ENTER** to terminate **MLINE**.

5. Save your work in a file named **multi.dwg**.

Let's add the interior walls shown in Fig. 15-10.

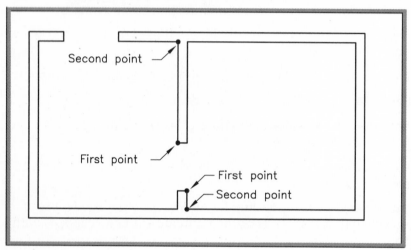

Fig. 15-10

1. Display the **Multiline Styles** dialog box.

2. Create a new multiline style named **S2** with the following description: **This style has a start end cap.** After you save it, be sure to load it.

3. Pick **OK** to exit the Multiline Styles dialog box, if you haven't already.

4. Redisplay the same dialog box and pick the **Multiline Properties...** button.

5. Uncheck the **End** check box located to the right of Line and pick **OK**; pick **OK** again.

6. Using **MLINE**, draw the interior walls as shown in Fig. 15-10. (Press **ENTER** to complete each wall.)

Draw

You will edit the wall intersections in the following section.

Editing Intersections

Using the Multiline Edit Tools dialog box, it is possible to edit the intersection of multilines.

1. From the **Modify** pull-down menu, pick the **Multiline...** item.

This enters the MLEDIT command and displays the Multiline Edit Tools dialog box (Fig. 15-11). The twelve squares in the dialog box are called *image tiles*.

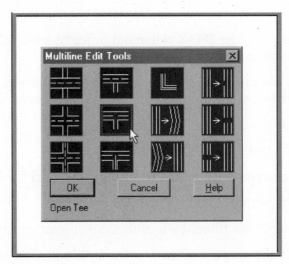

Fig. 15-11

2. Pick the image tile located in the second row and second column, as shown in the previous illustration.

Open Tee appears in the lower left corner.

3. Pick the **OK** button.

Focus your attention on either of the two interior walls.

4. In reply to Select first mline, pick one of the two interior walls.

5. In reply to Select second mline, pick the adjacent exterior wall.

This causes AutoCAD to break the exterior wall.

6. Edit the second interior wall by repeating Steps 4 and 5.

7. Press **ENTER** to terminate the **MLEDIT** command.

8. On your own, explore the remaining options found in the **Multiline Styles** and **Multiline Edit Tools** dialog boxes.

9. Save your work and exit AutoCAD.

Chapter 15
Review & Activities

Review Questions

1. What is the function of the CHAMFER command?

2. How is using the BREAK command different from using the ERASE command?

3. If you want to break a circle or arc, in which direction do you move when specifying points: clockwise or counterclockwise?

4. In what toolbar is the Fillet button found?

5. How do you set the fillet radius?

6. What can you accomplish by setting either FILLET or CHAMFER to 0?

7. Explain the purpose of the OFFSET command.

8. How might multilines be useful? Give at least one example.

9. Explain the purpose of the Multiline Styles and Multiline Edit Tools dialog boxes.

Challenge Your Thinking

1. Explore the Angle option of the CHAMFER command. How is it different from the Distance option? Discuss situations in which each option (Angle and Distance) may be useful.

2. To help potential clients understand floor plans, the architectural firm for which you work draws its floor plans showing the walls as solid gray lines the thickness of the wall. Exterior walls are 6″ thick, and interior walls (those that make up the room divisions) are 5″ thick. Explain how you could achieve these walls using MLINE and the Multiline Styles dialog box.

3. AutoCAD allows you to save multiline styles and then reload them later. Explore this feature and write a short paragraph explaining how to save and restore multilines.

Chapter 15
Review & Activities

Applying AutoCAD Skills

1. Create the first drawing on the left in Fig. 15-12. Don't worry about exact sizes and locations, but do use snap and ortho. Then use FILLET to change it to the second drawing. Set the fillet radius at .2 unit.

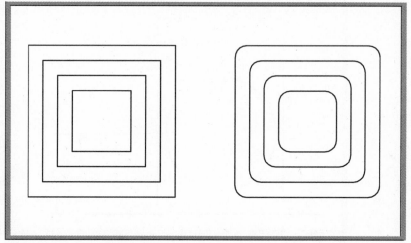

Fig. 15-12

2. Create the triangle shown in Fig. 15-13. Then use CHAMFER to change it into a hexagon. Set the chamfer distances at .66 unit.

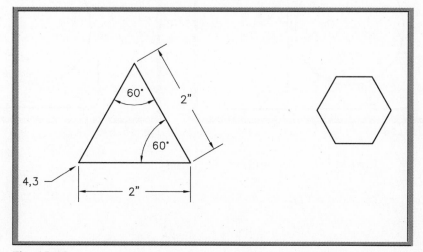

Fig. 15-13

Chapter 15
Review & Activities continued

3. Create the roller and cradle for a conveyor. Begin by drawing a circle and three lines as shown in Fig. 15-14A. Specify a fillet radius of .25 and create the fillets on the left side as shown in Fig. 15-14B. Then fillet the right side. The finished roller and cradle should look like the one in Fig. 15-14C.

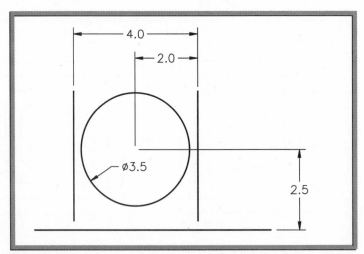

Fig. 15-14A

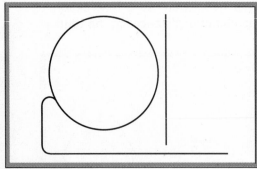

Fig. 15-14B

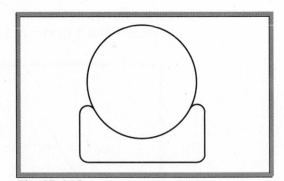

Fig. 15-14C

4. Draw the steel base shown in Fig. 15-15. Use the FILLET command to place fillets in the corners as indicated.

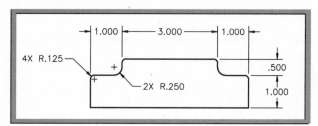

Fig. 15-15

5. Draw the shaft shown in Fig. 15-16. Make it 2 units long by 1 unit in diameter. Use the CHAMFER command to place a chamfer at each corner. Specify .125 for both the first and second chamfer distances.

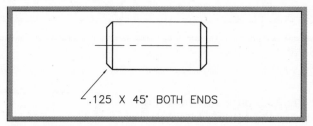

Fig. 15-16

6. Using the ARC and CHAMFER commands, construct the detail of the flat-blade screwdriver tip shown in Fig. 15-17.

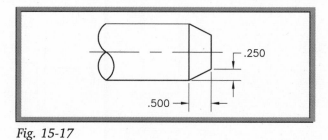

Fig. 15-17

Chapter 15
Review & Activities continued

7. Create the picture frame shown in Fig. 15-18. First create the drawing on the left using the LINE and OFFSET commands. Then modify it to look like the drawing on the right using the CHAMFER command.

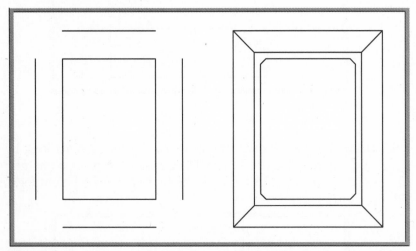

Fig. 15-18

8. Create the A.C. plug shown in Fig. 15-19 using the MLINE, MLSTYLE, CHAMFER, and FILLET commands. Measure a real electrical plug to obtain the sizes.

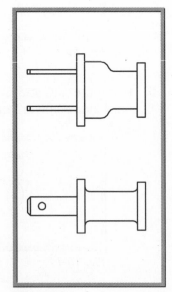

Fig. 15-19

Problems 7 and 8 courtesy of Dr. Joseph K. Yabu, San Jose State University

9. Draw the front view of the rod guide end cap shown in Fig. 15-20. Use the OFFSET command to create the outer circular shapes and the inner horizontal lines. Use the BREAK and FILLET commands to create the inner shape with four fillets. Do not dimension the drawing.

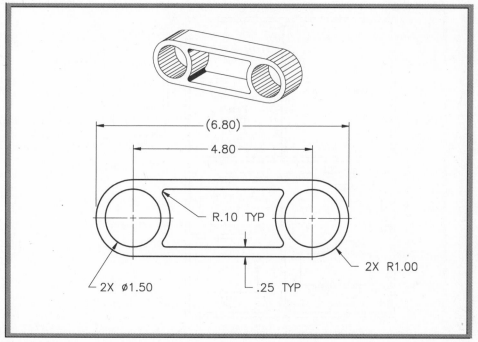

Fig. 15-20

Problem 9 courtesy of Gary J. Hordemann, Gonzaga University

10. Draw the objects and rooms shown in Fig. 15-21. Use multilines to produce the walls. Then use the MOVE and COPY commands to move the office furniture into the rooms. Omit the lettering on the drawing.

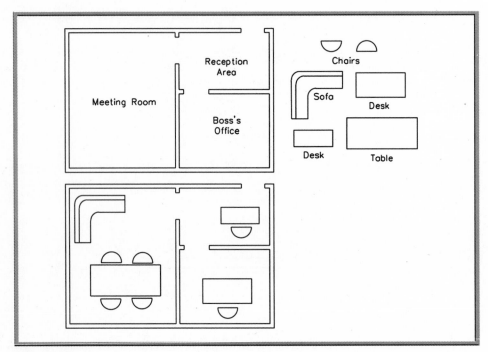

Fig. 15-21

11. Draw the front view of the ring separator shown in Fig. 15-22. Start by using the CIRCLE and BREAK commands to create a pair of arcs. Then use OFFSET to create the rest of the arcs. Do not dimension the drawing.

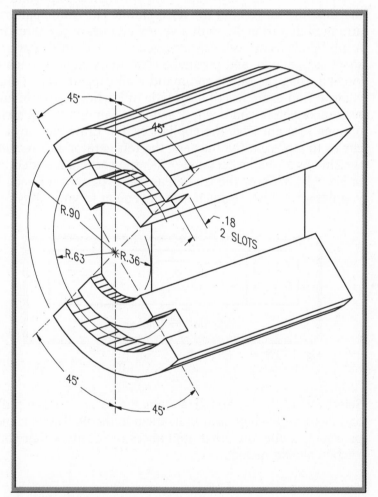

Fig. 15-22

Problem 11 courtesy of Gary J. Hordemann, Gonzaga University

Chapter 15
Review & Activities continued

Using Problem-Solving Skills

Complete the following activities using problem-solving skills and your knowledge of AutoCAD.

1. Kitchens are arranged around their three most important elements: the sink, the range, and the refrigerator. The traffic pattern is arranged in a triangle, with a vertex at each of the three items. As a result, kitchens are usually arranged in one of three patterns: U-shaped, L-shaped, and parallel. Draw an example of each type of kitchen using the MLINE command with capped ends. (You may want to find examples of each type of kitchen layout in architectural references before you begin drawing to familiarize yourself with the options.) Do not label the appliances, but draw the range using four circles to represent the burners, the refrigerator as a rectangle, and the sink as a smaller rectangle with the corners rounded, as shown in Fig. 15-23. Show the triangular traffic pattern for each kitchen you design. Which seems most convenient to you? Why?

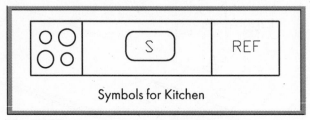

Symbols for Kitchen

Fig. 15-23

2. Select one of the kitchen arrangements you created in problem 1, and design a kitchen around it. Include the walls, door and window placements, and any other appliances or furniture that you think a kitchen should include.

Careers
Using
AutoCAD

Find the Right Amount of Light

People always notice the carpet on entering a room, but they seldom look at the ceiling, which is where lighting designers do much of their work. Proper lighting begins with the shape and size of the space to be lit. The hallway of a sleek professional building, for example, needs different lighting than a stairwell in a sports arena. Floor and wall treatments, with different light-reflecting qualities, also factor into the choice of which fixtures to use and where they should go.

Photo courtesy of Brent Phelps

To encourage energy efficiency in lighting, the Electric Power Research Institute sponsored development of LightCAD, a software package that creates a layer in an AutoCAD drawing to represent lighting. As you input details about fixture types and locations, LightCAD calculates totals for light output and energy consumption. You can try different combinations of fixtures and locations to find the most energy-efficient plan. A database in LightCAD provides data for standard fixtures. The program can also use files from fixture manufacturers.

LightCAD provides a report that shows how a lighting plan measures up to specifications of ASHRAE/IES 90.1, a national lighting standard. According to Tom McDougall of the Weidt Group (Minnetonka, MN), the architecture/energy organization that produced LightCAD, "this is an advanced lighting layout tool to improve basic commercial lighting."

Career Focus—Electrician

Electricians install and maintain a variety of electrical systems, including lighting as well as general power supply, climate control, security, and communications. Electricians also install boxes for switches and outlets. They test circuits after installation to ensure that they are properly grounded and have no breaks or shorts in them.

Electricians work with blueprints that specify locations for outlets, load centers, panel boards, and other components in electrical circuits. Electricians must follow the National Electric Code and comply with state and local requirements for electrical installations.

Education and Training

A four-year apprenticeship is the standard route to becoming qualified as an electrician. Apprenticeship programs are sponsored jointly by the International Brotherhood of Electrical Workers and the National Electrical Contractors Association through local affiliates. Apprenticeship programs generally require graduation from high school with coursework in mechanical drawing, mathematics, and electricity/electronics. Localities usually require electricians to obtain a license, issued upon passing a written examination.

Career Activity

- Interview someone who can tell you about employment of electricians in your area. Are apprenticeships available?
- Has employment in this field been steady over the past few years? What are the forecasts for future employment?

Chapter 16 — Moving and Duplicating Objects

Objectives:

- Change an object's properties
- Move and copy objects
- Mirror objects and parts around an axis
- Produce rectangular and polar arrays

Key Terms

array
circular array
polar array
rectangular array

Earlier in this book, you learned how to move and copy objects using the grips method. Now you will learn how to perform these and other operations using commands. Why would you want to? The commands give you additional options. Also, if you choose to write scripts or programs that automate certain AutoCAD operations, you could embed these commands into them, so knowing how they work is important.

You will also learn methods of arraying objects, both in rectangular and circular fashion. The arraying features are useful for many purposes, including architectural drafting, facilities planning, auditorium and stadium design, and countless machine design applications. The problems at the end of this chapter provide examples.

INFOLINK

You will learn more about creating scripts in Chapter 56.

Changing Object Properties

The Properties dialog box provides a method of changing several of an object's properties.

1. Start AutoCAD and open the drawing named **gasket.dwg**.

2. Without anything entered at the Command prompt, select one of the circles and then right-click.

3. Pick the **Properties** item from the shortcut menu.

This displays the Properties dialog box, as shown in Fig. 16-1. Review the list of object properties, including radius and diameter. Note that the properties listed in the dialog box depend on the object selected. In this case, the options presented are specific to the circle you selected in Step 2.

4. Change the diameter of the circle by entering a new number into the text box in the Properties dialog box and press **ENTER**.

The diameter of the circle changes.

You will use this dialog box many times throughout this book.

5. Change the diameter back to its original value, and close the dialog box.

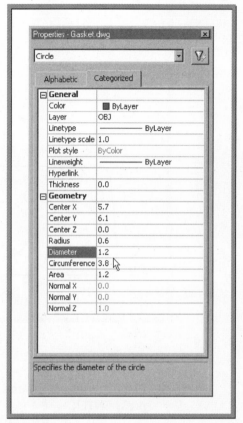

Fig. 16-1

Moving Objects

Let's move the gasket to the top of the screen using the MOVE command.

1. From the **Modify** toolbar, pick the **Move** button, or enter **MOVE** at the keyboard.

Modify

 HINT: ───────────────────────────

M is the command alias for MOVE.

AutoCAD displays the Select objects prompt.

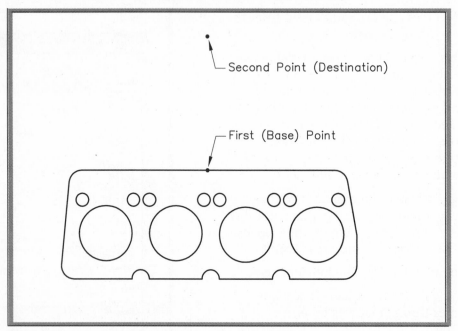

Fig. 16-2

2. Place a window around the entire gasket drawing and press **ENTER**.

3. In reply to Specify base point or displacement, place a base point somewhere on or near the gasket drawing as shown in Fig. 16-2.

4. Move the pointing device in the direction of the second point (destination).

5. Pick the second point.

The gasket should move as shown in Fig. 16-3.

6. Practice using the **MOVE** command by moving the drawing to the bottom of the screen.

7. Enter the **MOVE** command.

Modify

8. In response to the Select objects prompt, pick two of the four large holes (circles) and press **ENTER** (or right-click).

9. Specify the first point (base point). Place the point anywhere on or near either of the two circles.

10. In reply to Specify second point or displacement, move the pointing device upward. (Polar tracking should be on.)

11. Enter **3** to move the circles vertically 3 units.

12. Undo the move.

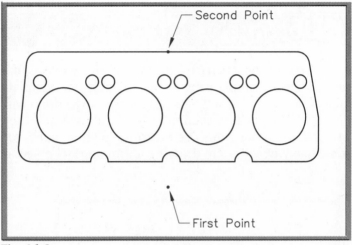

Fig. 16-3

Copying Objects

The COPY command is almost identical to the MOVE command. The only difference is that the COPY command does not move the object; it copies it.

1. Erase all of the large holes in the gasket except for one.

2. From the **Modify** toolbar, pick the **Copy Object** button.

Modify

This enters the COPY command.

3. Select the remaining large hole and press **ENTER**.

4. Specify the Multiple option by entering **M**, and select the center of the circle for the base point.

 HINT: ——————————————————————————

Use the Center object snap. Also, turn on ortho or polar tracking before completing the following step.

5. Move the crosshairs and place the circle in the proper location.

6. Repeat Step 5 until all four large circles are in place; then press **ENTER**.

7. Practice using the **COPY** command by erasing and copying the small circles.

8. Save your work.

Mirroring Objects

There are times when it is necessary to produce a mirror image of a drawing, detail, or part. A simple copy of the object is not adequate because the object being copied must be reversed, as was done with the butterfly in Fig. 16-4. One side of the butterfly was drawn and then mirrored to produce the other side. The mirroring feature can save you a large amount of time when you are drawing a symmetrical object or part.

Fig. 16-4

Mirroring Vertically

The same is true if the engine head gasket we developed is to be reproduced to represent the opposite side of an eight-cylinder engine. In this case, we will mirror the gasket vertically (around a horizontal plane).

Modify

1. Move the gasket to the bottom of the screen to allow space for another gasket of the same size.

2. From the **Modify** toolbar, pick the **Mirror** button.

Modify

This enters the MIRROR command.

3. Select the gasket by placing a window around it, and press **ENTER**.

4. Create a horizontal mirror line near the gasket by selecting two points on a horizontal plane as shown in Fig. 16-5.

AutoCAD asks if you want to delete the source objects.

5. Enter **N** for No, or press **ENTER** since the default is No.

AutoCAD mirrors the gasket as shown in Fig. 16-5.

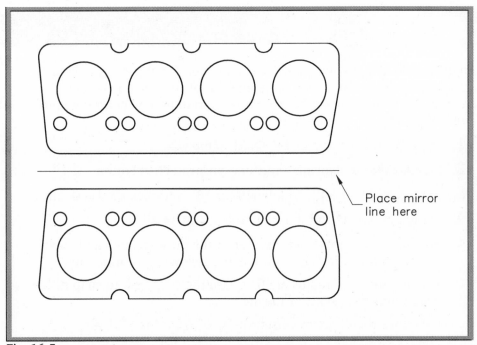

Fig. 16-5

Mirroring at Other Angles

You can also create mirror images around an axis (mirror line) other than horizontal.

Modify

1. Draw a small triangle and mirror it with an angular (*e.g.*, 45°) mirror line.

NOTE:

You can also mirror objects using grips. Without a command entered, pick an object, pick (activate) one of the grip handles, and then right-click. From the shortcut menu, pick Mirror. Notice that if you mirror using this method, the angular mirror function is the default. To enter a vertical or horizontal mirror line, you must first enter B (for Base point) at the Command line.

2. Erase the triangles and save your work.

Producing Arrays

An *array* is an orderly grouping or arrangement of objects. AutoCAD's arraying capability can save a large amount of time when you are creating a drawing that includes many copies of an object arranged in a rectangular or circular fashion.

1. Start AutoCAD and pick the **Quick Setup** wizard.

2. Pick **Architectural** for the units and pick the **Next** button.

3. Enter a width of **24'** and a length of **18'**, and pick the **Finish** button.

4. Set the grid at **1'** and the snap at **3"**; **ZOOM All**.

5. Zoom in on the lower left quarter of the drawing area and create the chair shown in Fig. 16-6. Use the following information.

   • Create the 1'-1" radius arc first. Use the snap grid to snap to the start and end points of the arc.

   • Produce the 7" arc by offsetting by 6" from the 1'-1" arc.

   • The 10" arc passes through the centers of the 3" arcs.

6. **ZOOM All** and save your work in a file named **chair.dwg**.

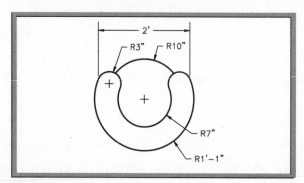

Fig. 16-6

Creating Rectangular Arrays

A *rectangular array* is one in which the objects are arranged in rows and columns. An example is shown in Fig. 16-7.

1. Pick the **Array** button from the docked **Modify** toolbar.

2. In reply to Select objects, select the chair and press **ENTER**.

Modify

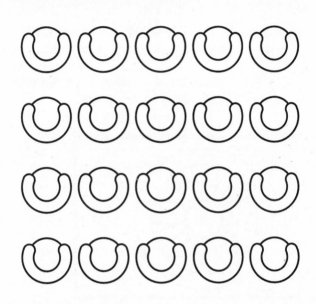

Fig. 16-7

3. Enter **R** for Rectangular.

4. In reply to Enter the number of rows, enter **4**.

5. In reply to Enter the number of columns, enter **5**.

6. Specify that you want **3'** between the rows.

The distance is measured from the center of one chair to the center of the next.

7. Specify that you want **30"** between the columns.

AutoCAD produces an array of 20 chairs, as shown in Fig. 16-7.

8. Save your work.

 NOTE: ————————————————————————

You can produce rectangular arrays at any angle by first changing the snap rotation angle. Also, you can specify the distances between rows and columns by picking the opposite corners of a rectangle with the pointing device.

Creating Circular Arrays

Next, we're going to produce the bicycle wheel shown in Fig. 16-8. Notice the arrangement of the spokes. We can create them quickly by drawing only two lines and then using a circular array to create the remaining lines. A *circular array* (also called a *polar array*) is one in which the objects are arranged radially around a center point.

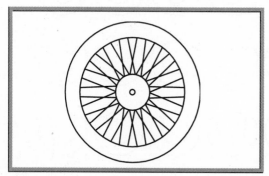

Fig. 16-8

1. Begin a new drawing from scratch.

2. Draw a wheel similar to the one in Fig. 16-9. Make the wheel large enough to fill most of the screen. Use the Center object snap to make the circles concentric.

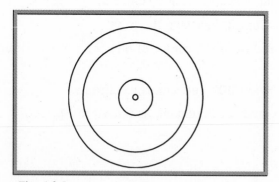

Fig. 16-9

3. Draw two crossing lines similar to the ones in Fig. 16-10. Use the **Nearest** object snap to begin and end the lines precisely on the appropriate circles.

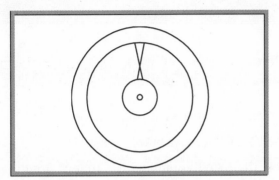

Fig. 16-10

4. Save your work in a file named **wheel.dwg**.

5. Pick the **Array** button or enter **AR**, the command's alias.

Modify

6. Use the following information when responding to each step. (Snap should be off.)

   • Select each of the two crossing lines for the array and press **ENTER**.

   • Enter **P** for Polar.

   • Make the center of the wheel the center point for the array.

   • Specify **18** for the number of items.

   • Enter **360** for the angle to fill by pressing **ENTER**.

   • Rotate the spokes as they are copied.

NOTE: ───────────────────────────────────

During the polar array sequence, the prompt Angle to fill (+=ccw, -=cw) appears. AutoCAD is asking for the angle (in degrees) to fill the array. +=ccw means that a positive number will produce an array in a counterclockwise direction, and -=cw means that a negative number will produce an array in a clockwise direction.

The wheel should look similar to one in Fig. 16-8.

7. Save your work.

8. Practice creating polar arrays using other objects. At least once, specify less than 360 degrees when replying to Angle to fill.

Modify

9. Save your work and exit AutoCAD.

Chapter 16
Review & Activities

Review Questions

1. What is the purpose of the Properties dialog box?

2. In what toolbar is the Move button located?

3. Explain how the MOVE command is different from the COPY command.

4. Describe a situation in which the MIRROR command would be useful.

5. During a MIRROR operation, can the mirror line be specified at any angle? Explain.

6. Name the two types of arrays.

7. State one practical application for each type of array.

8. What is the effect on a polar array if you specify an angle to fill of less than 360 degrees? Explain.

9. Explain how a rectangular array can be created at any angle.

Challenge Your Thinking

1. Experienced AutoCAD users take the time necessary to analyze the object to be drawn before beginning the drawing. Discuss the advantages of doing this. Keep in mind what you know about the AutoCAD commands presented in this chapter.

2. You have been asked to complete a drawing of the dialpad for a pushbutton telephone. You have decided to create one of the pushbuttons and use the ARRAY command to create the rest. Each pushbutton is a 13mm × 8mm rectangle, and the buttons are arranged as shown in Fig. 16-11. What should you answer when AutoCAD prompts you for the space between rows and space between columns?

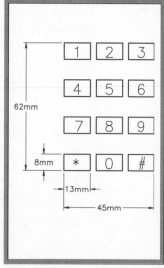

Fig. 16-11

Chapter 16
Review & Activities

Applying AutoCAD Skills

Work the following problems to practice the commands and skills you learned in this chapter.

1. Create the shim shown in Fig. 16-12. Begin by drawing the left half of the shim according to the dimensions shown, using construction lines for center lines. Then mirror the objects to create the right half of the shim. Erase any unneeded lines. Save the drawing as shim.dwg.

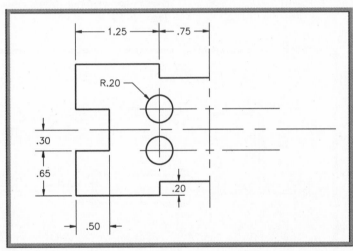

Fig. 16-12

2. Draw the chisel shown in Fig. 16-13 using the ARC, FILLET, LINE, COPY, and MIRROR commands.

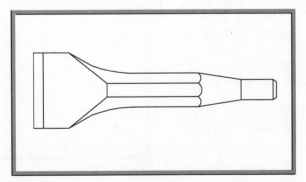

Fig. 16-13

Problem 2 courtesy of Joseph K. Yabu, Ph.D., San Jose State University

Chapter 16
Review & Activities continued

3. Draw the key shown in Fig. 16-14 using the ARC, FILLET, LINE or PLINE, COPY, and MIRROR commands.

4. Apply the commands introduced in this chapter to create the piping drawing shown in Fig. 16-15.

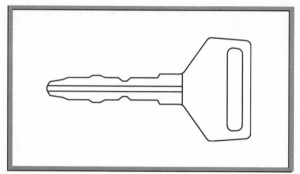

Fig. 16-14

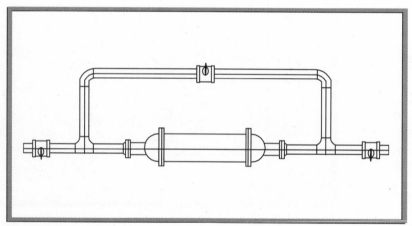

Fig. 16-15

5. Develop a plan for an auditorium with rows and columns of seats. Design the room any way you'd like. Save your work as auditorium.dwg.

6. Create the cooling fan shown in Fig. 16-16.

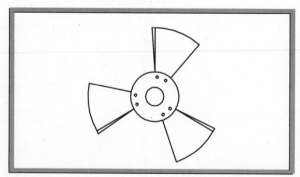

Fig. 16-16

Problems 3 and 6 courtesy of Joseph K. Yabu, Ph.D., San Jose State University
Problem 4 courtesy of Dr. Kathleen P. King, Fordham University, Lincoln Center

7. Draw the front view of the bar separator shown in Fig. 16-17. Draw one fourth of the bar separator using the LINE, ARC, and FILLET commands with a grid setting of .1 and a snap of .05. Use the MIRROR command to make half of the object; then use it again to create the other half.

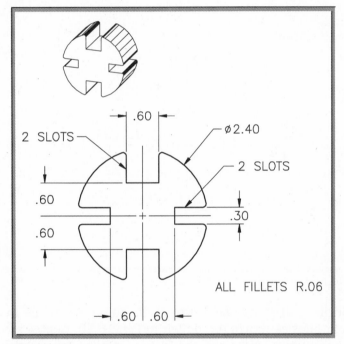

Fig. 16-17

8. Draw the air vent shown in Fig. 16-18 using the ARRAY command. Other commands to consider are POLYGON and COPY.

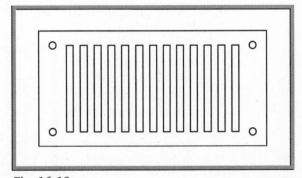

Fig. 16-18

Problem 7 courtesy of Gary J. Hordemann, Gonzaga University
Problem 8 courtesy of Joseph K. Yabu, Ph.D., San Jose State University

Chapter 16
Review & Activities continued

9. Draw the front view of the wheel shown in Fig. 16-19.

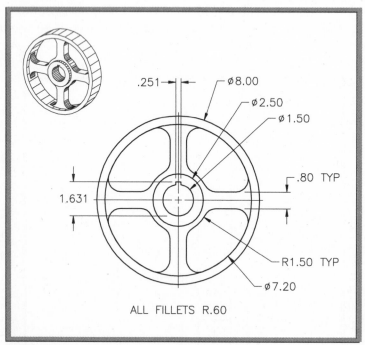

Fig. 16-19

Begin by drawing one-fourth of the object. Draw the three arcs and four lines as shown on the left in Fig. 16-20. Center the arcs on a known point. Next, use the FILLET command to create the cavity shown on the right. Erase the outside lines and the outside arc. Then use the MIRROR command to make three more copies of the cavity. Complete the drawing by adding arcs and circles. Try constructing the keyway by first drawing half of the shape, then using MIRROR to generate the other half. For more practice, try drawing with one-eighth of a wheel.

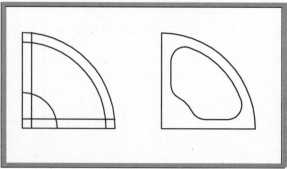

Fig. 16-20

Problem 9 courtesy of Gary J. Hordemann, Gonzaga University

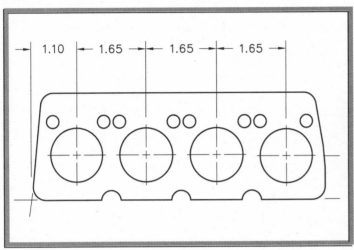

Fig. 16-21

10. Load the drawing named gasket.dwg. Erase each of the circles. Replace the large ones according to the locations shown in Fig. 16-21, this time using the ARRAY command. Reproduce the small circles using the ARRAY command, but don't worry about their exact locations. The diameter of the large circles is 1.25; the small circles have a diameter of .30.

11. Draw the snowflakes in Fig. 16-22 using the ARRAY command. Since snowflakes are six-sided, you may want to begin by drawing a set of concentric circles. Then use the ARRAY command to insert 6, 12, or more construction lines. The example in Fig. 16-23 shows the beginning of one of the snowflakes. Note that you only need to array three lines at 15° intervals for a total of 45° (two lines if you use the MIRROR command).

Fig. 16-22

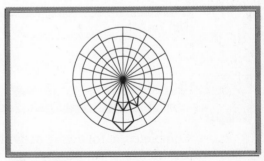

Fig. 16-23

Problem 11 courtesy of Gary J. Hordemann, Gonzaga University

Chapter 16
Review & Activities continued

12. Draw the top view of the power saw motor flywheel shown in Fig. 16-24. Use the ARRAY command to insert the 24 fins and arcs. You may find the OFFSET command useful in creating the first fin.

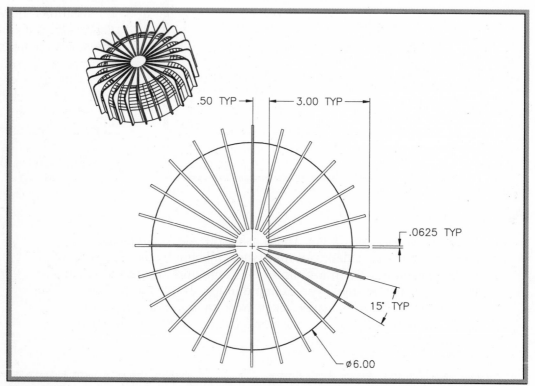

Fig. 16-24

Using Problem-Solving Skills

Complete the following activities using problem-solving skills and your knowledge of AutoCAD.

1. You are creating problems for an AutoCAD textbook and need to simulate the grid displayed with a 0.5 spacing. With the POINT command on the Draw toolbar, create a rectangular array within a drawing space of 12,9 so that the grid completely fills the space. If you find the point is too small to show on the screen, obtain on-line help to find out how to make it larger. When you have completed the grid, turn AutoCAD's grid on and off to see how your grid corresponds to the real grid. Save the drawing as grid.dwg.

2. You have received a design change for the shim you created in Fig. 16-12. Open your drawing and use the MOVE and COPY commands to arrange the holes in a pattern approximately as shown in Fig. 16-25. Save the drawing as shimrev.dwg.

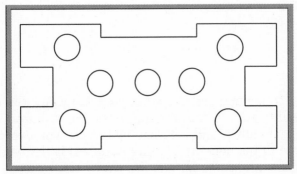

Fig. 16-25

3. Draw the wheel shown in Fig. 16-26 (you decide how).

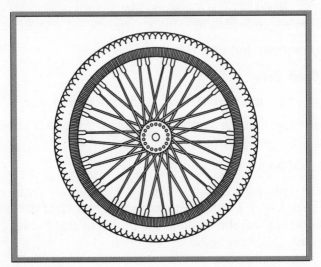

Fig. 16-26

Applying AutoCAD Skills problem 12 courtesy of Gary J. Hordemann, Gonzaga University
Using Problem-Solving Skills problem 2 adapted from a drawing by Bob Weiland.

Chapter 17 Modifying and Maneuvering

Objectives:

- Stretch objects to change their overall shape
- Scale objects using a scale factor or reference length
- Rotate objects to exact angles
- Trim and extend lines to specific boundaries

Key Terms

rotating

scale

The more you create and edit drawings with AutoCAD, the more you will discover a need to edit objects in many ways. You will find several uses for trimming, extending, and lengthening lines as you experiment with new ideas. One of the benefits of using AutoCAD is that you are not penalized for drawing something at the wrong size or in the wrong place. You can easily correct it using one of the commands covered in this chapter or another chapter.

Stretching Objects

1. Start AutoCAD and pick **Use a Wizard**.

2. Pick **Quick Setup** and **OK**.

3. For the units, use **Architectural**, and for the area, enter **120'** and **90'**.

Be sure to include the foot symbols (') so that AutoCAD knows that these are feet, not inches.

4. Enter **ZOOM All**.

5. Draw the site plan shown in Fig. 17-1 according to the dimensions shown. With snap set at **2'**, place the lower left corner of the property line at absolute point **10,10**. Omit dimensions. All points in the drawing should fall on the snap grid.

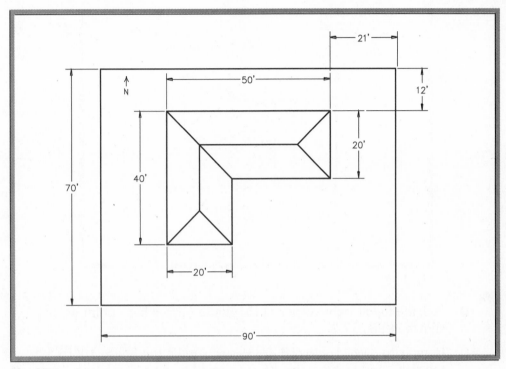

Fig. 17-1

6. Save your work in a file named **site.dwg**.

7. Pick the **Stretch** button from the **Modify** toolbar or enter **S**, the STRETCH command's alias.

Modify

This enters the STRETCH command.

8. Select the east end of the house as shown in the next drawing and press **ENTER**. Use the **Crossing** object selection procedure. (See the following hint.)

HINT: ─────────────────────

In reply to Select objects, pick a point to the right of the house and move the pointing device to the left to create a crossing window (Fig. 17-2).

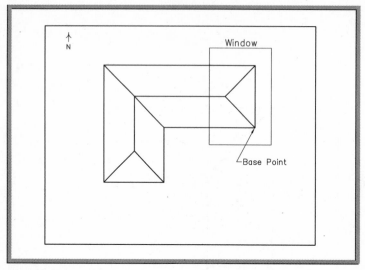

Fig. 17-2

9. Pick the lower right corner of the house for the base point as shown in Fig. 17-2.

10. In reply to Second point of displacement, stretch the house **10'** to the right, and pick a point. (Snap should be on.) The house stretches dynamically.

The house should now be longer.

11. Stretch the south portion of the house **10'** so that it looks similar to the house shown in Fig. 17-3.

12. Save your work.

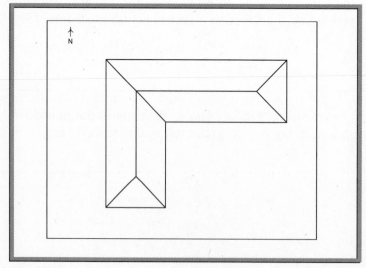

Fig. 17-3

Scaling Objects

Since the house is now slightly too large, let's scale it down to fit the lot. When you *scale* an object, you increase or decrease its overall size without changing the proportions of the parts to each other. You can use the SCALE command to scale by a specific scale factor, or you can use the existing size of the object as a reference length for the new size.

Using a Scale Factor

Modify

1. Pick the **Scale** button from the **Modify** toolbar or enter **SC**, the command's alias.

2. Select the house by placing a window around it and press **ENTER**.

3. Pick the lower left corner of the house as the base point.

At this point, you can dynamically drag the house into place or enter a scale factor.

4. Turn snap off. Move the crosshairs and notice the dynamic scaling.

5. Enter **.5** in reply to Scale factor.

Entering .5 reduces the house to half (.5) of its previous size.

Scaling by Reference

Suppose the house is now too small. Let's scale it up using the **SCALE Reference** option.

Modify

1. Enter **SCALE**, select the house as before, and press **ENTER**.

2. Pick the lower left corner of the house again for the base point.

3. Enter **R** for Reference.

4. For the reference length, pick the lower left corner of the house, and then pick the corner just to the right of it as the second point.

This establishes the length of the south wall (10′) as the reference length.

5. In response to Specify new length, move the crosshairs **4** feet to the right of the second point (snap should be on) and pick a point.

The south wall becomes 14′ long, and the rest of the house is scaled by reference to match.

Rotating Objects

Let's rotate the house on the site. Unlike scaling, *rotating* doesn't change the actual size of the house. It merely places it at a different angle. Using the ROTATE command, you can rotate objects by a specific number of degrees. You can also drag the objects to a new angle if the precise angle is not important.

Rotating by Dragging

Let's rotate the house by dragging it into place.

Modify

1. Enter **ROTATE**, select the house, and press **ENTER** after you have made the selection.

2. Pick the lower left corner of the house for the base point.

3. Drag the house in a clockwise direction a few degrees.

Standard

4. Pick the **Undo** button to undo the rotation.

Specifying an Angle of Rotation

AutoCAD also allows you to specify the angle of rotation. In practice, this is a more useful option than dragging because you can control the placement of objects more precisely.

Modify

1. Pick the **Rotate** button from the **Modify** toolbar, select the entire house, and press **ENTER**.

2. Pick the lower left corner of the house for the base point.

3. Turn off ortho and snap. As you move the crosshairs, notice that the object rotates dynamically.

4. For the rotation angle, enter **25** (degrees).

The house rotates 25° counterclockwise. Your drawing should now look similar to the one in Fig. 17-4.

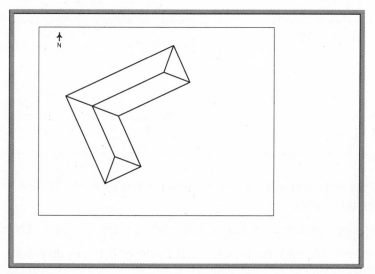

Fig. 17-4

Trimming Lines

The TRIM command allows you to trim lines that extend too far. You can use TRIM to clean up intersections and to convert construction lines and rays into useful parts of a drawing.

Draw

1. Draw two horizontal xlines **3′** apart to represent a sidewalk, as shown in Fig. 17-5. Make the south edge of the sidewalk **2′** from the south property line.

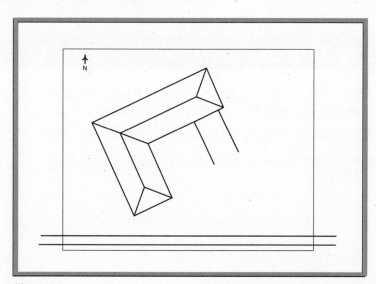

Fig. 17-5

2. With ortho off, draw a partial driveway as shown in Fig. 17-5. Make the driveway **13'** wide.

HINT:

Use the OFFSET command to create the second line of the driveway exactly 13' from the first line.

Modify

3. Pick the **Trim** button from the **Modify** toolbar or enter **TR**, the command's alias.

4. Select the east property line as the cutting edge and press **ENTER**.

5. Select the sidewalk lines on the right side of the property line.

The sidewalk should now end at the property line.

6. Press **ENTER** to return to the Command prompt.

7. Trim the west end of the sidewalk to meet the west property line using the **TRIM** command.

8. Obtain on-line help to learn about the **Project** and **Edge** options. (The UCS feature will be covered in an upcoming chapter.)

INFOLINK

You will learn more about UCSs in Chapter 38.

Extending Lines

The most accurate way to extend the driveway lines so that they meet the south property line is to use the EXTEND command.

Modify

1. Pick the **Extend** button from the **Modify** toolbar.

2. Select the south property line as the boundary edge and press **ENTER**.

3. In reply to Select object to extend, pick the ends of the two lines that make up the driveway and press **ENTER**.

The driveway extends to the south property line.

Modify

4. Using the **TRIM** command, remove the short intersecting lines so that the sidewalk and driveway look like those in Fig. 17-6.

HINT: —————————————————————————————

Use a crossing window to select all four objects and then select the parts you want to remove.

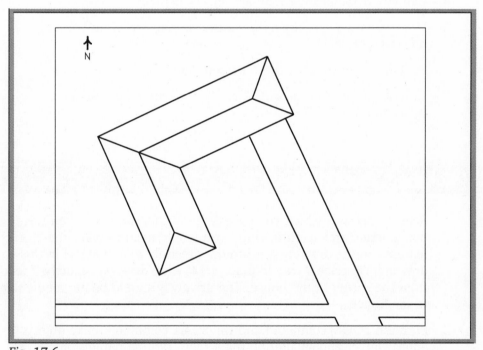

Fig. 17-6

5. Save your work and exit AutoCAD.

Chapter 17
Review & Activities

Review Questions

1. Explain the purpose of the STRETCH command.

2. Using the SCALE command, what number would you enter to enlarge an object by 50%? to enlarge it to 3 times its present size? to reduce it to ½ its present size?

3. Explain how you would dynamically scale an object up or down.

4. Can you dynamically rotate an object? Explain.

5. How would you specify a 90° clockwise rotation of an object accurately?

6. Explain the purpose of the TRIM command.

7. Describe a situation in which the EXTEND command would be useful.

Challenge Your Thinking

1. Refer to site.dwg, which you completed in this chapter. If you were doing actual architectural work, you would need to place the features on the drawing much more precisely than you did in this drawing. Describe a way to place a 14'-wide driveway exactly 2 feet from the corner of the house. The driveway should be perpendicular to the house.

2. Both the ZOOM command and the SCALE command make objects appear larger and smaller on the screen. Write a paragraph explaining the differences between the two commands. Explain the circumstances under which each command should be used.

Chapter 17
Review & Activities

Applying AutoCAD Skills

Work the following problems to practice the commands and skills you learned in this chapter.

1. Create a new drawing and draw the kitchen range symbol shown in Fig. 17-7A. The square is 5 units on each side, the large circles are R.8, and the small circles are R.6. Locate the circles approximately as shown: equidistant from the adjacent sides, with the large circles in the upper left and lower right corners.

   Notice that this is an incorrect symbol of a kitchen range. The large and small circles should be reversed. You can accomplish this by rotating the circles 90°. To make sure the circles stay in their same relative locations, you will rotate them around the center of the square. To do this, first draw two diagonal lines as shown in Fig. 17-7B. Then enter the ROTATE command and select the four circles and two diagonal lines. For the base point of the rotation, snap to the intersection of the two lines. After you have completed the rotation, erase both of the diagonal lines. Then scale the range to 25% of its current size. Save the drawing as range.dwg.

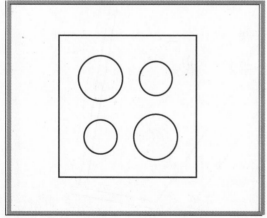

Fig. 17-7A

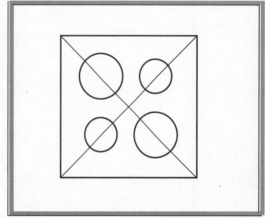

Fig. 17-7B

Chapter 17
Review & Activities continued

2. Load site.dwg, the drawing you created in this chapter. Perform each of the following operations on the drawing.

   • Stretch the driveway by placing a (crossing) window around the house and across the driveway. Stretch the driveway to the north so that the house is sitting farther to the rear of the lot.

   • Add a sidewalk parallel to the east property line. Use the TRIM and BREAK commands to clean up the sidewalk corner and the north end of the new sidewalk.

   • Reduce the entire site plan by 20% using the SCALE command.

   • Stretch the right side of the site plan to the east 10'.

   • Rotate the entire site plan 10° in a counterclockwise direction.

   • Place trees and shrubs to complete the site plan drawing. Use the ARRAY command to create a tree or shrub as shown in Fig. 17-8. Duplicate and scale the tree or shrub using the COPY and SCALE commands.

   When you finish, your drawing should look similar to the one in Fig. 17-8.

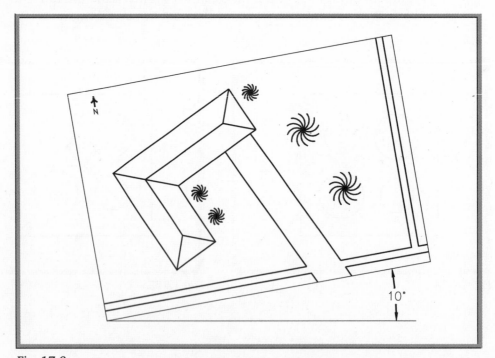

Fig. 17-8

Chapter 17
Review & Activities

3. Marketing wants to use the original design of your TV remote control device, but you must change it to meet competitive standards. Create the first drawing of the device (shown on the left in Fig. 17-9). Then copy the drawing to make the following changes using the commands you learned in this chapter.

• Scale the device down to 90% of its original size.

• Change the two buttons at the bottom into one large button.

• Reduce the overall length.

The final device should look similar to the drawing on the right.

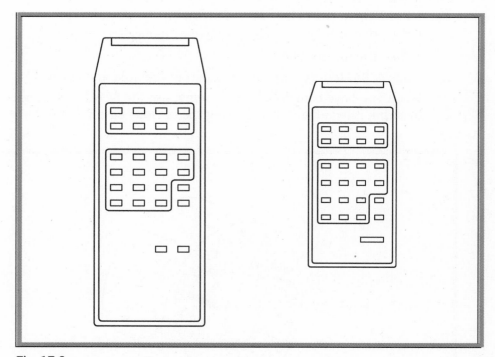

Fig. 17-9

Chapter 17
Review & Activities continued

4. Draw the front view of the tube bundle support shown in Fig. 17-10. Proceed as follows:

   • Draw the outside circle and one of the holes.

   • Offset both circles to create the width of the material.

   • Draw a line connecting the center of the outside circle and the center of the hole.

   • Offset the line to create the internal support structure that connects the hole with the center of the support.

   • Draw another line from the center of the large circle at an angle of 30° to the first line.

   • Use TRIM to create half of one cavity.

   • Use MIRROR to create the other half of the cavity.

   • Fillet the three corners.

   • Use ARRAY to insert five more copies of the hole and cavity.

   When you have completed the drawing, use the SCALE command to shrink it by a factor of .25 using the obvious base point.

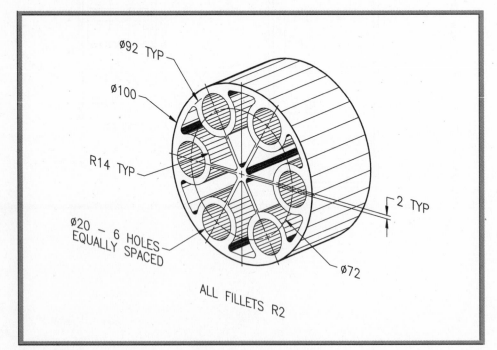

Fig. 17-10

Problem 4 courtesy of Gary J. Hordemann, Gonzaga University

Chapter 17
Review & Activities

 Using Problem-Solving Skills

Complete the following activities using problem-solving skills and your knowledge of AutoCAD.

1. You have a full single garage on your lot. The garage measures 16' × 25' and has a 9' door opening. Draw a floor plan of the garage with 6"-thick walls. Show the opening, but do not include the door. Your spouse's car necessitates a larger garage. Use the EXTEND command to convert the full single garage to a full double garage measuring 25' × 25'. Include two 9' door openings separated by a 1' center support. Trim where necessary. Save the drawing as garage.dwg.

2. Local zoning laws will not permit you to have a full double garage on your narrow lot. Use the TRIM command to convert the proposed full double garage to a small double garage measuring 20' × 20' with one large 16' door opening. Maintain the 6"-thick walls. Save the drawing as garage2.dwg.

Chapter 18 Notes and Specifications

Objectives:

- Create and edit text dynamically
- Specify the position and orientation of text
- Create and use new text styles
- Review options for fonts and adjust the quality of TrueType fonts
- Produce multiple-line text using AutoCAD's text editor
- Import text from word processors and other software

Key Terms

font

justify

mtext

notes

specifications

title block

Notes and specifications are a critical part of most engineering drawings. Although the two terms are often used interchangeably, notes and specifications are technically two different types of text. *Notes* generally refer to the entire drawing, rather than to any one specific feature. For example, a note might describe the thickness of a part, if the thickness is uniform throughout the part. *Specifications,* on the other hand, provide information about size, shape, and surface finishes that apply to specific portions of an object or part.

Without text, the drawings would be incomplete. The drawing in Fig. 18-1 shows the amount of text that is typical in many drawings. Some drawings contain even more text.

As you can see, the text is an important component in describing the drawing. With traditional drafting, the text is placed by hand, consuming hours of tedious work. With CAD, you can place the words on the screen almost as fast as you can type them.

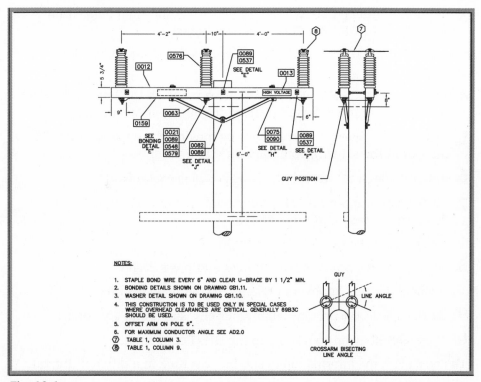

Fig. 18-1

Types of Text

AutoCAD supplies two different types of text objects. The one you use should depend on the type and amount of text you are inserting. The DTEXT command inserts one or more lines of text as single-line objects. The MTEXT command, on the other hand, treats one or more paragraphs of text as a single object. This chapter presents both types of text.

Dynamic Text

The DTEXT command enables you to display AutoCAD text dynamically in the drawing as you type it. It also allows you to *justify,* or align, the text in several ways, including left, right, and center justification.

1. Start AutoCAD and start a new drawing from scratch.

2. Set snap at **.25**.

3. From the **Draw** pull-down menu, select Text and then **Single Line Text**, or enter **DT** (the DTEXT command alias).

This enters the DTEXT command. The Command line prompts for the start point.

4. In response to Specify start point, place a point on the left side of the screen. The text will be left-justified beginning at this point.

5. Reply to the Specify height prompt by moving the crosshairs up **.25** unit from the starting point and picking a point.

6. Enter **0** (degrees) in reply to Specify rotation angle of text.

7. At the Enter text prompt, type your name using both upper- and lowercase letters and press **ENTER**.

You should again see the Enter text prompt.

8. Type your P.O. box or street address and press **ENTER**.

Notice where the text appears in relation to the previous line.

9. At the Specify text prompt, enter your city, state, and zip code.

10. Press **ENTER** again to terminate the DTEXT command.

Let's enter the same information again, but this time in a different format.

11. Enter the **DTEXT** command (press the space bar), and **J** for the Justify option.

The justification options appear at the Command line.

12. Enter the **C** (Center) option.

13. Place the center point near the top center of the screen and set the text height by entering **.2** at the keyboard. Do not insert the text at an angle.

14. Repeat Steps 7 through 10. Be sure to press **ENTER** twice at the end.

When you're finished, your text should be centered like the example in Fig. 18-2. If it isn't, try again.

15. Save your work in a file named **text.dwg**.

```
          Mr. John Doe
      601 West 29th Street
      Caddsville, CA  09876
```

Fig. 18-2

Multiple-Line Text

The MTEXT command creates a multiple-line text object called *mtext*. AutoCAD uses a text editor to create mtext objects.

1. Erase the text from the upper half of the screen.

2. From the **Draw** toolbar, pick the **Multiline Text** icon, or enter **T** at the keyboard.

Draw

This enters the MTEXT command and displays the text style and height at the Command line.

3. Pick a point in the upper left corner of the screen.

Notice the new list of options at the Command line.

4. Enter **W** for Width.

5. Enter **4** in reply to Specify width or, with ortho and snap on, pick a point about **4** units to the right of the first point.

AutoCAD displays the Multiline Text Editor dialog box. Notice that the font is Txt and the default text height is .2000 unit.

NOTE:

A *font* is a set of characters, including letters, numbers, punctuation marks, and symbols, in a particular style. Arial and Romans are examples of fonts used in AutoCAD.

6. Pick the **Properties** tab.

The style is Standard and the justification is Top left. This was selected automatically when you picked a point in the upper left area of the screen. The width is 4.0000, and the rotation is 0.

7. Type the following sentence.

   I'm using AutoCAD's text editor to write these words.

8. Pick the **Character** tab.

9. Highlight the word **using** by double-clicking and then pick the **B** (bold) button.

10. Highlight the word **words** and pick the **U** (underline) button.

You will learn more about the AutoCAD's text editor in Chapter 19.

The sentence should now look like the one in Fig. 18-3.

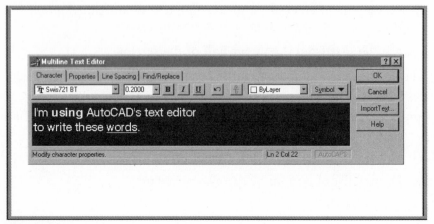

Fig. 18-3

11. Pick the **OK** button.

The new text appears on the screen.

12. Save your work.

Importing Text

Instead of typing text within AutoCAD, you can create it using a standard word processing program and then import it. Many people find it easier to create lengthy text passages using word processors with which they are familiar.

1. Minimize AutoCAD.

2. From the Windows **Start** menu, pick **Programs, Accessories,** and **Notepad.**

3. Enter the following text.

AutoCAD's text editor permits you to import text from a file. This text will appear in AutoCAD shortly.

4. From the **File** pull-down menu, pick **Save As...,** select the folder with your name, and name the file **import.txt.**

5. Pick the **Save** button, exit Notepad, and maximize AutoCAD.

6. From the **Draw** toolbar, pick the **Multiline Text** button or enter **T** at the keyboard.

Draw

7. Near the top center of the screen, pick the first corner.

8. Produce a rectangle that measures about **1** unit tall by **3** units wide.

The Multiline Text Editor dialog box appears.

9. Pick the **Import Text...** button on the right side of the dialog box.

10. Find and double-click **import.txt**.

The text appears in the text editor.

11. Pick the **OK** button.

The text appears in the drawing area.

12. Save your work.

Text Styles and Fonts

It's possible to create new text styles using the STYLE command. During their creation, you can expand, condense, slant, and even draw the characters upside-down and backwards.

Creating a New Text Style

1. From the **Format** pull-down menu, select **Text Style**, or enter **ST** at the keyboard for the STYLE command.

The Text Style dialog box appears as shown in Fig. 18-4.

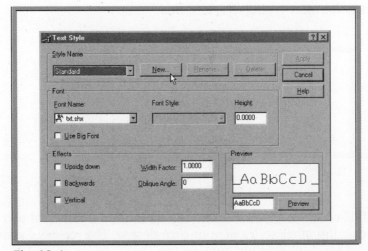

Fig. 18-4

2. Pick the **New...** button, enter **comp1** for the new text style name, and pick the **OK** button.

3. Under Font Name, display the list of fonts by picking the down arrow.

4. Using the scrollbar, find the font file named **complex.shx** and single-click it.

Notice that the text sample in the Preview box changes to show the appearance of the complex.shx font.

5. Study the other parts of the dialog box and then pick the **Apply** and **Close** buttons.

You are ready to use the new comp1 text style.

1. Enter the **DTEXT** command and **Justify** option.

HINT:

You can enter Justify options without first entering Justify. Just enter the capitalized letter(s) in the option names directly at the Command prompt.

2. Right-justify the text by entering **R** (for Right).

3. Place the endpoint near the right side of the screen.

4. Set the height at **.3** unit.

5. Set the rotation angle at **0**.

6. For the text, type the three lines shown in Fig. 18-5. Be sure to press **ENTER** twice after typing the third line.

Computer–aided
Design and Drafting
Saves Time

Fig. 18-5

With the STYLE command, you can develop an infinite number of text styles. Try creating other styles of your own design. The romans.shx font is recommended for most drafting applications.

As you create more text styles within a drawing file, you may occasionally want to check their names.

1. Enter the **STYLE** command.

2. Under Style Name, display the list of text styles by picking the down arrow.

3. Pick **STANDARD**.

Notice that the STANDARD style was developed using the txt.shx font. This is the default text style.

Text Fonts

AutoCAD supports TrueType fonts and AutoCAD compiled shape (SHX) fonts.

1. Reenter the **STYLE** command and pick the **New** button.

2. Enter **tt** for the new text style and pick the **OK** button or press **ENTER**.

3. Under **Font Name**, find **Swis721 BT** and select it.

INFOLINK

See Appendix E for examples of AutoCAD's standard and TrueType fonts.

The Swis721 BT font displays in the preview area. Notice the overlapping T's located at the left of the font name. This indicates that it is a TrueType font. Notice also that Roman displays under Font Style.

4. Under **Font Style**, display the list of options, pick **Italic**, and notice how the font changes in the Preview area.

Note the 0.0000 value in the text box under Height. This 0 value indicates that the text is not fixed at a specific height, giving you the option of setting the text height when you enter the DTEXT command.

5. Pick the **Apply** and **Close** buttons.

6. Enter the **DTEXT** command and pick a point.

7. Enter **.2** for the height and **0** for the rotation angle.

8. Type **This is TrueType.** and press **ENTER** twice.

The TrueType text should look like the text in Fig. 18-6.

This is TrueType.

Fig. 18-6

Text Quality

The TEXTQLTY system variable sets the resolution of text created with TrueType fonts. A value of 0 represents no effort to refine the smoothness of the text. A value of 100 represents a maximum effort to smooth the text. Lower values decrease resolution and increase plotting speed. Higher values increase resolution and decrease plotting speed.

INFOLINK

You will learn more about plotting in Chapter 23.

1. Enter the **TEXTQLTY** system variable at the keyboard.

2. Press the **ESC** key.

3. Experiment with other TrueType fonts on your own.

4. Save your work.

Setting the Current Style

Let's set a new current text style.

1. Enter the **DTEXT** command and enter **S** for Style.

At this point, you can generate a list of styles or enter a new current style. Let's bring back the Standard text style.

2. Type **standard** and press **ENTER**.

NOTE:

You can also enter the STYLE command and use the Text Style dialog box to change the text style.

3. Place some new text on the screen and then terminate the DTEXT command.

4. Change back to the ♯ text style and terminate the DTEXT command.

5. Save your work.

Applying Text in Drawing Files

Now that you have learned many ways of creating text, let's apply it to the creation of a title block. A *title block* is a portion of a drawing that is set aside to give important information about the drawing, the drafter, the company, and so on.

1. Begin a new drawing from scratch and save it as **title.dwg**.

2. Set the grid at **.125** and snap at **.0625**.

3. Create the title block shown in Fig. 18-7 using the following information.

   - For Dynamic Design, Inc., create a new style named **Swiss** using the **Swis721 Ex BT** font. Make the text **.25** tall.

   - Create a new text style named **Roms** using the **romans.shx** font, and use it to produce the small text, which is **.06** tall.

   - Use the **Justify** options of the **DTEXT** command to position the text accurately.

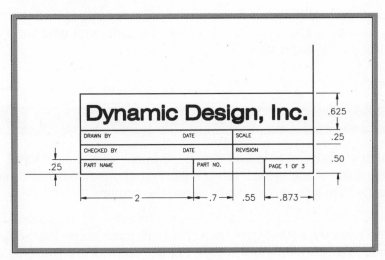

Fig. 18-7

4. Save your work and exit AutoCAD.

 NOTE: —————————————————

Many companies use their own customized title blocks. Autodesk also supplies several variations of title blocks in the Templates folder. If you need a ready-to-use title block for a drawing, you may want to use the Select File dialog box to find one that suits your needs.

Chapter 18
Review & Activities

Review Questions

1. What is the most efficient way to enter a single line of text? What command should you use?

2. What might be a benefit of using the MTEXT command?

3. What is the purpose of the DTEXT Style option?

4. Name at least six fonts provided by AutoCAD.

5. What command do you enter to create a new text style?

6. Briefly describe how you would create a tall, thin text style.

7. Explain how to import text from a word processor into AutoCAD. Why might you choose to do this?

8. What is the purpose of a title block?

9. Explain the differences between the DTEXT command and the MTEXT command. Describe a situation in which each command would clearly be a better choice.

Challenge Your Thinking

1. Obtain on-screen help and then experiment with the COMPILE command. What are the advantages and disadvantages of compiling fonts? When might you want to use this option?

2. In addition to standard AutoCAD and TrueType fonts, AutoCAD can also work with PostScript fonts. Find out more about PostScript fonts. Write a paragraph comparing and contrasting TrueType and PostScript fonts. Are there any situations in which one type might be preferable to the other? Explain.

Chapter 18
Review & Activities

Applying AutoCAD Skills

Work the following problems to practice the commands and skills you learned in this chapter.

1. Type your name and today's date in Windows Notepad. Import this text into the drawings you created in Chapter 17. Save your work.

2. Create a new text style using the following information.
   Style name: cityblueprint
   Font file: CityBlueprint
   Height: .25 (fixed)
   Width factor: 1
   Oblique angle: 15

3. Create a new text style using the following information.
   Style name: ital
   Font file: italic.shx
   Height: 0 (not fixed)
   Width factor: .75
   Oblique angle: 0

4. Use the DTEXT command to place the text shown in Fig. 18-8. Use the cityblueprint text style you created in problem 2. Right-justify the text. Do not rotate the text.

> Someday,
> perhaps in the near future,
> drafting boards will
> be obsolete.

Fig. 18-8

Chapter 18
Review & Activities continued

5. Use the MTEXT command and the ital text style you created in problem 3 to create the text shown in Fig. 18-9. Set the text height at .3 unit. Rotate the text 90°. Using mtext, set the width of the text line to 4 units.

6. Create the television remote shown in Fig. 18-10 with text using the STYLE and DTEXT commands. Use all commands from Chapters 1 through 18 to your advantage.

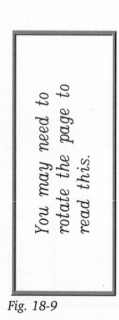

Fig. 18-9

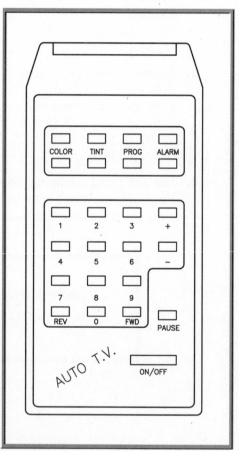

Fig. 18-10

7. Shown in Fig. 18-11 is a block diagram algorithm for a program that sorts numbers into ascending order by a method known as *sorting by pointers*. Use the LINE and OFFSET commands to draw the boxes. Then insert the text using the romans.shx font. For the words MAIN, SORT, and SWAPINT, use the romanc.shx font and make the text larger. For the symbol >, use the symath.shx font, character N.

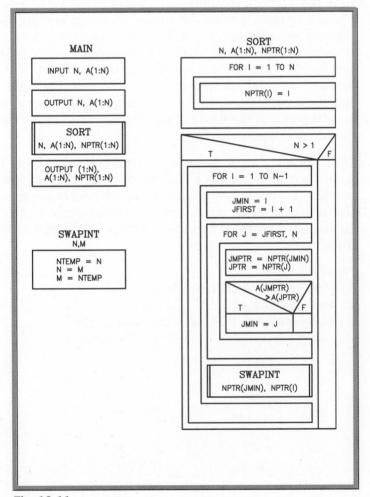

Fig. 18-11

Chapter 18
Review & Activities continued

 ## Using Problem-Solving Skills

Complete the following activities using problem-solving skills and your knowledge of AutoCAD.

1. Most companies have standard operating procedures for creating engineering drawings. Your company is new, and the procedures have not yet been formalized; however, it has become standard practice to place a rectangle around drawings and enclose the drawing identification in a title block. Open the shaftbearing.dwg drawing you created in Chapter 9 (Applying AutoCAD Skills problem 2). Draw a rectangle around the periphery of the drawing. Across the bottom, draw a title block .75 units high and divide it into three equal parts. In the first part, put your name; in the second part, put the name of the drawing; and in the third part, put today's date. Save the drawing.

2. Open the shim.dwg or shimrev.dwg drawing from Chapter 16. Make a rectangle and title block and complete the information as in problem 1, above. Add the following information to the drawing in the form of drawing notes. Use the text style and font of your choice, or as specified by your company's procedures. Save the drawing.

   Notes:
   1. Material: SAE 1010 Carbon Steel
   2. Break sharp edges
   3. All dimensions + - 0.01 unless otherwise specified

Careers
Using
AutoCAD

Freelance with AutoCAD

Over the past decade, companies have been increasing their use of independent contractors, or freelancers. Responding to this trend, many AutoCAD operators have gone into business for themselves to get a share of the new market for expert services.

Freelance drafting is an active area of enterprise, notably in facilities management. For example, when companies reorganize their office space, they may need a lot of floor plans in a hurry.

Often, facilities drawings depict equipment with special requirements for spacing, venting, electrical supply, or drainage. Facilities drawings may include information about lighting, electrical outles, plumbing, and HVAC. Many companies have discovered that having even a simple space plan in AutoCAD makes it much easier to accomplish asset management, energy control, motion studies, product flow studies, and so on.

Another considerable market for freelance drafters is technical illustration. Every new product needs documentation, and new products continue to be developed, especially peripheral devices for computers and specialized electronic and mechanical equipment.

Career Focus—AutoCAD Specialist

Apart from creating and modifying AutoCAD drawings, a successful AutoCAD freelancer must spend time finding and cultivating clients. Word of mouth is one important way clients find out about a freelancer's services. Freelancers also make valuable contacts by attending trade shows or other business gatherings. Another good way to find AutoCAD-related work is to stay in touch with AutoCAD dealers, who may need someone to call on for help when regular personnel are busy.

Based on a story by Pete Karaiskos in CADENCE. © 1994, Miller Freeman, Inc. Revised 1999.

Education and Training

In addition to familiarity with AutoCAD, a freelancer may benefit from a background in architecture, engineering, or construction. A two-year program at a community college or technical school can provide knowledge of AutoCAD and applications. Freelancers can also become certified in using AutoCAD by taking the certification exams offered periodically by an independent testing company.

To be successful, a freelancer typically develops a specialty and relies on repeat business from a handful of good customers. Understanding of a customer's area of business increases over time and becomes an important asset. Active efforts to develop expertise through work and continuing education help steer the specialization to an area that the freelancer finds rewarding.

Career Activity

- Interview someone who offers freelance AutoCAD services. What kinds of work are available in your area? How much should an AutoCAD freelancer charge?
- Find out how freelancers deal with times when work is scarce.

Chapter 19 Text Editing and Spell Checking

Objectives:

- **Edit text using AutoCAD's text editor**
- **Create special text characters**
- **Find and replace text**
- **Check the spelling of text in a drawing using AutoCAD's spell checker**

Key Terms

special characters

text editor

AutoCAD offers several commands and features for editing and enhancing text and mtext objects. Its text editor and spell checker help you produce error-free notes and specifications. With AutoCAD, it's possible to find and replace text using a feature similar to that in word processors such as Microsoft Word. For drawings that contain an abundance of text, AutoCAD offers a command that speeds the display of text and mtext objects.

Editing Text

The DDEDIT command provides a quick and easy way to change the contents of text. DDEDIT can edit text created with the DTEXT and MTEXT commands. The dialog box that displays depends on the type of text object you select for editing. In each case, the dialog box provides a *text editor*, or edit box, that allows you to edit the text you have selected.

1. Start AutoCAD and open the file named **text.dwg**.

2. Create a new text style named **roms** using the **romans.shx** text font. Apply all of the default values.

3. Using the **DTEXT** command, enter the following text at a height of **.15**.

   NOTE: ALL ROUNDS ARE .125

Editing Dynamic Text

The Edit Text dialog box allows you to edit text that was created with the DTEXT command.

1. Without anything entered at the Command prompt, select the text you just created and right-click.

2. From the shortcut menu, pick the **Text Edit...** item.

This displays the Edit Text dialog box, as shown in Fig. 19-1.

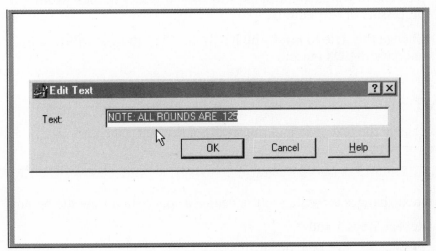

Fig. 19-1

3. Move the pointer to the area containing the text and pick a point between the words **ALL** and **ROUNDS**.

4. Enter **FILLETS AND** so that the line now reads NOTE: ALL FILLETS AND ROUNDS ARE .125.

5. Press **ENTER** or pick the **OK** button.

Notice the prompt Select an annotation object at the Command line.

6. Pick the text again.

AutoCAD immediately displays the Edit Text dialog box.

7. Pick the **Cancel** button and press **ENTER** to terminate the command.

Editing Mtext Objects

You can use a similar method to edit mtext.

1. Select the mtext object located in the upper left corner of the screen and then right-click.

2. Pick **Mtext Edit...** from the shortcut menu.

This displays the Multiline Text Editor.

3. With the pointing device, position the cursor in the text and change a couple of words in the same way as you would in most word processors or text editors.

4. Change the style to **roms**, the height to **.125**, and the width to **5**, and pick the **OK** button.

HINT:

To change the text style and the width of the text, pick the Properties tab.

The mtext changes according to the changes you made in the dialog box.

5. Repeat Steps 1 and 2.

6. Change the style to **tt**, the justification to **Top Right**, and the width to **4**.

7. Pick the **OK** button.

AutoCAD right-justifies the text.

Creating Special Characters

The text editor permits you to insert special characters such as the degree (°) and plus/minus (±) symbols. *Special characters* are those that are not normally available on the keyboard or in a basic text font.

1. Enter the **MTEXT** command.

2. In an empty area of the screen, create a rectangle measuring about **1** unit tall by **2** units wide.

Draw

3. With the help of the **Symbol** drop-down box, write the following specification.

Drill a ø.5 hole at a 5° angle using a tolerance of ±.010.

NOTE: ——————————

Selecting Diameter from the Symbol drop-down box causes the %%c characters to appear in the text editor. These are special codes that represent the diameter symbol in AutoCAD. When using DTEXT, you can enter these codes, which appear at the right in the Symbol drop-down box. When you select Degrees and Plus/Minus from the Symbol drop-down box, the actual symbols appear in the text editor.

4. Pick the **OK** button.

5. Pick the new mtext, right-click, and pick the **Mtext Edit...** item.

6. From the **Symbol** drop-down box, pick the **Other...** item.

This displays a character map, giving you access to many additional characters.

7. Close the character map and pick the **OK** button to close the text editor.

Finding and Replacing Text

Suppose you have lengthy paragraphs of text and you want to find and replace certain words.

1. Rest the crosshairs on top of the mtext object located in the upper left corner of the screen.

2. Right-click and pick the **Find...** item.

This displays the Find and Replace dialog box, as shown in Fig. 19-2.

3. In the Find text string text box, enter **write**, and in the Replace with text box, enter **draft**.

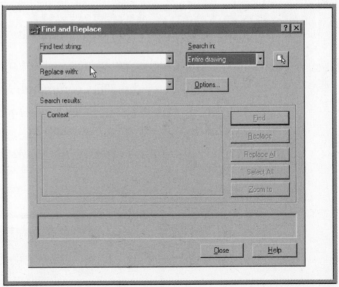

Fig. 19-2

4. Pick the **Find** button.

AutoCAD displays the text in the dialog box and finds and highlights the word write.

5. Pick the **Replace** button.

AutoCAD replaces the word.

6. Pick the **Options...** button.

This displays the Find and Replace Options dialog box.

If you check the Match case check box, AutoCAD finds the text only if the case of all characters in the text match the text in the Find text box. If you check the Find whole words only check box, AutoCAD matches the text in the Find text box only if it is a single word. If the text is part of another text string, AutoCAD ignores it.

The Include area of the dialog box specifies the type of objects you want to include in the search. By default, all options are selected.

7. Pick the **Cancel** button.

8. In the upper right area of the dialog box, pick the down arrow under **Search in**.

As you can see, you can search in the current selection or in the entire drawing.

9. Pick the **Close** button.

Using AutoCAD's Spell Checker

AutoCAD's spell checker examines text for misspelled words.

1. Enter **sp** to enter the SPELL command.

2. In reply to Select objects, pick the text **NOTE: ALL FILLETS AND ROUNDS ARE .125** and press **ENTER**.

If you spelled the words correctly, AutoCAD displays the message box shown in Fig. 19-3.

3. Pick **OK**.

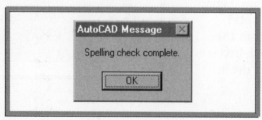

Fig. 19-3

NOTE: _____

If AutoCAD displays the Check Spelling dialog box, pick the
Cancel button.

4. If the term **Computer-aided** is no longer on the screen, add it now.

5. Enter the **SPELL** command, enter **All**, and press **ENTER**.

AutoCAD displays the Check Spelling dialog box because it found what
could be a misspelled word. "Computer-aided" is not included in the spell
checker's dictionary. As you can see, it has suggested "Computer" to
replace "Computer-aided."

6. Pick the **Lookup** button to review alternative suggestions. Then pick
the **Ignore** button to leave Computer-aided as it is.

The Context area of the dialog box displays the phrase in which AutoCAD
located the current word. The functions of the remaining buttons in this
dialog box are described in Table 19-1.

7. If AutoCAD finds other words that might be misspelled, pick the
Ignore All button.

8. Pick the **OK** button.

9. Save your work and exit AutoCAD.

Button	Function
Ignore All	Skips all remaining words that match the current word
Change	Replaces the current word with the word highlighted in the Suggestions box
Change All	Replaces the current word in all selected text objects
Add	Adds the current word to the current custom dictionary. (The maximum word length is 63 characters.)
Change Dictionaries	Displays the Change Dictionaries dialog box, which permits you to change the dictionary against which AutoCAD checks spelling.

Table 19-1

Chapter 19
Review & Activities

Review Questions

1. How does right-clicking selected text permit you to edit standard (dynamic) text? mtext?

2. Why might the character map be useful?

3. How would you add a diameter symbol to mtext?

4. Explain a situation in which AutoCAD's spell checker may find a word that is spelled correctly. What might cause this?

5. Suppose you have paragraphs of text that contain several words or phrases that need to be replaced with new text. What is the fastest way of replacing the text?

Challenge Your Thinking

1. Look again at the mtext object you edited in this chapter. If you changed the word text to test and the word write to rite, what misspellings do you think the spell checker would find? Try it and see. Then write a short paragraph explaining the proper use of a spell checker.

2. As you read in this chapter, AutoCAD allows you to change the dictionary against which it checks the spelling of words. AutoCAD also allows you to create one or more custom dictionaries. Under what circumstances might you want to use a custom dictionary? Might you ever need more than one? Explain.

Chapter 19
Review & Activities

Applying AutoCAD Skills

Work the following problems to practice the commands and skills you learned in this chapter.

1. Open shaftbearing.dwg, which you updated in the previous chapter to include text. Change the date to be the date you actually created the original drawing. Add your middle initial to your name. If you already have your middle initial, remove it. Spell-check the title block. Did your name appear for correction? If it did, add it to the dictionary. Save the drawing as shaft-rev1.dwg.

2. Edit the drawing notes in shimrev.dwg. Remove note number 2 (Break sharp edges). Renumber note 3, and add periods at the end of the notes. Spell-check the notes and the title block. Your name should not appear for correction. Save the drawing.

3. Refer again to shimrev.dwg. Your method of entering + – for a plus/minus value in the original note 3 is not acceptable to your supervisor. You are asked to change to the standard format of ±. Erase your notes, reenter the text using the MTEXT command, and insert the proper symbol. Save the drawing.

Using Problem-Solving Skills

Complete the following activities using problem-solving skills and your knowledge of AutoCAD.

1. Open truetype.dwg located in AutoCAD's Sample folder and save it in your named folder as truetype2.dwg. Zoom in so that you can read the individual lines and notice that the file lists the alphabet in each of the TrueType fonts offered within AutoCAD. Search and replace each occurrence of the alphabetic listing with the phrase ALL DIMENSIONS ARE IN INCHES UNLESS OTHERWISE SPECIFIED. This will show you how the phrase would look in each of the available TrueType fonts. Plot the drawing or zoom around the drawing as necessary to determine which fonts are best for notes and specifications on drawings. Select the five most appropriate fonts. Then create a new drawing, create a new text style for each of the fonts, and create example text in each style. Save the drawing as techtext.dwg and plot it for future reference.

Chapter 19
Review & Activities continued

2. The border and title block shown in Fig. 19-4 are suitable for a paper size of 8.5" × 11". The border allows .75" white space on all four sides.

Draw the border and title block using the dimensions shown in the detail drawing shown in Fig. 19-5. The three sizes of lettering in the title block are .0625", .125", and .25". Use the romans.shx font. Replace "Bulldog Engineering" with your own name and logo. Position the text precisely and neatly by using the suitable Justify options of the DTEXT command. Use the snap and grid as necessary. Intentionally omit words, such as BY in DRAWN BY, and misspell a few words. Then, using AutoCAD's editing and spell-checking capabilities, fix the text.

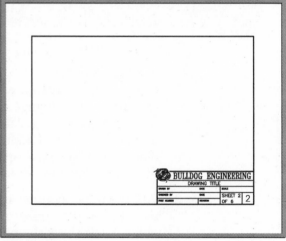

Fig. 19-4

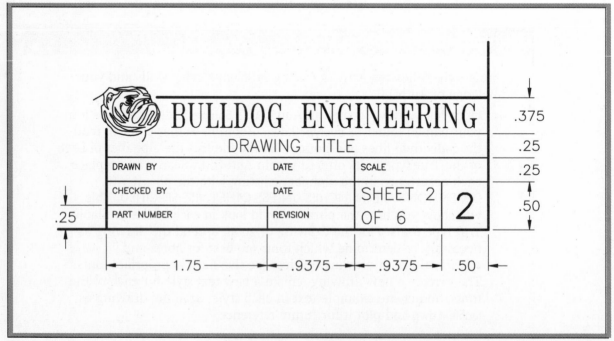

Fig. 19-5

Problem 2 courtesy of Gary J. Hordemann, Gonzaga University

Chapter 20 | Hatching and Sketching

Objectives:

- Hatch objects and parts according to industry standards to improve the readability of drawings
- Edit a hatch by changing its boundary
- Change the characteristics of an existing hatch
- Use sketching options to create objects with irregular lines

Key Terms

associative hatch

crosshatch

full section

hatch

section drawings

AutoCAD's hatching and sketching features can help you enhance the visual appearance and readability of a drawing. For example, on the map shown in Fig. 20-1, hatching makes it easy to tell which parts of the area are national forest and which are urban areas. The irregular lines representing the roads were created using the SKETCH command.

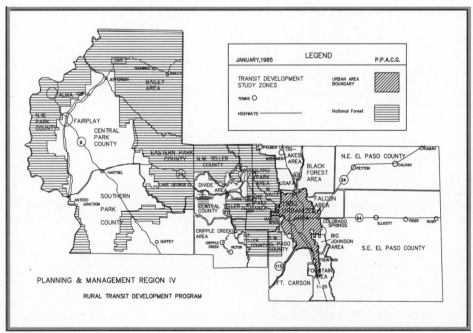

Fig. 20-1   Courtesy of David Salamon, Pikes Peak Area Council of Governments

Hatching and sketching can enhance the visual appearance and readability of a drawing. They also increase the size of the drawing file, so use them wisely.

Hatching

In drafting, a *hatch* or *crosshatch* is a repetitive pattern of lines or symbols that shows a related area of a drawing. Hatching is used extensively in engineering drawing and production drafting to show cut surfaces. For example, when you draw a section of a mechanical part, you use hatching to reflect the cut, or internal, surface. *Section drawings* are used to show the interior detail of a part. Figure 20-2 shows an example of a *full section* (one that extends all the way through the part). The drawing on the left is the front view, and the drawing on the right is the section view.

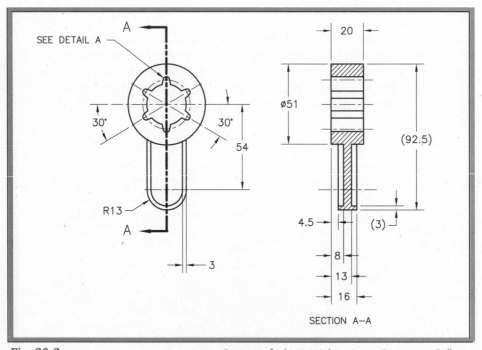

Fig. 20-2   Courtesy of Julie H. Wickert, Austin Community College

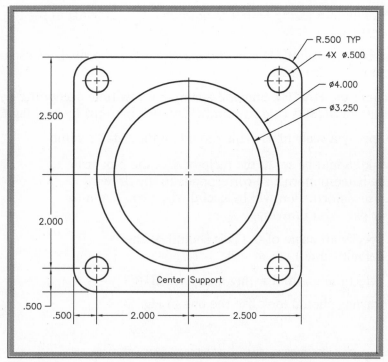

R.500 TYP
4X ⌀.500
⌀4.000
⌀3.250

2.500

2.000

Center Support

.500
.500
2.000
2.500

Fig. 20-3

Using the Command Line

AutoCAD has two hatching commands: HATCH and BHATCH. HATCH is a text-oriented command, while BHATCH is more sophisticated and provides more options. Let's start by working with the HATCH command.

1. Start AutoCAD and begin a new drawing from scratch.

2. Create the sectional view of the center support shown in Fig. 20-3. (You will add the hatch later in this exercise.) Use the dimensions shown in the figure to create it accurately. Include the text (Center Support), but do not include the dimensions.

3. Save your work in a file named **hatch.dwg**.

4. Enter the **HATCH** command at the keyboard.

The hatching options appear in the Command window.

5. Enter **?** to create a listing of all hatch patterns. Press **ENTER** a second time.

AutoCAD presents the list one screen at a time.

6. Continue pressing **ENTER** as prompted by AutoCAD to see the entire list.

As you can see, AutoCAD provides many hatch patterns. Let's use one of them in our drawing.

7. Enter the **HATCH** command again, and this time accept the hatch pattern **ANSI31** (the standard crosshatch pattern) by pressing **ENTER**.

8. Specify a scale of **1**. (This should be the default value.)

The scale should be set at the reciprocal of the plot scale so that the hatch pattern size corresponds to the drawing scale. This information will be useful when you begin to print or plot your drawings.

9. Specify an angle of **0**. (This should be the default value.)

10. Select the entire drawing and press **ENTER**.

Your drawing should look like the one in Fig. 20-4.

INFOLINK

You will learn more about plotting and plot scales in Chapter 23.

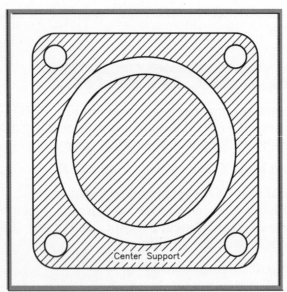

Fig. 20-4

Note the areas that received hatching.

11. Erase the hatching by picking any one of the hatch lines and pressing **ENTER**.

NOTE:

The entire hatch pattern is treated as a single object. If you want the freedom to edit small pieces from a hatch pattern, precede the hatch pattern name with an asterisk (*) when you insert the hatch.

12. Erase the hatching.

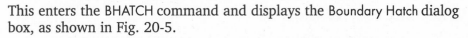
Creating a Boundary Hatch

BHATCH, an expanded version of the HATCH command, creates associative hatches. An *associative hatch* is one that updates automatically when its boundaries are changed.

Draw

1. From the docked **Draw** toolbar, pick the **Hatch** button.

This enters the BHATCH command and displays the Boundary Hatch dialog box, as shown in Fig. 20-5.

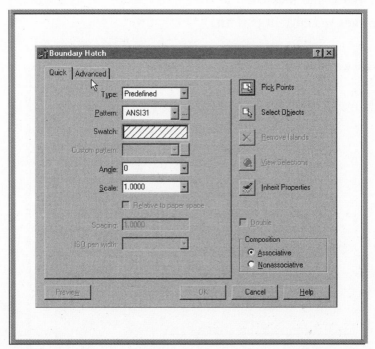

Fig. 20-5

2. Pick the **Advanced** tab.

The Advanced options appear, as shown in Fig. 20-6.

3. In the Island detection style area of the dialog box, pick **Normal**.

4. Pick the **Quick** tab.

5. Pick the **Pattern** down arrow.

A list of hatch patterns appears.

6. Pick the **...** button located to the right of the one you just picked.

The Hatch Pattern Palette appears with ANSI hatch patterns. ANSI stands for the American National Standards Institute, an industry standards organization in the United States.

7. Pick the **ISO** tab.

ISO hatch patterns appear. ISO stands for the International Standards Organization, which also produces industry standards.

8. Pick the **Other Predefined** tab.

AutoCAD displays other predefined hatch patterns that are available to you.

9. Pick **SOLID** and pick **OK**.

SOLID becomes the current pattern.

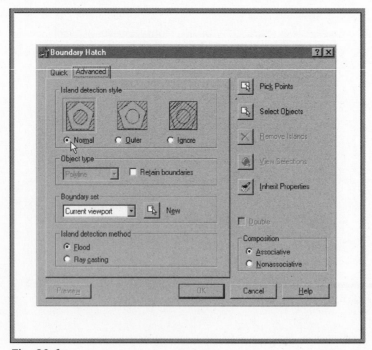

Fig. 20-6

10. Pick the **Select Objects** button, select the entire drawing, and press **ENTER**.

The Boundary Hatch dialog box reappears.

11. Pick the **Preview** button located in the lower left area of the dialog box to preview the hatching.

12. As instructed by AutoCAD, press **ENTER** or right-click to return to the dialog box.

13. Pick the **OK** button to create the associative hatch.

Editing a Hatch

By default, hatches created with the BHATCH command are associative. Therefore, you can change the hatch automatically by changing the size or shape of the hatched object.

Changing the Boundary

1. Select the entire drawing.

2. Select one of the four grip boxes located on the inner cylindrical part of the center support.

3. Reduce its size by **.250** using the **Stretch** option.

HINT:

With the POLAR button depressed in the status bar, move the crosshairs toward the center and enter .25.

AutoCAD updates the hatch pattern according to the new hatch boundary.

4. Press **ESC** twice to remove the selection.

5. Increase the diameter of the lower right hole by **.2**.

HINT:

Stretch the radius of the hole by .1.

Once again, AutoCAD updates the hatch pattern according to the new hatch boundary.

6. Undo your last operation and save your work.

Changing Hatch Characteristics

The HATCHEDIT command permits you to change the pattern and other characteristics of an associative hatch.

1. Display the **Modify II** toolbar and pick the **Edit Hatch** button.

Modify II

This enters the HATCHEDIT command.

2. Pick the hatch.

This causes the Hatch Edit dialog box to display. As you can see, it is very similar to the Boundary Hatch dialog box.

3. Make the following changes.
   Pattern: **ESCHER**
   Scale: **1.3**
   Angle: **30**

4. Pick the **OK** button.

AutoCAD applies the changes to the drawing.

5. Experiment further with creating and changing associative hatch objects on your own.

6. Using the **Edit Hatch** button, change the hatch pattern to **ANSI31**, the scale to **1**, and the angle to **0**, and pick **OK**.

Modify II

7. Close the Modify II toolbar and save your work.

Sketching

Sketching in AutoCAD is rarely used to create "sketches" because AutoCAD contains many other commands that make sketching (in its traditional sense) a faster, more accurate process. However, the SKETCH command is commonly used for many other purposes, such as to show irregular lines on maps like the one shown in Fig. 20-1.

Let's try some freehand sketching.

1. Using **Save As...** from the **File** pull-down menu, create a new drawing named **sketch.dwg**, and erase the current drawing.

2. Enter **SKPOLY** at the keyboard and specify a value of **1**.

Option	Function
Pen	Raises/lowers pen
eXit	Records all temporary lines and exits
Quit	Discards all temporary lines and exits
Record	Records all temporary lines
Erase	Selectively erases temporary lines
Connect	Connects to an existing line endpoint
. (period)	Line to point

Table 20-1

SKPOLY is a system variable that controls whether the SKETCH command creates lines (0) or polylines (1). Setting SKPOLY to 1 allows you to smooth sketched curves using the PEDIT command.

3. Enter the **SKETCH** command.

4. Specify **.1** unit for the Record increment. (This should be the default value.)

The sketching options appear at the Command line. Table 20-1 provides a brief description of each of the SKETCH options.

5. To begin sketching, pick a point where you'd like the sketch to begin. The pick toggles the pen down.

6. Move the pointing device to sketch a short line.

7. Pick a second point (to toggle the pen up), and enter **X** to exit.

8. Move to an open area on the screen and sketch the golf course sand trap shown in Fig. 20-7. Set the **SKETCH Record increment** at **.2**.

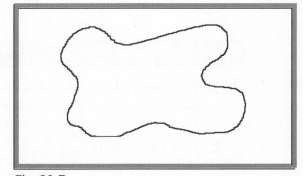

Fig. 20-7

It's okay if your sketch doesn't look exactly like the one in Fig. 20-7.

9. When you're finished sketching the sand trap, type **R** for Record or X for eXit.

10. Practice sketching by using the remaining **SKETCH** options. Draw anything you'd like.

11. When you're finished, save and exit.

Chapter 20
Review & Activities

Review Questions

1. Explain why hatch patterns are useful.

2. What is the most significant difference between a hatch created with the HATCH command and a hatch created with the BHATCH command?

3. With what command can you change a hatch pattern after the hatch has been applied to an object?

4. Briefly describe the purpose of each of the following options of the SKETCH command.
   Pen
   eXit
   Quit
   Record
   Erase
   Connect
   . (period)

5. SKETCH requires a Record increment. What does this increment determine? (If you're not sure, specify a coarse increment such as .5 or 1 and notice the appearance of the sketch lines.)

6. What is the purpose of the SKPOLY system variable?

Challenge Your Thinking

1. Find out what the Inherit Properties button on the Boundary Hatch dialog box does. Explain how it can save you time when you create a complex drawing that has several hatched areas.

2. Investigate the differences and similarities between ANSI and ISO standards. Which (if either) do companies in your area prefer? Why?

Chapter 20
Review & Activities

Applying AutoCAD Skills

Work the following problems to practice the commands and skills you learned in this chapter.

1. Draw the wall shown in Fig. 20-8. Replace the angled break line on the right with an irregular line drawn with the SKETCH command to indicate a continuing edge. Hatch the wall in the brick pattern. Save the drawing as brickwall.dwg.

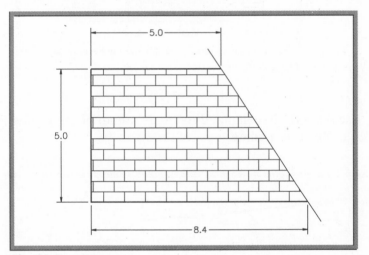

Fig. 20-8

2. Construct the map fragment shown in Fig. 20-9. Use the SKETCH and LINE commands to define temporary boundaries for the hatch patterns. Use HATCH, not BHATCH, to create the hatches. Then erase the boundaries.

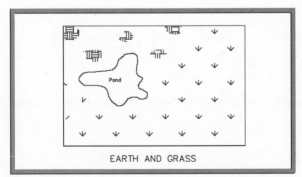

Fig. 20-9

Chapter 20
Review & Activities continued

3. Create the simplified house elevation shown in Fig. 20-10 using the same techniques you used in problem 2.

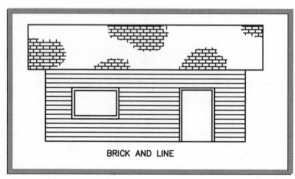

BRICK AND LINE

Fig. 20-10

4. Use the SKETCH command to create a map of the United States. Use Fig. 20-11 as a guide.

Fig. 20-11

5. Use the SKETCH command to draw a logo like the one shown in Fig. 20-12. Before sketching the lines, be sure to set SKPOLY to 1. Use the HATCH or BHATCH Solid pattern to create the filled areas.

Fig. 20-12

6. Draw the right-side view of the slider shown in Fig. 20-13 as a full section view. Assume the material to be aluminum. See Appendix D for a table of dimensioning symbols. For additional practice, draw the top view as a full section view.

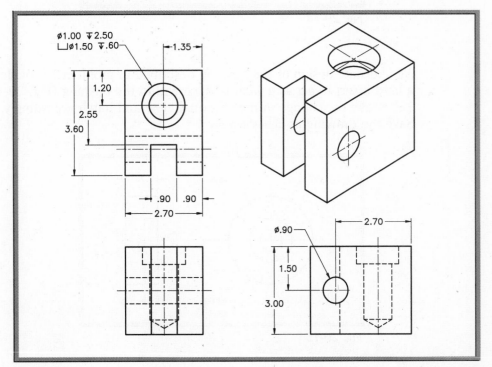

Fig. 20-13

Problems 5 and 6 courtesy of Gary J. Hordemann, Gonzaga University

Chapter 20
Review & Activities continued

Using Problem-Solving Skills

Complete the following activities using problem-solving skills and your knowledge of AutoCAD.

1. Your company is trying some new bushings for the swings in the child's playground set it manufactures. Draw the front view and full section of the bushing shown in Fig. 20-14. Hatch the section as appropriate. Save the drawing as bushing.dwg.

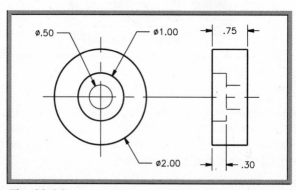

Fig. 20-14

2. An alternate design of the bushing described in problem 1 provides a hard insert to act as a wear surface within the bushing (Fig. 20-15). Make the change in the sectional view and hatch accordingly. Save the drawing as altbushing.dwg.

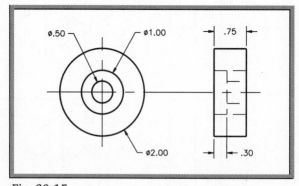

Fig. 20-15

3. The drawing shown in Fig. 20-16 was used with a computer-driven high pressure water cutter to cut a shield out of a piece of ¼″ brass. Use the SKETCH command with SKPOLY set to 1 to draw a similar shield.

Fig. 20-16

Part 3 Project

Creating a Floor Plan

As you have discovered in Part 3 of this textbook, AutoCAD can be used to depict many types of items, from small individual objects to large, complex systems. Even when you create large-scale drawings, AutoCAD's extensive zooming capabilities allow you to insert details as needed. Figure P3-1 shows an example of a single classroom detail. Use this figure as a guide only; your drawing should reflect your school building as closely as possible.

Description

Draw the actual floor plan of a classroom in your school. The level of detail and other specifications will be determined by your instructor, but the finished floor plan must be accurate.

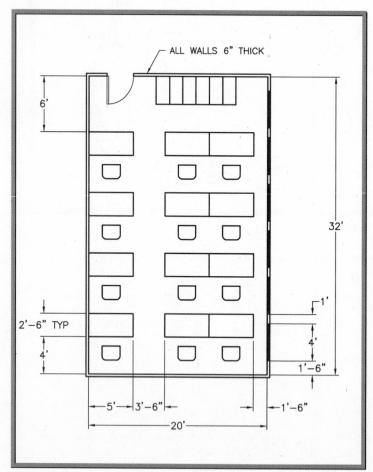

Fig. P3-1

Part 3 Project

Keep the following suggestions in mind as you work.
- Draw everything at its actual size.
- Include openings for doors and closets, and show windows as a line in the middle of and parallel to the window wall.
- Include furniture, if so instructed (outlines are adequate).
- Use multilines to show wall thicknesses. As a general rule, inside walls are 4″ thick, and outside walls are 6″ thick.

Hints and Suggestions

1. The drawing area will need to be increased to accommodate the size of your school building. Use AutoCAD's Quick Setup wizard to set up an appropriate drawing area. The drawing process will be easier if you specify Architectural units.

2. Fillet or chamfer the corners of furniture to resemble the furniture in use at your school. (Note: If you create a piece of furniture using a closed polyline or a polygon, you can fillet all of its corners in a single operation.)

3. Use COPY, MIRROR, and ARRAY as appropriate. Avoid drawing the same detail more than once.

4. Use SCALE and STRETCH when possible to modify basic furniture units into new proportions. For example, you can create a 5′ instructor's desk by stretching a 4′ desk.

5. Label the rooms and parts of the floor plan. (Remember to check your spelling.)

You are not required to include dimensions on the drawing, but your drawing should be precisely to scale.

Summary Questions/Self-Evaluation

1. What is the purpose of a floor plan? Why does it need to be precise?

2. Explain how the wall thickness could have been drawn using a wide polyline or a regular line and the OFFSET command. What is the advantage of using multilines?

3. List several reasons why it is beneficial to use MIRROR, COPY, and ARRAY to create details on a drawing of this type.

Careers
Using
AutoCAD

Draw Me a Map, Minneapolis

Unreadable maps bristle with information—a dozen styles of dots and lines, a beehive array of icons and micro-labels. A map like that, by trying to show everything, makes it hard to find anything. The ideal is a custom map, one that includes only the information that is relevant to the user. In Minneapolis, the city government uses AutoCAD maps, with selectable layers of information, to produce custom maps at a low cost for a variety of users.

Bus riders, tourists, city workers—everyone looks for different things in a map. They find the information they need in custom maps provided by the Transportation Division of the Department of Public Works. A map for bicyclists, for example, shows not only bike lanes and recommended routes but also bike parking and lockers. This and other maps of public interest are posted on the World Wide Web.

Within city government, many departments rely on detailed, up-to-date maps for managing municipal property and operations, such as roads, parking lots and structures, street lights, signs, sidewalks, and skyways. The city's maps are produced and maintained by an engineering graphic analyst, Jim Dahlseid, who says that with AutoCAD his office has multiplied its productivity. AutoCAD has shrunk the interval between updates for important maps from every four years to every six months.

Career Focus—Graphic Analyst

An engineering graphic analyst is a specialist within the field of civil engineering, which includes design and construction of roads, airports, tunnels, bridges, and water and sewage systems. More than 40% of civil engineers work for federal, state, or local government.

In the transportation sector, an engineering graphic analyst is responsible for producing and maintaining maps of city-owned facilities such as roads and traffic lights. The engineering graphic analyst interacts with various city departments to collect information for maps and to determine the needs of map users.

An engineering graphic analyst produces maps for diverse users, such as police, maintenance, and tourism agencies. Traditional drafting skills may be used, but an increasing share of the work is done using CAD.

Education and Training

The basic requirement for an engineering graphic analyst is a bachelor's degree in engineering. Additional courses in urban planning may prove valuable. Advancement to managerial positions within government may require a master's degree or higher.

Career Activity

- Write a report about other occupations related to mapping. What salaries do people make in different kinds of jobs in mapping? What kinds of training do their jobs require?

Part 4

Preparing and Printing a Drawing

Chapter 21 Drawing Setup

Objectives:

- Explain the purpose of a template file and list settings that are commonly included
- Choose the appropriate unit of measurement for a drawing
- Determine the appropriate sheet size and drawing scale for a drawing
- Check the current status of a drawing file

Key Terms

drawing area

limits

template file

The first 20 chapters of this book have provided a foundation for using AutoCAD. Now it's time to apply many of the pieces. As part of this effort, you will identify the scale and sheet size for a particular drawing. In doing this, you will determine the drawing units and drawing area—key elements of the drawing setup process. You will also consider settings of features such as the grid and snap that you can store in a template file and use over and over again.

Template Files

In AutoCAD, a *template file* is a file that contains drawing settings that can be imported into new drawing files. This feature is helpful to people who need to create the same types of drawings (drawings with similar settings) frequently. The purpose of a drawing template file is to minimize the need to change settings each time you begin a new drawing. Some users and companies choose to include a border and title block in their template files. This further shortens the time they have to spend on drawing preparation.

Once you have set up a template file, subsequent drawing setups for similar drawings are fast. When you use a template, its contents are automatically loaded into the new drawing. The template's settings thus become the settings for the new drawing.

Considerations for Template Development

Template development can include the following steps. Note that the first three steps are common to the planning of manual drawings using drafting boards.

1. Determine what you are going to draw (*e.g.,* mechanical detail, house elevation, etc.).

2. Determine the drawing scale.

3. Determine the sheet size. (Steps 2 and 3 normally are done simultaneously.)

4. Set the drawing units.

5. Set the drawing area. (You will learn more about the drawing area later in this chapter.)

6. Set the grid.

7. ZOOM All. (This will zoom to the new drawing area.)

8. Set the snap resolution.

9. Enter STATUS to review the settings.

10. Determine how many layers you will need and what information will be placed on each layer; establish the layers with appropriate colors, linetypes, lineweights, etc.

11. Set the linetype scale (LTSCALE).

12. Create new text styles.

13. Set DIMSCALE, dimension text size, arrow size, etc.

14. Store as a drawing template file.

This and the next several chapters will provide an opportunity to practice these steps in detail.

Using a Predefined Template File

AutoCAD provides many standard templates that can be used as is or modified to fit individual needs.

1. Start AutoCAD and pick the **Use a Template** button in the **Startup** dialog box.

AutoCAD displays a list of template files available to you.

2. Scroll down the list, and as you single-click several of them, notice those with borders and title blocks.

Notice also that the template files have a DWT file extension.

3. Find and select **Ansi_a.dwt** and pick the **OK** button.

The contents of Ansi_a.dwt display on the screen with the Layout1 tab in the foreground. This template is set up to fit an A-size drawing sheet according to standards established by the American National Standards Institute (ANSI).

4. Pick the **Model** tab.

As you can see, the Model tab does not display the border and title block.

INFOLINK

You will learn more about the Layout1 and Model tabs in Chapter 24.

Initial Template Setup

Many AutoCAD users choose to create custom template files. We will create one also. The first step is to identify the type of object (mechanical part, building, etc.) for which the new template will be used. For this exercise, let's create a template that we could use for stair details in drawings for homes and commercial buildings.

Next, we will determine the drawing scale for the stair detail. In this sense, a scale is a means of reducing or enlarging a representation of an actual object or part that is too small or to large to be shown on a drawing sheet. This information will give us a basis for establishing the drawing area, linetype scale, and the scale for the dimensions. Let's use a ½" = 1' scale, and let's base the template on a sheet size of 11" × 17".

1. Close the current drawing (do not save changes), and pick the **New** button.

This begins our setup of the template file. We will create and save it as a regular drawing file. Then, when we have finished defining the settings, we will save it as a template file.

2. Pick **Use a Wizard**.

3. Pick **Advanced Setup**, but do not pick OK at this time.

Read the Wizard Description. Advanced Setup uses the template file acad.dwt. So does Quick Setup.

4. Pick the **OK** button.

Advanced Setup consists of five steps:

- Units
- Angle
- Angle Measure
- Angle Direction
- Area

Completing these steps as part of the process of creating a new file can simplify part of the template setup process.

Specifying the Unit of Measurement

Prior to Release 14, AutoCAD users were required to use either the UNITS or the DDUNITS command to set the drawing units. Both commands are still available, but now AutoCAD provides the additional option of setting the drawing units using either Quick Setup or Advanced Setup.

1. Pick the **Architectural** radio button and review the sample drawing located at the right.

2. Under **Precision**, pick the down arrow and scroll up and down the list.

3. Choose the **0'0-1/16"** default setting.

This means that 1/16" is the smallest fraction that AutoCAD will display.

4. Pick the **Next** button to go to the next step.

Setting Angle Measurements

Decimal Degrees is the default setting for angle measurements. This is what we want to use for the template, so no selection is required from you. When Decimal Degrees is selected, AutoCAD displays angle measurements using decimals.

1. Under **Precision**, pick the down arrow to adjust the angle precision, and pick **0.0**.

A setting of 0.0 means that AutoCAD will carry out angle measurements to one decimal place, as shown in the sample.

2. Pick the **Next** button to proceed to the next step.

This dialog box permits you to control the direction for angle measurements. As discussed earlier in the book, AutoCAD assumes by default that 0 degrees is to the right (east). Let's not change it.

3. Pick the **Next** button.

We also established that angles increase in the counterclockwise direction by default. Do not change this, either.

4. Pick the **Next** button to proceed.

Setting the Drawing Area

The next step is to set the *drawing area,* or *limits,* of the drawing. The drawing area defines the boundaries for constructing the drawing, and it should correspond to both the drawing scale and the sheet size. The default drawing area is 12 × 9 units. Using architectural units, the drawing area is 1' × 9". (See Appendix C for a chart showing the relationships among sheet size, drawing scale, and drawing area.)

NOTE:

Actual scaling does not occur until you plot the drawing, but you should set the drawing area to correspond to the scale and sheet size. The drawing area and sheet size can be increased or decreased at any time using the LIMITS command. The plot scale can also be adjusted prior to plotting. For example, if a drawing will not fit on the sheet at ¼" = 1', you can enter a new drawing area to reflect a scale of ⅛" = 1'. Likewise, you can enter ⅛" = 1' instead of ¼" = 1' when you plot.

As mentioned before, the drawing area should reflect the drawing scale and sheet size. Let's look at an example.

If the sheet size is 11" × 8.5" and the drawing scale is ¼" = 1', what is the drawing area? Since each plotted inch on the sheet would occupy 4 scaled feet, it's a simple multiplication problem: 11 × 4 = 44 and 8.5 × 4 = 34. The drawing area, therefore, would be 44' × 34' because each plotted inch represents 4'.

Since our scale is ½" = 1', what should the drawing area be?

HINT:

How many ½" units would 11" occupy, and how many would 8.5" occupy?

1. Enter **22'** for the width and **17'** for the length. (See the following note.)

NOTE: ───────────────────────────────

When entering 22' and 17', type the numbers exactly as you see them here; use an apostrophe for the foot mark. As discussed in Chapter 12, if you do not use a foot mark, AutoCAD assumes that the numbers are inches. You can specify inches using " or no mark at all. Note also that if you need to change the drawing area later, you can do so using the LIMITS command.

2. Pick the **Finish** button.

You have completed AutoCAD's Advanced Setup process.

3. Save your work in a file named **tmp1.dwg**.

Establishing Other Settings

1. Enter the **GRID** command and set it at **1'**. (Be sure to enter the apostrophe.)

The purpose of setting the grid is to give you a visual sense of the size of the objects and the drawing area. After you enter ZOOM All in the next step, the grid will fill the drawing area, with a distance of 1' between grid dots.

2. Enter **ZOOM All**.

The grid reflects the size of the sheet.

3. Enter **SNAP** and set it at **6"**.

4. Position the crosshairs in the upper right corner of the grid and review the coordinate display.

It should read exactly 22'-0", 17'-0", 0'-0". The 0'-0" is the *z* coordinate.

INFOLINK

You will learn more about the *z* coordinate in Chapters 36 and 37.

Status of the Template File

1. To review your settings up to this point, select the **Tools** pull-down menu and pick **Inquiry** and **Status**.

This enters the STATUS command.

The text on the screen should be similar to that in Fig. 21-1. Note each of the components found in STATUS.

Several of the numbers in the status information, such as Display shows, Free dwg disk, and Free temp disk, will differ.

2. Press **ENTER** and close the AutoCAD Text Window.

Tmp1 now contains several settings specific to creating architectural drawings at a scale of $\frac{1}{2}'' = 1'$. It will work well for beginning the stairway detail mentioned earlier.

We have developed the basis for a template file. Technically, we have not yet created the template, because we have not stored the contents of our work in a template (DWT) file. We will do this later when we continue its development.

The next steps in creating a template file deal with establishing layers. You will create layers in the next chapter and complete the final steps for creating a template file in a subsequent chapter.

3. Save your work and exit AutoCAD.

4. Produce a backup copy of the file named **tmp1.dwg**.

Backup copies are important because if you accidentally lose the original (and you may, sooner or later), you will have a backup. You can produce a backup in seconds, and it can save you hours of lost work.

INFOLINK

See Chapter 8 for details on producing copies of files.

Command: status
42 objects in C:\Program Files\ACAD2000\Your Name\tmp1.dwg
Model space limits are X: 0'-0"   Y: 0'-0"   (Off)
          X: 22'-0"   Y: 17'-0"
Model space uses   *Nothing*
Display shows X: -5'-0 5/16"   Y: -0'-1 9/16"
          X: 27'-0 5/16"   Y: 17'-1 9/16"
Insertion base is   X: 0'-0"   Y: 0'-0"   Z: 0'-0"
Snap resolution is   X: 0'-6"   Y: 0'-6"
Grid spacing is X: 1'-0"   Y: 1'-0"

Current space: Model space
Current layout: Model
Current layer: "0"
Current color: BYLAYER – 7 (white)
Current linetype: BYLAYER – "Continuous"
Current lineweight: BYLAYER
Current plot style: ByLayer
Current elevation: 0'-0"   thickness: 0'-0"
Fill on Grid on Ortho off Qtext off Snap on Tablet off
Object snap modes: Center, Endpoint, Intersection, Midpoint,
          Perpendicular,
Free dwg disk (C:) space: 692.6 MBytes
Free temp disk (C:) space: 692.6 MBytes
Free physical memory: 0.1 Mbytes (out of 31.5M).
Free swap file space: 696.6 Mbytes (out of 744.6M).

Fig. 21-1

Chapter 21
Review & Activities

Review Questions

1. Explain the purpose and value of template files.

2. What settings are commonly included in a template file?

3. If you select architectural units, what does a precision of 0′0–¼″ mean?

4. What determines the drawing area?

5. If the sheet measures 22″ × 16″, and the scale is 1″ = 10′, what should you enter for the drawing area?

6. Describe the information displayed as a result of entering the STATUS command.

Challenge Your Thinking

1. Discuss unit precision in your drawings. If necessary, review the precision options listed in the Units Control dialog box for units and angles. *Hint:* You can display this dialog box by entering the DDUNITS command.) Then describe applications for which you might need at least three of the different settings.

2. For a drawing to be scaled at ⅛″ = 1′ and plotted on a 17″ × 11″ sheet, what drawing area should you establish?

Applying AutoCAD Skills

Work the following problems to practice the commands and skills you learned in this chapter.

1. Create a new drawing using the Ansi layout templates.dwt template file. Use the STATUS command to review its settings. Create another new drawing using the adadiso.dwt template file and review its settings also. What differences do you see between AutoCAD's templates for ANSI and ISO standards?

Chapter 21
Review & Activities

2. Establish the settings for a new drawing based on the information below, setting each of the values as indicated. Save the file as prb21-2.dwg.

   Drawing type: Mechanical drawing of a machine part
   Scale: 1″ = 2″
   Sheet size: 17″ × 11″
   Units: Engineering
   Drawing area: (You determine it.)
   Grid value: .5″
   Snap resolution: .25″

   Be sure to ZOOM All.

   Review settings with the STATUS command.

3. Establish the settings for a new drawing based on the information below, setting each of the values as indicated. Save the file as prb21-3.dwg.

   Drawing type: Architectural drawing of a house and site plan
   Scale: ¹/₈″ = 1′
   Sheet size: 24″ × 18″
   Units: Architectural (You choose the appropriate options.)
   Drawing area: (You determine it.)
   Grid value: 4′
   Snap resolution: 2′

   Reminder: Be sure to ZOOM All.

   Review settings with the STATUS command.

4. Consider the following drawing requirements:

   Drawing type: Architectural drawing of a detached garage
   Approximate dimensions of garage: 32′ × 20′

   Other considerations: Space around the garage for dimensions, notes, specifications, border, and title block

   Based on this information, write the missing data for drawing setup on a separate sheet of paper. Suggest values for scale, paper size, units, drawing area, grid, and snap resolution.

Chapter 21
Review & Activities continued

5. Consider the following drawing requirements:

   Drawing type: Mechanical drawing of the wheel and bearings for an in-line skate.

   Approximate diameter of wheel: 2.5″

   Other considerations: Space for dimensions, notes, specifications, border, and title block.

   Based on this information, write the missing data for drawing setup on a separate sheet of paper. Suggest values for scale, paper size, units, drawing area, grid, and snap resolution.

 ## Using Problem-Solving Skills

Complete the following activities using problem-solving skills and your knowledge of AutoCAD.

1. Your architectural office needs a drawing template setup that will accommodate the floor plans of houses measuring 16,000 mm by 8400 mm. Create the template. In addition to the units, angles, and area, specify reasonable settings for both snap and grid. Save the template as arch1.dwt.

2. Electronic engineering requires large drawings of small parts. Set up a drawing for electronic circuits measuring 1/4 × 3/32 inches. Use decimal settings, and set the appropriate snap and grid. Save the template as elec.dwt.

Objectives:

- Create layers with appropriate characteristics for the current drawing or template
- Use layers to control the visibility and appearance of objects in a drawing
- Change an object's properties
- Use object properties to filter object selections
- Apply a custom template file

Key Terms

center lines

filtering

freeze

hidden lines

layers

palette

phantom lines

viewport

AutoCAD gives you the option of separating classes of objects into layers. It may be helpful to think of AutoCAD's *layers* as transparency films that help you organize your drawing. One of the benefits of using layers is the ability to make them visible and invisible. For example, you can place construction lines and reference notes on a layer and then turn it off when you're not using them. A house floor plan could be drawn on a layer called Floor and displayed in red. The dimensions of the floor plan could be drawn on a layer called Dimension and displayed in yellow. Furthermore, a layer called Center could contain blue center lines.

Table 22-1 on the next page shows an example set of layers. Note the layer names, colors, linetypes, and lineweights. We will create these layers in this chapter.

NOTE:

Be sure to make a backup copy of tmp1.dwg before you begin working with it in this chapter if you have not already done so.

Layer Name	Color	Linetype	Lineweight
0	white	CONTINUOUS	Default
Border	cyan	CONTINUOUS	0.50 mm
Center	magenta	CENTER	0.20 mm
Dimensions	blue	CONTINUOUS	0.20 mm
Hidden	green	HIDDEN	0.30 mm
Notes	magenta	CONTINUOUS	0.30 mm
Objects	red	CONTINUOUS	0.40 mm
Phantom	yellow	PHANTOM	0.50 mm

Table 22-1

Creating New Layers

1. Start AutoCAD and open the drawing named **tmp1.dwg**.

2. Pick the **Layers** button from the docked Object Properties toolbar, or enter the letters **la**, the command alias for the LAYER command.

Object Properties

This enters the LAYER command and displays the Layer Properties Manager dialog box, as shown in Fig. 22-1.

First, let's create layer Objects.

3. Pick the **New** button.

This creates a new layer with a default name of Layer1. Notice that the name Layer1 is highlighted and a cursor appears in the edit box.

4. Using upper- or lowercase letters, type **Objects** and press **ENTER**.

The name Objects replaces Layer1. If you make a mistake, you can single-click the layer name and edit it.

5. Create the layers **Border**, **Center**, **Dimensions**, **Hidden**, **Notes**, and **Phantom** on your own, as listed in Table 22-1.
(You will set the color, linetype, and lineweight of each layer later in this chapter.)

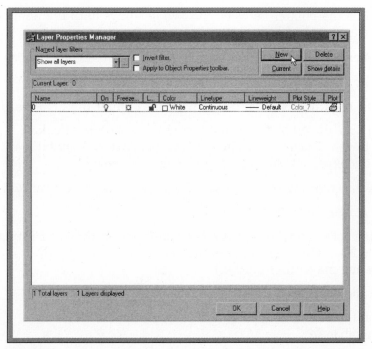

Fig. 22-1

Changing the Current Layer

Let's change the current layer to Objects.

1. Pick **Objects** and pick the **Current** button.

Objects is now the current layer.

2. Pick the **OK** button.

Notice that Objects now appears in the docked Object Properties toolbar.

3. Save your work.

Assigning Colors

The Layer Properties Manager dialog box permits you to assign screen colors to the layers.

Object Properties

1. Pick the **Layers** button or enter **la.**

2. Find the **Color** or (**C...**) heading, which is located at the left of the Linetype column heading. If the Color column heading and the color names are visible, skip to Step 6. If not, proceed to Steps 3, 4, and 5.

3. Position the pointer between the two headings until the pointer changes to a double arrow.

4. Click and drag to the right until the names of the colors appear.

5. Adjust the column to its original size.

6. At the right of Objects, pick the white box under the Color heading.

The Select Color dialog box appears. It contains a palette of colors available to you. *Palette* is a term used in AutoCAD to describe a selection of colors, similar to an artist's palette of colors.

7. Pick the color red from Standard Colors and pick the **OK** button.

AutoCAD assigns the color red to Objects.

8. Assign colors to the other layers as indicated in the layer listing in Table 22-1. (Use the colors in the Standard Colors area. Cyan is light blue and magenta is purple.)

9. Pick the **OK** button and save your work.

Notice that the color red appears beside Objects in the Object Properties toolbar.

 NOTE:

AutoCAD offers commands called COLOR and DDCOLOR that allow you to set the color for subsequently drawn objects, regardless of the current layer. Therefore, you can control the color of each object individually.

The ability to set the color of objects individually or by layer gives you a great deal of flexibility, but it can become confusing. It is recommended that you avoid use of the COLOR and DDCOLOR commands and that their settings remain at ByLayer. The ByLayer setting means that the color is specified by the layer on which you draw the object.

The Color Control drop-down box in the Object Properties toolbar allows you to set the current color. The recommended setting is ByLayer.

10. Draw a circle of any size on the current Objects layer.

It should appear in the color red.

11. Set **Hidden** as the current layer and draw a concentric circle inside the first circle.

It should appear in the color green.

Assigning Linetypes

Various types of lines are used in drafting to show different elements of a drawing. By convention, for example, *hidden lines* (those that would not be visible if you were looking at the actual object) are shown as dashed or broken lines. *Center lines* (imaginary lines that mark the exact center of an object or feature) are shown by a series of long and short line segments. Let's look at and load the different linetypes AutoCAD makes available to you.

Object Properties

1. Display the **Layer Properties Manager** dialog box.

2. At the right of layer Hidden, and under the Linetype heading, pick **CONTINUOUS**.

This displays the Select Linetype dialog box.

3. Pick the **Load...** button.

This displays the Load or Reload Linetypes dialog box, as shown in Fig. 22-2.

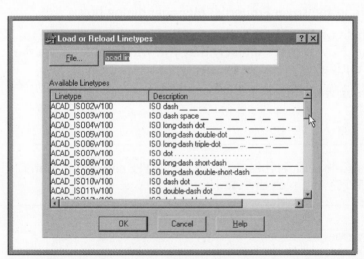

Fig. 22-2

AutoCAD stores the list of linetypes in a file named acad.lin, as listed in the box at the right of the File... button.

These linetypes conform to International Standards Organization (ISO) standards. This is important because ISO is the leading organization for the establishment of international drafting standards.

At this point, you can select individual linetypes that you want to load and use in the current drawing.

4. Using the scrollbar, review the list of linetypes.

5. Find and select each of the linetypes listed in the Key Terms on the first page of this chapter.

HINT:

Press the CTRL key when selecting them. This allows you to make multiple selections.

Hidden lines are used to show invisible edges on drawings. This is why they are called "hidden." Center lines are used to show the centers of holes, cylinders, rounded corners, and fillets. *Phantom lines,* drawn using a thick lineweight, are used for cutting planes in sectional views.

6. Pick the **OK** button and notice the new list of linetypes.

7. Pick **HIDDEN** for the linetype and pick **OK**.

8. Pick **OK** in the **Layer Properties Manager** dialog box.

The inner circle on layer Hidden changes from a continuous line to a hidden line. However, it's unlikely that you will be able to see the hidden line until after we scale the linetypes.

9. Save your work.

NOTE:

You can also load linetypes using the LINETYPE command. Enter lt (for LINETYPE) and pick the Load button.

Scaling Linetypes

The LTSCALE command permits you to scale the linetypes so that they are appropriate for the overall drawing scale.

1. Enter **lts** for the LTSCALE command.

Let's scale the linetypes to correspond to the scale of the prototype drawing. This is done by setting the linetype scale at $\frac{1}{2}$ the reciprocal of the plot scale. When you do this, broken lines, such as hidden and center lines, are plotted to the ISO standards.

NOTE:

As you may recall, we are creating a template file based on a scale of $\frac{1}{2}'' = 1'$. Another way to express this is $1'' = 2'$ or $1'' = 24''$. This can be written as $\frac{1}{24}$. The reciprocal of $\frac{1}{24}$ is 24, and half of 24 is 12. Therefore, in this particular case, you should set LTSCALE at 12.

2. In reply to New scale factor, enter **12**.

The inner circle should now be made up of hidden lines.

3. If you're not sure, zoom in on it.

4. Open the **Layer Properties Manager** dialog box.

5. Assign the **CENTER** linetype to layer Center and the **PHANTOM** linetype to layer Phantom.

6. Pick **OK** to close the dialog box.

NOTE:

AutoCAD permits you to set the linetype for subsequently drawn objects, regardless of the current layer. The Linetype Control drop-down box, located in the Object Properties toolbar, permits you to change the current linetype. The default setting is ByLayer, which means that the linetype is specified by the layer on which you create the object. This is the recommended approach—and the approach that you've used in this chapter. If you change ByLayer to a specific linetype, all subsequently drawn objects will use this linetype regardless of the layer on which you create them. This gives you a great deal of flexibility, but it can become confusing.

Setting Lineweights

Lineweights are important in technical drawing. For example, object lines should be thicker so they stand out more than dimensions. Normally, object lines are thick, hidden lines and text are of medium thickness, and center lines, dimensions, and hatch lines are thin. The exact lineweights vary from industry to industry and from company to company. In AutoCAD, the default linewidth is a thickness of .25 mm.

1. Open the **Layer Properties Manager** dialog box.

2. At the right of layer Objects, and under the Lineweight heading, pick the word **Default**.

This displays the Lineweight dialog box as shown in Fig. 22-3.

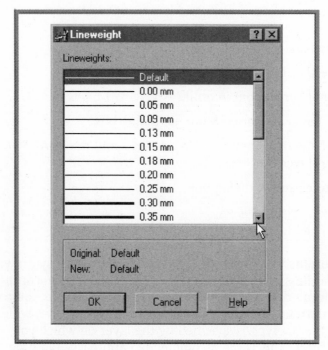

Fig. 22-3

3. Scroll down the list to review the lineweight options.

4. Pick **0.40 mm** and pick **OK**.

This assigns a lineweight of .4 mm to layer Objects. Notice that this lineweight replaces the word default in the Layer Properties Manager dialog box.

5. Pick the **LWT** button in the status bar.

LWT is short for "lineweight." This turns on the display of lineweights in the drawing area.

6. Pick the **LWT** button to toggle it off; toggle it back on.

7. Assign lineweights to the other layers as shown in Table 22-1 on p. 314.

8. Pick **OK** to close the dialog box.

9. Save your work.

NOTE:

In the docked Object Properties toolbar, you can set the lineweight for subsequently drawn objects, regardless of the current layer. The Lineweight Control drop-down box permits you to change the current lineweight. The default setting is ByLayer, which means that the lineweight is controlled by the layer on which you create the object. This is the recommended approach and the approach that you've used so far. If you change ByLayer to a specific lineweight, all subsequently drawn objects will use this lineweight, regardless of the layer on which you create them. This gives you a lot of flexibility, but it can become confusing, so it is not recommended.

Working with Layers

Setting up the appropriate layers for a drawing gives you much more control over the drawing. It also gives you more flexibility. For example, suppose you are working on the plans for a new residence. You can set up separate layers for plumbing, electrical, and other work that is often subcontracted. By placing the plans for these elements on separate layers, you can use the same drawing to generate plans for use by each of the subcontractors. For the electrical plans, you can turn off or *freeze* the plumbing layer so that the plumbing specifications don't clutter up the electrical drawing, and so on.

Turning Layers On and Off

Object Properties

1. Display the **Layer Properties Manager** dialog box.

2. At the right of Objects, click the light bulb.

This toggles off layer Objects.

3. Pick the **OK** button and notice that the red circle disappears.

4. Press the spacebar to display the dialog box again.

5. Toggle on layer **Objects** by clicking the darkened light bulb.

6. Pick the **OK** button and notice that the red circle reappears.

Freezing and Thawing Layers

Object Properties

1. Display the **Layer Properties Manager** dialog box.

2. Adjust the headings so that you can read the heading title Freeze in all VP.

VP stands for viewport. A *viewport* is a single-window drawing area.

3. Adjust the column to its original size.

4. In this column, and at the right of Objects, pick the symbol representing the sun.

This changes the symbol to a snowflake and freezes the layer.

5. Pick the **OK** button and notice what happens to the circle.

It disappears.

6. Display the dialog box again, change the snowflake into a sun by picking it, and pick the **OK** button.

The circle reappears. As you can see, freezing and thawing layers is similar to turning them off and on. The difference is that AutoCAD regenerates a drawing faster if the unneeded layers are frozen rather than turned off. Therefore, in most cases, Freeze is recommended over the Off option. Note that you cannot freeze the current layer.

INFOLINK

You will learn more about <u>viewports</u> in Chapter 24.

Locking Layers

AutoCAD permits you to lock layers as a safety mechanism. This prevents you from editing objects accidentally in complex drawings.

1. Display the **Layer Properties Manager** dialog box.

2. At the right of Objects, click the symbol that looks like a padlock.

The lock closes, indicating that the layer is now locked.

3. Pick the **OK** button and try to edit the circle.

AutoCAD will not permit you to edit the circle because it resides on a locked layer.

 NOTE:

AutoCAD does allow you to create new objects on a locked layer. However, once you have created them, you cannot edit them.

4. Reopen the dialog box and unlock layer Objects by clicking the locked padlock.

5. Pick **OK** to close the dialog box.

Controlling Layers

There are several ways to control layers, layer properties, and the current layer in AutoCAD. The Make Object's Layer Current button allows you to change the current layer based on a selected object. The Layer Properties Manager dialog box allows you to change layer properties and delete unused layers.

1. Pick the **Make Object's Layer Current** button on the docked **Object Properties** toolbar.

Notice that the Command line reads Select object whose layer will become current.

2. Pick the green circle.

Hidden is now the current layer.

3. Pick the down arrow in the upper left area to display a list of options.

These options, along with the two check boxes, determine which layers are displayed. You can filter layers based on whether they contain objects and on their name, state, color, and linetype. These filters are especially helpful in complex drawings that contain a large number of layers.

 NOTE: _____

> To filter layers based on their name, color, or other characteristics, you can pick the [...] button to the right of the down arrow. This displays the Named Layer Filters dialog box, from which you can set or remove multiple filters.

The Delete button, located at the right, permits you to delete selected layers. However, AutoCAD will not delete any layer that contains objects.

4. Highlight layer **Phantom** by single-clicking the layer name.

5. Pick the **Show Details** button.

This displays more information about the Phantom layer. You can change the layer name, color, lineweight, linetype, and various other settings from the Details portion of the box.

6. Pick the Hide Details button to remove the details from the dialog box.

Focus your attention on the Plot Style and Plot headings. Colors appear under Plot Style. These are the colors you see in the drawing area. Color 1 is red, color 2 is yellow, and so on, as shown in the dialog box. They are grayed out and unavailable because this drawing does not use plot styles.

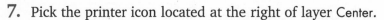

Under the Plot heading, you can pick the printer icon to turn off plotting for a layer. If you turn off plotting for a layer, AutoCAD still displays the objects on that layer, but they do not plot.

INFOLINK

You will learn more about plotting in Chapter 23.

7. Pick the printer icon located at the right of layer Center.

A small red circle with a line through it appears on top of the printer, indicating that plotting has been turned off for this layer. You will learn more about plotting in the following chapter.

8. Turn on plotting for layer **Center** and pick **OK** to close the dialog box.

Changing and Controlling Layers

AutoCAD offers a convenient way of changing and controlling certain aspects of layers.

1. In the docked Object Properties toolbar, pick the **Layer Control** drop-down box.

This displays the list of layers in the current drawing, as shown in Fig. 22-4. By picking the symbols, you can quickly change the current status of the layers.

2. Pick **Objects**.

Layer Objects becomes the current layer.

3. Display the list again.

4. Beside Objects, click the padlock.

The padlock changes to its locked position, and layer Objects becomes locked.

5. Pick it again to unlock layer Objects.

6. Beside Hidden, pick the sun.

The sun changes to a snowflake.

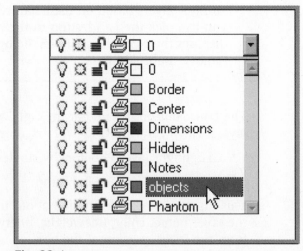

Fig. 22-4

7. Pick a point anywhere in the drawing area.

Layer Hidden is now frozen.

8. Display the list, pick the snowflake, and pick a point anywhere in the drawing area.

This thaws layer Hidden.

Working with Objects

Now that you have created a number of layers and assigned properties to them, you have much more flexibility in working with objects in the drawing. You can control their appearance and style by placing them on appropriate layers. You can also change an object's properties and even filter objects according to their characteristics.

Changing an Object's Properties

The Properties dialog box provides a quick method of reviewing and changing the properties of an object.

1. Pick the green circle, right-click, and pick **Properties** from the menu.

This displays the Properties dialog box, as shown in Fig. 22-5.

The Properties dialog box provides a lot of information on the circle, including its color, linetype, lineweight, and layer name. Suppose you created the object on the wrong layer—something you will certainly do in the future—and want to move it to a different layer.

2. At the right of Layer, pick **Hidden**.

This produces a down arrow.

3. Pick the down arrow.

This displays the list of layers you created.

4. Pick **Center**, close the dialog box, and press **ESC** twice to remove the selection.

The circle now resides on layer Center.

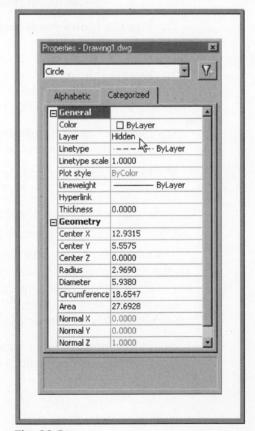

Fig. 22-5

5. Using the same method, change the circle back to layer **Hidden**.

The Properties dialog box is a handy way of quickly making changes to the objects in the drawing.

Matching Properties

The Match Properties button provides a quick way to copy the properties of one object to another.

Standard

1. From the **Standard** toolbar, pick the **Match Properties** button.

This enters the MATCHPROP command.

2. In reply to Select source object, pick the red circle.

3. In reply to Select destination object(s), pick the green circle.

This copies the properties of the first object to the second object.

4. Press **ENTER** to terminate the command.

5. Undo the last operation.

The circle reverts to its original properties.

Standard

6. Pick the **Match Properties** button again and pick the green circle.

7. Enter **S** for Settings.

This displays the Property Settings dialog box. The checked items are copied from the source object to the destination object(s), allowing you to control the properties that are copied.

8. Pick the Cancel button in the dialog box and press ENTER to terminate the command.

9. Erase both circles and save your work, but do not close the drawing.

Filtering by Object Property

The Quick Select dialog box allows you to filter a selection set based on an object's properties. *Filtering* means to selectively include or exclude using specific criteria. You can use the Quick Select dialog box with the Properties dialog box to make changes quickly to a large number of objects.

INFOLINK

See Chapter 5 for other methods of object selection.

1. Start a new drawing using **Quick Setup**. Specify **Architectural** units and a drawing area of **17′ × 11′**.

Layer Name	Color	Linetype	Lineweight
Risers	yellow	CONTINUOUS	0.30 mm
Treads	green	CONTINUOUS	0.60 mm

Table 22-2

2. Create two new layers as shown in Table 22-2.

3. Create the stair detail shown in Fig. 22-6. Place the risers on layer Risers and the treads on layer Treads. (The *risers* are the vertical lines, and the *treads* are the horizontal lines.)

4. Zoom in so that the stairs fill most of the screen.

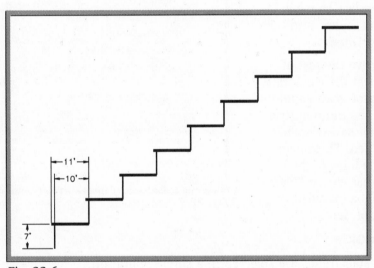

Fig. 22-6

5. Save your work in a file named **stairs.dwg**.

Suppose your supervisor reviews the drawing and asks you to place both the treads and the risers on the stairs layer. However, you must keep the differences in lineweight and color for display purposes.

6. Select the **Tools** pull-down menu and pick the **Quick Select...** item.

This displays the Quick Select dialog box, as shown in Fig. 22-7.

7. Pick the down arrow by the Object type box and select **Line**.

The Properties box lists several properties by which you can select objects.

8. Pick **Layer**.

9. Pick a point in the **Value** box to see a list of layers, and pick **Treads**.

We will leave the rest of the values at their defaults. The operator is an equal sign, and the radio button for Include in new selection set is selected. This means that all the lines in the drawing that are on layer Treads will be included in the selection set.

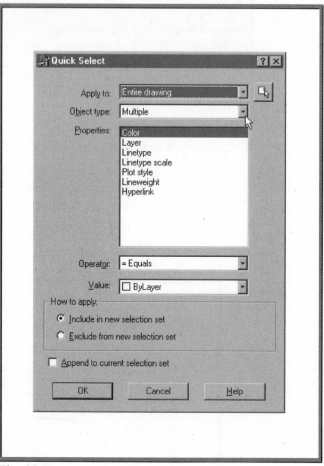

Fig. 22-7

10. Pick **OK**.

All of the stair treads are selected in a single operation.

Standard

11. Pick the **Properties** button on the **Standard** toolbar and pick **Layer** in the Properties dialog box.

12. Pick the down arrow to see a list of layers, and choose **Risers**.

13. Close the Properties dialog box and press **ESC** twice to remove the grips.

All of the stair treads are now located on layer Stairs, but they are now all the same lineweight. Because they are all on the same layer, you can't select according to layer this time. However, you know that all of the treads are exactly 11 inches long. The Quick Select dialog box allows you to select them based on their length.

14. Select the **Tools** pull-down menu and pick **Quick Select...** again.

15. Select **Line** in the Object type box.

16. Scroll down the **Properties** selection box and pick **Length**.

17. In the **Value** box, enter **11** and pick **OK**.

All of the 11″ lines in the drawing—all of the treads—are selected.

Standard

18. Pick the **Properties** button on the **Standard** toolbar and pick **Lineweight** in the Properties dialog box.

19. Pick the down arrow and select **0.60 mm** from the drop-down list.

20. Pick **Color**, pick the down arrow, and choose green.

21. Close the dialog box, save your work, and close the file.

Applying the Custom Template

The tmp1.dwg file should still be open on your screen. This template file is now nearly complete. The last few steps (12, 13, and 14 from page 303) typically involve creating new text styles and setting the dimensioning variables and DIMSCALE. All of this is covered in Chapters 25 and 26.

The template file concept may lack meaning to you until you have actually applied it. Therefore, let's convert tmp1.dwg to a template file and use it to begin a new drawing. First we will create the template file.

1. Be sure to save your work if you haven't already.

2. Select **Save As...** from the **File** pull-down menu.

3. At the right of Save as type, pick the down arrow and select **AutoCAD Drawing Template File (*.dwt)**.

4. Find and select the folder with your name and pick the **Save** button.

The Template Description box appears.

5. Enter **Created for stair details** and pick **OK**.

6. Pick the **New** button in the **Standard** toolbar to display the Create New Drawing dialog box and pick the **Use a Template** button.

7. Pick the **Browse...** button.

8. Find and double-click the **tmp1.dwt** template file.

AutoCAD loads the contents of tmp1.dwt into the new drawing file.

9. Review the list of layers.

Look familiar? Do you see why drawing templates are of value?

10. Save the new drawing file as **staird.dwg** and exit AutoCAD.

Chapter 22
Review & Activities

Review Questions

1. Name at least two purposes of layers.

2. Using the Layer Properties Manager dialog box, how do you change the current layer?

3. Describe the purpose of the LTSCALE command and explain how to set it.

4. Describe a situation in which you would want to freeze a layer.

5. Name five of the linetypes AutoCAD makes available.

6. What is the purpose of locking layers?

7. If you accidentally draw on the wrong layer, how can you correct your mistake without erasing and redrawing?

8. Explain how to freeze a layer using the Layer Control drop-down box in the docked Object Properties toolbar.

9. Describe a way to select several similar objects using a single operation.

Challenge Your Thinking

1. When might you use the Layer Control drop-down box in the Object Properties toolbar instead of using the Layer Properties Manager dialog box? Explain.

2. AutoCAD allows you to use linetypes in your drawings that correspond to ISO standards. Find out more about ISO standards. When and where are they used? What is the purpose of having such standards?

Chapter 22
Review & Activities

Applying AutoCAD Skills

Work the following problems to practice the commands and skills you learned in this chapter.

1–2. Create two new drawings using the tmp1.dwt template file. Change the layers to match those shown in Tables 22-3 and 22-4. Name the drawings prb22-1.dwg and prb22-2.dwg. Make Objects the current layer in prb22-1.dwg, and make 0 the current layer in prb22-2.dwg.

Layer Name	State	Color	Linetype	Lineweight
0	Frozen	white	CONTINUOUS	Default
Border	On	cyan	CONTINUOUS	0.50 mm
Center	On	yellow	CENTER	0.20 mm
Dimensions	On	green	CONTINUOUS	0.20 mm
Hidden	On	yellow	HIDDEN	0.30 mm
Objects	On	red	CONTINUOUS	0.40 mm
Phantom	On	blue	PHANTOM	0.50 mm
Text	Frozen	magenta	CONTINUOUS	0.30 mm

Table 22-3

Layer Name	State	Color	Linetype	Lineweight
0	On	white	CONTINUOUS	Default
Center	Frozen	blue	CENTER	0.20 mm
Dimensions	On	yellow	CONTINUOUS	0.20 mm
Electrical	On	cyan	CONTINUOUS	0.30 mm
Found	On	magenta	DASHED	0.40 mm
Hidden	On	blue	HIDDEN	0.30 mm
Notes	On	yellow	CONTINUOUS	0.30 mm
Plumbing	Frozen	white	CONTINUOUS	0.30 mm
Title	Frozen	magenta	CONTINUOUS	0.30 mm
Walls	On	red	CONTINUOUS	0.40 mm

Table 22-4

Chapter 22
Review & Activities continued

3. Create a new drawing from scratch. Set up the layers shown in Table 22-5. Then create the slide shown in Fig. 22-8 on layer Object. Place the center lines on layer Center, positioning them as shown in the illustration. Change back to the Object layer and create the hole (circle) for the slide so that its center point is at the intersection of the two center lines. To finish the drawing, offset the hole by .2 unit to the outside and trim the center lines to the outside circle. Erase the temporary trim circle, and save the drawing as slide.dwg.

Layer Name	Color	Linetype	Lineweight
Object	white	CONTINUOUS	0.40 mm
Center	blue	CENTER2	0.20 mm
Dims	red	CONTINUOUS	(default)

Table 22-5

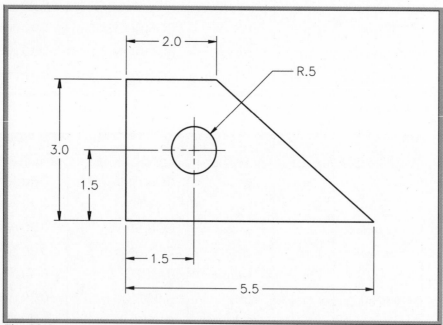

Fig. 22-8

4. Create a drawing template for use with an A-size drawing sheet. Use a scale of ¼″ = 1′. Use architectural units with a precision of ¹/₁₆″. Set up the following layers: Floor, Dimensions, Electrical, Plumbing, and Furniture. Call it ch22tmp.dwt.

5. Using the tmp2.dwt template file you created in the previous problem, create a simple floor plan for a house. Make the floor plan as creative and as detailed as you wish. Place all the objects on the correct layers.

6. The graph in Fig. 22-9 shows the indicated and brake efficiencies as functions of horsepower for a small engine. Reproduce the graph as follows: Using a suitable scale, draw the grid and plot the given points as shown; then draw a spline through each set of points. Place the border, title border, and curves on a layer named Visible; the grid and point symbols on layer Grid; and the text on layer Text. Trim the grid around the text and arrows, and trim the curves and grid out of the symbols. Use the appropriate text justification to align the axis numbers and titles properly.

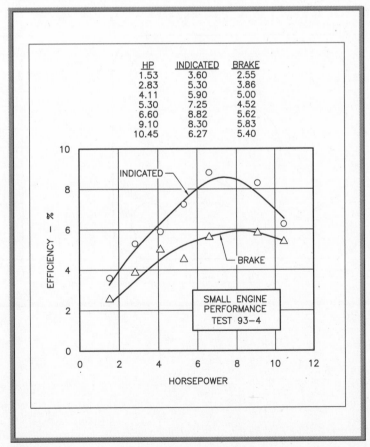

HP	INDICATED	BRAKE
1.53	3.60	2.55
2.83	5.30	3.86
4.11	5.90	5.00
5.30	7.25	4.52
6.60	8.82	5.62
9.10	8.30	5.83
10.45	6.27	5.40

Fig. 22-9

Problem 6 courtesy of Gary J. Hordemann, Gonzaga University

Chapter 22
Review & Activities continued

Using Problem-Solving Skills

Complete the following activities using problem-solving skills and your knowledge of AutoCAD.

1. Create the drawing of the washing machine spacer from the graph-paper sketch sent to the drafting department by engineering (Fig. 22-10). Each square represents .25 inch. Add a title to the drawing and include a note saying that the material is AISI 1010 carbon steel. Make the appropriate drawing setup, and create the necessary layers. Add center lines for all holes. Save the drawing as ch22spacer.dwg.

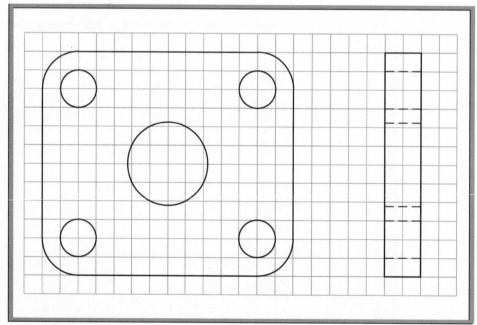

Fig. 22-10

2. The chief engineer created a sketch of the step block for a bowling alley pin-setter (Fig. 22-11). Create the drawing. Each square represents 5 cm. Add a title to the drawing and these notes: 1) to break all sharp edges, 2) fillets and rounds 5cm, and 3) material is UNS S30451 stainless steel. Make the appropriate drawing setup, and create the necessary layers. Save the drawing as ch22pinsetter.dwg.

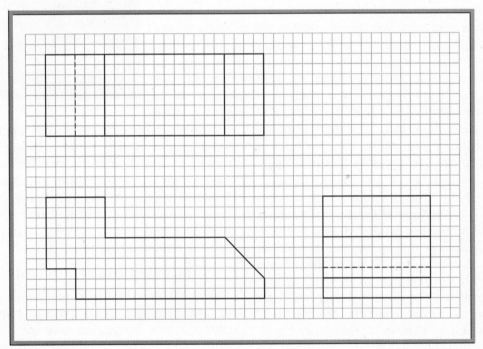

Fig. 22-11

Careers Using AutoCAD

The Stock Market/Jon Feingersh

Tool for Sales Engineers

The product catalogs lugged around by sales engineers are going out of date faster than ever before, in part because CAD has shortened design cycles, getting new product ideas to the factory floor sooner. Now CAD can also be applied to help sales engineers keep pace.

CustomWorks is AutoCAD-based software for sales engineers who have to give presentations about a product line that is complex or frequently changing—hydraulic pumps, for example. It is especially useful in situations in which you need to show a particular model configured for a particular customer. With CustomWorks running on a laptop computer, the sales engineer drafts a layout diagram or assembly drawing for the customer's site. He or she uses built-in parametric controls to generate accurate, easy-to-understand technical drawings in minutes.

Developed by The Premisys Corporation (Chicago), CustomWorks is third-party software for AutoCAD users. For people who do not have AutoCAD, Premisys makes another version of the software with AutoCAD embedded. Sales engineers can use this system in virtually any business in which the price quote is based on measurements from the customer's site. Companies using CustomWorks range from makers of materials-handling equipment to custom window manufacturers.

Career Focus—Sales Engineer

Manufacturer's sales engineers sell to businesses, governments, wholesalers, and other organizations. During a sales call, they show samples and catalogs that describe their company's products. They explain how a particular product can improve the customer's business by reducing costs or increasing productivity. After writing up the sales order and taking information for installation, a sales engineer calls on the customer periodically to make sure the installation is working as planned.

To find new customers and opportunities, sales engineers gather information about companies within their territory. This information may come as "leads" from the home office. Satisfied customers may recommend the sales engineer to other companies, or the sales engineer might develop sales leads from reading articles and advertisements in trade journals.

Education and Training

A bachelor's degree in engineering, science, or mathematics is a typical requirement, though some employers hire on the basis of sales experience in a related field. Many companies have formal training programs for sales engineers, lasting up to two years. In some programs, trainees rotate through various jobs to learn all the phases of the business. Training may include formal classes followed by a period of work with an experienced sales engineer.

Career Activity

- Interview someone who can tell you about the advantages and disadvantages of a job based on sales commissions.
- Do the rewards and demands of engineering sales fit with your goals? Explain.

Chapter 23 Plotting and Printing

Objectives:

- Preview a plot
- Adjust plotter settings
- Plot an AutoCAD drawing to scale

Key Terms

landscape
plotter
plot scale
plot style
portrait
printer

CAD users plot and print drawings using devices called plotters and printers. Many years ago, there was a distinct difference between a plotter and a printer, but today, the distinction has blurred to the extent that there's now little difference between the two. Both use the same basic technology—usually inkjet or laser—and both print black, grayscale, or color onto paper and other sheet materials. Some people still refer to devices that handle large sheets, such as 36″ × 48″ and even larger, as *plotters*. They think of printers as smaller, tabletop devices handling sheets up to 11″ × 17″. Note that this is not a universally accepted difference, however. The terms *plotter* and *printer* and *plotting* and *printing* are used interchangeably in this book.

Let's create and plot a set of stairs.

1. Start AutoCAD and open the file named **staird.dwg.**

This file, which is based on the tmp1.dwt template file, was created in the previous chapter.

2. On layer **Objects**, draw the stair step shown in Fig. 23-1. Use polar tracking and enter the lengths at the keyboard. (The LWT button should *not* be depressed.)

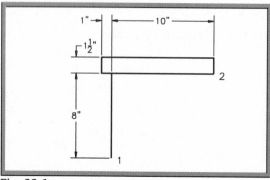

Fig. 23-1

3. Copy the objects five times to produce the stairs as shown in Fig. 23-2.

 HINT:

Use the COPY command's Multiple option. When copying, use point 1 as the base point and point 2 as the second point.

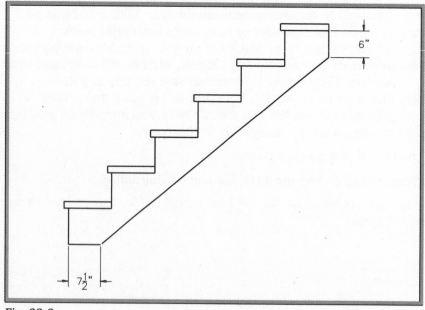

Fig. 23-2

4. **ZOOM All** and save your work.

Previewing a Plot

AutoCAD allows you to preview a plot before you send it to the plotter. This feature allows you to catch mistakes before you spend the time and supplies to create the actual plot.

Standard

1. From the docked **Standard** toolbar, pick the **Plot** button or enter the **PLOT** or **PRINT** command at the keyboard.

The Plot dialog box appears, as shown in Fig. 23-3.

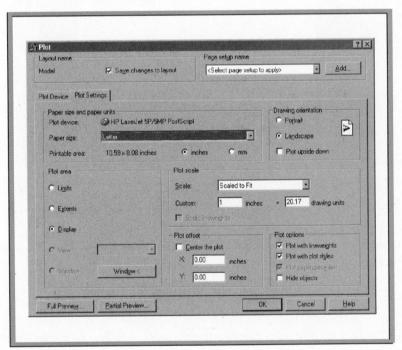

Fig. 23-3

In the illustration, notice that HP LaserJet 5P_5MP Postscript appears in the upper left area of the dialog box. This means a device was configured previously. The device you see in this area will probably be different. AutoCAD stores plotter settings for each configured plotter so other settings may also differ, depending on how they were last set.

NOTE:

If no plotter names appear in the upper left area, you must configure one before you proceed with the following steps. Chapter 24 provides information on how to configure plotter devices.

Note the preview buttons located in the lower left area of the dialog box.

2. Pick the **Full Preview...** button.

This feature gives you a preview of how the drawing will appear on the sheet, as shown in Fig. 23-4.

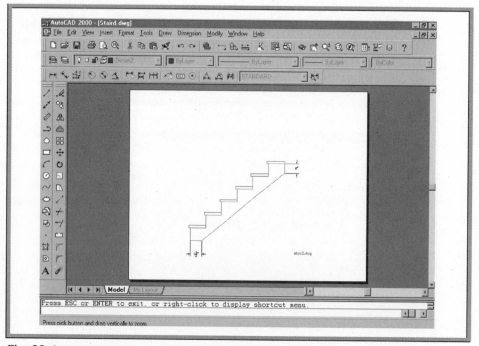

Fig. 23-4

3. Use the pick button to zoom up on the drawing.

4. Right-click to display a pop-up menu, and pick **Pan** to move around in the drawing.

This enables you to examine the drawing closely before you plot it. Large drawings can take considerable time to plot, especially on pen plotters, so spotting errors before you plot can save time as well as plotting supplies.

5. Press **ESC** or **ENTER**.

6. If your computer is connected to the configured plotter or printer and it is turned on and ready, continue with Step 7. If it is not, prepare the device. If you do not have access to an output device, pick the **Cancel** button and skip to the next section, titled "Tailoring the Plot."

7. Pick **OK**.

Since we did not make changes to the plot settings before plotting, the plot may be incorrect.

Tailoring the Plot

The following steps consider sheet size, drawing scale, and other plotter settings and parameters.

Standard

1. Pick the **Plot** button.

2. Near the center of the dialog box, pick the inches radio button, if it is not already selected.

Specifying Sheet Size

1. At the right of Paper size, pick the down arrow.

2. Review the list of sizes available for the configured printer.

3. Pick **Letter** if it is available. If not, pick the option that is closest to 11″ × 8.5″.

Setting the Drawing Orientation

Notice the illustration to the right of the Portrait and Landscape radio buttons in the Drawing orientation area of the Plot dialog box. In *portrait* orientation, the drawing is positioned so that "north" or "up" falls on the narrower edge of the paper. In *landscape,* the wide edge of the paper is north.

1. Pick each radio button while watching the illustration to see the difference between portrait and landscape.

2. In the Drawing Orientation area, pick **Landscape**.

Defining the Plot Area

1. Under Plot area, select the **Limits** radio button.

We selected Limits because we want to plot the entire drawing area as defined by the 22′ × 17′ drawing limits.

In the future, you may choose to select one of the other plotting options. Table 23-1 on page 342 explains what each option means.

Option	Function
Limits	Plots the entire drawing area
Extents	Similar to ZOOM Extents; plots the portion of the drawing that contains objects
Display	Plots the current view
View	Plots a saved view
Window	Plots a window whose corners you specify; use the Window... button to specify the window to plot.

Table 23-1

Setting the Drawing Scale

Focus your attention on the Plot scale area of the dialog box. The *plot scale* is the scale at which the drawing is plotted to fit on a drawing sheet.

1. At the right of Scale, pick the down arrow and review the list of options.

2. Pick **Scale to Fit**.

When Scaled to Fit is selected, the values entered at the right of Custom reflect the actual scale used to fit the drawing on the sheet. Focus on the values located at the right of Custom. As you may recall, the scale for the stairs drawing is ½″ = 1′, which is the same as .5 = 12, or 1 = 24.

3. At the right of Custom, enter **.5** for inches and **12** for drawing units.

When Scale lineweights is available and checked, AutoCAD scales the lineweights in proportion to the plot scale. This setting controls the LWSCALE system variable.

Reviewing Other Settings

The Plot offset area specifies an offset of the plotting area from the lower left corner of the sheet.

1. Pick the **Plot Device** tab.

Much of this information is self-explanatory. The Plotter configuration area provides information on the configured device. The Properties button presents information and options relevant to the device.

The Plot style table (pen assignments) allows you to create and edit plot style tables. A *plot style* is a collection of property settings saved in a plot style table. The What to plot area permits you to select what you want to plot and the number of copies.

When the Plot to file check box is checked in the Plot to file area, AutoCAD sends plot output to a file rather than to a device and creates a PLT file type.

2. Pick the Plot Settings tab and review all of the settings.

Plotting the Drawing

Let's preview and plot the drawing.

1. Pick the **Full Preview...** button.

2. Review the preview and then press **ESC**.

3. Prepare the device for plotting.

4. Pick the **OK** button to initiate plotting.

After plotting is complete, examine the output carefully. The dimensions on the drawing should measure correctly using a $1/2'' = 1'$ scale.

Adding and Configuring a Plotter

Adding a new plotter for AutoCAD is an elaborate process that should involve an instructor or systems administrator.

1. Pick **Options...** from the **Tools** pull-down menu to display the Options dialog box.

2. Select the **Plotting** tab.

3. Pick the down arrow located in the upper left area to display a list of currently available output devices, but do not select a new one.

4. Pick the **Add or Configure Plotters...** button.

5. Double-click **Add-A-Plotter Wizard** and read the information that displays in the Add Plotter–Introduction Page.

6. Pick the **Next** button.

7. Read the information in the box and then pick the **Cancel** button if an instructor or systems administrator is not available to assist you.

8. Close the window containing Add-A-Plotter Wizard.

9. If you are interested in learning more about adding and configuring plotters, pick the **Help** button and read the information presented in the Help window.

10. Close any windows or dialog boxes that are still open and exit AutoCAD. Do not save any changes.

Chapter 23
Review & Activities

Review Questions

1. Explain why a full plot preview is useful.

2. In the Plot dialog box, what is the purpose of the Plot to file check box?

3. Briefly describe each of the following plot options.
   Limits
   Extents
   Display
   View

4. The drawing plot scale for a particular drawing is 1 = 4″. What does the 1 represent and what does the 4″ represent?

5. Describe how you would add a new printer or plotter device.

Challenge Your Thinking

1. Explore ways of sending PLT files to a printer or plotter.

2. How is a full plot preview different from a partial preview? Write a short paragraph explaining the difference.

 ## Applying AutoCAD Skills

Work the following problems to practice the commands and skills you learned in this chapter. In problems 1 and 2, prepare to plot a drawing using the information provided. Choose any drawing to plot.

1. Paper size: Letter
   Units: Inches
   Drawing orientation: Landscape
   Plot scale: Custom 1 inch = 2 drawing units
   Plot offset: Do not center the plot
   Plot with lineweights
   Do not plot with plot styles
   Do not hide objects
   Plot current tab
   Number of copies: 1
   Do not plot to file
   Perform a full preview

Chapter 23
Review & Activities

2. Paper size: A4
   Units: Millimeters
   Drawing orientation: Landscape
   Plot area: Limits
   Plot scale: Custom 1 mm = 10 drawing units
   Plot offset: Do not center the plot
   Plot with lineweights
   Do not plot with plot styles
   Plot current tab
   Number of copies: 1
   Perform a full preview
   Plot to file

3. Open slide.dwg (problem 3 in Chapter 22). Zoom in so that the slide fills most of the screen. Enter the PLOT command and choose the Limits radio button in the Plot area portion of the dialog box. Plot the drawing. Then create two more plots, one with the Extents radio button selected, and once with the Display radio button selected. Compare the drawings.

4. Choose and plot a drawing that you created in an earlier chapter. Consider the scale and drawing area so that dimensions measure correctly on the plotted sheet. Text and linetypes should also measure correctly on the sheet. For example, ¹⁄₈″ text should measure ¹⁄₈″ in height.

 ## Using Problem-Solving Skills

Complete the following activities using problem-solving skills and your knowledge of AutoCAD.

1. The R & D department has requested a full-size plot of the step block for the bowling alley pin-setter you created in Chapter 22. Adjust the plot area and scale accordingly, and plot the drawing.

2. The engineering department has requested a half-size plot of the washing machine spacer that you created in Chapter 22. Make the necessary adjustments and plot the drawing.

Chapter 24 Multiple Viewports

Objectives:

- Create and use multiple viewports in model space
- Create and use viewports in paper space
- Position objects in paper space viewports
- Edit, position, and plot paper space layouts

Key Terms

layout
model space
paper space
viewports

Viewports in AutoCAD are portions of the drawing area that you define to show a specific view of a drawing. By default, AutoCAD begins a new drawing using a single viewport. By defining and using multiple viewports, you can see more than one view of a drawing at once.

AutoCAD includes two different environments—*model space* and *paper space*—for working with objects. When you create a new drawing, AutoCAD presents a single viewport in model space, because this is where most drafting and design work is done.

Each new drawing also has two default layout tabs. A *layout* is an arrangement of one or more views of an object on a single sheet. By default, AutoCAD provides two layout tabs for every new drawing you create. When you pick a layout tab, AutoCAD automatically switches to paper space. In general, paper space is used to lay out, annotate, and plot one or more views of an object or 3D model. After you have set up the paper space parameters, AutoCAD positions model space objects against a white "paper" background to show you how they will appear on the printed sheet.

Viewports in Model Space

You can apply viewports to both model space and paper space. In model space, you can use viewports to draw and edit in more than one view at a time. The magnification of each viewport can be set individually, so viewports provide capabilities that would otherwise be impossible.

After creating a drawing in model space, you can create floating viewports in paper space to display different views of the drawing. Because they float, you can easily position them for plotting. Depending on your needs, you can set options that determine what is plotted and how the viewports fit on the sheet.

Figure 24-1 gives an example of applying multiple viewports in model space to zoom.dwg. Notice that each viewport is different both in content and in magnification.

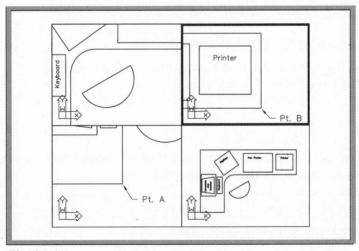

Fig. 24-1

Creating Viewports in Model Space

Viewports are controlled with the VPORTS (or VIEWPORTS) command.

1. Open the drawing file named **zoom.dwg**.

2. Enter **ZOOM All** to make the drawing fill the screen.

3. Enter the **VPORTS** command or select **Viewports** and **New Viewports...** from the **View** pull-down menu.

4. Under **Standard viewports**, click on each of the options.

Each viewport option appears in the Preview area.

5. Pick **Four: Equal** and pick the **OK** button.

AutoCAD produces four viewports of equal size, each with an identical view of the drawing.

6. Move the pointer to each of the four viewports.

The crosshairs appear only in the viewport with the bold border. This is the current (active) viewport.

7. Move to one of the three nonactive viewports.

An arrow appears in place of the crosshairs.

8. Press the pick button on the pointing device.

This viewport becomes the current one.

Using Viewports in Model Space

Let's modify zoom.dwg using the viewports.

1. Refer to Fig. 24-1 and create four similar viewports using AutoCAD's zoom and pan features.

 HINT: ─────────────────────────────────

> Make one of the viewports current, and then zoom and pan. Repeat this process for the other three viewports.

2. Make the lower left viewport current.

3. Enter the **LINE** command and pick point **A**. (Refer to Fig. 24-1 for point A.)

4. Move to the upper right viewport and make it current.

Notice that the LINE command is now active in this viewport.

5. Pick point **B** and press **ENTER**.

View all four viewports. The line represents the edge of a hard surface for the chair.

So you see, you can easily begin an operation in one viewport and continue it in another. Any change you make is reflected in all viewports. This is especially useful when you are working on large drawings with lots of detail.

Let's move the printer from one viewport to another.

NOTE:

You may need to shrink the printer a small amount so that it will fit in its new location. Use the grips Scale option to scale the printer.

6. With the upper right viewport current, select the printer, and pick a grip box.

7. Right-click and pick **Move** from the shortcut menu.

8. Move to the lower left viewport and make it current. (Make sure ortho is off.)

9. Place the printer in the open area on the table by picking a point at the appropriate location.

Notice that the printer location changed in the other viewports.

Viewport Options

Let's combine two viewports into one.

1. From the **View** pull-down menu, pick **Viewports** and **Join**.

2. Choose the upper left viewport in reply to Select dominant viewport.

3. Now choose the upper right viewport in reply to Select viewport to join.

As you can see, the Join option enables you to expand—in this case, double—the size of a viewport.

4. Enter **VPORTS**, pick **Single**, and pick **OK**.

The screen changes to single viewport viewing. This single viewport is inherited from the current viewport at the time you selected Single.

AutoCAD also allows you to subdivide current viewports into two or more additional viewports.

5. Pick the new **Viewports** tab, pick **Three: Right**, and pick **OK**.

6. Make the upper left viewport the current one.

7. Redisplay the **Viewports** dialog box and pick **Two: Vertical**.

8. Under Apply to, located in the lower left corner, pick the down arrow and select **Current Viewport**, and pick **OK**.

As you can see, AutoCAD applies Two: Vertical to the current viewport.

9. Try the remaining viewport options on your own. Practice drawing and editing using the different viewport configurations.

NOTE:

The REGEN command affects only the current viewport. If you are using multiple viewports and you want to regenerate all of them, you can use the REGENALL command.

10. Save your work and close the drawing file, but do not exit AutoCAD.

Viewports in Paper Space

When you are creating a layout in AutoCAD, you can consider viewports as objects with a view into model space that you can move and resize. By default, AutoCAD presents a single viewport in the paper-space layout.

1. Open the file named **staird. dwg**.

2. Right-click the **Layout1** tab.

This produces a menu that allows you to name and rename the current layout.

3. Click anywhere to make the menu disappear.

4. With the pick button, pick the **Layout1** tab.

When you pick a layout tab for the first time, a sheet with margins displays, reflecting the paper size of the currently configured plotter and printable area of the sheet. AutoCAD also displays the Page Setup dialog box as shown in Fig. 24-2.

5. Under Layout name, double-click **Layout1** and then enter **My Layout**.

This dialog box permits you to establish all of the settings for plotting. For now, do not change any of these settings.

6. After reviewing the dialog box, pick the **OK** button.

AutoCAD renames the tab to My Layout.

In paper space, you can view the exact size of the sheet and see how the drawing will appear on the sheet when you plot. The dashed line represents the plotting boundary, and the solid line represents the single viewport. (AutoCAD's paper space defaults to a single viewport if you do not specify more than one.)

The paper space icon replaces the standard coordinate system icon (Fig. 24-3). The paper space icon is present whenever paper space is the current space. The coordinate system icon is present whenever model space is the current space.

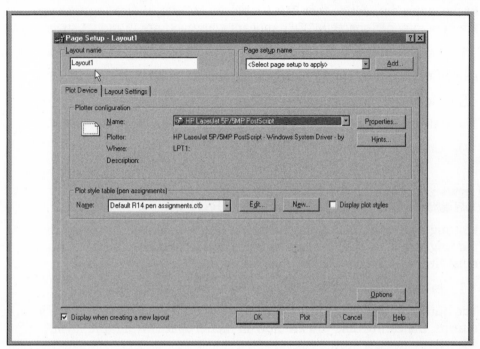

Fig. 24-2

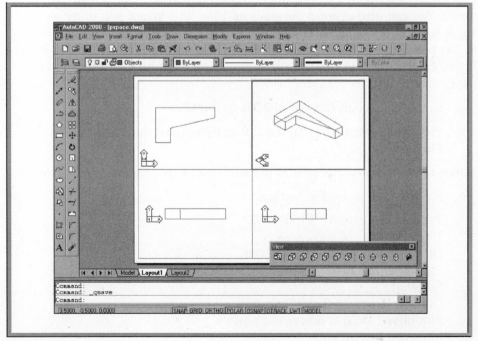

Fig. 24-3

In the status bar, notice that PAPER replaced MODEL, indicating that you are now in paper space. MODEL appears when you are in model space.

7. Pick the viewport (solid border).

As you can see, a viewport in paper space is an object.

8. Move the viewport a short distance.

The stair detail moves too, because it belongs to the viewport.

Standard

9. Undo the last step so that the viewport reappears.

Plotting a Single Viewport in Paper Space

Plotting a viewport containing an object in paper space is different than plotting the same object in model space. Before plotting to scale, you must fit the objects in the viewport using the ZOOM command.

1. In the status bar, click **PAPER** so that it now reads **MODEL**.

The outline of the viewport becomes bold and the coordinate system icon returns, indicating that you are now in model space within the layout.

2. Enter the **ZOOM** command and enter the **Scale** option.

3. For the scale factor, enter **1/24xp**.

The "1 to 24" reflects the scale of the drawing. As you may recall, we established a scale of $1/2'' = 1'$ for tmp1.dwt, which we used for the stair detail. You can also express the scale as $1'' = 24''$. When you enter a value such as 1/24 followed by xp, AutoCAD specifies the scale relative to the paper space scale. (The term xp means "times paper space scale.") If you enter .5xp, AutoCAD displays model space at half the scale of the paper space scale.

4. In the status bar, change **MODEL** to **PAPER**.

Standard

5. Pick the **Plot** button from the docked **Standard** toolbar and pick the **Plot Settings** tab.

Notice under Plot area that Limits is not available; Layout appears in its place.

6. Produce the following settings:

Paper size: Letter or 11" × 8.5"
Units: Inches
Drawing orientation: Landscape
Plot area: Layout
Scale: 1:1
Custom: 1 inch = 1 drawing unit
Plot offset: Do not center the plot
Plot with lineweights

The plot scale is 1=1 because we do not want to scale the layout up or down. We did that in Step 3. This is one distinct difference between plotting from model space and plotting from paper space.

7. Perform a full preview and then press **ESC**.

8. Make sure that your plotter is ready, and pick the **OK** button to initiate plotting.

AutoCAD plots the stairs to scale.

9. Save your work and close the drawing file.

Adding Viewports in Paper Space

You can add viewports in paper space using a method similar to the one you used in model space, but you must be in paper space. Let's set up a drawing with multiple paper space viewports.

1. Create a new drawing and pick the **Use a Wizard** button.

2. Pick **Quick Setup** and **OK**.

3. Pick **Next** to accept Decimal units.

4. Enter **11"** for the width and **8.5"** for the length, pick the **Finish** button, and **ZOOM All**.

5. Create the layers shown in Table 24-1, making **Vports** the current layer.

6. Turn the grid off, and set snap at **.5"** , and create a new text style named **romans** using the **Romans.shx** font. Use the default settings.

7. Save your work in a file named **pspace.dwg**.

8. Pick the **Layout1** tab to enter paper space, and pick **OK** to accept all of the default settings.

9. Erase the viewport that appears on layer Vports.

10. From the **View** pull-down menu, pick **Viewports** and **4 Viewports**, and press **ENTER** to accept the **Fit** default value.

AutoCAD inserts four equally-sized viewports on layer Vports.

Layer Name	Color	Linetype	Lineweight
Objects	red	CONTINUOUS	0.40 mm
Border	blue	CONTINUOUS	0.60 mm
Vports	magenta	CONTINUOUS	0.20 mm

Table 24-1

NOTE: ─────────────────────────

You can insert viewports of any polygonal shape. From the View pull-down menu, select Viewports and Polygonal Viewport.

11. On the status bar, change **PAPER** to **MODEL**.

Coordinate system icons appear in each of the four viewports, showing that you are now in model space.

Objects in Viewports

You will better understand the benefits of using viewports in paper space when you create an object.

1. Make **Objects** the current layer.

2. Enter the **THICKNESS** system variable and set it to **1**.

The THICKNESS system variable enables you to specify the thickness of an object in the *z* direction, resulting in a three-dimensional object.

3. Make the upper left viewport the current viewport by picking a point inside it.

4. In the upper left viewport, draw the object shown in Fig. 24-4.

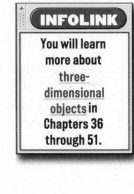

INFOLINK

You will learn more about three-dimensional objects in Chapters 36 through 51.

An identical view of the object appears in all four viewports.

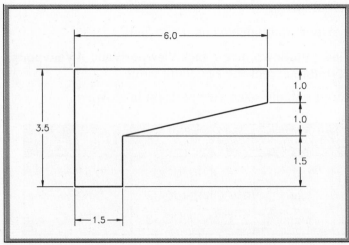

Fig. 24-4

Creating Four Individual Views

Taking advantage of AutoCAD's viewports, let's create four different views of the solid object.

1. Make the lower left viewport the current one by picking a point inside the viewport.

2. Display the **View** toolbar.

3. View the object from the front by picking the **Front View** button from the toolbar.

4. Enter **ZOOM** and **1**.

This scales the viewport to an apparent size of 1. Entering 2 would make it twice the size.

5. Make the lower right viewport the current one.

6. View the object from the right side by picking the **Right View** button.

7. Enter **ZOOM** and **1**.

8. Make current the upper right viewport and view the object from above, in front, and to the right by picking **SE Isometric View**.

9. Enter **ZOOM** and **1**, and save your work.

The screen should look similar to the one in Fig. 24-5.

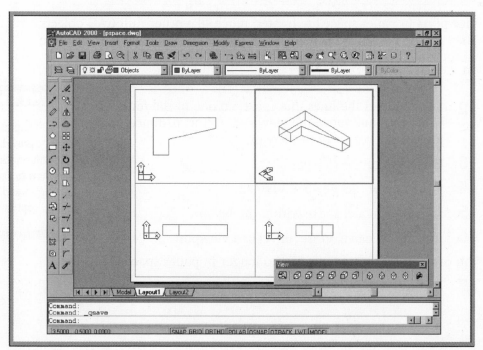

Fig. 24-5

Editing Objects

When you are working in paper space, you cannot edit objects created in model space. Likewise, when you are in model space, you cannot edit objects created in paper space.

1. In the status bar, change **MODEL** to **PAPER**.

Little appears to change except for the coordinate system icons.

2. Attempt to select the object in any of the four viewports.

As you can see, you cannot select it because it was created in model space.

Working with Paper Space Viewports

One of the big advantages of working with paper space viewports is that they allow you to print more than one view of an object on a single sheet of paper. Paper space is used to arrange views and embellish them for plotting, while model space is used to construct and modify objects that make up the model or drawing.

Editing Viewports

Viewports in paper space are treated much like other AutoCAD objects. They can be moved and even erased, but not in model space. Only the views (objects) themselves can be edited in model space.

1. Enter the **UNDO** command and the **Mark** option.

2. Select one of the lines that make up one of the four viewports, and move the viewport a short distance toward the center of the screen.

3. Erase one of the viewports.

4. Scale one of the viewports by **.75**.

5. Switch to model space within the layout.

6. Attempt to move, erase, or scale a viewport.

You can't do it, because you are no longer in paper space.

7. Enter **UNDO** and the **Back** option.

INFOLINK

Chapter 10 provides information on how to use the Mark and Back options.

Plotting Multiple Viewports in Paper Space

One of the benefits of paper space is multiple viewport plotting. (You cannot plot more than one model space viewport at a time.)

1. Freeze layer **Vports** and make **Border** the current layer.

The lines that make up the viewports should now be invisible.

2. Switch to model space within the layout.

The viewport lines are invisible in model space also.

3. Switch to paper space and draw a border and basic title block similar to those in Fig. 24-6.

4. Plot the layout at a scale of **1 = 1**.

INFOLINK

Refer to Chapter 23 for more information about plotting a drawing.

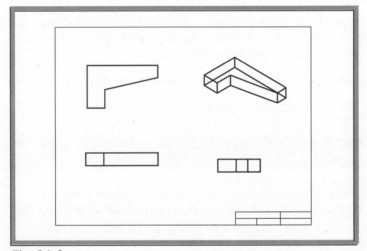

Fig. 24-6

Positioning Viewports

It is possible that the current position of the border and views did not plot perfectly. Even if it did, make adjustments to the location of the border, title block, and views by following these steps.

1. Thaw layer **Vports**.

2. Move the individual viewports to position them better in the drawing. It is normal for them to overlap.

3. Edit the size and location of the border and title block if necessary.

4. Freeze layer **Vports**, save your work, and replot the drawing.

5. Close the View toolbar and exit AutoCAD.

Chapter 24
Review & Activities

Review Questions

1. Explain the difference between model space and paper space.

2. How can using multiple viewports in model space help you construct drawings?

3. How do you make a viewport in model space the current viewport?

4. Explain how you would join two viewports in model space.

5. What option should you enter to obtain two viewports in the top half of the screen and one viewport in the bottom half of the screen?

6. If you are working in paper space and you discover that you need to change an object that was created in model space, what must you do before you can make the change? Why?

7. Explain why you may want to edit viewports in paper space.

8. What is the main benefit of plotting in paper space?

Challenge Your Thinking

1. Find out how many viewports you can have at one time in AutoCAD. Would you want to use that many? Why? Explain the advantages and disadvantages of using multiple model space viewports in your drawings.

2. Experiment with viewports created in model space and paper space. Is it possible to create more than one viewport in model space, then import the model space viewports into a paper space viewport? Explain.

Chapter 24
Review & Activities

Applying AutoCAD Skills

Work the following problems to practice the commands and skills you learned in this chapter.

1. Create each of the viewport configurations shown in Fig. 24-7.

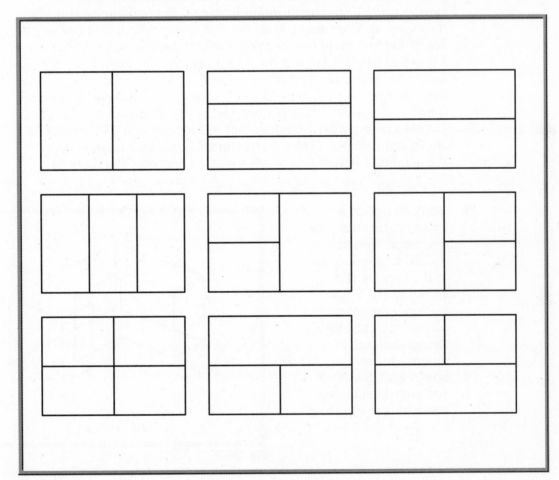

Fig. 24-7

Chapter 24
Review & Activities continued

2. Open the db_samp.dwg file in AutoCAD's Sample folder and save it in your named folder as Fremont.dwg. The current tenant of this building wants to reconfigure the office by moving a printer island to an area of unused offices. Create two viewports side by side (use the Vertical option). Make the right viewport active, and zoom in so that the building almost fills the viewport. Then zoom in further on the printer island on the right side of the building. Make the right viewport active, and again zoom in so that the building almost fills the viewport. Then zoom in on the five empty offices in a horizontal line at the bottom of the viewport (office numbers 6156, 6158, 6162, and 6164). This will be the new location of the printer island. Erase the five offices, including the wall that lies against the outside wall. Using the two viewports as necessary, move the entire printer island to the space formerly occupied by the offices. (Note: When you select the printer island, do not include the gray H or the lines above and below it. This is a structural I-beam that helps support the building.) Return to a single viewport and save the drawing.

3. Using viewports in paper space, create the top, front, right side, and SE isometric view of the security clip shown in Fig. 24-8. (The clip fastens to a planter designed for the agricultural industry.) Create a border and title block, and plot the multiple views on a single sheet.

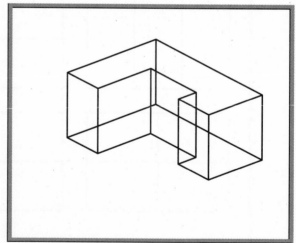

Fig. 24-8

Chapter 24
Review & Activities

Using Problem-Solving Skills

Complete the following activities using problem-solving skills and your knowledge of AutoCAD.

1. Draw the locking receptacle for the alarm system to be installed in the administration building (Fig. 24-9). Use three viewports (large right) in paper space, and show the top and front views. In the large right viewport, show the SW isometric view. Plot the views.

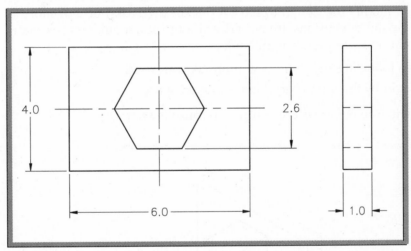

Fig. 24-9

2. Draw the hexagonal locking pin that fits into the alarm system receptacle in the administration building (problem 1). The pin, shown in Fig. 24-10, is 2″ long and measures 2.5″ across the flats. Use three viewports in paper space, and show the same views that you used for the locking receptacle in problem 1. Plot the views.

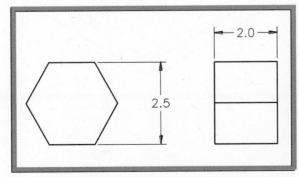

Fig. 24-10

Part 4 Project

Designing a Bookend

Creating and working with multiple viewports allows the AutoCAD user greater flexibility in preparing, viewing, and plotting drawings. This exercise requires the use of both model space and paper space.

Description

Design a bookend based primarily on one or more initials of your name. Figure P4-1 shows an example of a bookend created using the initials L.B.

Create or modify the design to have a thickness (Z axis) of 2 or more inches. Actual dimensions are not critical, but they should approximate the height and width of a small book.

Keep the following suggestions in mind as you work.

1. Begin by drawing everything in model space.

2. Switch to paper space, and set up four viewports.

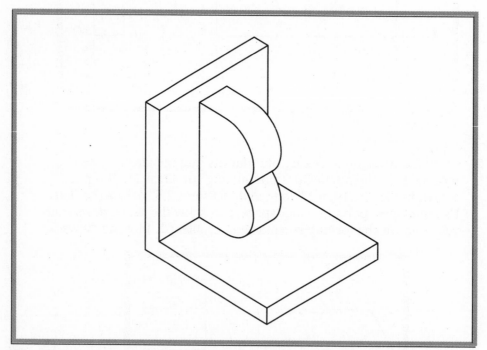

Fig. P4-1

3. Show the plan view in the lower left viewport, the top view in the upper left viewport, the right view in the lower right viewport, and an isometric view in the upper right viewport.

4. Place each view on its own layer. Assign each layer a different color.

5. Zoom each viewport appropriately.

Hints and Suggestions

1. Thickness gives the piece "mass," or a three-dimensional look.

2. Remember as you work that lines and objects created in model space cannot be altered in paper space, and vice versa.

3. Only paper space allows adjustment and manipulation of multiple viewport plotting. Some trial and error may be required to achieve the desired look.

4. You may wish to try several potential designs on paper before beginning the AutoCAD drawing. The drawing task will be easier if you decide on a design and plan how to draw it in AutoCAD before beginning the drawing.

Summary Questions/Self-Evaluation

Your instructor may direct you to answer these questions orally or in writing, or simply to think about your responses as a means of self-evaluation.

1. What factors determine the drawing setup? (How do you determine the "right" choices?) Why can't all drawings successfully use the same setup?

2. What variables can be modified on each layer?

3. Discuss some advantages of being able to draw on and individually control separate layers.

4. Compare and contrast model space with paper space.

5. What is useful about working in multiple viewports?

Careers
Using
AutoCAD

The Stock Market/Lester Lefkowitz

Accident Recreations

From long experience talking to juries, lawyers know that a visual helps people understand and retain information better. In recent years, they have even turned to moviemakers for help in reconstructing events at the scene of a crime or accident. However, movies are expensive.

The Focal Group, a litigation-support firm in Sacramento, uses AutoCAD and 3D Studio Max to generate *forensic animation,* which lets a jury see what happened but costs much less to produce than a movie. The animation can show a scene from different points of view. For example, it might show everything within the driver's field of view during the moments before an accident.

Moreover, computer-generated animation may offer a more accurate picture than a movie. If construction or other changes have altered an accident scene, animation can show things the way they were.

To re-create an accident scene, the forensic animator begins with a topographic survey of the site, scanning in and applying photographic images, defining the path of the vehicle, and constructing before-and-after models of the vehicle. Using information from AutoCAD, 3D Studio Max puts the sequence in motion. The result is an accurate visual that allows investigators, jurors, and others to understand events moment by moment.

Career Focus—Paralegal

A forensic animation specialist is one type of paralegal. Paralegals assist litigation-support firms to provide specialized services such as research or jury analysis. A paralegal assists a lawyer, usually with research, contracts, and traditional legal work, but sometimes with a specialty not directly related to law.

In addition to making drawings, a forensic animator may help gather and analyze maps, photographs, reports, and other information to be portrayed in an animation. A forensic animator must maintain documentation so that the content of the animation and its sources can be verified.

Education and Training

Education and training for paralegals vary widely, but two-year programs are most common. There is no required level of education for paralegals; law firms hire according to their needs and preferences. A forensic animator, in addition to CAD skills, would benefit from paralegal curriculum, because there are many technical and ethical constraints on the presentation of evidence in court.

Forensic animation may involve analysis of causes and effects of physical events, so a broad background in the sciences and mathematics will improve the forensic animator's ability to contribute to the portrayal of an accident or crime scene.

 ## Career Activity

- Write a report about the various occupations that may be involved in analysis of traffic accidents—from police to insurance specialists to traffic engineers. What kinds of training does each specialty require?

Part 5

Dimensioning and Tolerancing

Chapter 25 Basic Dimensioning

Objectives:

- **Set up a text style for dimensions**
- **Produce linear dimensions using dimensioning commands and shortcuts**
- **Dimension round shapes, curves, and holes**
- **Dimension angles**
- **Determine the need for and use baseline and ordinate dimensioning when appropriate**

Key Terms

baseline dimensions
datum
datum dimensions
dimensioning
dimension line
extension lines
linear dimensions
ordinate dimensions

Technical drawings lack meaning without information that communicates size. Drafters and designers use a method known as *dimensioning* to describe the size of features on a drawing. In this and the two chapters that follow, you will discover the wide range of dimensioning options, settings, and styles that AutoCAD makes available to you for mechanical, architectural, and other types of drafting and design work.

We will dimension the following drawing of a part that fits into an injection mold, but first we need to prepare the drawing (Fig. 25-1).

1. Start AutoCAD and use the **Quick Setup** wizard to establish decimal units and a drawing area of 11″ × 8.5″.

2. Create two new layers using the information in Table 25-1.

Layer Name	Color	Linetype	Lineweight
Objects	red	CONTINUOUS	0.50 mm
Dimensions	green	CONTINUOUS	0.20 mm

Table 25-1

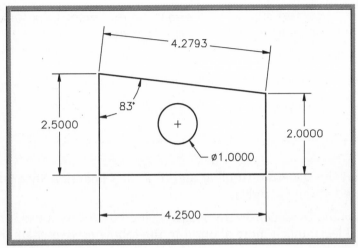

Fig. 25-1

3. Set layer **Objects** as the current layer.

4. Set snap at **.25** and grid at **.5**; **ZOOM All**.

5. Produce the drawing shown in Fig. 25-1 on layer **Objects**, but omit the dimensions at this point. From the lower left corner of the drawing, position the hole **2"** in the positive *x* direction and **1.25"** in the positive *y* direction.

HINT:

Begin in the upper left corner and work counterclockwise as you specify the line endpoints.

6. Save your work in a file named **dimen.dwg**.

Setting the Dimension Text Style

AutoCAD uses text styles for its dimension text. AutoCAD's "romans" text font is suitable for dimension text and is popular among companies that use AutoCAD. This text font resembles the Gothic lettering used extensively in hand-produced drawings. For most drafting applications, a text height of .125" (1/8") is standard practice in industry. Some companies use taller text, such as .1875" (3/16"). Whichever height you choose to use, it's important that you use a consistent text height throughout the drawing or set of drawings.

1. Create a new text style named **rom** using the **romans.shx** font. Accept the default settings for the new style.

HINT:

Enter the STYLE command by entering ST.

Dimension

2. Display the **Dimension** toolbar and pick the **Dimension Style** button (located at the far right).

This displays the Dimension Style Manager dialog box. We will use it here, but you will learn much more about it in the following two chapters.

3. Pick the **Modify...** button.

4. Pick the **Text** tab.

5. At the right of Text style, pick the down arrow and select **rom**.

6. Pick **OK** and then pick **Close**.

Creating Linear Dimensions

Linear dimensions are those with horizontal, vertical, or aligned dimension lines. A *dimension line* is the part of the dimension that typically contains arrowheads at each of its ends, as shown in Fig. 25-2.

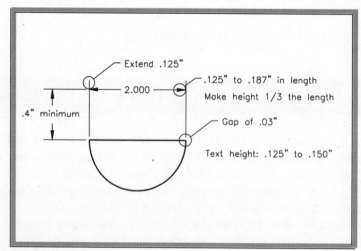

Fig. 25-2

1. Set layer **Dimensions** as the current layer.

2. Pick the **Linear Dimension** button from the Dimension toolbar.

Dimension

This enters the DIMLINEAR command.

3. In reply to Specify first extension line origin, pick one of the endpoints of the mold insert's horizontal line.

The dimension's *extension lines* are the lines that extend from the object to the dimension line.

4. In reply to Specify second extension line origin, pick the other end of the horizontal line.

5. Move the crosshairs downward to locate the dimension line **1″** away from the object and press the pick button.

Dimensioning Overall Size

Let's dimension the vertical edges of the object.

1. Press the spacebar to reenter the **DIMLINEAR** command.

2. This time, when AutoCAD asks for the first extension line origin, press **ENTER** or the spacebar.

The crosshairs change to a pick box.

3. Pick any point on the left edge of the object.

4. Move the crosshairs to the left and pick a point **1″** from the edge to place the dimension.

Let's dimension the other vertical edge, but this time we will use AutoCAD's Quick Dimension feature.

Dimension

5. Pick the **Quick Dimension** button from the Dimension toolbar or enter the **QDIM** command.

As you can see, a pick box replaces the crosshairs. AutoCAD asks that you select geometry to dimension.

6. Pick the right edge of the object and right-click for **ENTER**.

7. Move the crosshairs to the right and pick a point for the dimension.

Notice that adding the last two dimensions required only three clicks for each dimension.

NOTE:

You can create more than one dimension at a time using the Quick Dimension method. However, the results are often unpredictable.

Dimensioning Inclined Edges

Let's dimension the inclined edge by aligning the dimension to the edge.

Dimension

1. Pick the **Aligned Dimension** button from the Dimension toolbar.

This enters the DIMALIGNED command.

2. Press **ENTER**, pick the inclined edge, and place the dimension.

Dimensioning Round Features

Now let's dimension the hole. For mechanical drafting, you should use diameter dimensions for features such as holes and cylinders. Use radius dimensions for features such as fillets and rounds (rounded outside corners).

Dimension

1. Pick the **Diameter Dimension** button from the Dimension toolbar.

This enters the DIMDIAMETER command.

2. Pick a point anywhere on the hole. (You may need to turn off snap.)

The dimension appears.

3. Move the crosshairs around the circle and watch what happens.

Notice that you have dynamic control over the dimension.

4. Pick a point down and to the right as shown in Fig. 25-1 (page 367).

Dimensioning Angles

Let's dimension the angle as shown in the drawing on page 367.

Dimension

1. Pick the **Angular Dimension** button from the Dimension toolbar.

This enters the DIMANGULAR command.

2. Pick both edges that make up the angle.

3. Move the crosshairs outside the mold insert and watch the different possibilities that AutoCAD presents.

4. Pick a location for the dimension arc inside the drawing as shown in the illustration.

5. Save your work as a template file in your named folder. Name the file **dimen.dwt** and enter **Dimensioning Practice** for the template description.

6. Close the file.

Using Other Types of Dimensioning

The dimensioning process you have followed so far in this chapter works for most objects. However, other types of dimensioning are better suited for some objects and drafting applications. In many cases, they produce a cleaner, less confusing drawing.

Baseline Dimensioning

Baseline dimensions are progressive, each starting at the same place, as shown in Fig. 25-3. Baseline dimensioning is most useful on machine drawings where precision is critical. Measuring from a single reference dimension reduces the chance of errors that can accumulate from a "stack" of dimensions.

1. Using **dimen.dwt** as the drawing template, begin a new drawing.

2. Save the drawing and name it **base.dwg**.

3. Erase the drawing and dimensions and create the sheet metal drawing shown in Fig. 25-3. Place the sheet metal part on layer **Objects**. Fill the top half of the screen, and ignore all dimensions.

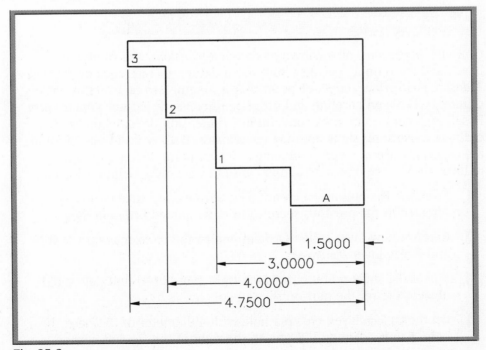

Fig. 25-3

4. Make **Dimensions** the current layer.

Dimension

5. Pick the **Linear Dimension** button and dimension line A. It is important that you pick line A's right endpoint first.

6. Pick the **Baseline Dimension** button.

Dimension

This enters the DIMBASELINE command.

7. Pick points 1, 2, and 3 in order.

8. Press **ENTER** twice and save your work.

The Quick Dimension method also works with baseline dimensioning.

9. Erase the dimensions on the drawing.

Dimension

10. Pick the **Quick Dimension** button or enter **QDIM**.

11. Select the entire drawing, press **ENTER**, and read the options at the Command line.

12. Move the crosshairs downward, but do not pick a point.

AutoCAD assumes that you want to do baseline dimensioning. Now you need to tell AutoCAD from which baseline you want to dimension.

13. Enter **P** for datumPoint, and select the right edge of the drawing.

14. Place the dimensions.

Ordinate Dimensioning

Ordinate dimensions, also known as *datum dimensions,* are similar to baseline dimensioning because both use a datum, or reference dimension. A *datum* is a surface, edge, or point that is assumed to be exact. A basic difference between baseline and ordinate dimensions is their appearance. Baseline dimensioning uses conventional dimension lines, whereas ordinate dimensions show absolute coordinates. Both types of dimensioning are especially useful when producing machine parts. Using a datum reduces the chance of error buildup caused by successive dimensions.

1. From the **File** pull-down menu, pick **Save As...** to save base.dwg (created in the previous section) to a file named **ordinate.dwg**.

2. Create a new layer called **Orddim**, assign the color magenta to it, and freeze layer Dimensions.

3. Mirror the sheet metal part to the right (Fig. 25-4) and delete old objects. Center the part on the screen.

4. On the Objects layer, create a hole with a diameter of **.5**. Place the hole **.5** unit down and **.5** unit to the right of the top left corner.

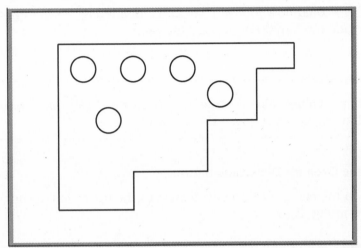

Fig. 25-4

5. Use **COPY Multiple** to create four additional holes. Place the holes as shown in Fig. 25-4.

6. On layer Center, add center marks to each of the holes. Now we're ready to define the datum, or reference dimension.

7. Make **Orddim** the current layer.

Dimension

8. Pick the **Ordinate Dimension** button.

This enters the DIMORDINATE command.

9. At the Specify feature location prompt, pick the upper left corner of the sheet metal part.

The Xdatum option measures the absolute X ordinate along the X axis of the drawing. Also, it determines the orientation of the leader line and prompts you for its endpoint. This defines the location of the datum.

10. Enter **X** for Xdatum.

11. Pick a point about **1″** above the object to make the ordinate dimension appear.

12. Reenter the command and pick the upper left corner of the sheet metal part again.

13. Enter **Y** for Ydatum and pick a point about **1″** to the left of the object to make the ordinate dimension appear.

The Ydatum option measures the absolute Y ordinate along the Y axis of the drawing. This defines the location of the datum.

Your drawing should now look similar to the one in Fig. 25-5, although it will not include the 3.0000 ordinate dimension.

NOTE: ————————————————————————————————

The ordinate values may differ in your drawing, depending on where in the drawing area you place the drawing.

Dimension

14. Pick the **Ordinate Dimension** button.

15. Snap to the center of the hole nearest to the upper left corner, as shown in Fig. 25-5.

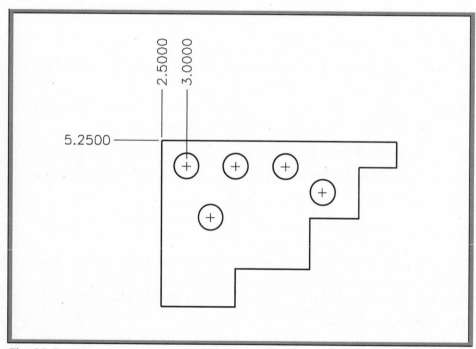

Fig. 25-5

16. Enter **X** for Xdatum and pick a point about **1** unit from the top of the sheet metal part. (Snap should be on.)

Dimension

The X-datum ordinate dimension appears.

17. Repeat Steps 14 through 16 to create X-datum ordinate dimensions for the remaining holes.

18. Create the Y-datum ordinate dimensions on your own using Fig. 25-6 as a guide.

When you finish, your drawing should look similar to the one shown in Fig. 25-6.

19. Close the Dimension toolbar, save your work, and exit AutoCAD.

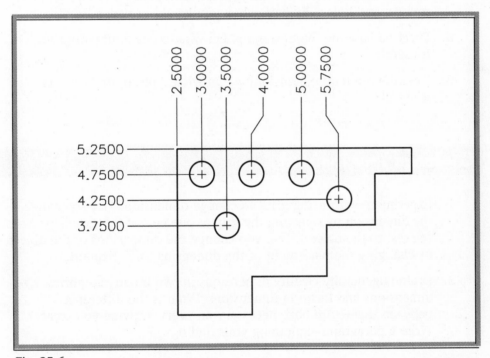

Fig. 25-6

Chapter 25
Review & Activities

Review Questions

1. How do you specify a text style for dimension text?

2. Describe the alternative to specifying both endpoints of a line when dimensioning the entire length of the line.

3. Which dimension button do you use to dimension inclined lines? Angles?

4. Which dimension buttons do you use to dimension fillets, rounds, and holes? Explain when you would use each button.

5. Describe baseline dimensioning. On what types of drawings is it useful?

6. Describe ordinate dimensioning. On what types of drawings is it useful?

Challenge Your Thinking

1. Experiment with using grips to change dimensions. Try to change the dimension by selecting the grip at one end of the extension lines (closest to the object). Can you change the dimensions of the object by changing the placement of the dimension line? Explain.

2. Drafters generally classify dimensions in two broad categories: size dimensions and location dimensions. What is the difference between them? Are both necessary on every drawing you create? Write a paragraph explaining your findings.

Review & Activities

Applying AutoCAD Skills

Work the following problems to practice the commands and skills you learned in this chapter. Begin a new drawing for each problem.

1–2. Create the blocks shown in Fig. 25-7 and 25-8. Place object lines on a layer named Objects and dimensions on a layer named Dimensions. Create a new text style using the romans.shx font. Approximate the location of the holes. If you have access to the drawing files on the Instructor's Resource CD-ROM, refer to block1.dwg and block2.dwg and the instructions that apply.

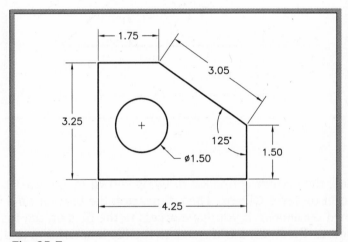

Fig. 25-7

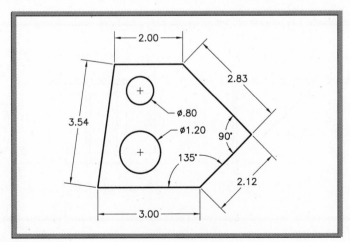

Fig. 25-8

Chapter 25
Review & Activities continued

3. Open the slide.dwg file that you created at the end of Chapter 22 and dimension it. When you finish, your drawing should look like the one in Fig. 25-9. Save the drawing as slide2.dwg.

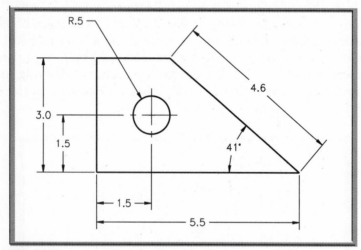

Fig. 25-9

4. Create the mounting bracket drawing shown in Fig. 25-10. Place the bracket on layer Objects. Then dimension the bracket on a layer named Dimensions. If you have access to the files on the Instructor's Resource Guide CD-ROM, refer to the file bracket1.dwg and the instructions that accompany it.

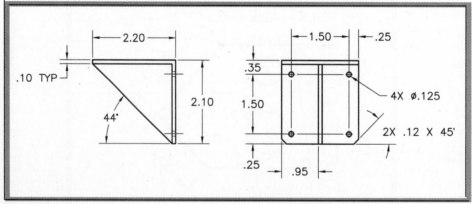

Fig. 25-10

Problem 3 courtesy of Joseph K. Yabu, Ph.D., San Jose State University

5. Draw and dimension the top and front views of the clamp support shown in Fig. 25-11. Assume the spacing between dots to be .20.

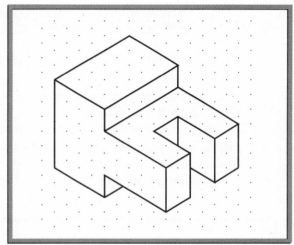

Fig. 25-11

6. Draw the front view of the shaft lock shown in Fig. 25-12. Create appropriate layers for the dimensions and various linetypes. Use the romans.shx font to create a new text style named Roms for the dimension text. Dimension the front view, placing the dimensions on layer Dimensions. Show all pertinent dimensions except the location of the holes. Also, draw and dimension the right-side view. Change the drawing area as necessary to accommodate the dimensions of the part. If you have access to the files on the Instructor's Resource CD-ROM, refer to the file shaftloc.dwg and the accompanying instructions.

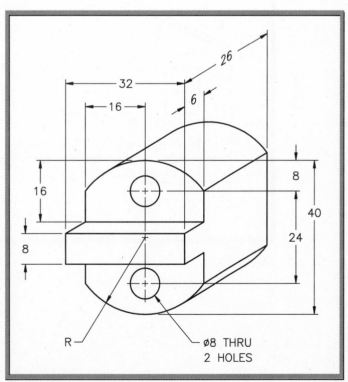

Fig. 25-12

Problems 5 and 6 courtesy of Gary J. Hordemann, Gonzaga University

Chapter 25
Review & Activities continued

Using Problem-Solving Skills

Complete the following activities using problem-solving skills and your knowledge of AutoCAD.

1. The architect for your firm has given you the preliminary elevation drawing shown in Fig. 25-13. Select an appropriate template, complete the title block, and draw the front elevation; include dimensions. Since the drawing was made by an architect and not a drafter, make any changes necessary to comply with correct dimensioning practices. Save the drawing as elev1.dwg.

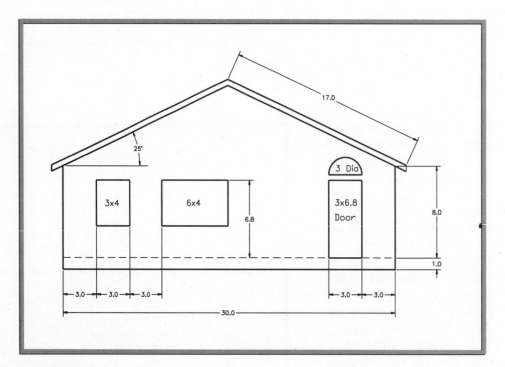

Fig. 25-13

2. Your manager has given you the sketch of a shaft support shown in Fig. 25-14 and told you that it is not to scale (obviously); however the following description applies. The base is 2″ square and ¼″ thick. The boss (lipped area) is in the center of the base, has an OD of 0.5″, a thickness of ¼″, and a through hole with a .35″ radius. The four corner holes are each .15″ in diameter and their center lines are offset from the horizontal and vertical sides by .35″. Draw two views of the shaft support from this description and dimension the drawing appropriately.

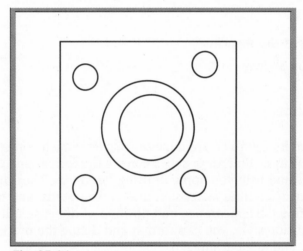

Fig. 25-14

Chapter 26 Advanced Dimensioning

Objectives:

- **Create a dimension style and apply it to a drawing and template file**
- **Adjust the dimension text, arrowheads, center marks, and scale in a dimension style**
- **Edit individual dimensions using a shortcut menu**
- **Use AutoCAD's associative dimensioning feature to update dimensions on a rescaled drawing**
- **Apply dimension styles to a drawing template file**

Key Terms

associative dimensions

dimension styles

leaders

spline leader

AutoCAD permits you to create *dimension styles,* which consist of a set of dimension settings. Dimension styles control the format and appearance of dimensions and help you apply drafting standards. They define the format of dimension lines, extension lines, arrowheads, and center marks. They also define the appearance and position of dimension lines and text. Using a dimension style, you can format and define the precision of the linear, radial, and angular units.

Before we can apply a dimension style to something meaningful, we need to produce a drawing.

1. Start AutoCAD, pick **Use a Template**, and select **tmp1.dwt**.

2. On layer **Objects**, create the top view drawing of the redwood deck shown in Fig. 26-1. Do not add dimensions yet.

The round object represents an outdoor table. It was included in the drawing to ensure that the deck is of sufficient size and shape to accommodate a table of this size.

3. Save your work in a file named **deck.dwg**.

 HINT:

For best results, begin in the upper right corner and work counterclockwise to draw the edges of the deck. Add the inside rounded corner using the FILLET command.

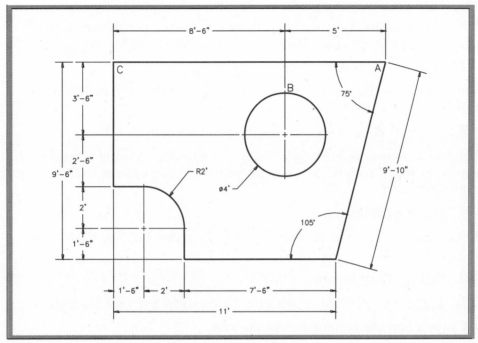

Fig. 26-1

Creating a Dimension Style

Dimension styles provide an easy method of managing the appearance of dimensions. This saves time and effort and helps you produce professional drawings. Note that drafting standards vary from industry to industry and even among companies in the same industry.

1. Display the **Dimension** toolbar.

2. On layer Dimensions, place a horizontal dimension at the bottom of the drawing.

As you can see, the dimension text, arrowheads, etc., are much too small. This is because we need to make some adjustments.

3. Erase the dimension.

4. From the Dimension toolbar, pick the **Dimension Style** button.

Dimension

This displays the Dimension Style Manager dialog box. As you can see, STANDARD is the default dimension style. The example drawing shows how the dimensions will appear using this style.

5. Pick the **New...** button.

6. For New Dimension Style, enter **My Style**.

STANDARD appears after *Start With*. This means that you can start with all of the settings saved in the STANDARD style. That's what we will do.

7. After *Use for*, pick the down arrow.

You can apply a dimension style to the types of dimensions included in this list.

8. Select **All dimensions** and pick the **Continue** button.

The New Dimension Style: My Style dialog box appears, showing several tabs across the top. This is where you change the settings to store in the dimension style.

9. Pick the **OK** button.

Notice that My Style now appears in the list of styles, but STANDARD is still the current style.

10. Pick the **Close** button.

11. In the Dimension toolbar, pick the down arrow and pick **My Style**.

My Style is now the current dimension style.

12. Save your work.

Changing a Dimension Style

Currently, My Style is identical to the STANDARD style. We will need to customize the new My Style dimension style for our particular drawing. This means changing the size of the arrowheads and center marks, the style and size of the text, and the overall scale of the dimensions.

Arrowhead Size

Dimension

1. Pick the **Dimension Style** button and pick the **Modify...** button.

2. Pick the **Lines and Arrows** tab.

You can make adjustments to the arrowheads in the lower right area of the dialog box.

3. Using the edit box at the right of *Arrow size*, change the value to ¹/₈″ and watch closely at how the arrowheads in the sample drawing change to reflect the new value.

Sizing Center Marks or Center Lines

Notice that center marks currently appear at the center of the circle in the dialog box's sample drawing. Below Arrow size, in the area titled Center Marks for Circles, you can make changes to the center marks. Use center marks when space is not available for complete center lines. You may also need to make changes in this area when a company style requires one or the other.

1. At the right of Type, pick the down arrow and select **Line**, watching closely at how the sample drawing changes to reflect the new center lines.

This causes complete center lines to appear at the centers of circles and arcs, as shown in the sample drawing.

2. At the right of Size, click the down arrow until it drops to a value of **0″** and then click the up arrow and stop at ¹⁄₁₆″.

As you can see, the sample drawing changes from displaying in increments of 32nds to 16ths.

Tailoring Dimension Text

Let's create a new text style and height for our dimension style.

1. Pick the **Text** tab.
2. In the upper left area titled Text Appearance, pick the button containing the three dots.

This causes the familiar Text Style dialog box to appear.

3. Create a new style named **roms** using the **romans.shx** font and the default settings.
4. After closing the Text Style dialog box, pick the down arrow at the right of Text style and select **roms**.

Roms is now the text style for My Style, as shown in the sample drawing.

5. In the Text height edit box, change the value to ¹⁄₈″.

As a reminder, the dimension text should be .125″ to .150″ in height.

Setting the Units

1. Pick the **Primary Units** tab.
2. At the right of Unit format, change the setting to **Architectural**.
3. At the right of Precision, change the value to **0**.

The sample drawing changes to reflect the unit format and precision.

Scaling the Dimensions

1. Pick the **Fit** tab and focus your attention on the area titled Scale for Dimension Features.

This is where you set the scale for dimensions. The value should be the reciprocal of the plot scale. When creating tmp1.dwt, we determined that the scale would be $\frac{1}{2}'' = 1'$, which is the same as $\frac{1}{24}$. The reciprocal of $\frac{1}{24}$ is 24.

2. Double-click the number in the edit box and change it to **24**.

As you have discovered by now, there are many other options. We have changed only those needed for the dimension style we will use to dimension the redwood deck.

3. Pick **OK**; pick **Close**.

4. Save your work.

> **INFOLINK**
>
> You will learn more about dimension style options in Chapters 27 and 28.

Applying a Dimension Style

Now that you have made several adjustments to the dimension style, you are ready to dimension the drawing.

1. Make layer **Dimensions** the current layer.

2. Begin by dimensioning the table and rounded corner. Use the **Diameter Dimension** button for the table and the **Radius Dimension** button for the corner. Place the dimensions as shown in Fig. 26-1 (page 383).

Now that you have center lines, you can use them to locate the centers of the table and rounded corner using dimensions.

3. Pick the **Linear Dimension** button from the Dimension toolbar.

4. For the first and second extension line origins, snap to points A and B, and then place the dimension **1.5'** from the edge of the deck.

Dimension

Dimension

Dimension

 HINT: ——————————————————————————

Use grid or snap to determine the 1.5' distance.

Dimension

5. Pick the **Continue Dimension** button, snap to point C, and press **ENTER** twice to terminate the command.

6. Using a similar method, add the remaining horizontal and vertical dimensions.

HINT:

When locating the center of the rounded corner with dimensions, use the end of the center lines for the extension line origins, similar to the way you picked point B. Separately, using the grips method of moving dimension text, move the 9'-6" text of the vertical dimension so that it is lower than the 2'-6" dimension beside it.

7. Add the aligned and angular dimensions.

Your dimensioned drawing of the redwood deck should now look similar to the one at the beginning of this chapter.

8. Save your work.

9. If you have access to a plotter or printer, plot the drawing.

Editing a Dimension

Using AutoCAD's grip feature, you can change the location of dimension lines and dimension text.

1. Pick one of the linear dimensions.

Five grip boxes appear.

2. Pick the grip box located at the center of the dimension text.

3. Move the text up and down and back and forth, and then pick a new point.

4. Press **ESC** twice.

By selecting a dimension and right-clicking it, you can make other useful changes.

5. Select one of the linear dimensions and right-click.

A shortcut menu appears.

6. Rest the pointer on **Dim Text position**.

These options permit you to adjust the location of the dimension text.

7. Try each of them, and pick **Home text** last.

As you can see, it's easy to tailor individual dimensions, regardless of the current dimension style.

8. Select one of the linear dimensions and right-click.

9. Rest the pointer on **Dim Style**.

This allows you to save a new style or apply an existing one.

10. Select **STANDARD**.

The dimension uses settings stored in the STANDARD dimension style. The arrowheads and dimension text are there, but you can't see them because they are so small. The value of the dimension scale in STANDARD is 1, and we set it to 24 in My Style. This means that the text, arrowheads, and other dimensioning features are 24 times larger in My Style.

11. Change the dimension's style back to **My Style**.

Leaders

Leaders are lines with arrowheads at one end that are used to point out a particular feature in a drawing. Radial and diameter dimensions usually use leaders to show the radius and diameter, respectively. The plastic retainer in Fig. 26-2 provides an example of a leader.

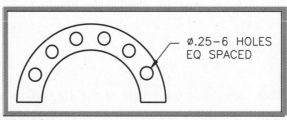

Fig. 26-2

Dimension

1. Pick the **Quick Leader** button on the Dimension toolbar.

This enters the QLEADER command.

2. Pick point A.

3. Move the crosshairs up and to the right about **1.5′** and pick a point.

4. Press **ENTER**.

AutoCAD now asks for the text width.

5. Enter **1'**.

6. In reply to Enter first line of annotation text, enter **USE REDWOOD**.

7. Press **ENTER** to display the text and terminate the command.

A leader appears with the note USE REDWOOD. Notice that the width of the note does not exceed 1'.

Using the Text Editor with Leader Text

You can also use AutoCAD's Multiline Text Editor to produce notes for leaders.

1. Undo the addition of the leader.

2. Repeat Steps 1 through 5 from the previous section.

3. When AutoCAD asks you to enter the first line of annotation text, press **ENTER**.

4. Enter **USE REDWOOD** and pick **OK**.

The leader and note appear.

Spline Leaders

A *spline leader* is similar to a regular leader, except the leader line can consist of a spline curve, giving you flexibility with its shape. Spline leaders permit you to produce leaders where space is tight on a drawing.

Dimension

1. Pick the **Quick Leader** button and enter **S** for Settings.

The Leader Settings dialog box appears.

2. On your own, review all of the parts of this dialog box, including each of the three tabs.

3. Pick the **Leader Line & Arrow** tab.

4. In the Arrowhead area, pick the down arrow to review the list of arrowhead options.

5. In the Leader Line area, pick the **Spline** radio button.

6. In the Number of Points area, pick the **No Limit** check box.

Now AutoCAD will not restrict the number of points you can use to create a spline leader.

7. Pick **OK**.

8. Pick three points anywhere on the screen to form a complex leader.

9. Press **ENTER** and **3'** for the text width.

10. Enter **THIS IS A SPLINE LEADER**.

11. Press **ENTER** to display the text and terminate the command.

12. Undo the addition of the spline leader and save your work.

Working with Associative Dimensions

By default, AutoCAD creates associative dimensions. *Associative dimensions* are dimensions that update automatically when you change the drawing by stretching, scaling, and so on.

1. Using the **SCALE** command, scale the drawing by **1.1**.

HINT:

In reply to Select objects, enter all. Use 0.0 for the base point and 1.1 for the scale factor.

Notice that the dimensions changed automatically. This becomes useful when you need to change a drawing after you've dimensioned it.

2. Using the **STRETCH** command, stretch the rightmost part of the drawing a short distance to the right. (When selecting, use a crossing window.)

The dimensions change accordingly.

3. Undo Steps 1 and 2.

Adding Dimension Styles to Templates

Now you know how to perform the remaining steps (12 through 14) in creating a template file (see page 303). Let's apply these steps to the tmp1.dwt template.

1. Open **tmp1.dwt**.

HINT: ────────────────────────────────

When the Select File dialog box appears, you will need to select Drawing Template File (*.dwt) at the right of Files of type.

Dimension

2. Create a new dimension style using the following information.
   - Name the dimension style **Preferred**. In the New Dimension Style dialog box, start with **STANDARD** and apply the new style to all dimensions.
   - In the **Primary Units** tab, set Unit format to **Architectural** and Precision to **0'-0"**.
   - In the **Lines and Arrows** tab, set the dimension arrow size to ¹/₈". Use full center lines (versus center marks only) and set the size of the center marks to ¹/₁₆".
   - In the **Text** tab, create a new text style named **roms** using the **romans.shx** font. Use the default settings and make **roms** the current style for Preferred. Set the text height at ¹/₈".
   - In the **Fit** tab, set Scale for Dimension Features to **24**.

3. Close the dialog boxes.

You could make other changes to the dimension style, but let's stop here.

4. In the Dimension toolbar, pick the down arrow and select **Preferred**.

5. Save your work.

 HINT: ────────────────────────────────

AutoCAD is expecting you to save a drawing file. Therefore, you will need to select AutoCAD Drawing Template File (*.dwt) at the right of File name. Also, you will need to locate the folder with your name before saving the file.

The tmp1.dwt file is now ready for use with other new drawing files. As you use template files, feel free to tailor them to your specific needs.

6. Close the Dimension toolbar and exit AutoCAD.

Chapter 26
Review & Activities

Review Questions

1. Describe the purpose of dimension styles. Why might you need to create one?

2. How do you determine the dimensioning scale?

3. Explain the difference between center marks and center lines. When should you use each?

4. Explain the use of the Continue Dimension button in the Dimension toolbar.

5. When you select a dimension and right-click, AutoCAD displays a menu. Name and briefly describe the three dimension-related options that are available to you.

6. When might you use a spline leader?

7. What is the advantage of using associative dimensions?

8. What is the fastest way to change the position of a dimension line?

Challenge Your Thinking

1. In this chapter, you placed object lines on one layer and dimensions on another. Discuss the advantages of placing dimensions on a separate layer. Are there any disadvantages to placing dimensions on a separate layer? Explain.

2. Describe the changes you would need to make to tmp1.dwt to create a template for a B-size sheet at the same drawing scale.

Chapter 26
Review & Activities

Applying AutoCAD Skills

Work the following problems to practice the commands and skills you learned in this chapter.

1. Begin a new drawing and establish the following drawing settings. Store as a template file. (You could name it tmp2.dwt.)

    Units: Engineering
    Scale: 1″ = 10′ (or 1″ = 120″)
    Sheet size: 17″ × 11″
    Drawing Area: You determine the drawing area based on the scale and sheet size.
    Grid: 10′
    (Reminder: Be sure to enter ZOOM All.)
    Snap: 2′
    Layers: Set up as shown in Table 26-1.

Layer Name	Color	Linetype	Lineweight
Thick	red	CONTINUOUS	0.50 mm
Thin	green	CONTINUOUS	0.25 mm

Table 26-1

Create a new dimension style for the template using the following information.

- Name the dimension style D001. In the New Dimension Style dialog box, start with STANDARD and apply the new style to all dimensions.
- Set the unit format to Engineering and the precision to 0′-0.00″.
- Set the dimension arrow size to .125″.
- Use center marks and set their size to .1″.
- Create a new text style named ariel using the Ariel font. Use the default settings and make arial the current style for NewStyle. Set the text height at .125″.
- Set the dimension scale to the proper value.

Chapter 26
Review & Activities continued

2–3. Use the template file you created in problem 2 to create the hotel recreation areas shown in Fig. 26-3 and 26-4.

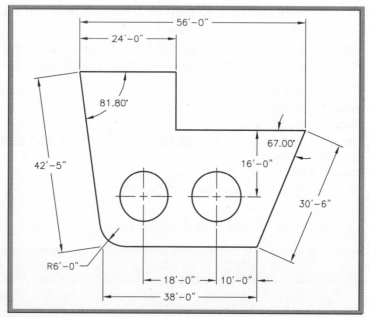

Fig. 26-3

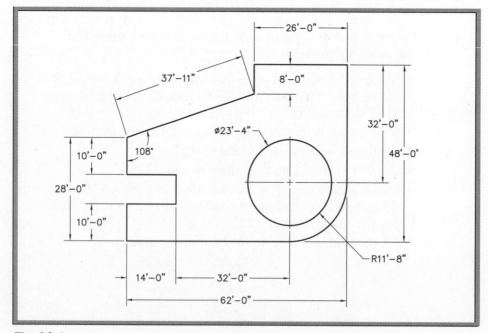

Fig. 26-4

4. Create and identify the radii of the sheet metal elbow shown in Fig. 26-5. First draw the perpendicular lines, and then create two arcs: one with a radius of 1.5 and the other with a radius of 3. Then dimension the 90° angle and add the notes. Before you create the leaders, open the Dimension Styles dialog box and change the arrowheads to dots. Add the text as shown, and save the drawing as elbow1.dwg.

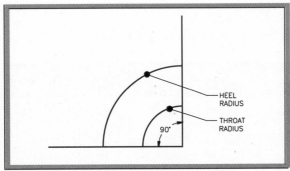

Fig. 26-5

5. Draw the front view of the shaft shown in Fig. 26-6. Use the Quick Setup method to change the drawing area to accommodate the dimensions of the drawing. Place the visible lines on a layer named Visible using the color green. Create a layer named Forces using the color red. Draw and label the four load and reaction forces, using leaders and a text style based on the romans.shx font. Create another layer named Dimensions using the color cyan. Dimension the front view, placing the dimensions on layer Dimensions.

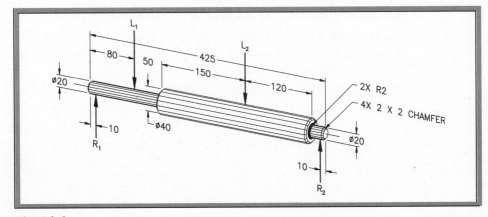

Fig. 26-6

Problem 5 courtesy of Gary J. Hordemann, Gonzaga University

Chapter 26
Review & Activities continued

6. Draw and dimension the front view of the metal casting shown in Fig. 26-7. Use the following drawing and dimension settings, and place everything on the proper layer.

Units: Decimal with two digits to the right of the decimal point
Drawing Area: 11 × 8.5
Scale: 1:1
Grid: .1
Snap: .05
LTSCALE: Start with .5
Layers: Set up as shown in Table 26-2.

Layer Name	Color	Linetype	Lineweight
Visible	red	CONTINUOUS	0.50 mm
Dimensions	blue	CONTINUOUS	0.25 mm
Text	magenta	CONTINUOUS	0.30 mm
Center	green	CENTER	0.20 mm

Table 26-2

Create a new dimension style using the following information.

- Name the dimension style using a name of your choice. Choose a name that will help you remember the purpose of this dimension style. In the New Dimension Style dialog box, start with STANDARD and apply the new style to all dimensions.
- Set the unit format to Decimal and the precision to 0.00.
- Set the dimension arrow size to .125.
- Use center lines and set the size of the center marks to .1.
- Create a new text style using a name of your choice. Use the romans.shx font, apply default settings, and make it the current style. Set the text height at .125.
- Set the dimension scale to the proper value.

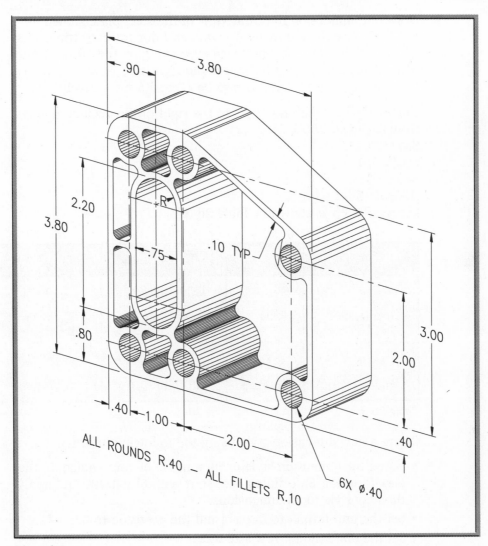

Fig. 26-7

Problem 6 courtesy of Gary J. Hordemann, Gonzaga University

Chapter 26
Review & Activities continued

7. Draw and dimension the front view of the steel plate shown in Fig. 26-8. Use the following drawing and dimension settings to set up the metric drawing, and place everything on the proper layers. Also draw and dimension the top and right-side views, assuming the plate and boss thicknesses to be 10 and 5 respectively.

Units: Decimal with no digits to the right of the decimal point
Drawing Area: 280 × 216
Scale: 1:1
Grid: 10
Snap: 5
LTSCALE: Start with 10
Layers: Set up as shown in Table 26-3.

Layer Name	Color	Linetype	Lineweight
Visible	red	CONTINUOUS	0.60 mm
Dimensions	blue	CONTINUOUS	0.20 mm
Text	magenta	CONTINUOUS	0.20 mm
Center	green	CENTER	0.20 mm
Hidden	magenta	HIDDEN	0.30 mm

Table 26-3

Create a new dimension style using the following information.

- Name the dimension style using a name of your choice. In the New Dimension Style dialog box, start with STANDARD and apply the new style to all dimensions.
- Set the unit format to Decimal and the precision to 0.
- Set the dimension arrow size to 3.
- Use center lines and set the size of the center marks to 1.5.
- Create a new text style using a name of your choice. Use the romans.shx font, apply default settings, and make it the current style. Set the text height at 3.
- Set the dimension scale to the proper value.

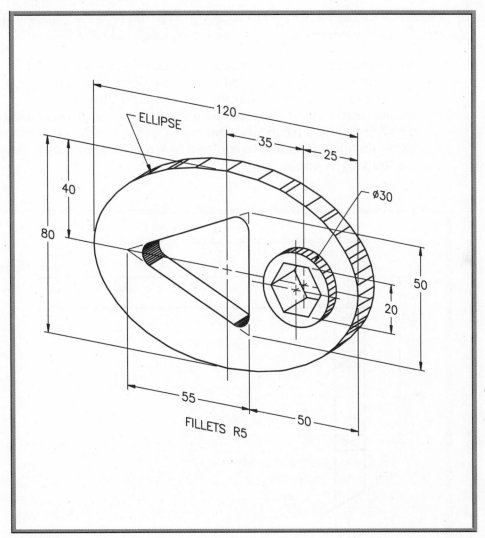

Fig. 26-8

Problem 7 courtesy of Gary J. Hordemann, Gonzaga University

Chapter 26
Review & Activities continued

Using Problem-Solving Skills

1. Your architectural drafting trainee showed you the wall corner shown in Fig. 26-9. Notice that the trainee dimensioned the drawing in a mechanical drafting style. Redraw the corner and set the correct dimension size and units, change the arrowheads to architectural ticks (short diagonal lines), and move the dimensions above the dimension line to conform to your company's style. Save the drawing as framing1.dwg.

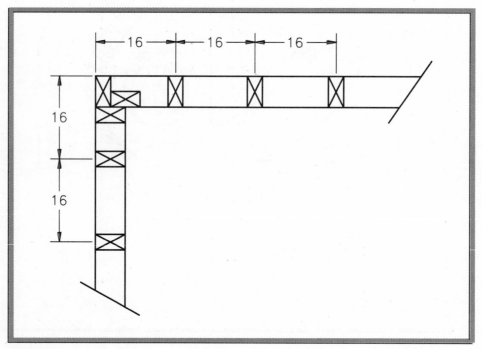

Fig. 26-9

2. As the set designer for a theatrical company, you are creating a template for a cityscape. The director has given you a sketch of the general building landscape she would like (Fig. 26-10). Draw the template and dimension it from the datum plane. Each square equals 9 inches. Save and plot the drawing using the Extents plotting option.

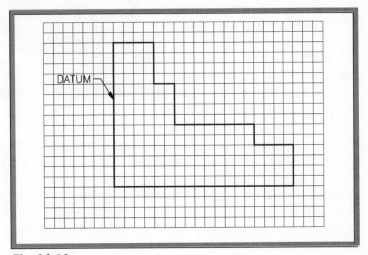

Fig. 26-10

Careers
Using
AutoCAD

Mapping Underground Mines

In underground mining, you can't see where you're going until you get there. A miner needs good maps to show any previous tunneling, which could affect the strength of a new tunnel's walls. One thing that a tunneler does not want to stumble across unexpectedly is somebody else's mine.

Through the last decade, AutoCAD has come into widespread use as a mapping tool. Companies that act as suppliers or contractors to mines have adopted AutoCAD as well. The result is not only increased safety, but reduced costs. In mining, the line between profit and loss is determined by costs.

As an example, if the cost of getting gold out of the tunnel exceeds the market price for gold, the mine shuts down. However, if mining engineers can find a way to extract the gold at a lower cost, then the mine can resume operations. Continually reducing costs relative to payback is especially important in mining because it is a finite-resource industry.

Continued supplies depend on mining ore that is less pure or is harder to reach, and it has to be done with less waste and lower costs. The same dilemma faces all finite-resource industries, such as agriculture (limited arable soil) and energy (limited fossil fuels).

Mining engineers keep trying to find ways to do their job better; they have to because the demand for minerals continues. As miners say: "Everything in your world either comes from a farm or out of the ground."

Career Focus—Mining Engineer

Mining engineers find, extract, and prepare minerals for manufacturers to use. Mining engineers design open-pit and underground mines, supervise construction and excavation, and devise systems for delivering ore to a processing plant. Mining engineers have overall

The Stock Market/Lester Lefkowitz

responsibility for the safe and economical operation of mines, including ventilation, water, power, communications, and equipment maintenance. Mining engineers often specialize by mineral; for example, an engineer may work only in coal or metal mining.

Education and Training

A bachelor's degree in engineering is required for most jobs as a mining engineer. A degree from a school that emphasizes mining will improve readiness, as will work experience with a mining or related company. After learning the characteristics of a site and the mining methods used there, a mining engineer takes on more responsibility in designing systems for new or expanded operations.

 Career Activity

- Interview someone who can tell you about different jobs held by people in a mining company.
- What kinds of work must be done before the ore is dug?
- What kinds of work are required to deliver the ore for processing?

Chapter 27 Fine-Tuning Dimensions

Objectives:

- **Adjust the appearance of dimension lines, arrowheads, and text**
- **Control the placement of dimension text**
- **Adjust the format and precision of primary and alternate units**
- **Edit a dimension's properties**
- **Explode a dimension**

Key Terms

alternate units

primary units

Most industries that use AutoCAD apply dimensioning differently. The architectural industry, for example, uses architectural units and fractions, whereas manufacturers of consumer and industrial products typically use decimal units. Within each industry, companies apply standards, some more rigidly than others. In manufacturing, most use a variation of ANSI or ISO standards and enforce the standard company-wide. Because thousands of companies use AutoCAD worldwide, the software must accommodate a wide range of industries and company standards. This chapter focuses on how you can fine-tune dimensions to meet a particular company's style or standard.

Dimension Style Options

AutoCAD offers a wealth of options for tailoring a dimension style to a specific need.

- Start AutoCAD and open **deck.dwg**.
- From the **File** pull-down menu, select **Save As...** and enter **deck2.dwg** for the file name.
- Open the **Dimension** toolbar and pick the **Dimension Style** button.

Dimension

The current dimension style is My Style, as shown in the upper left corner of the dialog box. In the last chapter, we covered the purpose of the New... and Modify... buttons. The Set Current button permits you to make another style current. The Override... button enables you to override one or more settings in a dimension style temporarily. The Compare... button allows you to compare the properties of two dimension styles or view all the properties of one style.

Lines and Arrows

1. Pick the **Modify...** button and pick the **Lines and Arrows** tab (Fig. 27-1).

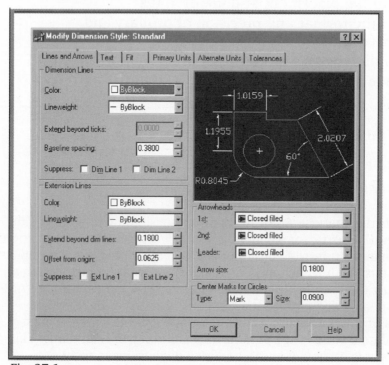

Fig. 27-1

Focus your attention on the Dimension Lines area. Color and Lineweight permit you to change the color and lineweight of dimension lines. In most cases, it is best not to change them. Extend beyond ticks specifies a distance to extend the dimension line past the extension line when you use oblique, architectural, tick, integral, and no marks for arrowheads. Some companies and individual architects use ticks instead of arrows. Oblique is a less bold variation of a tick. Integral is a name for a curved tick. It would be rare to use None, but AutoCAD's flexibility gives you that option. Baseline spacing sets the spacing between the dimension lines when they're outside the extension lines. Dim Line 1 suppresses the first dimension line, and Dim Line 2 suppresses the second dimension line, as shown in Fig. 27-2.

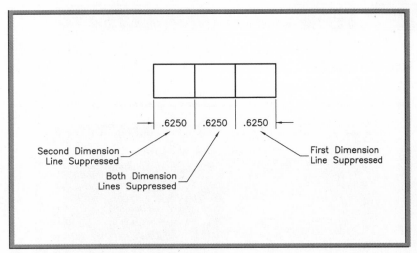

Fig. 27-2

2. Make changes to the settings in the Dimension Lines area and notice how the sample drawing changes.

Focus on the area labeled Extension Lines. Color and Lineweight allow you to change the color and lineweight of the extension lines. Extend beyond dim lines specifies a distance to extend the extension lines above the dimension line. Offset from origin specifies the distance to offset the extension lines from the origin points that define the dimension. Suppress prevents the display of extension lines. Ext Line 1 suppresses the first extension line, and Ext Line 2 suppresses the second extension line. Suppressing an extension line is useful when you want to dimension to an object line. In this case, the object line would serve the same purpose as the extension line.

3. Make changes to the settings in the Extension Lines area and notice how the sample drawing changes.

Concentrate on the area titled Arrowheads. 1st sets the arrowhead for the first end of a dimension line, while 2nd sets the arrowhead for the second end of a dimension line. In most cases, you would use the same arrowhead style for both ends, but AutoCAD gives you flexibility in case you need to make them different. Leader sets the arrowhead for leaders. As you learned in the previous chapter, Arrow size displays and sets the size of the arrow.

4. Make changes to the settings in the Arrowhead area and notice how they change the sample drawing.

Text

1. Pick the Text tab (Fig. 27-3).

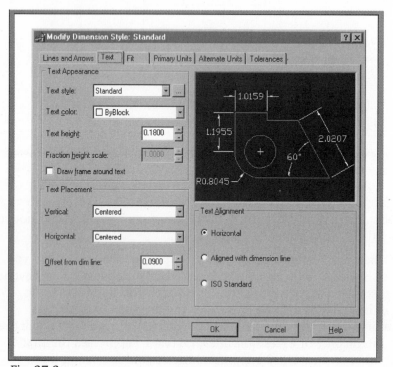

Fig. 27-3

In the Text Appearance area, you are familiar with Text style and Text height. Text color permits you to change the color of the dimension text. In most circumstances, it is best to leave it alone. Fraction height scale sets the scale of fractions relative to dimension text by multiplying the value entered here by the text height. Draw frame around text draws a frame around the dimension text. This option would be useful to companies that frame certain notes in a drawing.

2. Make changes to the settings in the Text Appearance area and notice the changes in the sample drawing.

The Text Placement area allows you to control the placement of text. Vertical controls the vertical justification of dimension text along the dimension line. Horizontal controls the horizontal justification of dimension text along the dimension line and the extension line. Offset from dim line sets the text gap (distance) around the dimension text when the dimension line is broken to accommodate the dimension text. If the space between the dimension text and the end of the dimension lines becomes too large or small, use this option to adjust the space. It is rare for a company or even a government agency to require a specific distance.

3. Make changes to the settings in the Text Placement area and notice the changes in the sample drawing.

The Text Alignment area permits you to change the orientation (horizontal or aligned) of dimension text, whether it is inside or outside the extension lines. Horizontal places text in a horizontal position. Aligned with dimension line aligns the text with the dimension line. ISO Standard aligns text with the dimension line when text is inside the extension lines. If the text is outside the extension lines, it aligns it horizontally.

4. Experiment with these options.

Fit Tab

1. Pick the **Fit** tab (Fig. 27-4).

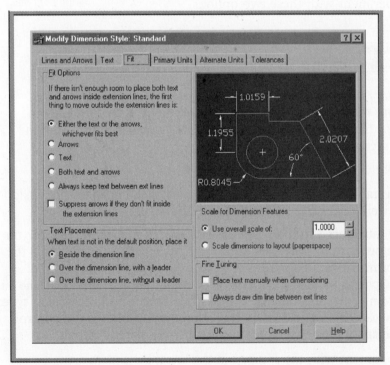

Fig. 27-4

Focus on the Fit Options area. The Fit options permit you to control the placement of dimension text and arrows when there is not enough space to place them inside the extension lines.

2. Select each of the options and notice how they affect the placement of dimension text and arrows.

The Text Placement area allows you to control the placement of text when it's not in the default position.

3. Try each of these options.

As you discovered in Chapter 26, the Scale for Dimension Features area controls the scaling of dimensions. When you pick Scale dimension to layout (paperspace), AutoCAD determines a scale factor based on the scaling between the current model space viewport and paper space.

In the Fine Tuning area, if you check Place text manually when dimensioning, AutoCAD allows you to place the text at the position you specify. If you check Always draw a dim line between ext lines, AutoCAD draws dimension lines between the measured points even when the arrowheads are outside the extension lines.

4. Experiment with these two options.

Primary Units

The *primary units* are those that appear by default when you add dimensions to a drawing. The term *primary* is used by AutoCAD to distinguish them from alternate dimensions, which are discussed in the next section.

1. Pick the **Primary Units** tab (Fig. 27-5).

This tab establishes the format and precision of the primary dimension units, as well as prefixes and suffixes for dimension text.

You were introduced to the Linear Dimensions area in Chapter 26, so you know the purpose of Unit format and Precision. Fraction format controls the format for fractions. Decimal separator sets the separator for decimal formats. Options include a period (.), comma (,), or space( ). Round off sets rounding rules for dimension measurements for all dimension types except Angular. Prefix and Suffix permit you to enter text or use control codes to display special symbols such as the diameter symbol.

2. Experiment with the settings in the Linear Dimensions area.

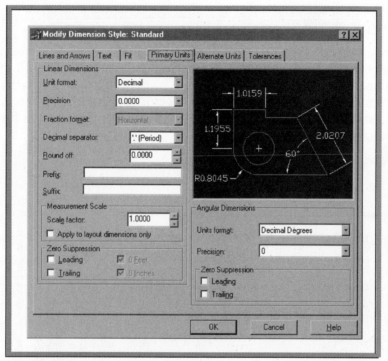

Fig. 27-5

Under Measurement Scale, Scale factor establishes a scale factor for all dimension types except Angular.

Zero Suppression controls the suppression of zeros. Some companies do not use a leading zero, for example. One widely accepted rule of thumb is to include the leading zero in drawings dimensioned using the metric system, but to exclude it when dimensioning using the English system (feet and inches). Space (or lack of it) for the dimension may also be a consideration.

A distance of 0.2000 becomes .2000 when you check Leading. A distance of 2.0000 becomes 2 when you check Trailing. A distance of 0'-2 ¹/₄" becomes 2 ¹/₄" when you check 0 Feet, and a distance of 2'-0" becomes 2' when you check 0 Inches.

Concentrate on the area labeled Angular Dimensions. Units format sets the angular units format, and Precision sets the precision for the units. The Zero suppression area controls the suppression of zeros in angular dimensions.

3. Make changes to the settings in the Angular Dimensions area.

Alternate Units

Alternate units are a second set of distances inside brackets in dimensions. The most common method of using alternate units is to show both inches and millimeters, so millimeters are shown by default when you activate alternate units. However, the conversion factor is customizable, as you will see, so that you can change the alternate units to meet needs for specific drawings.

1. Pick the **Alternate Units** tab (Fig. 27-6).

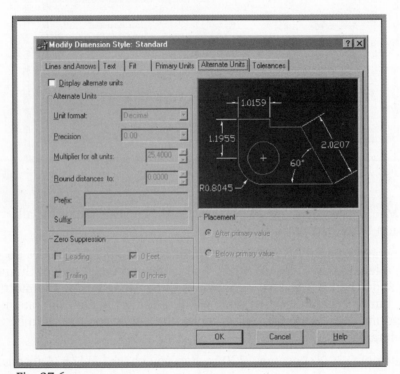

Fig. 27-6

2. Pick the **Display alternate units** check box in the upper left corner.

This displays alternate units in the sample drawing. The alternate units, in millimeters, are in brackets.

Unit format and Precision in the Alternate Units area are self-explanatory. Multiplier for alt units sets the multiplier to use as the conversion factor between primary and alternate units. Since an inch is equal to 25.4 millimeters, the default value is 25.4. Round distances to establishes the rounding rules for alternate units for all dimension types except Angular. Prefix and Suffix permit you to include a prefix and suffix in the alternate dimension text.

3. Experiment with the settings in the Alternate Units area.

The Zero Suppression area is the same as the one for Primary Units.

In the Placement area, After primary value places the alternate units after the primary units. Below primary value places alternate units below the primary units.

4. Pick **Below primary value** to compare the difference between the two options.

You will learn about the Tolerance tab in Chapter 28.

5. Pick the **Cancel** button; pick the **Close** button to close the Dimension Style Manager dialog box.

Editing a Dimension

AutoCAD offers additional options for editing a dimension.

Dimension

1. Pick the **Dimension Edit** button from the Dimension toolbar.

AutoCAD enters the DIMEDIT command and displays several Command line options.

Home moves dimension text to its default position. New allows you to change dimension text using the Multiline Text Editor. Rotate rotates dimension text. Oblique makes the extension lines of linear dimensions oblique instead of perpendicular to the dimension line.

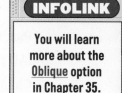

INFOLINK

You will learn more about the Oblique option in Chapter 35.

2. Press **ESC** to end the command.

Changing a Dimension's Properties

1. Select any linear dimension.

2. Right-click and select **Properties** at the bottom of the shortcut menu.

AutoCAD displays the Properties dialog box, as shown in Fig. 27-7.

Fig. 27-7

3. Pick the **Categorized** tab.

AutoCAD provides a list of the dimension's properties.

4. Change the color and lineweight to new values of your choice.

5. Click the small box with the plus sign in it at the left of Text.

This displays all of the settings associated with the dimension's text.

6. Change the height to $^3/_{16}''$.

7. Click the small box located at the left of Primary Units, Alternate Units, Fit, and Lines & Arrows.

As you can see, you can easily custom-tailor the properties of an individual dimension. In most circumstances, all of the dimensions on a drawing should use a consistent style, but AutoCAD gives you the option of changing the individual elements of a particular dimension if necessary.

8. Pick the **Alphabetic** tab.

This displays a detailed listing of the same properties, but this list is in alphabetical order.

9. Close the dialog box.

Exploding a Dimension

1. Attempt to erase a single element of any dimension on the drawing, such as the dimension text or an extension line.

AutoCAD selects the entire dimension because it treats an associative dimension as a single object.

2. Pick the **Explode** button from the docked Modify toolbar.

This enters the EXPLODE command.

Modify

3. Select any dimension and press **ENTER**.

AutoCAD explodes the dimension, enabling you to edit (move, erase, trim, etc.) the individual parts of the dimension.

4. Select the dimension.

As you can see, it's now possible to select individual parts of the dimension. Sometimes, it's necessary to explode a dimension to edit it because it is impossible to change it any other way. Use this method, however, as the last alternative because exploded dimensions lose their associativity and you cannot control their appearance using a dimension style.

5. Close the Dimension toolbar.

6. Exit AutoCAD without saving your changes.

Chapter 27
Review & Activities

Review Questions

1. Why is it important for AutoCAD to offer so many settings that control the appearance of dimensions?

2. What are alternate units?

3. Can you rotate dimension text? Describe a drawing situation in which you might need to do this.

4. What is the purpose of the New option of the DIMEDIT command?

5. How do you adjust the angle of a dimension's extension lines? Why might you choose to make this adjustment?

6. Will the Home option of the DIMEDIT command work with a dimension that is not an associative dimension? Explain.

7. What is the fastest way of reviewing and changing the properties of an individual dimension?

8. Under what circumstances would you want to suppress one or both dimension lines? Explain.

Challenge Your Thinking

1. A drawing of a complex automotive part has been completed with AutoCAD using the English inch system. Without redimensioning the drawing, how could you change it to show metric units only?

2. Contact two different companies in your area that use CAD and find out what their company dimensioning standards are. How do the two companies compare? How might any differences be explained?

Chapter 27
Review & Activities

Applying AutoCAD Skills

Work the following problems to practice the commands and skills you learned in this chapter.

1. Create a new drawing named prb27-1.dwg. Plan for a drawing scale of 1″ = 1″ and sheet size of 11″ × 17″. After you apply the following settings, create and dimension the drawing of an aircraft door hinge shown in Fig. 27-8. In addition to saving prb27-1.dwg as a drawing file, save it as a template file.

Snap:	.25
Grid:	1
Layers:	Create layers to accommodate multiple colors, linetypes, and lineweights.
Linetype scale:	.5
Font:	romans.shx
Dimension style name:	Heather
Dimension scale:	1
Dimension text height:	.16
Arrowhead size:	.16
Center mark size:	.08
Mark with center lines?	Yes

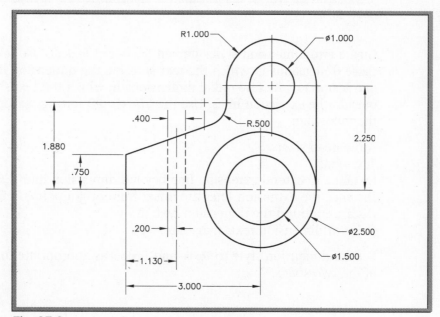

Fig. 27-8

Chapter 27
Review & Activities continued

2. Draw and dimension the front view of the oscillating follower shown in Fig. 27-9. Use the drawing settings shown below. Draw the top and right-side views, assuming the follower to have a thickness of 8. This is a metric drawing.

Units:	Decimal with no digits to the right of the decimal
Drawing area:	280 × 216
Grid:	10
Snap:	5
Linetype scale:	Start with 10
Text style:	Use the romans.shx font to create a new text style and set the height at 3.
Layers:	Set up layers as shown in Table 27-1.

Layer Name	Color	Linetype	Lineweight
Visible	red	CONTINUOUS	0.60 mm
Dimensions	blue	CONTINUOUS	0.20 mm
Text	magenta	CONTINUOUS	0.35 mm
Center	green	CENTER	0.20 mm
Hidden	magenta	HIDDEN	0.35 mm

Table 27-1

Create two dimension styles named Textin and Textout. Use Textin for those dimensions in which the text is inside the dimension lines, and Textout for the few radial dimensions in which the text should be outside the extension lines. Use the STANDARD settings except for the following:

Arrowhead size: 3
Text size: 3
Distance to extend extension line beyond dimension line: 1.5
Distance the extension lines are offset from origin points: 1.5
Distance around the dimension text: 1.5
Color of dimension text: magenta

Set the dimension style to Textin and Textout as appropriate for each of the two styles.

Problem 2 courtesy of Gary J. Hordemann, Gonzaga University

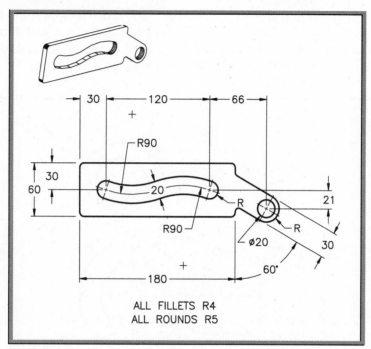

Fig. 27-9

3. Dimension the gasket as shown in Fig. 27-10. Set the arrowheads to Integral and the text font to Times New Roman. Display alternate units below the primary values and in the same unit format. Save the drawing as ch27gasket.dwg.

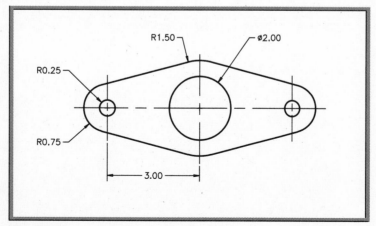

Fig. 27-10

Chapter 27
Review & Activities continued

4. Create a new drawing using the prb27-1.dwt template file and make the changes shown in Fig. 27-11. Note that the 2.250 dimension is now 2.500. Stretch the top part of the hinge upward .250 unit and let the associative dimension text change on its own.

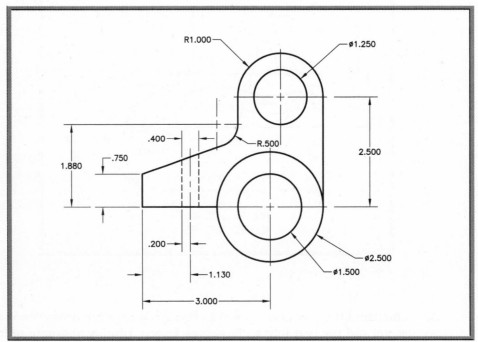

Fig. 27-11

Chapter 27
Review & Activities

Using Problem-Solving Skills

Work the following problems to practice the commands and skills you learned in this chapter.

1. Dimension the shim shown in Fig. 27-12. Set the font for Times New Roman. Apply Override as necessary. Save the drawing as ch27shim.dwg.

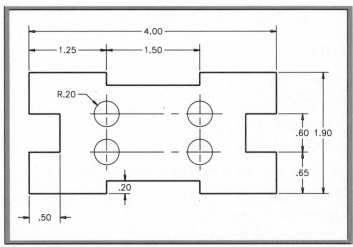

Fig. 27-12

2. Architects are noted for their individuality. Open the elevation drawing from Chapter 25. Redimension the elevation, including text, with the following style modifications. Save the drawing as Ch27elev-rev1.dwg.

   Text font name: Trekker
   Arrowheads: Architectural tick
   Dimension Lines: Extend beyond ticks ¹/₈
   Vertical Text Placement: Outside

Objectives:

- Apply the symmetrical, deviation, limits, and basic methods of tolerancing
- Insert surface controls on a drawing
- Create geometric characteristic symbols and feature control frames that follow industry standard practices for geometric dimensioning and tolerancing
- Add material condition symbols, datum references, data identifiers, and projected tolerance zones to drawings in accordance with industry standards

Key Terms

basic dimension

deviation tolerancing

feature control frames

geometric characteristic symbols

geometric dimensioning and tolerancing (GD&T)

limits tolerancing

material condition symbols

Maximum Material Condition (MMC)

projected tolerance zone

surface controls

symmetrical deviation

symmetrical tolerancing

tolerance

AutoCAD permits you to add tolerances to dimensions. A *tolerance* specifies the largest variation allowable for a given dimension. Tolerances are necessary on drawings that will be used to manufacture parts because some variation is normal in the manufacturing process.

This chapter applies AutoCAD's tolerancing capabilities and provides a basic introduction to *geometric dimensioning and tolerancing (GD&T)*. This is a system by which drafters and engineers define tolerances for the location of features (such as holes) of a part. The guidelines in this chapter reflect the standards that have been adopted by the American National Standards Institute (ANSI).

Methods of Tolerancing

Tolerances can be shown in more than one way, so AutoCAD provides many options for different kinds and styles of tolerances. The method you use depends on the individual drawing. In some cases, you can place the tolerance information in a note or title block. However, when many different tolerances apply to different parts of the drawing, you may need to specify them individually with the dimension information.

1. Start AutoCAD and use the **dimen.dwt** template file to create a new drawing.

2. Display the **Dimension** toolbar.

Dimension

3. Pick the **Dimension Style** button from the toolbar and pick the **New...** button.

4. Create a new dimension style named **toler**. Start with **STANDARD** and use it for all dimensions.

5. Pick the **Text** tab and change the text height to **0.125**.

6. Pick the **Primary Units** tab and change Precision to **0.000**.

7. Pick the **OK** button; pick the **Close** button.

8. In the Dimension toolbar, make **toler** the current dimension style.

9. Save your work in a file named **toler.dwg**.

10. Pick the **Dimension Style** button from the toolbar and pick the **Modify...** button.

11. Pick the **Tolerances** tab to display the tolerance options shown in Fig. 28-1.

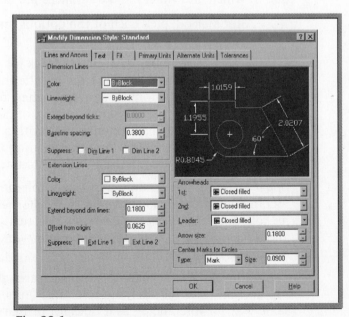

Fig. 28-1

Focus your attention on the area titled Tolerance Format. Notice that many of the settings are grayed out and not available. This is because Method is set to None. Method refers to the approach used to calculate the tolerance. The four methods that AutoCAD makes available, along with an example, are shown in Fig. 28-2.

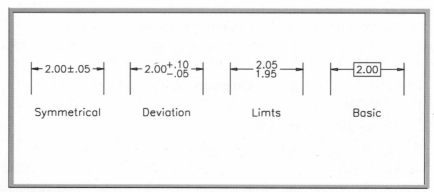

Fig. 28-2

The Symmetrical option permits you to create a plus/minus expression tolerance. AutoCAD applies a single value of variation to the dimension measurement. In other words, the upper and lower tolerances are equal. Deviation allows you to add different values for the upper and lower tolerance for cases in which the two are not equal. Limits adds a limit dimension with a maximum and a minimum value—one on top of the other. Basic creates a plain dimension inside a box. A *basic dimension* is one to which allowances and tolerances are added to obtain limits.

12. After Method, pick each of the options and notice how the sample drawing changes.

Precision sets the number of decimal places for tolerances. Precision varies with the tolerance precision and the numbering system you use. If you are using the U.S. customary system (inches, feet, etc.), the basic dimension should have the same number of decimal places as the tolerance. For example, a 1.5-inch hole with a tolerance of ±.005 should be written as 1.500 ±.005. Note that leading zeros (zeros to the left of the decimal point) are not used with customary measurements.

Upper value sets the maximum or upper tolerance value, and Lower value sets the minimum or lower value. Scaling for height sets the current height for tolerance text. Vertical position controls text justification for symmetrical and deviation tolerances. Zero Suppression works the same way as it does for primary and alternate units.

The settings in the Alternate Unit Tolerance area sets the precision and zero suppression rules for alternate tolerance units. Zero Suppression works the same way for tolerancing as it does for primary and alternate units.

13. Experiment with each of the options in the Tolerance tab and notice how their settings affect the sample drawing.

14. Pick the **Cancel** button to clear all changes that you made.

Symmetrical Tolerancing

In *symmetrical tolerancing* (also called *symmetrical deviation*), the dimension specifies a plus/minus (±) tolerance in which the upper and lower limits are equal.

1. Pick the **Modify...** button and pick the **Tolerances** tab.

2. Set Method to **Symmetrical** and Precision to **0.000**.

3. Set Upper value to **0.005**.

Lower value is not available because it is the same as the upper value by definition in symmetrical tolerancing.

4. Under Zero Suppression, check both **Leading** and **Trailing**, and do not change anything else.

5. Pick **OK**; pick **Close**.

Tolerances do not appear because the toler dimension style was created after the drawing was completed, and toler has not yet been applied to the drawing.

6. Select the drawing and dimensions.

7. In the Dimension toolbar, make **toler** the current dimension style and press **ESC** twice to remove the grips.

Your drawing should look similar to the one in Fig. 28-3.

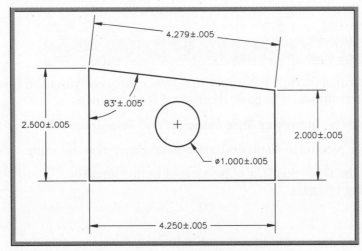

Fig. 28-3

Deviation Tolerancing

Sometimes the upper and lower tolerances for an object are unequal. AutoCAD's *deviation tolerancing* method allows you to set the upper and lower values separately.

1. Display the **Tolerances** tab of the Dimension Style Manager dialog box.

2. Change Method to **Deviation**, set Lower value to **0.007**, and leave everything else the same.

3. Close the Dimension Style Manager and notice how the drawing changes.

Your drawing should now look similar to the one in Fig. 28-4.

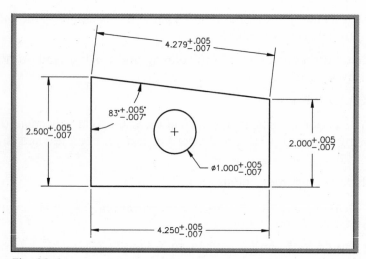

Fig. 28-4

Limits Tolerancing

In *limits tolerancing,* only the upper and lower limits of variation for a dimension are shown. The basic dimension is not shown.

1. Display the **Dimension Style Manager** and **Tolerances** tab.

2. Change Method to **Limits** and leave everything else the same.

3. Close the Dimension Style Manager and notice how the drawing changes.

Your drawing should now look similar to the one in Fig. 28-5.

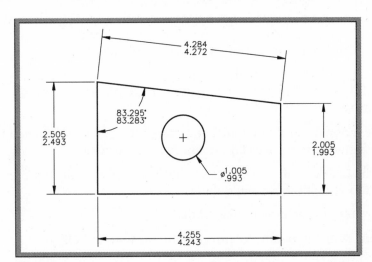

Fig. 28-5

Basic

When you select the Basic method, AutoCAD draws a box around the dimension. A *basic dimension* is one to which allowances and tolerances are added to obtain limits. Therefore, you should use the Basic method to present theoretically exact dimensions.

1. Display the **Tolerances** tab of the Dimension Style Manager dialog box.

2. Change Method to **Basic**.

3. Close the Dimension Style Manager and notice how the drawing changes.

GD&T Practices

With AutoCAD, you can create geometric characteristic symbols and feature control frames that follow industry standard practices for geometric dimensioning and tolerancing (GD&T). *Geometric characteristic symbols* are used to specify form and position tolerances on drawings. *Feature control frames* are the frames used to hold geometric characteristic symbols and their corresponding tolerances.

Surface Controls

When a feature control frame is not associated with a specific dimension, it usually refers to a surface specification regardless of feature size. For this reason, such frames are often known as *surface controls.*

In Chapter 26, you created simple and moderately complex leaders using the QLEADER command. You can also use QLEADER to create surface controls.

1. Create a new layer named **GDT** and assign the color magenta to it. Make it the current layer and freeze layer Dimensions.

2. Pick the **Quick Leader** button from the Dimension toolbar and press **ENTER** to accept the Settings default.

This displays the Leader Settings dialog box.

3. Pick the **Annotation** tab, **Tolerance** radio button, and **OK**.

4. Press **ENTER**.

5. In reply to Specify first leader point, pick the midpoint of the inclined line.

HINT:

Use the Midpoint object snap.

6. Pick a second point about **1** unit up and to the right of the first point.

7. Press **ENTER**.

The Geometric Tolerance dialog box appears, as shown in Fig. 28-6.

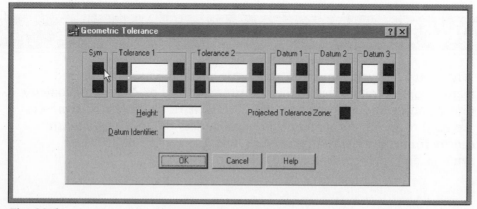

Fig. 28-6

This dialog box permits you to create complete feature control frames by adding tolerance values and their modifying symbols. Focus your attention on the area labeled Sym.

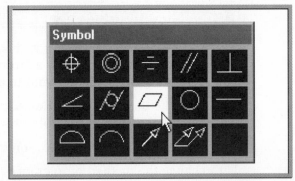

Fig. 28-7

8. Pick the top black box under Sym.

This displays the Symbol dialog box, as shown in Fig. 28-7.

The dialog box contains 14 geometric symbols that are commonly used in GD&T.

9. Pick the flatness symbol as shown in Fig. 28-7.

AutoCAD adds the symbol to the black box you picked under Sym.

10. Under Tolerance 1, in the top white box, enter **.020** for the tolerance value.

11. Pick **OK**.

INFOLINK

See Appendix D for more information about geometric symbols.

The feature control frame appears as a part of the leader, as shown in Fig. 28-8. The feature control frame specifies that every point on the surface must lie between two parallel planes .020 apart.

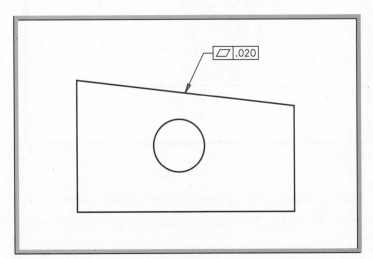

Fig. 28-8

Geometric Characteristic Symbols

To associate geometric symbols with specific dimensions, you must first create the dimension using the appropriate dimensioning command. Then you use the TOLERANCE command to add the tolerancing information.

Dimension

1. Pick the **Diameter Dimension** button from the toolbar and pick a point anywhere on the hole.

2. Pick a second point to position the dimension as shown in Fig. 28-9.

Let's change the method of tolerance from Basic to Limits.

3. Select the new dimension, right-click, and pick **Properties** from the menu.

4. In the Properties dialog box, pick the small box at the left of Tolerances, change Tolerance display from **Basic** to **Limits**, and close the menu.

5. Press **ESC** twice to remove the selection.

Your drawing should now look similar to the one in Fig. 28-9.

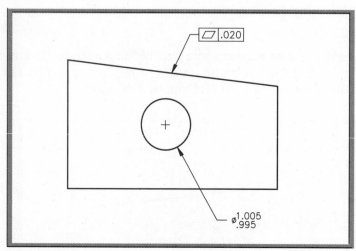

Fig. 28-9

Dimension

6. Pick the **Tolerance** button from the Dimension toolbar.

This enters the TOLERANCE command and displays the Geometric Tolerance dialog box.

7. Under Sym, pick the top black box and pick the true position symbol located in the upper left corner of the Symbol box.

8. Under Tolerance 1, pick the black box located in the upper left corner.

The diameter symbol appears.

9. Pick it again, and again, causing it to disappear and reappear.

As you can see, this serves as a switch, toggling the diameter symbol on and off.

10. In the top white box under Tolerance 1, enter **.010** for the tolerance value.

Material Condition Symbols

The black box located at the right of the white text box specifies material condition. *Material condition symbols* are used in GD&T to modify the geometric tolerance in relation to the produced size or location of the feature. In this case, we will use the symbol for *Maximum Material Condition (MMC)*. MMC specifies that a feature, such as a hole or shaft, is at its maximum size or contains its maximum amount of material.

11. At the right of the same white box, pick the black box.

This displays the Material Condition dialog box, as shown in Fig. 28-10.

12. Pick the first symbol, which is the symbol for Maximum Material Condition.

The symbol appears in the Geometric Tolerance dialog box.

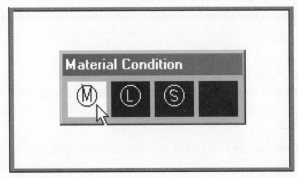

Fig. 28-10

Datum References and Datum Identifiers

AutoCAD allows you to specify datum reference information on three different levels (primary, secondary, and tertiary).

13. In the top white boxes under Datum 1, Datum 2, and Datum 3, enter **A**, **B**, and **C**, respectively.

You can also associate a material condition with any or all of the datum references. To do this, you would pick the black box to the right of the datum boxes and select the appropriate material condition symbol. For the current drawing, do not use material condition symbols.

Datum Identifier near the bottom left of the dialog box allows you to enter a feature symbol to identify the datum.

14. In the box located to the right of Datum Identifier, enter **-D-**.

The Geometric Tolerance dialog box should now look like the one in Fig. 28-11.

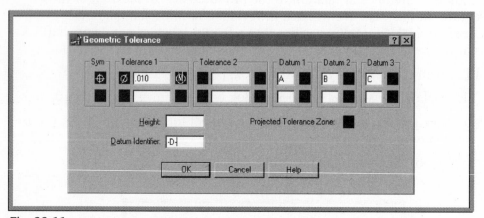

Fig. 28-11

The second row of black and white boxes permit you to create a second feature control frame. You would need to use a second feature control frame when the feature has two geometric characteristics that are of dimensional importance.

15. Pick the **OK** button and position the feature control frame as shown in Fig. 28-12.

16. Save your work.

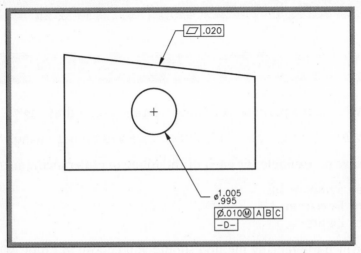

Fig. 28-12

Projected Tolerance Zone

You can specify projected tolerances in addition to positional tolerances
to make a tolerance more specific. The *projected tolerance zone* controls
the height of the extended portion of a perpendicular part. To display
a projected tolerance, you would pick the black box next to Projected
Tolerance Zone in the Geometric Tolerance dialog box. Then, under the
Tolerance 1 area of the dialog box, enter a value for the height. Doing
so creates a projected tolerance zone in the feature control frame.

NOTE:

To experiment with the projected tolerance zone, first save your
work. Then enter the UNDO command and Mark option. Create
another feature control frame for practice. Then experiment with the
projected tolerance zone by picking the black box next to Projected
Tolerance Zone and entering a height in the Height box. Place a
dimension to see how the projected tolerance appears. When you
have finished experimenting, enter the UNDO command and specify
the Back option.

17. Close the Dimension toolbar and exit AutoCAD.

Chapter 28
Review & Activities

Review Questions

1. What is the purpose of adding tolerances to a drawing?

2. When is it necessary to include tolerances on a drawing?

3. Give an example for each of the following tolerancing methods.

   a. Symmetrical
   b. Deviation
   c. Limits

4. Which tolerancing method should you choose in AutoCAD if the upper tolerance value of a dimension is different from the lower value?

5. What is a surface control? How can you create a surface control in AutoCAD?

6. How is the TOLERANCE command similar to the QLEADER command?

7. What is a geometric characteristic symbol?

8. Explain how you would include material condition symbols in feature control frames.

9. How would you include a second feature control frame below the first one?

Challenge Your Thinking

1. With AutoCAD, you can quickly produce feature control frames, complete with geometric characteristic symbols. Investigate ways of editing them.

2. When a second feature control frame is added to a dimension, how does this affect the meaning of the first feature control frame?

3. Investigate material condition symbols and their meanings. In what way do material conditions affect dimensions?

Chapter 28
Review & Activities

Applying AutoCAD Skills

Work the following problems to practice the commands and skills you learned in this chapter.

1. Create the two-view drawing of the plug according to the dimensions shown in Fig. 28-13A. Use a drawing area of 100,80, and name the drawing plug.dwg. Dimension it as shown in Fig. 28-13B to show the runout of the plug. (*Runout* is the form and location of a feature relative to a datum.) Notice that the limits are different on the two holes.

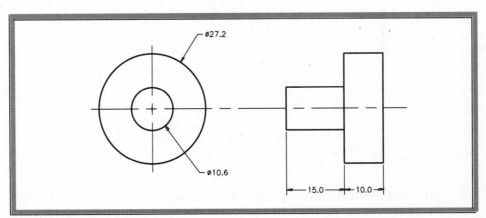

Fig. 28-13A

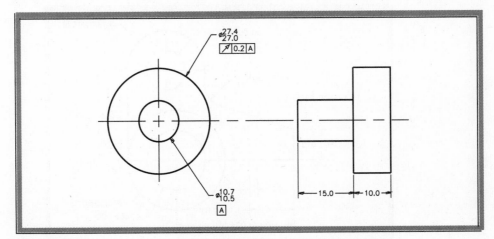

Fig. 28-13B

Chapter 28
Review & Activities continued

2. Create a new drawing named Ch28gasket.dwg and draw the gasket shown in Fig. 28-14. Dimension the gasket using limits tolerancing, and specify an upper value of .02. Create a feature control frame to show a true position diameter tolerance of .025 at maximum material condition.

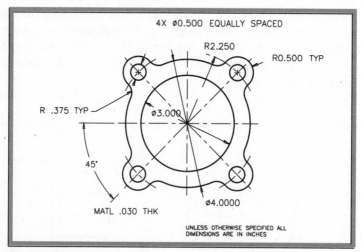

Fig. 28-14

3. Create a new drawing named hinge.dwg and draw the aircraft door hinge shown in Fig. 28-15. Dimension the hinge using symmetrical tolerancing and specify tolerances as shown.

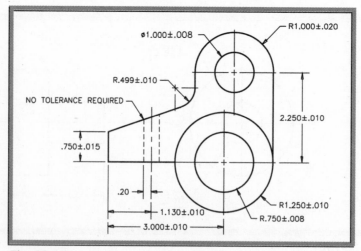

Fig. 28-15

Chapter 28
Review & Activities

 Using Problem-Solving Skills

Complete the following activities using problem-solving skills and your knowledge of AutoCAD.

1. The drawing you dimensioned in Chapter 25, slide2.dwg, will be used in a new CAD textbook to illustrate dimensioning styles. Save the drawing with a new name of slide3.dwg. Create two new layers named Symmetrical and Deviation. Freeze the layer on which the current dimensions were drawn. Make the Symmetrical layer current and redimension the drawing with tolerances expressed in the symmetrical style. Then freeze Symmetrical, make Deviation current, and redimension using the deviation style. The tolerance on the angle is ±15 units, on the radius +0.02° and −0.00°, and all others ±.05 units. Plot the drawing twice to show the two different dimensioning methods.

2. Your editor also wants an example of GD&T with limits dimensioning. Save slide3.dwg as slide4.dwg. Redimension the drawing to maintain the 3.0-unit vertical side perpendicular to the 5.5-unit base within 0.02 units. Save the drawing.

Chapter 29 A Calculating Strategy

Objectives:

- Find the coordinates of points using AutoCAD's **Inquiry** feature
- Calculate the distance between specific points in a drawing file
- Calculate area and circumference of objects in a drawing
- Use AutoCAD's online geometry calculator to place objects at precise points in a drawing
- Display information from AutoCAD's drawing database on objects and entire drawings
- Divide an object into equal parts
- Place markers or points at specified intervals on an object

Key Terms

area

circumference

delta

drawing database

integer

perimeter

vector

With the power of the computer, AutoCAD can perform measurement and calculation tasks that would take significant time to do manually. An example is calculating miles of a chain-link fence on a drawing.

The drawing in Fig. 29-1 shows an apartment complex with parking lots, streets, and trees. With AutoCAD, you can calculate the square footage of the parking lot and the distance between parking stalls on such a drawing.

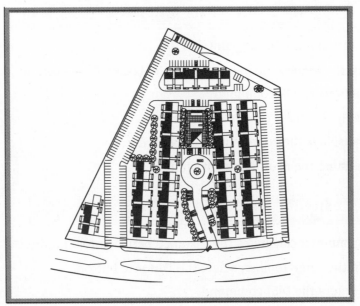

Fig. 29-1

Calculations

AutoCAD includes commands that help you find the coordinates of specific points on a drawing, calculate the distance between two points, calculate the area of an object, and perform other geometric calculations.

1. Start AutoCAD and start a new drawing using the **dimen.dwt** template file.

2. Erase the drawing and dimensions.

3. Create the drawing of the end view of a shaft with a square pocket machined into it (Fig. 29-2). Use the following guidelines.

   • Use the sizes shown, but omit dimensions.

   • The shaft and pocket share the same center point.

   • Use **RECTANG** to create the square pocket.

   • Create the end view on the layer named Objects.

4. Save your work in a file named **calc.dwg**.

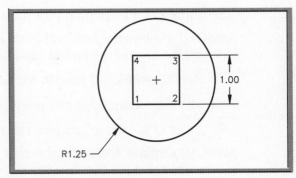

Fig. 29-2

Locating Points

Let's find the coordinates of point 1.

1. Open the **Inquiry** toolbar and pick the **Locate Point** button.

AutoCAD enters the ID command.

2. In reply to Specify point, pick point 1.

AutoCAD displays the coordinates of point 1.

3. Determine the coordinates of point 2.

Inquiry

Calculating Distances

The DIST command calculates the distance between two points.

1. From the same toolbar, select the **Distance** button.

AutoCAD enters the DIST command.

2. Pick points 1 and 3.

In addition to the distance, AutoCAD calculates angle and delta information. If you have studied geometry, you may be familiar with the use of the term *delta* to mean "change." Delta X is the distance, or change, in the X direction, and Delta Y is the distance in the Y direction from one point to the other.

Inquiry

Calculating Area

The AREA command determines the area and perimeter of several different object types. *Area* is the number of square units (inches, acres, miles, etc.) needed to cover an enclosed two-dimensional shape or surface. *Perimeter* is the distance around a two-dimensional shape or surface.

1. Select the **Area** button from the Inquiry toolbar.

AutoCAD enters the AREA command.

Inquiry

2. Enter **O** for Object and pick the square pocket.

AutoCAD displays the area and perimeter of the pocket. Suppose we want to know the area of the end of the shaft minus the pocket.

3. Reenter **AREA** and enter **A** for Add.

This puts the command in Add mode.

4. Enter **O** for Object and pick the shaft.

AutoCAD displays 4.9087 for the area and 7.8540 for the circumference. (*Circumference* is the distance around a circle.) Notice that AutoCAD is still in Add mode and is asking you to select objects.

5. Press **ENTER**.

6. Enter **S** for Subtract.

Now the command is in Subtract mode.

7. Enter **O** for Object and select the pocket.

AutoCAD subtracts the area of the pocket from the area of the end of the shaft and displays the result (3.9087).

8. Press **ENTER** twice to terminate the command.

Using the Online Geometry Calculator

With the CAL command, AutoCAD offers an online geometry calculator that evaluates vector, real, and integer expressions. A *vector* is a line segment defined by its endpoints or by a starting point and a direction in 3D space. An *integer* is any positive or negative whole number or 0. You can use the expressions in any AutoCAD command that requests points, vectors, or numbers.

1. Enter the **CAL** command.

2. Type **(3*2)+(10/5)** at the keyboard and press **ENTER**.

As in basic algebra, the parts in parentheses are calculated first. The asterisk (*) means to multiply and the forward slash means to divide the numbers.

AutoCAD calculates the answer as 8.0.

You can also use CAL with object snaps to calculate the exact placement of a point. For example, suppose you wanted to add a feature to the shaft that required starting a polyline halfway between the center of the shaft and the lower left corner of the machined pocket.

3. Enter the **PLINE** command.

4. Enter **'CAL**. (Note the leading apostrophe.)

By preceding the command with an apostrophe, you have entered the command transparently. In other words, when you complete the CAL command, the PLINE command will resume.

5. Type **(cen+end)/2** and press **ENTER**.

6. Pick any point on the shaft.

7. Pick either line near point 1.

AutoCAD calculates the midpoint between the shaft's center and point 1 and places the first point of the polyline at that location.

8. Press **ENTER** to terminate the PLINE command.

Displaying Database Information

AutoCAD maintains an internal database on every drawing. The *drawing database* contains information about the types of objects in the drawing, defined layers, and the current space (model or paper). It also contains specific numerical information about individual objects and their placement in the drawing. This information can be useful to drafters and programmers using AutoCAD for a variety of tasks.

Listing Information About Objects

Inquiry

1. Pick the **List** button from the Inquiry toolbar.

2. Pick any point on the shaft and press **ENTER**.

The AutoCAD Text Window displays the object type, layer, space, center point, radius, circumference, and area.

3. Close the window.

4. Select the shaft.

5. Right-click and pick **Properties** from the shortcut menu.

AutoCAD displays some of the same information, but does not give the object type or space. The benefit of this list is that you can make changes to the object, such as changing its circumference and area.

6. Change Area to **3**.

AutoCAD redraws the shaft with an area of 3 units.

7. Close the Properties dialog box and undo Step 5.

Viewing the Drawing Database

1. Enter **DBLIST**.

AutoCAD displays database information on all objects in the drawing database. In this case, there are only two objects.

2. Close the window.

Editing Objects Mathematically

You can apply the same power that AutoCAD uses to calculate coordinates, areas, and circumferences to edit objects with mathematical precision.

Dividing an Object into Equal Parts

The DIVIDE command divides an object, such as the end of the shaft, into a specified number of equal parts.

1. From the **Draw** pull-down menu, pick **Point** and **Divide**.

This enters the DIVIDE command.

2. Select the shaft.

3. Enter **20** for the number of segments.

It appears as though nothing happened. Something did happen: the DIVIDE command divided the end of the shaft into 20 equal parts using 20 points; you just can't see them. Here's how to use them.

4. Turn off object snap and enter the **LINE** command.

5. Enter the **Node** object snap. (Node is used to snap to the nearest point.)

6. Move the crosshairs along the shaft and snap to one of the nodes.

7. Snap to the center of the shaft.

8. Enter the **Node** object snap again and snap to another point on the shaft.

Your drawing should now look similar to the one in Fig. 29-3.

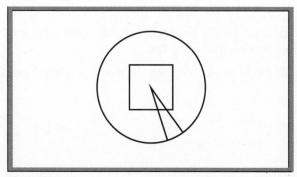

Fig. 29-3

9. Terminate the **LINE** command.

10. Save your work.

Displaying the Points

AutoCAD's DDPTYPE command is used to control the appearance of points. Let's use it to make the points on the shaft visible.

1. Enter **DDPTYPE**.

AutoCAD displays the dialog box shown in Fig. 29-4.

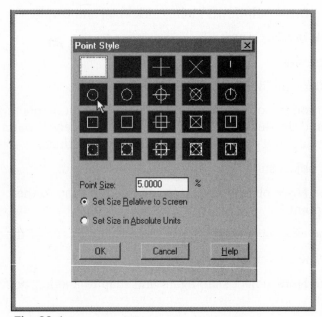

Fig. 29-4

2. Pick the first point style in the second row—the one with a circle and a point at its center—and pick **OK**.

AutoCAD places a small circle at each of the 20 equally spaced points as shown in Fig. 29-5.

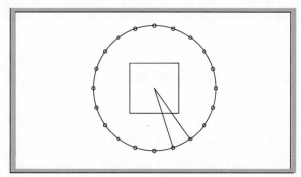

Fig. 29-5

3. Display the same dialog box (press the spacebar).

4. Experiment with other point styles and adjust the point size.

5. Display the **Point Style** dialog box and set the style to the last one in the top row.

6. Pick **OK** to make the points invisible.

Placing Markers at Specified Intervals

The MEASURE command is similar to DIVIDE except that MEASURE does not divide the object into a given number of equal parts. Instead, the MEASURE command allows you to place markers along the object at specified intervals.

1. From the **Draw** pull-down menu, pick **Point** and **Measure**.

This enters the MEASURE command.

2. Select one of the two lines.

3. Enter **.3** unit.

AutoCAD adds points spaced .3 unit apart.

4. Further experiment with MEASURE.

5. Set the point style to a single dot.

6. Close the Inquiry toolbar, save your work, and exit AutoCAD.

Chapter 29
Review & Activities

Review Questions

1. Which AutoCAD command is used to find coordinate points?

2. What information is produced with the AREA command?

3. What information is produced with the LIST command?

4. Describe the difference between LIST and DBLIST.

5. How do you calculate the perimeter of a polygon?

6. How do you find the circumference of a circle?

7. Explain how you control the appearance of points.

8. Explain the difference between the DIVIDE and MEASURE commands. Under what conditions would you use each of these commands?

Challenge Your Thinking

1. Describe one drafting situation in which you might need to use the DIVIDE command and one in which you would prefer to use the MEASURE command.

2. Investigate the difference between AutoCAD's points created when you use the POINT command and those created when you use the DIVIDE and MEASURE commands. Write a paragraph comparing and contrasting them.

Applying AutoCAD Skills

Work the following problems to practice the commands and skills you learned in this chapter.

In problems 1 through 3, draw the views of the fasteners in Figs. 29-6 through 29-8 at the sizes indicated, omitting all letters. Use the appropriate commands to find the information requested above the fasteners. Write their values on a separate sheet of paper. If you have access to the Instructor's Resource CD-ROM, refer to the files calcprb1.dwg, calcprb2.dwg, and calcprb3.dwg. These files correspond to Figs. 29-6, 29-7, and 29-8, respectively.

Chapter 29
Review & Activities

1. Draw the top view of the Phillips-head screw shown in Fig. 29-6 according to the dimensions shown, omitting all letters. Then find the following information:

    - location of point A
    - distance between points A and B
    - area of the polygon
    - perimeter of the polygon

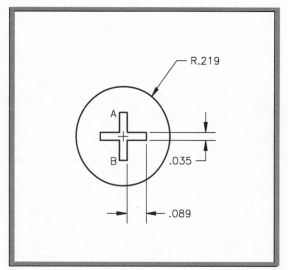

Fig. 29-6

2. Draw the side view of the screw shown in Fig. 29-7 according to the dimensions shown, omitting all letters. Then find the following information:

    - distance between A and B
    - distance between B and C
    - area of hole
    - circumference of hole
    - area of bolt head
    - perimeter of bolt head
    - area of bolt head minus hole

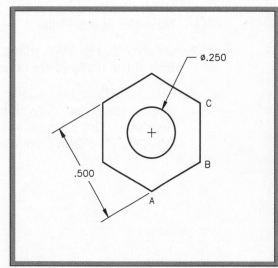

Fig. 29-7

3. What information does AutoCAD list for arc A on the screw shown in Fig. 29-8?

4. List the information AutoCAD provides for line B on the screw.

5. What information does DBLIST provide for the screw? How does this differ from the LIST information?

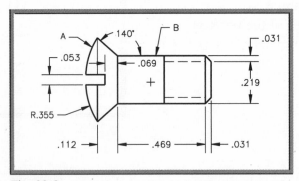

Fig. 29-8

6. On the screw shown in Fig. 29-8, divide line B into five equal parts. Make the points visible.

7. On arc A in Fig. 29-8, place markers along the arc at intervals of .02 unit. If the markers are invisible, make them visible.

8. Create the floor plan shown in Fig. 29-9 according to the dimensions given. Then find the information requested below the floor plan. This floor plan is also available on the Instructor's Resource CD-ROM as flplan.dwg.

Chapter 29
Review & Activities

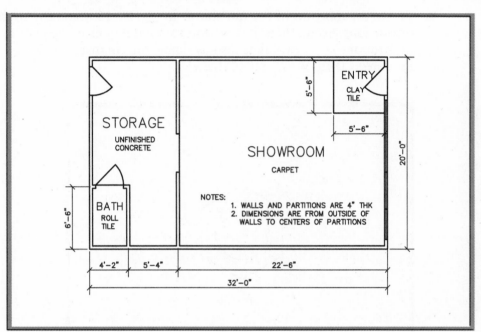

STORAGE
UNFINISHED CONCRETE

ENTRY
CLAY TILE

5'-6"

SHOWROOM

CARPET

5'-6"

20'-0"

NOTES:
1. WALLS AND PARTITIONS ARE 4" THK
2. DIMENSIONS ARE FROM OUTSIDE OF
   WALLS TO CENTERS OF PARTITIONS

BATH
ROLL TILE

6'-6"

4'-2"   5'-4"   22'-6"

32'-0"

Fig. 29-9

Find the following information in square feet:

- area of showroom carpet
- area of entry clay tile
- area of bathroom roll tile

As you may know from your math courses, 1 square yard contains 9 square feet. Calculate the area of the carpet, entry clay tile, and bathroom roll tile as square yards by dividing your answers above by 9 using the CAL command.

- square yards of carpet
- square yards of clay tile
- square yards of roll tile

Calculate the distance between opposite corners of the following areas.

- showroom
- entry area
- bathroom

Problem 8 courtesy of Mark Schwendau, Kishwaukee College

Chapter 29
Review & Activities continued

9. Create the top view of the nut shown in Fig. 29-10 according to the dimensions given. Then find the information requested below the illustration. This drawing is also available on the Instructor's Resource CD-ROM as 1-8UNC2B.dwg.

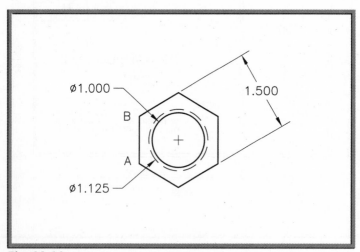

Fig. 29-10

Find the following information:

- distance between A and B
- area of the minor diameter of the thread
- area of the major diameter of the thread
- circumference of the minor diameter
- circumference of the major diameter
- area of top surface of the nut
- area of top surface of the nut minus the minor diameter
- area of top surface of the nut minus the major diameter

Calculate the average of the last two answers using the CAL command.

Problem 9 courtesy of Mark Schwendau, Kishwaukee College

Chapter 29
Review & Activities

 Using Problem-Solving Skills

Complete the following activities using problem-solving skills and your knowledge of AutoCAD.

1. Create a new drawing using the Quick Setup wizard. Specify a drawing area of 300′ × 200′. Draw two horizontal lines of different lengths. Place one near the top of the drawing area and the other near the bottom. Then connect their ends to form a rectangular area.

2. The area you created in problem 1 represents a real estate plot. As a paralegal for an attorney, you have been asked to divide this plot into five equal lots. Determine how to do this and carry out your plan. (Hint: consider using the DIVIDE command and the Node object snap.) To avoid future conflict, the attorney also wants you to verify the size of each lot. Determine the area and perimeter of each lot, and prepare a statement for the attorney explaining why each of the areas are the same, but the perimeters are not. Save the drawing as realestate.dwg.

Part 5 Project

Designer Eyewear

Section 5 reviewed the extensive dimensioning and calculating abilities of AutoCAD. In this exercise, you will be asked to design and draw an object and then to dimension it as specified. To review, traditional dimensions generally inform of size and location. Tolerance dimensions also yield information about characteristics or features of the item (e.g., How round is the hole? How flat is the surface?).

Description

Your task is to design a pair of eyeglasses. The actual style (appearance) and size are up to you. Whether latest fashion fad or very old-fashioned, the eyeglasses should be designed to be appropriate for the human face. (Feel free to inquire whether your instructor will allow a design for some other creature, real or imagined.) In any event, the chosen design must be revealed in two or three views and fully dimensioned. Of particular interest will be dimensioning information concerning the minimum and maximum tolerance permitted for the frame and lenses (such as ± .01). The nose pad(s) or nosepiece may be separate or part of the frame itself. You may wish to measure your own eyeglasses or those of family or friends, or you may have someone actually measure your face to gain some insight with respect to realistic sizes. Figures P5-1 and P5-2 may also give you some design ideas.

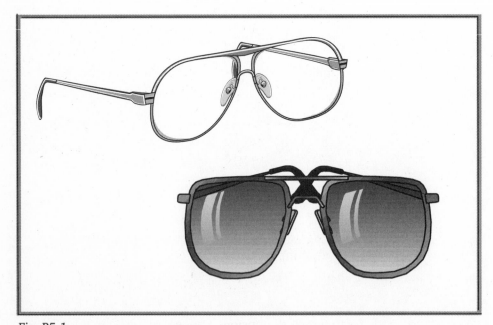

Fig. P5-1

Part 5 Project

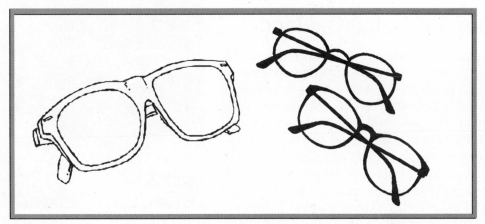

Fig. P5-2

Hints and Suggestions

- Consider showing a sketch of your idea to the instructor for feedback before investing significant time and effort into the 3-view CAD version. Guidance gained early is likely to save more time than constructive criticism given upon completion.

- Don't spend too much time on exact sizes for the eyeglasses, but once drawn, dimension details thoroughly and accurately.

- Include dimensions for the angles of the earpieces (end piece) as well as radii for the curvature of the lenses.

- Consider drawing one half of the eyeglasses and mirror the drawing to complete the pair.

Summary Questions/Self-Evaluation

Your instructor may direct you to answer these questions orally or in writing, or simply to think about your responses as a means of self-reflection.

1. Why might accurate dimensions be needed for one drawing and not for another?

2. Name some purposes for which tolerance dimensioning (min/max) might be considered of vital importance.

3. Once created, a dimension style can be a great timesaver. In what ways might it save time?

4. How do you determine which dimensions are needed on a drawing?

Careers
Using
AutoCAD

Dana White/Dana White Productions

No Waiting in Virtual Oakland

Reams of routine business transactions, from real estate to construction, depend on information kept on file at city hall. To speed its response to information requests, Oakland, California, has put a site on the World Wide Web that lets you point and click your way to public records on zoning, property titles, and more.

Eighty percent of the data in city records can be referenced through a geographical or spatial address, according to Corbin deRubertis of the consulting firm Fourth Dimension Interactive (4Di). These addresses make it convenient to store information in vector-based maps. Using Autodesk MapGuide, 4Di put the Virtual Oakland site on-line in just six weeks.

In addition to quicker access, there are other benefits. As city staff spend less time processing requests, they will have more time to keep records complete and up to date. With information now organized into map layers, it is easier for city departments to access and review information in their layer. "I think the ideal function of the World Wide Web is to democratize information and present it in a way that is pleasing and easy to use," says deRubertis. Virtual Oakland exemplifies that ideal.

Career Focus: Systems Analyst

Robert deRubertis performs as a computer systems analyst. He works for a consulting firm that helps companies and other organizations start up and maintain new computer systems.

The job of a computer systems analyst is first to understand how the client organization conducts its business. The computer systems analyst then uses his or her knowledge of available hardware and software to identify the system that will best suit the client's needs.

In some cases, the computer systems analyst may have to write code to adapt software for the client. Setting up a training program for users of the new system is often part of the job. Debugging also is a normal part of installing a new system.

Education and Training

A bachelor's degree in computer science is generally required for employment as a computer systems analyst. Organizations that hire their own permanent systems analysts may look for work experience or college courses in business management or a field related to their specialty.

An aptitude for mathematics and logical analysis is important to success as a computer systems analyst. This job requires long hours at computer terminals. It sometimes requires work under deadline pressure. A computer systems analyst needs the ability to explain technical material to others.

Career Activity

- Write a report about the work of a computer systems analyst. What are the advantages and challenges?
- Does this kind of work fit with your goals and hopes for the future?

Part 6

Groups and Details

Chapter 30 Groups

Objectives:

- **Create a group**
- **Add and delete objects from a group**
- **Edit the group name and description**
- **Make a group selectable or unselectable**
- **Reorder the objects in a group**

Key Terms

group
selectable

A *group* in AutoCAD is a named set of objects. A circle with a line through it, for instance, could become a group. Groups save time by allowing you to select several objects with just one pick. This speeds up editing operations such as moving, scaling, and erasing these objects.

Creating a Group

The GROUP command permits you to create a named selection set of objects.

1. Start AutoCAD and open the drawing named **calc.dwg**.

2. Using **Save As...**, create a new drawing named **groups.dwg** and erase all objects.

3. Create the drawing of the pulley wheel shown in Fig. 30-1 using the following guidelines.

   - Create a layer named **Hidden** and assign the hidden linetype and a light gray color to it. Place the hidden line on this layer.
   - Place the remaining objects on layer Objects.
   - Omit the dimensions, including the center line.
   - Set LTSCALE to **.5**.

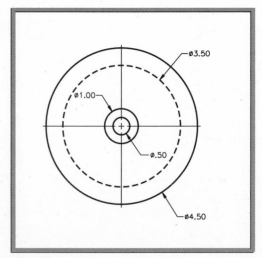

Fig. 30-1

4. Enter the **GROUP** command.

This displays the Object Grouping dialog box, as shown in Fig. 30-2.

5. For Group Name, enter **wheel** using upper- or lowercase letters.

6. Enter **Cast aluminum** for Description.

Focus your attention on the Create Group area of the dialog box. Notice that the Selectable check box is checked.

7. Pick the **New** button, select all four objects, and press **ENTER**.

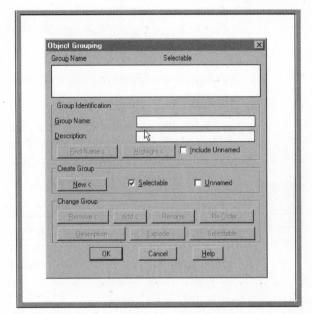

Fig. 30-2

When the dialog box returns, notice that WHEEL is listed under Group Name. Yes appears under Selectable because the Selectable box was checked when you picked the New button. *Selectable* means that when you select one of the group's members (*i.e.*, one of the circles), AutoCAD selects all members in the group.

8. Pick the **OK** button.

9. Pick any one of the objects.

Notice that AutoCAD selected all four circles.

10. Move the group a short distance.

11. Lock layer Hidden.

12. Pick one of the objects again and then move the group a short distance.

The gray hidden-line object did not move because it is on a locked layer.

13. Undo the move and unlock layer Hidden.

Changing Group Properties

The Object Grouping dialog box offers several features for changing a group's definition and behavior. For example, you can add and delete objects from the group, change the group's name or description, and re-order the contents (objects) in the group. The dialog box also allows you to explode the group, which deletes the group definition from the file.

Adding and Deleting Objects

1. Enter the **GROUP** command.

2. Select **WHEEL** in the Group Name list box.

3. Pick the **Highlight** button.

This highlights all members of the group to show you at a glance which objects are members of the group. This could be particularly useful when you're editing a complex drawing containing many objects and groups.

NOTE: ————————————————————

If the small Object Grouping box obscures a critical part of the drawing, you can move the box out of the way. Use the pointing device to pick the top of the box and drag it away.

4. Pick the **Continue** button.

Focus your attention on the Change Group area of the dialog box.

5. Pick the **Remove** button and pick the gray hidden line.

AutoCAD removes this circle from the group.

6. Press **ENTER**.

7. Select the **Highlight** button again.

Notice that the hidden-line circle does not highlight; it is no longer part of the group.

8. Pick **Continue**.

9. Pick the **Add** button, pick the gray hidden-line circle, and press **ENTER**.

This adds the circle back to the group.

Changing the Group Name and Description

1. In the Group Name edit box, change the group name to **pulley** and pick the **Rename** button.

PULLEY appears in the Group Name list box near the top of the dialog box.

2. Rename the group back to **WHEEL**.

3. In the Description edit box, change the group description from Cast aluminum to **Cast magnesium**.

4. Pick the **Description** button (just above the OK button).

This causes the group description to update, as indicated at the bottom of the dialog box.

Making a Group Selectable

1. In the Change Group area, pick the **Selectable** button.

In the upper right corner under Selectable, notice that Yes changed to No.

2. Pick the **OK** button and select one of the objects in the group.

As you can see, the group is no longer selectable. Grips appear only for the specific object you select.

3. Reenter the **GROUP** command, pick **WHEEL** in the Group Name list box, pick the **Selectable** button again, and pick **OK**.

The group is selectable once again.

Reordering the Objects

AutoCAD numbers the objects in the order in which you select them when you create the group. In the Order Group dialog box, AutoCAD allows you to change the numerical sequence of objects within a group. Reordering can be useful when you are creating tool paths for computer numerical control (CNC) machining.

1. Pick the **Re-order...** button to make the Order Group dialog box appear, as shown in Fig. 30-3, and pick **OK**.

2. Save your work and exit AutoCAD.

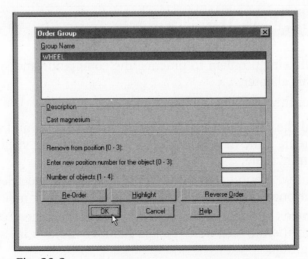

Fig. 30-3

Chapter 30
Review & Activities

Review Questions

1. What is a group?

2. How can groups help AutoCAD users save time?

3. When creating a new group, what is the purpose of entering the group description?

4. Explain how you would add and delete objects from a group.

5. Is it possible to remove the selectable property from a group but still retain the group definition in the drawing? Explain.

6. Under what circumstances might you need to reorder the objects in a group?

Challenge Your Thinking

1. Experiment with object selection using a group name. Notice that you can enter the name of a group at any Select objects prompt to select it for a move, copy, scale, or other operation. When might this feature be useful? Explain.

2. Under what circumstances might you want to make a group unselectable? Explain.

Applying AutoCAD Skills

Work the following problem to practice the commands and skills you learned in this chapter.

1. Perform the following tasks.

   a. Load the calc.dwg file from the previous chapter and use Save As... to create a file named stop.dwg. If the points are visible, make them invisible.

   b. Create a group named WIDGET. Enter New product design for the description. The group should consist of all four objects.

   c. Change the name of the group to WATCH and the group description to Stop watch design. Remove the two lines from the group.

d. Move the group to a new location, away from the two lines. Set the text height to .2 and add 12:30 inside the square. Add this text to the group. Your drawing should look similar to the one in Fig. 30-4.

Fig. 30-4

Using Problem-Solving Skills

1. The exterior trim design of your company's administration building has not yet been approved. Open db_samp.dwg in AutoCAD's Sample folder and save it as AdmBldg.dwg. Make a group of each of the three exterior entrances (two side in turquoise and the front in yellow). Move these groups to various locations to find a better arrangement. You may also remove the five-sided structure at the rear of the building to be replaced by a possible entry. Save the design you think is the best, and be prepared to explain your choice.

2. Your company has been hired to create a sales brochure for a new line of bicycles. The bicycle will be available in several colors, and your customer wants the brochure to show the bicycle in each color. Research typical bicycle shapes and designs. Then create a bicycle design of your own. Use commands such as PLINE, SOLID, and FILL to make the body (frame) of the bicycle solid green. Save the drawing as bikesales.dwg. Then, using the GROUP and COPY commands, create additional copies of the bicycle and change the body colors to show a version in six colors that you think would look good on a bicycle. Use the entire Color palette to make your color selections.

Chapter 31 Building Blocks

Building Blocks

Objectives:

- Create and insert blocks
- Rename, explode, and purge blocks
- Insert a drawing file into the current drawing
- Create a drawing file from a block
- Copy and paste objects from drawing to drawing

Key Terms

block
copy and paste
purge

With AutoCAD, you seldom need to draw the same object twice. Using blocks, you can define, store, retrieve, and insert symbols, components, and standard parts without the need to recreate them. A *block* is a collection of objects that you can associate together to form a single object. Blocks are especially useful when you are creating libraries of symbols and components.

Blocks are similar to groups, but the two are different. A block definition is stored as a single, selectable object in a drawing file that you can insert, scale, and rotate. A group is a named selection set of objects.

Working with Blocks

The BLOCK command allows you to combine several objects into one and then store and retrieve it for later use.

1. Start AutoCAD and start a new drawing from scratch.

2. Draw the top view of a flat-head screw as shown in Fig. 31-1. Omit dimensions.

3. Save your work in a file named **blks.dwg**.

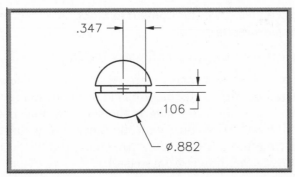

Fig. 31-1

Creating a Block

1. Pick the **Make Block** button from the docked Draw toolbar.

This displays the Block Definition dialog box, as shown in Fig. 31-2.

2. After Name, enter **flat**.

3. Under Name, pick the **Pick point** button.

4. Snap to the center point of the head of the screw.

This defines the base point for subsequent insertions of the screw head.

Draw

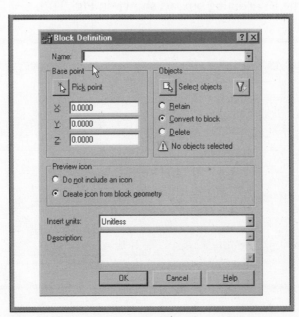

Fig. 31-2

5. In the Objects area, pick the **Convert to block** radio button and then pick the **Select objects** button.

6. Select the entire screw head and press **ENTER**.

In the same area, AutoCAD displays 6 objects selected. The Retain option retains the selected objects as distinct objects in the drawing after you create the block. Convert to block converts the selected objects to a block instance in the drawing after you create the block. Delete deletes the selected objects from the drawing after you create the block. (The block definition remains in the drawing database.)

7. In the Preview icon area, pick **Create icon from block geometry**.

At the right, AutoCAD shows you a view of the icon. You will use this icon in the following chapter.

Focus on Insert units and Description. Insert units specifies the units AutoCAD uses when you insert the block.

8. Set Insert units at **Unitless**, do not add a description, and pick the **OK** button.

You have just created a new block definition.

Inserting a Block

1. Pick the **Insert Block** button from the docked Draw toolbar.

This displays the Insert dialog box, as shown in Fig. 31-3.

Draw

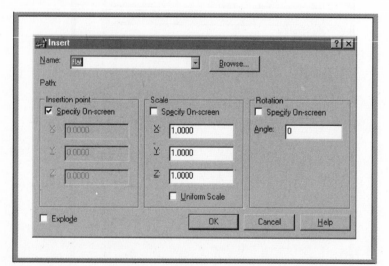

Fig. 31-3

Flat is listed after Name because it is the only block definition in the drawing. Picking the down arrow would list additional blocks, if any existed, in alphabetical order.

The Insertion point area permits you to enter specific coordinates for the block's insertion. Specify On-screen, which should be checked, allows you to specify the insertion point on screen. The Scale area allows you to scale the block now or on the screen when you insert it. The Rotation area enables you to specify a rotation angle in degrees now or when you insert the block. The Explode check box explodes the block and inserts the individual parts of the block.

2. Under Scale, pick the **Specify On-screen** check box.

3. Under Rotation, pick the **Specify On-screen** check box.

4. Pick the **OK** button.

AutoCAD locks the insertion point of the block onto the crosshairs.

5. Pick an insertion point anywhere on the screen.

6. Move the crosshairs up and down, back and forth.

As you can see, you can drag the X and Y scale factors.

7. Enter **1.5** for the X scale factor and **1.5** for the Y scale factor.

You can also drag the rotation into place, or you can enter a specific degree of rotation.

8. Enter **45** for the rotation angle.

9. Insert additional instances of the block using different settings in the Insert dialog box.

Editing an Inserted Block

1. Try to erase one of the lines from the first inserted block.

You cannot, because a block is a single object.

Modify

2. Pick the **Explode** button from the docked Modify toolbar, pick the inserted block, and press **ENTER**.

3. Now try to erase a line from the inserted block.

As you can see, exploding a block returns it to its component parts. Exploding a block with the EXPLODE command is the same as checking the Explode check box in the Insert dialog box when inserting the block.

Renaming a Block

The RENAME command lets you rename previously created blocks.

1. Pick **Rename...** from the **Format** pull-down menu or enter **RENAME** at the keyboard.

This enters the RENAME command and displays the Rename dialog box, as shown in Fig. 31-4.

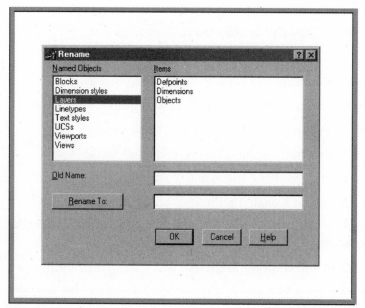

Fig. 31-4

The dialog box lists, under Named Objects, all of the object types in AutoCAD that you can rename.

2. Pick **Blocks**.

AutoCAD lists the only block, flat, under Items.

3. Pick **flat** and then enter **flathead** in the box at the right of Rename To.

4. Pick the **Rename To** button.

Flathead is now listed under Items.

5. Pick the **OK** button.

In the future, if you need to rename blocks, dimension styles, layers, linetypes, text styles, UCSs, viewports, or views, use the RENAME command.

INFOLINK

You will learn more about UCSs in Chapter 38.

Purging Blocks

The PURGE command enables you to selectively delete, or *purge,* any unused named objects, such as blocks. Purging named objects reduces drawing size.

1. Enter the **PURGE** command at the keyboard.

AutoCAD lists the different object types that you can purge.

2. Enter **B** for Blocks.

3. In reply to Enter names(s) to purge <*>, press **ENTER**.

4. Enter **Y** for Yes in reply to Verify each name to be purged?

AutoCAD displays the message No unreferenced blocks found. You have used the only block stored in the current drawing, so there are no blocks to purge. In future, you may choose to purge unused blocks, dimension styles, layers, and other object types.

Blocks and Drawing Files

It is possible to insert a drawing (DWG) file as if it were a block. This can be useful when you want to use part or all of another drawing in the current drawing. Also, it's possible to create a drawing file from a block. This can be especially helpful when you need to transport a block to another computer system.

Inserting a Drawing File

Draw

1. Pick the **Insert Block** button from the docked Draw toolbar and pick the **Browse** button.

2. Find and double-click the folder with your name.

3. Click once on the file named **toler.dwg** and pick the **Open** button.

Toler appears in the Name box.

4. Pick the **Explode** check box and pick **OK**.

5. Move the pointing device and notice that the drawing's insertion base point is 0,0.

6. Pick a point so that the drawing appears on the screen.

7. Pick one of the objects from the toler.dwg drawing.

This is possible because we selected the Explode check box to make the components of the block selectable individually.

Creating a Drawing File from a Block

WBLOCK, short for Write BLOCK, writes (saves) objects or a block to a new drawing file.

1. Enter the WBLOCK command.

AutoCAD displays the Write Block dialog box, as shown in Fig. 31-5.

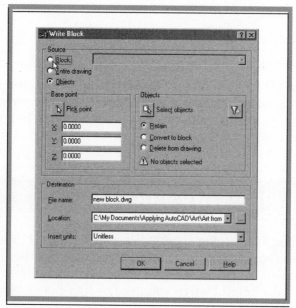

Fig. 31-5

2. Under Source, pick the **Block** radio button.

3. Click in the box at the right of Block.

4. Pick **flathead.**

The Base point and Objects areas are grayed out because both base point and objects are a part of the block definition.

5. Review the information listed under Destination, but do not change any of it.

Notice that the proposed file name is flathead.dwg.

6. Pick the **OK** button.

AutoCAD creates a new file named flathead.dwg.

7. Pick the **Open** button from the docked Standard toolbar.

8. Find flathead.dwg and pick it once to preview it, but do not open it.

9. Pick the **Cancel** button.

Standard

Copying and Pasting Objects

AutoCAD's Windows-standard *copy and paste* feature provides an alternative to using the WBLOCK and INSERT commands. Copying and pasting is a faster approach when you want to transfer a block or set of objects to another drawing on the same computer. Note that this approach does not work when you need to transfer a block or set of objects to another computer.

1. From the Standard toolbar, pick the **Copy to Clipboard** button.

This enters the COPYCLIP command.

Standard

2. Pick a couple of objects and press **ENTER**.

This copies the objects to the Windows Clipboard.

NOTE: ──────

You can select the object(s) first and then pick the Copy to Clipboard button.

3. Save your work.

4. Create a new drawing using the **Start from Scratch** method.

5. From the Standard toolbar, pick the **Paste from Clipboard** button.

This enters the PASTECLIP command.

Standard

6. Move the crosshairs and notice that the objects are attached to them.

7. Pick an insertion point.

As you can see, this is a fast and simple way of copying objects from drawing to drawing.

Chapter 31
Review & Activities

Review Questions

1. Briefly describe the purpose of blocks.

2. Explain how the INSERT command is used.

3. How can you list all defined blocks contained within a drawing file?

4. A block can be inserted with or without selecting the Explode check box. Describe the difference between the two.

5. Describe the function of the WBLOCK command.

6. When would WBLOCK be useful?

7. How can you rename blocks?

8. Why would you want to purge unused blocks from a drawing?

9. When copying an object from one drawing to another, why might you prefer the copy and paste method over the WBLOCK and INSERT approach?

Challenge Your Thinking

1. An electrical contractor using AutoCAD needs many electrical symbols in his drawings. He has decided to create blocks of the symbols to save time. Describe at least two ways the contractor can make the blocks easily available for all his AutoCAD drawings. Which method would you use? Why?

2. If you were to copy a block from one drawing and paste it into another, would AutoCAD recognize the pasted object as a block? Explain.

Chapter 31
Review & Activities

Applying AutoCAD Skills

Work the following problems to practice the commands and skills you learned in this chapter.

1. Open range.dwg, which you created in the "Applying AutoCAD Skills" section of Chapter 17. Make a block of the range and reinsert it into the drawing at a scale factor of .25, with a rotation of 90°. Save the drawing as range2.dwg.

2. Begin a new drawing named livroom.dwg. Draw the furniture representations shown in Fig. 31-6 and store each as a separate block. Then draw the living room outline. Don't worry about exact sizes or locations, and omit the text. Insert each piece of furniture into the living room at the appropriate size and rotation angle. Feel free to create additional furniture and to use each piece of furniture more than once. This file is also available on the Instructor's Resource CD-ROM as livroom.dwg.

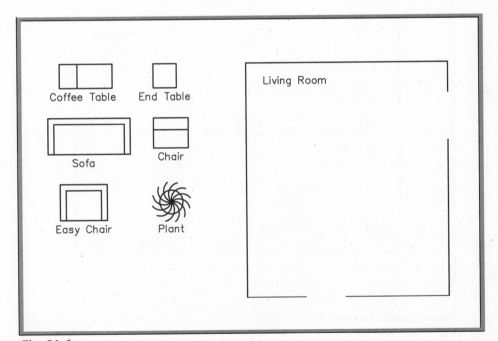

Fig. 31-6

Chapter 31
Review & Activities continued

3. After creating the blocks in problem 2, write two of them (of your choice) to disk using WBLOCK.

4. Copy the block of the easy chair and paste it into another drawing.

5. Explode the PLANT block and erase every fourth arc contained in it. Then store the plant again as a block.

6. Rename two of the furniture blocks.

7. Purge unused objects.

8. Create a new drawing named revplate.dwg. Create the border, title block, and revisions box according to the dimensions shown in Fig. 31-7. (Do not include the dimensions.) Block the revisions box using the insertion point indicated. Then insert the block in the upper right corner of the drawing. This drawing is also available as revplate.dwg on the Instructor's Resource CD-ROM.

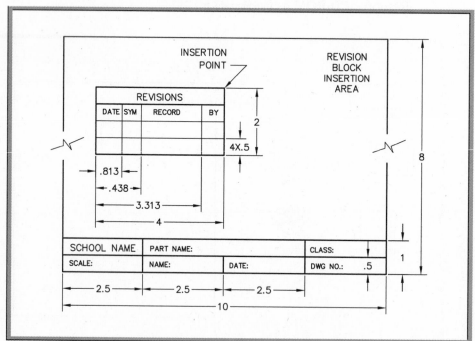

Fig. 31-7

Problem 8 courtesy of Mark Schwendau, Kishwaukee College

Chapter 31
Review & Activities

Using Problem-Solving Skills

Complete the following problem using problem-solving skills and your knowledge of AutoCAD.

1. Your company wants to make the kitchen elevations you worked with in Chapter 30 available to all the designers in your company. Open Ch30kitchen.dwg from your named folder, block the two elevations and save them as DWG files. Be sure to give the files descriptive names so that the designers will know what is in each drawing file.

2. Draw the electric circuit shown in Fig. 31-8 as follows: Draw the resistor using the mesh shown; then save it as a block. Draw the circuit, inserting the blocks where appropriate. Grid and snap are handy for drawing the resistor and circuit. Finish the circuit by inserting small donuts at the connection points. Add the text. The letter omega ([Ω]), which is used to represent the resistance in ohms, can be found under the text style GREEKC (character W).

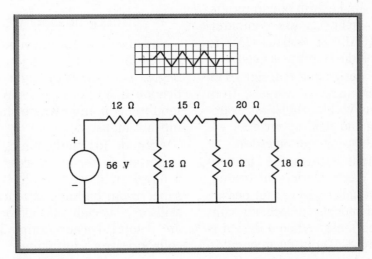

Fig. 31-8

Problem 2 courtesy of Gary J. Hordemann, Gonzaga University

Careers
Using
AutoCAD

The Stock Market/Ray Soto

Architecture that Floats

A good ship outlives its electronics. The service life of a well-designed vessel may include more than one return to the shipyard for remodeling. However, midlife changes to a ship's design have to be made with care, and usually a great many drawings. Onboard space is precious, with every square inch accounted for. Ship designers (called *naval architects,* even if the ship is civilian) are common users of AutoCAD.

Putting up an array of new antennas, for example, may also change a ship's center of gravity. Top-heaviness endangers stability, a ship's designed-in tendency to come back upright after heeling over in heavy seas. Warships often have concentrations of mass above the waterline—in the form of gun turrets, for example. Naval architects use computer-aided engineering (CAE) as well as CAD to ensure that a ship has the stability it needs.

CAD is used in the design and redesign of all kinds of vessels and parts of vessels, from supertankers to racing yachts. Hulls, for example, with their jutting and yielding surfaces, are rendered in CAD with speed and precision.

Another major benefit of CAD is the ability to use drawing data to create models for testing. Designers preview propeller and rudder performance by the use of models in specially controlled and monitored tanks. When a design is ready for the shipyard, CAD output may guide the cutting and shaping of steel plates. Naval architects say the day is coming when the whole job can be done in CAD, with a ship proceeding straight from CAD to the water.

Career Focus—Naval Architect

In the design of ships, naval architects consider stability, structure, power systems, and controls. In power systems, for example, the goal is to match engine capacity to the finished vessel's weight and handling. An underpowered ship will have obvious problems. But engines that are too powerful may burn more fuel than the ship can earn with its cargo.

Naval architects design ships for a specific kind of duty, whether it is to carry cargo or serve as a floating airport. A successful design optimizes a ship's function within constraints of cost and safety.

Education and Training

Qualifications for naval architects are similar to those for other engineers. A bachelor's degree is required to begin professional work. Schools that offer degrees in naval architecture tend to be in areas where shipbuilding is a significant industry.

Aptitude for mathematics and physics is an important factor for success. After graduation, a naval architect typically works under close supervision, building practical knowledge and skills in a specialty. Good communication skills are helpful, because shipbuilding is always a collaborative undertaking.

 Career Activity

- Write a report about today's shipbuilding industry. In what parts of the world are shipyards clustered?
- Are employment forecasts encouraging for naval architects?

Objectives:

- Create a library of symbols and details
- Insert symbols and details using **AutoCAD DesignCenter**
- Insert layers, dimension styles, and other contents from other drawings using **AutoCAD DesignCenter**

Key Terms

palette

symbol library

Figure 32-1 shows a collection of electrical substation schematic symbols in an AutoCAD drawing file. Each of the symbols was stored as a single block and given a block name. (In this particular case, numbers were used for block names rather than words.) The crosses, which show the blocks' insertion base points, and the numbers were drawn on a separate layer and frozen when the blocks were created. They are not part of the blocks; they are used for reference only. A drawing file such as this that contains a series of blocks for use in other drawings is known as a *symbol library*.

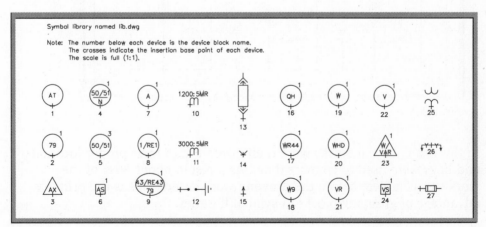

Fig. 32-1

After the symbols were developed and stored in a drawing file, the AutoCAD DesignCenter was used to insert the symbols into a new drawing for creation of the electrical schematic shown in Fig. 32-2.

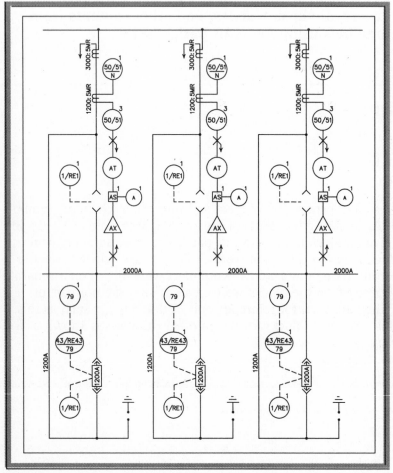

Fig. 32-2

In this example, the blocks were then inserted into their proper locations, and lines were used to connect them. As a result, about 80% of the work was complete before the drawing was started. This is the primary advantage of grouping blocks in symbol libraries.

Creating a Library

Let's step through a simple version of the procedures just described. Keep in mind that a symbol library is nothing more than a drawing file that contains a collection of blocks for use in other drawings.

1. Start AutoCAD and start a new drawing using the **tmp1.dwt** template file.

2. Using the **LIMITS** command, set the drawing area to **24′ × 14′** and **ZOOM All.**

3. Create the schematic representations of tools shown in Fig. 32-3. Set snap at **3″**. Construct each tool on layer **0**. Omit the text.

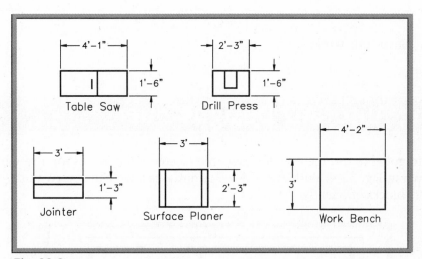

Fig. 32-3

4. Save your work in a file named **lib1.dwg.**

5. Create a block from each of the tools using the following information.
   • Use the names shown in Fig. 32-3.
   • Pick the lower left corner of each tool for the insertion base point.
   • Create an icon from the block geometry in the Block Definition dialog box.
   • For Insert units, use inches.
   • The block names adequately describe the blocks, so do not add a description for each block.

Draw

NOTE:

A block can be made up of objects from different layers, with different colors and linetypes. The layer, color, and linetype information of each object is preserved in the block. When the block is inserted, each object is drawn on its original layer, with its original color and linetype, no matter what the current drawing layer and object linetype are.

A block created on layer 0 and inserted onto another layer inherits the color and linetype of the layer on which it is inserted and resides on this layer. This is why it was important to create the tools on layer 0. Other options exist, but they can cause confusion. Therefore, creating the block on layer 0 is generally recommended if you want the block to take on the characteristics of the layer on which it is inserted.

6. Save your work.

Using AutoCAD DesignCenter

We're going to use the new lib1.dwg file to create the workshop drawing shown in Fig. 32-4. With AutoCAD DesignCenter, we can review and insert blocks efficiently.

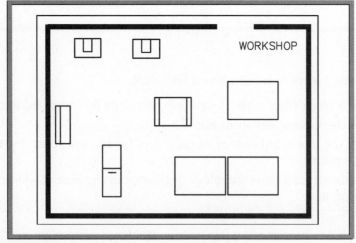

Fig. 32-4

1. Start a new drawing using the **Quick Setup** method and the following information.
   - Use **Architectural** units.
   - Make the drawing area **22′ × 17′**.
   - Set snap at **6″** and grid at **1′**.
   - Be sure to **ZOOM All**.

2. Create a new layer named **Objects**, assign color red to it, and make it the current layer.

3. Using **PLINE,** create the outline of the workshop as shown in the previous illustration. Make the starting and ending width **4″**.

4. Save your work in a file named **workshop.dwg**.

Standard

5. Pick the **AutoCAD DesignCenter** button from the docked Standard toolbar.

AutoCAD DesignCenter appears. How it appears depends on how the last user of the software left it. It will either be docked along the left edge of the AutoCAD window or it will appear as a floating window in the drawing area. If it is docked, you can drag it into the drawing area to make it a floating window. Likewise, if it is a floating window, you can drag it to the left and dock it along the left edge of the AutoCAD window. Notice the row of buttons along the top of the DesignCenter window.

6. Pick the **Load** button.

The Load DesignCenter Palette dialog box appears. The *palette* in AutoCAD DesignCenter is an organized, hierarchical tree that is structured in the same way files and folders are organized in Windows. In fact, the palette consists of your computer's files, folders, and devices.

7. Find the folder with your name and double-click on it.

8. Find and double-click the drawing file named **lib1.dwg**.

This displays several types of content stored in the lib1.dwg. As you can see, one of them is Blocks.

9. Double-click **Blocks.**

The icons of the five blocks you created appear in the palette.

Inserting Blocks

1. Right-click and drag the jointer into the drawing area and release the right button.

2. Pick **Insert Block...** from the shortcut menu.

The Insert dialog box appears.

3. Under Rotation, pick **Specify On-screen**, but leave everything else the same, and pick **OK**.

4. Insert the block in the position shown in Figure 32-4 (page 476).

5. Insert the remaining blocks and save your work.

 NOTE: ─────────────────────────────

> If you do not need to rotate the block when you insert it, you can save steps by left-clicking and dragging the block into position in the drawing. When you use this method however, the block is inserted on layer 0, rather than on the appropriate layer in the drawing.

You can insert symbols and details that are stored as blocks from any drawing file. AutoCAD provides several good examples.

1. Pick the **Load** button in the DesignCenter to display the Load DesignCenter Palette dialog box.

2. Find the **AutoCAD 2000** folder; then double-click the **Sample** folder, and then the **DesignCenter** folder.

3. Find the file named **House Designer.dwg** and double-click it.

4. Double-click **Blocks.**

5. Insert a **36″** right-swing door into the doorway of the workshop. (You may need to edit the size of the doorway.)

6. Experiment with the blocks contained in some of the other drawing files located in the DesignCenter folder, such as **Home Space Planner.dwg.**

7. Erase any blocks that are not appropriate for this drawing, and save your work.

Inserting Other Content

As you may have noticed, AutoCAD DesignCenter permits you to drag other content into the current drawing. Examples include layouts, text styles, layers, and dimension styles.

1. In the **Load DesignCenter** dialog box, find and double-click the **lib1.dwg** file.

2. Double-click **Dimstyles.**

The Preferred and STANDARD dimension styles appear in the palette because they are stored in lib1.dwg.

3. Left-click and drag **Preferred** into the drawing area.

AutoCAD adds the Preferred dimension style to the current drawing.

4. Display the **Dimension** toolbar and review the dimension styles.

As you can see, Preferred was indeed added to the current drawing.

5. Pick the **Up** button (the button that has a folder with an Up arrow in it) once in DesignCenter and double-click **Layers.**

This displays the layers contained in lib1.dwg.

6. Click and drag **Center** to the drawing area.

7. Check the list of layers in the current drawing.

Center is now among them. So, you see, it is very easy to add content such as blocks, dimension styles, layers, and text styles from another drawing to the current drawing.

Other Features

The buttons along the top of the AutoCAD DesignCenter window offer several Windows-standard features. The Tree View Toggle button displays and hides the tree view.

1. Pick the **Tree View Toggle** button.

This splits the window, with the left side consisting of an Explorer-style tree view of your computer's files, folders, and devices.

2. Pick the **Tree View Toggle** button again to toggle off the display of the tree view.

The Favorites button displays the contents of the Favorites folder. You can select a drawing or folder or another type of content and then choose Add to Favorites. AutoCAD creates a shortcut to that item, which is then added to the AutoCAD Favorites folder. The original file or folder does not actually move. Note that all the shortcuts you create using AutoCAD DesignCenter are stored in the AutoCAD Favorites folder.

3. Pick the **Up** button twice and right-click **lib1.dwg.**

The shortcut menu includes Add to Favorites. Picking it would add lib1.dwg to the Favorites folder. Do not pick it at this time.

The Find button displays the Find dialog box. You can enter search criteria and locate drawings, blocks, and nongraphic objects within drawings. The Preview button displays a preview of the selected item at the bottom of the palette if a preview image has been saved with the selected item. If not, the Preview area is empty.

The Description button displays a text description of the selected item at the bottom of the palette. When creating a new block, you are given the opportunity to enter a description; this is where you can view this information.

Chapter 32
Review & Activities

Review Questions

1. What is the primary purpose of creating a library of symbols and details?

2. When you create a library, on what layer should you create and store the blocks? Why?

3. What does the AutoCAD DesignCenter palette display?

4. From the palette, how do you insert a block into the current drawing?

5. Name three examples of other types of content that you can insert from the palette into the current drawing.

Challenge Your Thinking

1. Identify an application for creating and using a library of symbols and details. Identify the symbols the library should include. Discuss your idea with others and make changes and additions according to their suggestions.

2. Why are blocks, rather than groups, used in symbol libraries? Can you think of any applications for which you could use groups as well as blocks in a symbol library? Explain.

3. Explore reasons it might be useful to add shortcuts to the Favorites folder that appears in the DesignCenter. How might adding shortcuts to libraries of symbols and details be especially helpful?

Applying AutoCAD Skills

Work the following problems to practice the commands and skills you learned in this chapter.

1. Based on steps described in this chapter, create an entirely new symbol library specific to your area of interest. For example, if you practice architectural drawing, create a library of doors and windows. First create and/or specify a drawing template. If you completed the first "Challenge Your Thinking" question, you may choose to create the symbol library you planned.

Chapter 32
Review & Activities

2. After you have completed the library of symbols and details, begin a new drawing and insert the blocks. Create a drawing using the symbols and details.

3. The logic circuit for an adder is shown in Fig. 32-5. Draw the circuit by first constructing the inverter and the AND gate as blocks. Use the DONUT command for the circuit connections.

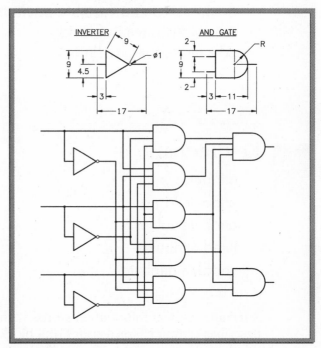

Fig. 32-5

 ## Using Problem-Solving Skills

Complete the following activities using problem-solving skills and your knowledge of AutoCAD.

1. As a building contractor, you want to make a customized library from AutoCAD's existing libraries to suit your own needs. Open DesignCenter, select the Sample folder, and then pick the DesignCenter folder. Find out what blocks are most suitable for a building contractor. Browse through AutoCAD's drawing libraries and select the blocks that are most appropriate for a building contractor. Adjust the scale so that they fit on the screen. Save the drawing as contractor.dwg.

2. As a drafter for an electronics firm, you need a library of electronic symbols. Consult reference books to determine the correct schematic symbols for a PNP transistor, NPN transistor, resistor, capacitor, diode, coil, and transformer. Draw the symbols and make blocks of them to create an electronics symbol library. Save it as electsymbol.dwg.

Problem 3 courtesy of Gary J. Hordemann, Gonzaga University

Objectives:

- **Create fixed and variable attributes**
- **Store attributes in blocks**
- **Edit individual attributes**

Key Terms

attributes
attribute tag
attribute values
fixed attributes
variable attributes

Attributes are text information stored within blocks. The information describes certain characteristics of a block, such as size, material, model number, cost, etc., depending on the nature of the block. The advantage of adding attribute information to a drawing is that it can be extracted to form a report such as a bill of materials. The attribute information can be made visible, but in most cases, you do not want the information to appear on the drawing. Therefore, it usually remains invisible, even when plotting.

The electrical schematic shown in Fig. 33-1 contains attribute information, even though you cannot see it. It's invisible. (The numbers you see in the components are not the attributes.)

Figure 33-2 shows a zoomed view of one of the components. Notice that in this example, the attribute information is displayed near the top of the component.

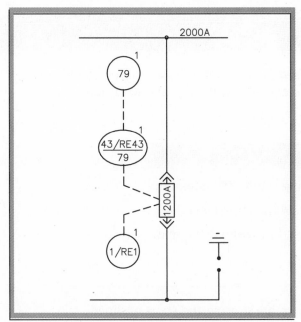

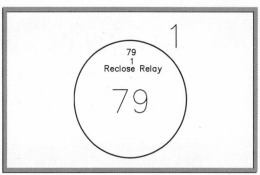

Fig. 33-2

Fig. 33-1

All of the attributes contained in this schematic were compiled into a file and placed into a program for report generation. The report (bill of materials) in Fig. 33-3 was generated directly from the electrical schematic drawing.

DESCRIPTION	DEVICE	QUANTITY/UNIT
Recloser Cut-out Switch	43/RE43/79	1
Reclose Relay	79	1
Lightning Arrester	–	3
Breaker Control Switch	1/RE1	1
1200 Amp Circuit Breaker	52	1

Fig. 33-3

Fixed Attributes

Attributes can be fixed or variable. *Fixed attributes* are those whose values you define when you first create the block. You will learn about variable attributes later in this chapter.

Creating Attributes

Attributes are created, or defined, using the ATTDEF command.

1. Start AutoCAD and open the drawing named **lib1.dwg**.

It should look like the one in Fig. 33-4, but without the text.

NOTE: ────────────────

If you do not have lib1.dwg on file, create it using the steps outlined in Chapter 32.

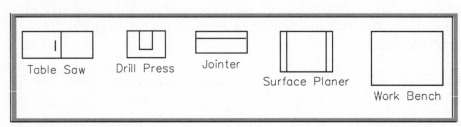

Table Saw   Drill Press   Jointer   Surface Planer   Work Bench

Fig. 33-4

Let's assign information (*attribute values*) to each of the tools so that we can later insert them and generate a bill of materials. We'll design the attributes so that the report will contain a brief description of the component, its model, and the cost.

2. Zoom in on the table saw. It should fill most of the screen.

3. From the **Draw** pull-down menu, select **Block** and the **Define Attributes...** item, or enter **ATTDEF** (short for "attribute definition") at the keyboard.

This enters the ATTDEF command and displays the Attribute Definition dialog box, as shown in Fig. 33-5.

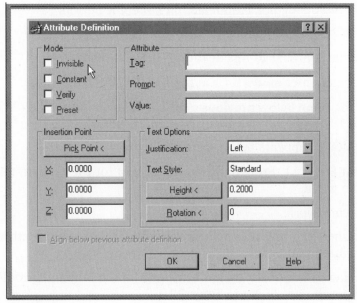

Fig. 33-5

4. In the Mode area, check **Invisible** and **Constant**.

Invisible specifies that attribute values are not displayed when you insert the block. Constant gives attributes a fixed value for block insertions. You will learn more about these two items when you insert blocks that contain attributes.

5. In the Attribute area, type the word **description** (in upper- or lowercase letters) in the box at the right of Tag.

The *attribute tag* identifies each occurrence of an attribute in the drawing. Once again, this will become clearer after you create and use attributes.

6. In the box at the right of Value, type **Table Saw** (exactly as you see it here).

Table Saw becomes the default attribute value.

7. Under Text Options, enter the following information.
   - Justification: **Left**
   - Text Style: **roms**
   - Height: **3"**
   - Rotation: **0.0**

8. Pick the **Pick Point** button.

9. In reply to Start point, pick a point inside the table saw, near the top.

10. When the dialog box reappears, pick the **OK** button.

The word DESCRIPTION appears on the table saw.

11. Press the spacebar to repeat the ATTDEF command.

12. This time, enter **model** for the tag and **1A2B** for the value.

13. Pick the **Align below previous attribute definition** check box in the lower left corner of the dialog box and pick **OK**.

The word MODEL appears below DESCRIPTION.

14. Repeat Steps 11 through 13, but enter **cost** for the tag and **$625.00** for the value.

15. Save your work.

You are finished entering the table saw attributes. Let's assign attributes to the remaining tools.

16. Zooming as necessary, assign attributes to the remaining tools. Use the information shown in Table 33-1.

Description	Model	Cost
Drill Press	7C-234	$590.00
Jointer	902-42A	$750.00
Surface Planer	789453	$2070.00
Work Bench	31-1982	$825.00

Table 33-1

 HINT: ————————————————————

Be sure to pick the Align below previous attribute definition check box in the Attribute Definition dialog box when adding the second and third attributes to each tool. This will save time and will automatically align the second and third attributes under the first one.

17. **ZOOM All** and save your work.

Storing Attributes

Let's store the attributes in the blocks.

Modify

1. Pick the **Explode** button from the docked Modify toolbar and explode each of the tools.

Exploding the blocks permits you to redefine them using the same name.

Draw

2. Pick the **Make Block** button from the docked Draw toolbar.

3. Pick the down arrow at the right of Name and pick **Table Saw.**

4. Pick the **Pick point** button and pick the lower left corner of the table saw for the insertion base point.

5. Pick the **Select objects** button, select the table saw and its attributes, and press **ENTER.**

6. Pick the **OK** button.

7. When AutoCAD asks whether you want to redefine Table Saw, pick **Yes.**

This redefines the Table Saw block, and the attribute information disappears.

INFOLINK

You will learn more about attribute extraction and bills of materials in Chapter 34.

8. Repeat Steps 2 through 7 to redefine each of the remaining tool blocks.

9. Save your work.

All of the tools in lib1.dwg now contain attributes. When you insert these tools into another drawing, the attributes will insert also.

Displaying Attributes

Let's display the attribute values using the ATTDISP (short for "attribute display") command.

1. Enter **ATTDISP** and specify **On.**

You should see the attribute values, similar to those shown in Fig. 33-6.

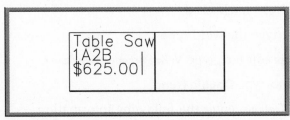

```
Table Saw
1A2B
$625.00
```

Fig. 33-6

2. Reenter **ATTDISP** and enter **N** for Normal.

The attribute values should again be invisible.

3. Save your work.

Variable Attributes

Thus far, you have experienced the use of fixed attribute values. With *variable attributes,* you have the freedom of changing the attribute values as you insert the block. Let's step through the process.

1. Using the **tmp1.dwt** template file, begin a new drawing.

2. Zoom in on the lower quarter of the display, make layer **0** the current layer, and set the snap resolution to **2″**.

3. Draw the architectural window symbol shown in Fig. 33-7. The dimensions not given are **2″** in length. Do not place dimensions on the drawing. (The symbol represents a double-hung window for use in architectural floor plans.)

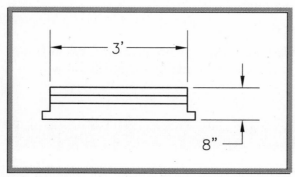

Fig. 33-7

4. Save your work in a file named **window.dwg**.

5. Enter the **ATTDEF** command and check **Invisible**, but leave Constant, Verify, and Preset unchecked.

6. For the tag, type the word **type**.

7. In the Prompt edit box, type **What type of window?**

8. For the value, type **Double-Hung**.

9. Under Text Options, enter the following information.
   - Justification: **Center**
   - Text Style: **roms**
   - Height: **3″**
   - Rotation: **0.0**

10. Pick the **Pick point** button and pick a point over the center of the window, leaving space for two more attributes.

11. When the dialog box reappears, pick the **OK** button.

The word TYPE appears.

12. Press the spacebar to redisplay the dialog box, and leave the attribute modes as they are.

13. Type **size** for the attribute tag, **What size?** for the attribute prompt, and **3′ × 4′** for the attribute value.

14. Pick the **Align below previous attribute definition** check box, and pick **OK**.

The word SIZE appears on the screen below the word TYPE.

15. Repeat Steps 12 through 14 using the following information.
    • Tag: **manufacturer**
    • Prompt: **What manufacturer?**
    • Value: **Andersen**

16. Store the window symbol and attributes as a block. Name it **DH** (short for "double-hung") and pick the lower left corner for the insertion base point.

The Edit Attributes dialog box appears.

17. We do not want to edit the attributes at this time, so pick the **Cancel** button and save your work.

18. Enter **ATTDISP** and **On**.

Did the correct attribute values appear?

Editing Attributes

ATTEDIT, short for "attribute edit," allows you to edit attributes in the same way as you edited the size attribute of the double-hung window. Your drawing should currently look similar to the one in Fig. 33-8.

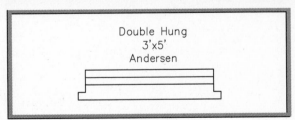
Fig. 33-8

1. Enter the **ATTEDIT** command and pick one of the two blocks.

This displays the Edit Attributes dialog box.

2. Change **Andersen** to **Pella** and pick **OK**.

3. Save your work and exit AutoCAD.

Chapter 33
Review & Activities

Review Questions

1. Explain the purpose of creating and storing attributes.

2. Briefly define each of the following commands.

   ATTDEF
   ATTDISP
   ATTEDIT

3. What are attribute tags?

4. What are attribute values?

5. Explain the attribute modes Invisible and Constant.

Challenge Your Thinking

1. Discuss the advantages and disadvantages of using variable and fixed attribute values. Why might you sometimes prefer one over the other?

2. Brainstorm a list of applications for blocks with attributes. Do not limit your thinking to the applications described in this chapter. Compare your list with lists created by others in your group or class. Then make a master list that includes all the ideas.

Chapter 33
Review & Activities

Applying AutoCAD Skills

Work the following problems to practice the commands and skills you learned in this chapter.

1. Load the drawing containing the furniture representations you created in Chapter 31. If this file is not available, create a similar drawing. Outline a simple plan for assigning attributes to each of the components in the drawing. Create the attributes and redefine each of the blocks so the attributes are stored within the blocks.

2. The small hardware shop in which you work is computerizing its fasteners inventory. The shop carries both U.S. and metric sizes. All of the available fasteners are saved as blocks in the DesignCenter folder in AutoCAD's Sample folder. Place all of the blocks from Fasteners – metric.dwg and Fasteners – U.S.dwg into a drawing named hardware.dwg. Place them in an orderly and logical fashion. Then assign attributes to each fastener to describe it. Use attribute tags of Type and Size. For the individual attribute values, use the information given in the name of each block.

3. Look in a computer hardware catalog to find the sizes and basic shapes of several printers from different manufacturers. Draw the basic shape of at least five different printers. Create attributes to describe the printers in terms of type (inkjet, laser, etc.), brand, and cost. Save the drawing as printers.dwg.

4. Refer to schem.dwg and schem2.dwg and instructions contained on the Instructor's Resource CD-ROM. Display and edit the attributes contained in the drawing files.

Chapter 33
Review & Activities continued

 ## Using Problem-Solving Skills

Complete the following problem using problem-solving skills and your knowledge of AutoCAD.

1. Attributes can be used for many things besides creating bills of materials. For example, you can use attributes to track items in a collection of anything from model cars to unique bottles to action figures. For this problem, suppose you are a dealer in and collector of Avon bottles that have a transportation theme. You have been asked to participate in a display of various collections being sponsored by the local library. Display space is limited, so you need to make a layout to determine your specific needs. Create a library of schematic symbols for the following car-shaped bottles. The cars vary in size from $2'' \times 5''$ for the Corvette and Studebaker to $2\frac{1}{2}'' \times 8''$ for the Pierce Arrow and $3'' \times 9''$ for the Cord and Deusenberg. Assign the attributes as shown in Table 33-2. Save the drawing as cars.dwg.

Bottle	Contents	Price
1937 Cord	Wild Country	$12
1951 Studebaker	Spicy	$8
1988 Corvette	Spicy	$8
Silver Deusenberg	Oland	$13
1933 Pierce Arrow	Deep Woods	$10

Table 33-2

Chapter 33
Review & Activities

 Using Problem-Solving Skills

2. You are setting up sprinklers for new landscaping for a residential plot. Create a new drawing with architectural units and a drawing area of 300′ × 200′. Draw the plot boundaries with corners at the following absolute coordinates:

   45′,170′
   235′,140′
   210′,45′
   35′,25′

   The house will be rectangular and will be 60′ × 32′. The northwest corner of the house is at absolute coordinates 90′,112′. Your client wants a sprinkler system with enough sprinkler heads to cover the entire property. Each sprinkler head can be set to turn a full 360° or any number of degrees greater than 30°. When set to turn 360°, each sprinkler can cover a radial area of approximately 30′. Decide on the number of sprinklers needed, their placement on the property, and the angle (number of degrees) that each should be set to cover. Then create the sprinkler heads, load them from your library, or load them from AutoCAD DesignCenter. Place them in the locations you have determined. On a separate layer, draw the range of each sprinkler to show the client that the sprinkler heads will indeed cover the entire property.

   Finally, define attributes with the following tags: Location, Angle, and Direction. Assign these attributes to each sprinkler head in your drawing and give them the appropriate values according to the plan you have developed. Save the drawing as sprinklers.dwg.

Chapter 34 Bills of Materials

Objectives:

- **Extract attributes from blocks for report generation**
- **Create a bill of materials**

Key Terms

attribute extraction

comma delimited file (CDF)

dialog control language file (DCL)

drawing interchange format file (DXF)

space delimited file (SDF)

template extraction file

After finishing the attribute assignment process (Chapter 33), you are ready to create a report such as a bill of materials. The first step in this process involves *attribute extraction* (gathering the information stored in the attributes and placing it into an electronic file that can be read by a computer program).

 NOTE: ────────────────────────

This chapter requires files that are located on the Instructor's Resource Guide CD-ROM.

1. Start AutoCAD and begin a new drawing using the **tmp1.dwt** template file.

2. Create the four walls of the workshop and insert each of the blocks as shown in Fig. 34-1.

 HINT: ────────────────────────

Pick the AutoCAD DesignCenter button from the docked Standard toolbar. Find the file named lib1.dwg located in the folder with your name and insert the five blocks contained in this drawing file.

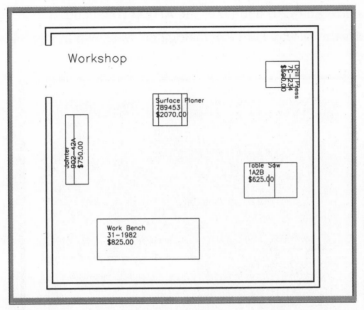

Fig. 34-1

3. Save your work in a file named **extract.dwg**.

4. Enter **ATTDISP** and **On**.

Each of the tools contains attributes, as shown in Fig. 34-1.

Extracting the Attributes

The ATTEXT command enables you to extract the attributes from the extract.dwg file. First, we'll need to copy files from the Instructor's Resource Guide CD-ROM.

1. Minimize AutoCAD.

2. Copy **attext.txt**, **extract.dcl**, and **extract.lsp** from the folder of drawing files from the Instructor's Resource Guide CD-ROM into the folder with your name.

NOTE:

It is important that these three files and extract.dwg (the one currently open in AutoCAD) are in the same folder. You will learn about these three files as you use them.

3. Maximize AutoCAD and enter the **ATTEXT** command.

This displays the Attribute Extraction dialog box, as shown in Fig. 34-2.

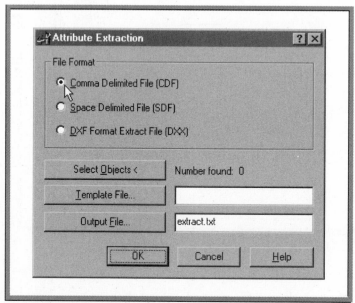

Fig. 34-2

Comma delimited file (CDF) and *space delimited file* (SDF) are file formats that allow you to write attributes to an ASCII text file.

A *drawing interchange format* (DXF) file is a variant of AutoCAD's DXF format used to import and export AutoCAD's drawing files from one CAD system to another. The variant creates a DXX file extension to distinguish it from normal DXF files.

INFOLINK

See Chapter 61 for more information about DXF files.

4. Pick the **Comma Delimited File** (CDF) radio button.

5. Pick the **Select Objects** button, enter **All**, and press **ENTER** a second time.

The dialog box reappears.

In order to use CDF, SDF, or DXX files to display attributes in a text form that is easy to read, you must use a *template extraction file* to define the format. In this case, the template extraction file has been created for you.

6. Pick the **Template File...** button and find and double-click the **attext.txt file** located in the folder with your name.

The dialog box displays Attext.txt as the template extraction file.

The file name extract.txt appears in the edit box located to the right of the Output File... button. AutoCAD suggests this name because the current drawing file is named extract.dwg.

7. Pick the **Output File...** button, double-click the folder with your name, and pick the **Save** button.

This saves the extract.txt file in the folder with your name.

8. Pick the **OK** button.

AutoCAD creates the extract.txt file consisting of the attributes. Notice that AutoCAD displays 5 records in extract file on the Command line.

Creating a Report

You can format and display the contents of CDF files such as extract.txt using special programs such as extract.lsp.

1. Select **Load Application...** from the **Tools** pull-down menu.

This displays the Load/Unload Applications dialog box, as shown in Fig. 34-3.

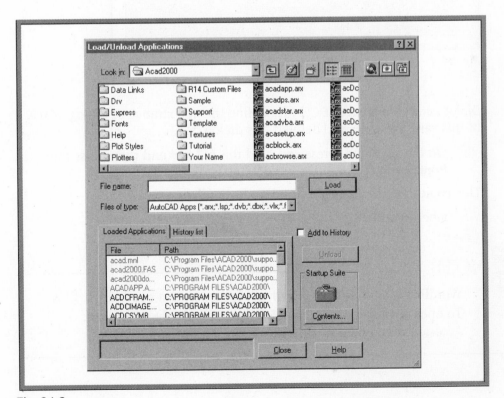

Fig. 34-3

2. In the upper area of the dialog box, double-click the folder with your name, and then double-click the **extract.lsp** file.

3. Pick the **Close** button.

This displays the attributes in a dialog box, as shown in Fig. 34-4.

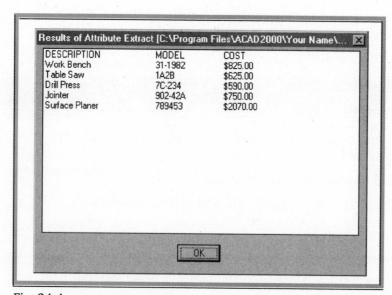

Fig. 34-4

Dialog control language (DCL) files define the appearance of dialog boxes. In this case, the extract.dcl defined this dialog box.

4. On the keyboard, press and hold the **ALT** key and then press the **Print Scrn** key.

This copies the dialog box to the Windows Clipboard.

5. Open **WordPad**.

NOTE:

WordPad is similar to Notepad, except that it can open larger files. To open it, pick the Windows Start button and select Programs, Accessories, and then WordPad.

6. Paste the contents of the Clipboard into it, and print the document.

NOTE: _____

You can use this procedure to copy and print any dialog box, as long as it is active on the screen.

7. Exit WordPad without saving the file.

8. Save your work in the extract.dwg file and exit AutoCAD.

Reviewing the Files

Let's review the contents of the files we used in this chapter.

1. Open the folder with your name and double-click the **extract.txt** file.

This opens the comma delimited file named extract.txt, as shown in Fig. 34-5.

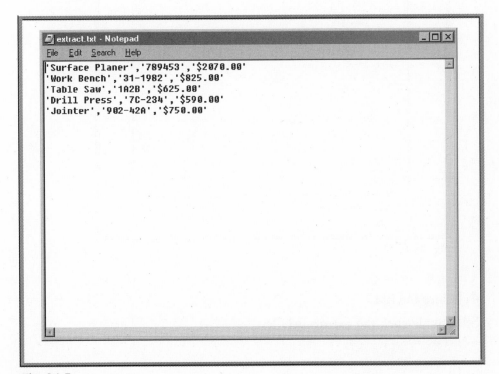

Fig. 34-5

2. Review the information in the file, and then close it.

NOTE:

If you have access to Microsoft Excel or a similar spreadsheet program, you could import extract.txt into it.

3. Double-click the file named **attext.txt**.

This displays the contents of the template extraction file as shown in Fig. 34-6. This is the file that you specified earlier in the Attribute Extraction dialog box.

Fig. 34-6

4. Close the file.

5. Double-click the **extract.dcl** file.

This displays the contents of the dialog control language file, as shown in Fig. 34-7.

```
Extract.dcl - Notepad
File  Edit  Search  Help

//
//   Extract Dialogue
//   Includes a dialog to display a text file.
//
//   Written By:  Donald F. Sanborn          June 1997
//
//   (c) Copyright 1997, Unique Solutions, Inc., All rights reserved
//


//
//   Text File display dialogue
//
frm_text : dialog {
  key = FRM_text ;
  value = "Text Display" ;
  :list_box {
    key = "LST_text" ;
    label = "" ;
    mnemonic = "" ;
    alignment = left ;
    fixed_width = true ;
    fixed_height = true ;
    multiple_select = false ;
    width = 70 ;
    height = 17 ;
    tabs = "25 40 55 70 80 90" ;
  }
  spacer ;
  ok_only ;
  spacer ;
}
```

Fig. 34-7

6. Close the file.

7. Double-click the **extract.lsp** file.

This displays the contents of the AutoLISP file used to format and display the attributes.

8. Close the file.

Chapter 34
Review & Activities

Review Questions

1. Describe the purpose of AutoCAD's ATTEXT command.

2. Describe the process of creating a bill of materials from a drawing that contains blocks with attributes.

3. What is the purpose of the extract.dcl file?

4. What is the purpose of the extract.lsp file?

Challenge Your Thinking

1. Find out more about SDF and DXF/DXX files. What must you do first if you want to use either of these formats for attribute extract files? After you have created the files, how can you display them on the screen or print them?

2. Research template extraction files. Write a short paragraph explaining what they do, how they work, and why they are necessary.

Applying AutoCAD Skills

Work the following problems to practice the commands and skills you learned in this chapter.

1. Open the livroom.dwg file from Chapter 31, which contains several pieces of furniture. Add attributes to the furniture and create blocks. Using the ATTEXT command, CDF option, and the files you used in this chapter, create a bill of materials.

2. Using the CDF file created in problem 1, insert its contents into Microsoft Excel or another spreadsheet if you have access to one.

3. Open sprinklers.dwg from Chapter 33. Create a list of the sprinkler heads for the landscaping contractor to use as reference when the sprinkler system is installed.

4. Refer to schem.dwg and schem2.dwg and the instructions on the Instructor's Resource Guide CD-ROM. Using the ATTEXT command, CDF option, and the files you used in this chapter, create a bill of materials from each of the two drawings.

Review & Activities

 Using Problem-Solving Skills

Complete the following activities using problem-solving skills and your knowledge of AutoCAD.

1. Open cars.dwg from Chapter 33. Using the schematic blocks, create a display to fit in a $2\frac{1}{2}' \times 3'$ display space. Before you begin, you may want to consult antique car or Avon references if you are unfamiliar with any of the car models. Knowing what the cars look like will help you position them attractively in the display. Then extract the attributes and create a list to distribute to potential customers.

2. Open printers.dwg from the previous chapter. Extract the attributes from the printer blocks you created and create a list of the printer attributes. Import the list into a spreadsheet or word processor and create an eye-catching one-page advertising flyer for distribution by mail. Be creative!

Part 6 Project

Computer Assembly Library

One rule of thumb for CAD drafters is not to draw anything twice. Although that may sound strange at first, what it really means is to use CAD commands and functions to reproduce the same details again and again without having to redraw them. You are already familiar with the MIRROR, COPY, and ARRAY commands. To extend the capabilities of these and other commands, AutoCAD also permits you to group two or more objects so that you can manipulate them as one. These blocks of objects may also be assigned characteristics or attributes, which can be retrieved and listed later if desired.

Description

A business that assembles computers has decided that it wishes to automate its advertising and billing practices. To do so, the CAD drafter has been assigned the task of drawing basic symbols to represent the various parts of the computer. Figure P6-1 shows four possible configurations each for CPUs, CD-ROM drives, and hard disk drives as an example, and Fig. P6-2 shows a possible symbol library based on these configurations.

Using the examples shown in Figs. P6-1 and P6-2, create a symbol library that contains various components of a computer. Do not limit the library to just the three parts shown in Figs. P6-1 and P6-2. Do research if necessary to find out what other components should be included in the symbol library and the configurations for each component. After you have created the symbols, assign appropriate attributes to them, as shown in Fig. P6-1, block the individual symbols and save the symbol library.

Create a new drawing and use the symbol library to configure three computers—a low-cost model, a high-end "supermodel," and a middle-of-the-road model. Extract the attributes and create a list of materials needed to assemble the three computers.

CPUs	
300 MegaHertz	$300
350 MegaHertz	$350
400 MegaHertz	$400
500 MegaHertz	$450
Hard Drives	
4.2 Gigabyte	$125
6.4 Gigabyte	$175
8.0 Gigabyte	$225
12.0 Gigabyte	$300
CD-ROM Drives	
24x	$35
42x	$75
Read/write	$300
DVD-ROM	$350

Fig. P6-1

Part 6 Project

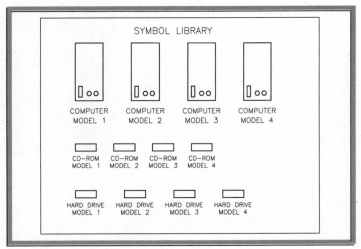

Fig. P6-2

Hints and Suggestions

- Create all blocks on layer 0.

- Create only one symbol for each type of component (CPU, hard disk drive, etc.). Assign attributes to it, and then copy it and edit the attributes to create the other models in that category.

- One way to allow people to distinguish quickly among types of symbols in a library is to use color to classify categories. For example, all CPU symbols could be blue, CD-ROM drives green, and so on.

Summary Questions/Self-Evaluation

1. Why is grouping objects a timesaver for drafters?

2. Distinguish between groups and blocks.

3. Describe the function of a symbol library.

4. What factors might determine the attributes assigned to a block?

5. Why is creating a bill of materials from a set of blocks a timesaving technique? Explain.

Careers
Using
AutoCAD

The Stock Market/Jean Miele

Cams, CAM, and Camco

CAM has come to mean computer-aided manufacturing, but long before the age of computers, cams were (and continue to be) a central technology of automated manufacturing. A cam is an attachment to a crankshaft or spindle. Somewhat like a gear, it is specially shaped to control the motion of a connected device. On a sewing machine, for example, cams convert the spinning motion of the drive wheel to the up-and-down motion of the needle.

Camco of Wheeling, Illinois, makes cam drives (multiple-cam assemblies) for factories. Their smallest cams would fit in the palm of your hand; their largest cam drive is as big as a merry-go-round. This large cam drive is part of an important change in the automobile industry, as some manufacturers have reconfigured the traditional assembly line into more efficient assembly circles. Each circle has a revolving table that carries cars from station to station, where robots perform assembly and welding tasks. The circular shape of the cam-driven tables allows assembly operations to be laid out in a smaller floorspace, which reduces costs such as heating, taxes, and maintenance.

Cam geometry, which can be curved or straight-line, begins with a drawing. Camco engineers design in AutoCAD and save the drawings as DXF files, which provide data to guide machines that grind or mill the cam.

Career Focus—Industrial Engineer

Industrial engineers determine the most effective ways for an organization to use the basic factors of production—people, machines, materials, information, energy, and time. Industrial engineers are more concerned with people and methods of business organization than are engineers in other specialties, who generally work more with products or processes.

Industrial engineers apply mathematical analysis to establish procedures and solve problems related to production. They also develop systems for cost analysis, design production systems, and plan for distribution of goods and services. In the setup of new manufacturing facilities, industrial engineers conduct surveys to find plant locations with the best combination of raw materials, transportation, and taxes.

Education and Training

A bachelor's degree in engineering is the basic requirement for industrial engineers. A curriculum that includes industrial engineering as a specialty or manufacturing practices within a specific industry improves employability.

Industrial engineers should be able to work as part of a management team. Industrial engineers need creativity, an analytical mind, and a capacity for detail. They should express themselves well, both orally and in writing.

 Career Activity

- Write a report about industrial engineering, including all processes that are part of the development of a new product.
- How many different kinds of jobs are typically involved in bringing a new product to market?

Part 7

3D Drawing and Modeling

Objectives:

- Set up a drawing file for isometric drawing
- Create an isometric drawing
- Dimension an isometric drawing correctly

Key Terms

isometric drawing
isometric planes
pictorial representation

An *isometric drawing* is a type of two-dimensional drawing that gives a three-dimensional (3D) appearance. Isometric drawings are often used to produce a *pictorial representation* of a machined part, injection-molded part, architectural structure, or some other physical object. Pictorial representations are useful because they provide a realistic view of the object being drawn, compared to orthographic views. The drawing in Fig. 35-1 is an example of an isometric drawing created with AutoCAD.

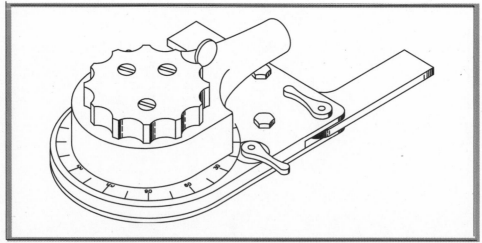

Fig. 35-1

Setting Up an Isometric Drawing

Because isometric drawings are two-dimensional, all the lines in an isometric drawing lie in a single plane parallel to the computer screen. The drawing achieves the 3D effect by incorporating three axes at 120° angles, as shown in Fig. 35-2. In AutoCAD, isometric drawing is accomplished by changing the SNAP style to Isometric mode.

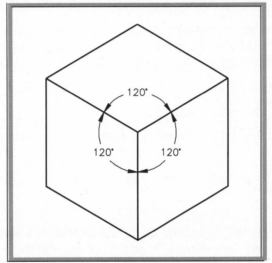

Fig. 35-2

Isometric Planes

As you can see from Fig. 35-2, the 120° angles produce left, right, and top imaginary drawing planes. These three planes are known as the *isometric planes.*

1. Start AutoCAD and begin a new drawing from scratch.

2. Set the grid at **1** unit and the snap resolution at **.5** unit.

3. Enter the **SNAP** command.

4. Enter **S** for Style.

5. Enter **I** for Isometric.

6. Enter **.5** for the vertical spacing. (This should be the default value.)

You should now be in Isometric drawing mode, with the crosshairs shifted to one of the three isometric planes.

7. Move the crosshairs and notice that they run parallel to the isometric grid.

8. Pick **Drafting Settings...** from the **Tools** pull-down menu and pick the **Snap and Grid** tab in the Drafting Settings dialog box.

Notice in the lower right area of this dialog box that you can turn isometric snap on and off. Currently, it is on because you turned it on with the SNAP command.

9. Pick the **Cancel** button.

Toggling Planes

When you produce isometric drawings, you will often use more than one of the three isometric planes. To make your work easier, you can shift the crosshairs from one isometric plane to the next.

1. Press the **CTRL** and **E** keys and watch the crosshairs change.

2. Press **CTRL** and **E** again, and again.

As you can see, this toggles the crosshairs from one isometric plane to the next.

 NOTE:

You can also toggle the crosshairs using the ISOPLANE command. Enter ISOPLANE and then enter L, T, or R to change to the left, top, and right planes, respectively.

Creating an Isometric Drawing

Let's construct a simple isometric drawing.

1. Create a layer named **Objects**, assign a color of your choice to it, and make it the current layer.

2. Enter the **LINE** command and draw the aluminum block shown in Fig. 35-3. Make it **3 × 3 × 5** units in size.

3. Save your work in a file named **iso.dwg**.

4. Alter the block so that it looks similar to the one in Fig. 35-4, using the **LINE**, **BREAK**, and **ERASE** commands.

5. Further alter the block so that it looks similar to the one in Fig. 35-5. Use the **ELLIPSE** command to create the holes.

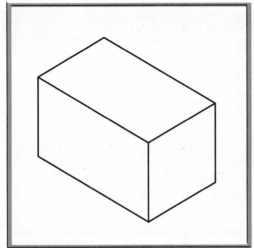

Fig. 35-3

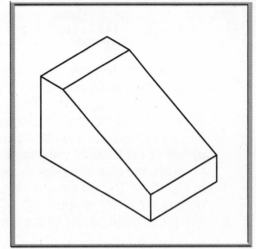

Fig. 35-4

HINT:

Since the ellipses are to be drawn on the three isometric planes, toggle the crosshairs to the correct plane. Then choose the ELLIPSE command's Isocircle option.

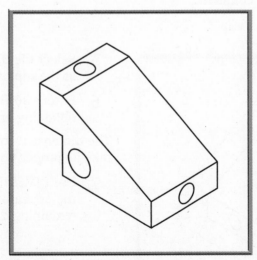

Fig. 35-5

Dimensioning an Isometric Drawing

To dimension an isometric drawing properly, you must rotate the dimensions so that they show more clearly what is being dimensioned. To do this, you can use the Oblique option of the DIMEDIT command. Compare the two drawings shown in Fig. 35-6. The drawing at the right is the result of applying oblique to the drawing at the left.

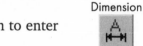

Fig. 35-6

Now let's dimension the aluminum block.

Dimension

1. Display the **Dimension** toolbar and use the **Aligned Dimension** button to create the dimensions as shown in Fig. 35-7.

2. In the lower left corner of the block, draw one horizontal and one vertical extension line.

Your drawing should now look similar to the one Fig. 35-7.

Notice that dimensions are not well-placed. This is because you have not yet applied oblique to them.

Dimension

3. From the Dimension toolbar, pick the **Dimension Edit** button to enter the DIMEDIT command.

4. Enter **O** for Oblique.

5. Select the **3.000** dimension at the top of the block.

6. Enter **–30** for the obliquing angle.

The dimension rotates –30° to take its proper position.

7. Use DIMEDIT to correct the appearance of the remaining dimensions.

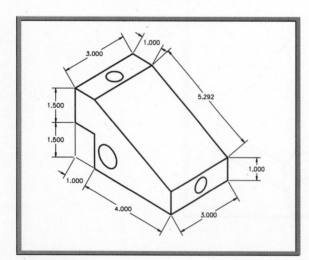

Fig. 35-7

HINT:

Enter an obliquing angle of 30 or −30, depending on the dimension you are editing.

8. Use grips if necessary to adjust the positions of the dimension lines.

9. Edit the **1.000** dimension at the top of the block to suppress the right arrow for a cleaner appearance.

When you finish, your drawing should look like the one in Fig. 35-8.

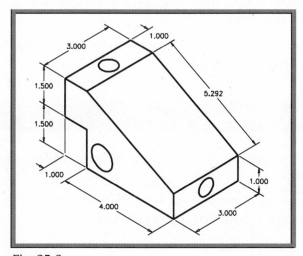

Fig. 35-8

NOTE:

In this drawing, we have not dimensioned the holes in the aluminum block. The completed drawing would include size and location dimensions for all three holes.

10. Return to AutoCAD's standard drawing format using the **SNAP** command or the **Drafting Settings** dialog box.

11. Close the Dimension toolbar, save your work, and exit AutoCAD.

Chapter 35
Review & Activities

Review Questions

1. Describe two methods of changing from AutoCAD's standard drawing format to isometric drawing.

2. Describe how to change the isometric crosshairs from one plane to another.

3. Explain how to create accurate isometric circles.

4. How do you correct the appearance of dimensions used on an isometric drawing?

5. How do you change from isometric drawing to AutoCAD's standard drawing format?

Challenge Your Thinking

1. Discuss the purpose(s) of isometric drawings. When and why are they used?

2. If you were to dimension the holes in the aluminum block created in this chapter, how would you go about it?

Applying AutoCAD Skills

Work the following problems to practice the commands and skills you learned in this chapter.

1-3. Create the machined parts shown in Figs. 35-9 through 35-11 using AutoCAD's isometric capability. Approximate their sizes.

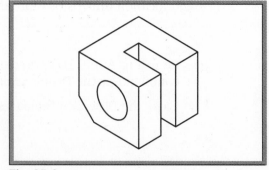

Fig. 35-9

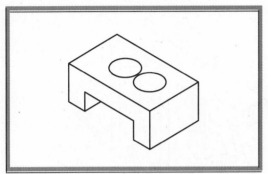

Fig. 35-10

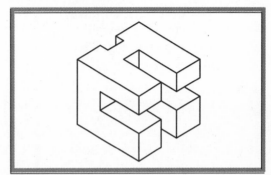

Fig. 35-11

4. Create a mating part for the plug you created in Chapter 28 ("Applying AutoCAD Skills," problem 1). Draw the isometric solid according to the dimensions shown in Fig. 35-12A. Then center the hole in the top surface of the part. Specify a diameter of 10.61. Your finished drawing should look like the one in Fig. 35-12B. Save the drawing as plug2.dwg.

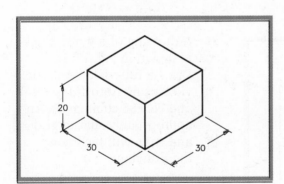

Fig. 35-12A

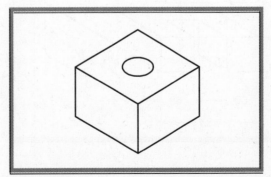

Fig. 35-12B

Chapter 35
Review & Activities continued

5. Create the cut-away view of the wall shown in Fig. 35-13.

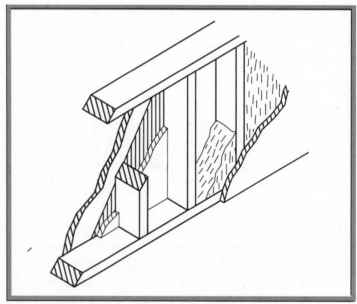

Fig. 35-13

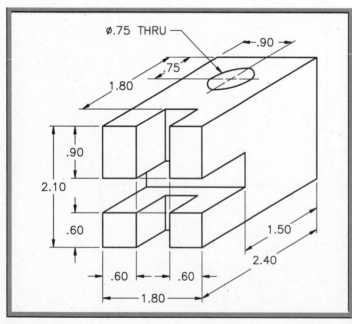

Fig. 35-14

6. A drawing of a guide block is shown in Fig. 35-14. Draw a full-scale isometric view using AutoCAD's SNAP and ELLIPSE commands. An appropriately spaced grid and snap will facilitate the drawing.

7. An exploded, sectioned orthographic assembly is shown in Fig. 35-15. Draw the assembly as a full-scale exploded isometric assembly. Use isometric ellipses to represent the threads as shown in the isometric pictorial.

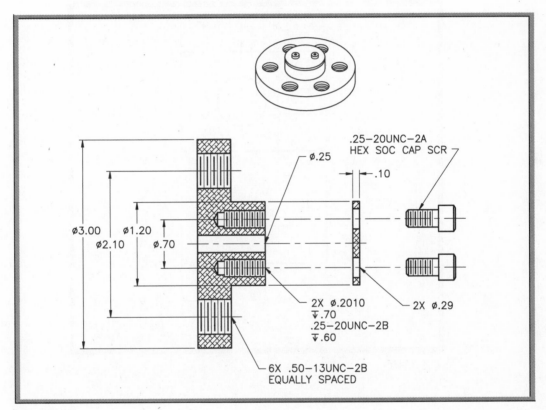

Fig. 35-15

Problem 7 courtesy of Gary J. Hordemann, Gonzaga University

8. Accurately draw an isometric representation of the orthographic views shown in Fig. 35-16. Draw and dimension the isometric according to the dimensions provided.

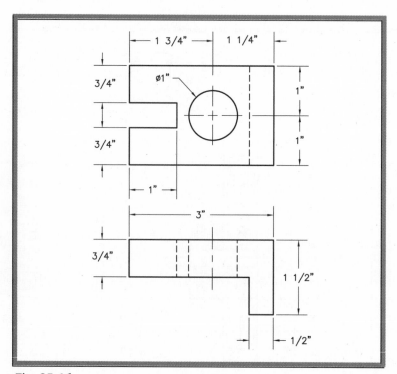

Fig. 35-16

Chapter 35
Review & Activities

Using Problem-Solving Skills

Complete the following problem using problem-solving skills and your knowledge of AutoCAD.

1. The design engineer for the quality assurance division of your company asked you for an isometric drawing of the dovetail block shown in Fig. 35-17, which is used in a milling machine setup. Draw the block, dimension it, and save it as ch35dove.dwg.

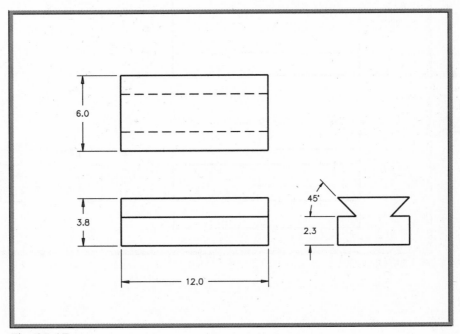

Fig. 35-17

Chapter 35
Review & Activities continued

2. The engineer also needs an isometric drawing of the slide shown in Fig. 35-18 for the same milling machine setup. This one, however, does not need to be dimensioned. Save the drawing as ch35slide.dwg.

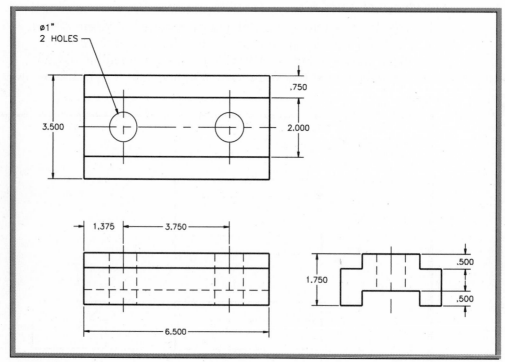

Fig. 35-18

Objectives:

- Create a basic 3D model
- Apply several 3D viewing options
- View and control a model interactively in 3D space

Key Terms

arcball

clipping plane

compass

extrusion thickness

flat shading

Gouraud shading

hidden line removal

parallel projection

perspective projection

roll

Z axis

Three-dimensional (3D) drawing and modeling is becoming increasingly popular due, in large part, to improvements in hardware and software. AutoCAD's 3D capabilities allow you to produce realistic views of car instrument panels, snow skis, office buildings, and countless other products and structures. A key benefit of creating 3D drawings is to communicate better a proposed design to people and groups inside and outside a company. Also, the availability of 3D data speeds many downstream processes such as engineering analysis, digital and physical prototyping, machining, and mold-making. In the field of architecture, 3D modeling can help companies persuade customers to move ahead with a proposed construction project such as a shopping mall.

This chapter introduces AutoCAD's 3D capabilities with four easy-to-use commands. These commands permit you to create simple 3D models and view them from any point in space.

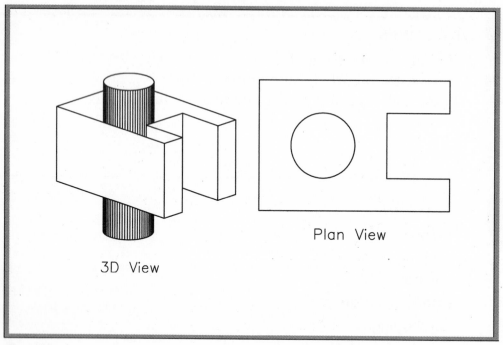

Fig. 36-1

An example of an AutoCAD-generated model is shown in Fig. 36-1, left. The model on the right is the same object, viewed from the top.

The model consists of a section of steel rod and a sheet-metal shroud that fasten into an aircraft fuel control system. The model shows how the shroud protects a wiring harness that runs through the square notch in the shroud next to the rod. The shroud protects the wires from the heat of the rod. This prevents the plastic insulation from melting and causing an electrical short.

Creating a Basic 3D Model

To work in three dimensions in AutoCAD, you will use a third axis on the Cartesian coordinate system. This axis, called the *Z axis*, shows the depth of an object, adding the third dimension. (In this context, the X axis shows the width of the object, and the Y axis shows the length or height.) The Z axis runs through the intersection of the X and Y axes at right angles to both of them.

Let's draw the steel rod and sheet-metal shroud similar to those shown in Fig. 36-1.

1. Start AutoCAD and create a new drawing using the following information.
   - Use the **Quick Setup** wizard.
   - Set the units to **Architectural**.
   - Set the area to **120′ × 90′**.
   - **ZOOM All**.

2. Create a **10′** grid and a **5′** snap.

3. Create a new layer named **Objects**, assign the color magenta to it, and make it the current layer.

4. Enter the **ELEV** command and set the new default elevation at **10′** and the new default thickness at **30′**.

An elevation of 10′ means the base of the object will be located 10′ above a baseplane of 0 on the Z axis. The thickness of 30′ means the object will have a thickness on the Z axis of 30′ upward from the elevation plane. This is called the *extrusion thickness*.

5. Draw the top (plan) view of the object using the **LINE** command. Construct the object as shown on the right in the previous drawing, but omit the circle (cylinder) at this time. The width and length of the object are 40′ × 60′, respectively. Approximate the remaining dimensions.

6. Save your work in a file named **3d.dwg**.

Viewing and Hiding

Let's view the object in 3D.

1. From the **View** pull-down menu, pick **3D Views** and **VPOINT**.

HINT:

Instead, you may choose to type the VPOINT command and press ENTER twice. You may find this to be faster than picking items from the pull-down menu.

2. Move the pointing device and watch what happens.

3. Place the crosshairs inside the globe representation, also referred to as the *compass,* as shown in Fig. 36-2 and pick that approximate point.

A 3D model of the object appears on the screen.

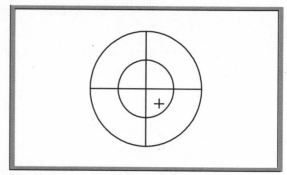

Fig. 36-2

Study the globe representation in Fig. 36-3 carefully. The placement of the crosshairs on the globe indicates the exact position of the viewpoint. Placing the crosshairs inside the inner ring (called the "equator") results in viewing the object from above. Placing the crosshairs outside the inner ring results in a viewpoint below the object.

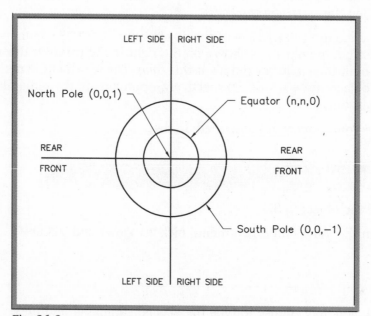

Fig. 36-3

If the crosshairs are on the right side of the vertical line, the viewpoint will be on the right side of the object. Similarly, if the crosshairs are in front of the horizontal line, the viewpoint will be in front of the object.

4. Enter **Z** (for ZOOM) and **.9x**.

The object should look somewhat like the one in Fig. 36-4.

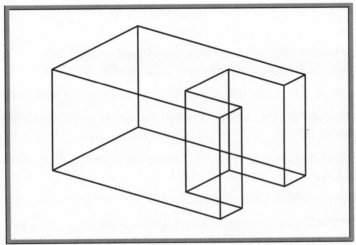

Fig. 36-4

The HIDE command removes edges that would be hidden if you were viewing a solid 3D object. Hiding these edges produces a more realistic view of the object. This is called *hidden line removal*.

5. From the **View** pull-down menu, select **Hide**, or enter the **HIDE** command at the keyboard.

The object should now look similar to the one in Fig. 36-5.

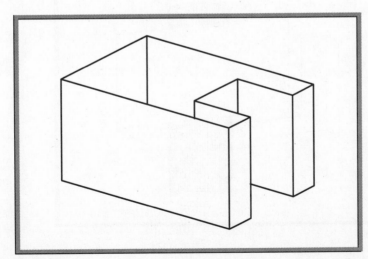

Fig. 36-5

6. Return to the plan view of the object by selecting the **View** pull-down menu and then selecting **3D Views** and **Top**.

7. **ZOOM All** to restore the original plan view of the model.

Changing Elevation and Thickness

Let's add the steel rod to the 3D model at a new elevation and thickness.

1. Enter the **ELEV** command and set the elevation at **−10′** and the thickness at **60′**.

2. Draw a **20′**-diameter circle in the center of the model, as shown in Fig. 36-1.

Visualize how the cylinder will appear in relation to the existing object.

3. Enter **VPOINT** and press **ENTER** twice. Place the crosshairs in approximately the same location as before and pick a point.

Does the model appear as you had visualized it?

4. Remove the hidden lines by entering the **HIDE** command.

The model should now look similar to the one in Fig. 36-1.

5. Experiment with **VPOINT** to obtain viewpoints from different points in space, and create a 3D view of the model that is similar to the one in Fig. 36-6.

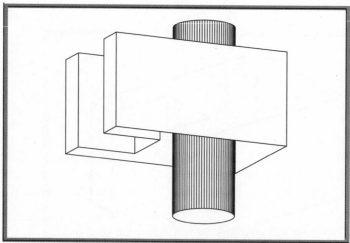

Fig. 36-6

HINT:

Type VPOINT and press ENTER twice. Select a point in the globe. Press the spacebar (or ENTER) to reissue the VPOINT command, and press ENTER to display the globe. Repeating this is a quick way to view the object from several different viewpoints.

6. Save your work.

Viewing Options

AutoCAD offers several options for viewing models in 3D space. One method, which you've practiced, is to use the VPOINT command and globe. Another is to use the 3D Views option in the Views pull-down menu. A third is to select buttons from a toolbar. Finally, you can make selections from a dialog box. In the future, the method you choose should be based on your personal preference. One method is only better than another if it's faster for you and if it accomplishes what you are trying to achieve.

Using the Predefined Viewpoint Buttons

AutoCAD offers several predefined viewpoint buttons in its View toolbar.

1. Display the **View** toolbar.

2. Rest the pointer on each of the buttons to view their names.

The cyan-colored side of the button's icon shows the view that will display when the button is picked.

3. Pick each of the buttons except for the first and last ones and enter **HIDE** after each one.

As you can see, this is a fast way of displaying preset views of a model. The SW Isometric View button displays a southwest isometric view of the model, as if the model were oriented on a map. Likewise, SE stands for southeast, NE stands for northeast, and NW stands for northwest.

Using a Dialog Box

1. Display the top view of the model and **ZOOM All**.

2. Enter **VP** at the keyboard.

This displays the Viewpoint Presets dialog box, as shown in Fig. 36-7.

Fig. 36-7

For now, ignore the Absolute to WCS and Relative to UCS radio buttons at the top of the dialog box.

Focus on the half-circle located at the right. It allows you to set the viewpoint height.

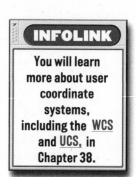

INFOLINK

You will learn more about user coordinate systems, including the WCS and UCS, in Chapter 38.

3. Pick a point (as shown in Fig. 36-7) so that you view the object from above at a **60°** angle and pick the **OK** button.

You should now be viewing the object from above at a 60° angle.

4. Enter **HIDE**.

5. Display the dialog box again.

The full circle (located in the left half of the dialog box) allows you to set the viewpoint rotation.

6. Pick a point so that you will view the object at a **135°** angle and pick **OK**.

7. Remove hidden lines.

8. Display the dialog box again.

The two edit boxes enable you to enter values for the rotation and height.

9. Pick the **Set to Plan View** button and pick **OK**.

10. **ZOOM All**.

11. Experiment further with the Viewpoint Presets dialog box.

12. Display the top view and **ZOOM All**.

3D Orbit

3D Orbit is a tool for manipulating the view of 3D models by clicking and dragging. It's especially useful because you can use it to manipulate 3D models while they are shaded or when hidden lines are removed. 3D Orbit's right-click shortcut menu offers options for removing hidden lines, rotating objects, panning, zooming, shading, perspective viewing, and clipping planes. You will explore these options in the following steps.

Standard

1. Pick the **3D Orbit** button from the docked Standard toolbar, or enter **3DO** at the keyboard.

This activates an interactive 3D Orbit view in the current viewport. 3D Orbit displays an object called an *arcball,* as well as a grid of lines (unless the grid was left off by the last person to use AutoCAD on your computer). The arcball is the green circle divided into four quadrants by smaller circles. When the 3DORBIT command is active, the target of the view stays stationary and the point of view (also called the *camera*) moves around the target. The target point is the center of the arcball, not the center of the objects(s) you are viewing.

2. Position the cursor inside the arcball and click and drag upward to rotate the model.

3. With the **3DORBIT** command active, right-click to display a shortcut menu.

4. In the menu, rest the pointer on **Shading Modes** and then select **Hidden**.

This removes the hidden lines from the model.

5. With the cursor inside the arcball, click and drag.

As you can see, the model remains in hidden-line mode.

A red, green, and blue icon made up of the X, Y, and Z axes appears in the lower left area (unless the last person to use AutoCAD on your computer turned it off). The icon serves as a visual aid to help you understand the direction from which you are viewing the model. If you do not see it now, that's okay.

Rotating Objects in 3D Space

When you move the cursor over different parts of the arcball, the cursor icon changes. When you click and drag, the appearance of the cursor indicates the rotation of the view.

1. Study the cursor; then move it outside the arcball and then back inside it.

The cursor changes to a small sphere encircled by two lines. When the cursor is a small sphere, as shown in Fig. 36-8A, you can manipulate the view freely.

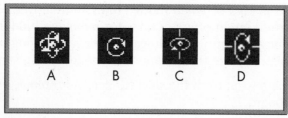

Fig. 36-8

2. With the cursor inside the arcball, click and drag up and down, back and forth.

3. Move the cursor outside the arcball.

The cursor changes to a circular arrow around a small sphere as shown in Fig. 36-8B.

When you click and drag outside the arcball, AutoCAD moves the view around an axis at the center of the arcball, perpendicular to the screen. This action is called a *roll*.

4. With the cursor outside the arcball, click and drag.

5. Move the cursor over one of the smaller circles on the left or right side of the arcball.

The cursor becomes a horizontal ellipse around a small sphere as shown in Fig. 36-8C.

When you click and drag from either of these points, AutoCAD rotates the view around the Y axis that extends through the center of the arcball.

6. With the cursor over one of the smaller circles at the left or right, click and drag.

7. Move the cursor over one of the smaller circles on the top or bottom of the arcball.

The cursor becomes a vertical ellipse around a small sphere as shown in Fig. 36-8D.

When you click and drag from either of these two points, AutoCAD rotates the view around the horizontal or X axis that extends through the center of the arcball.

8. With the cursor over one of the smaller circles at the top or bottom, click and drag.

Panning and Zooming

Standard

1. With the **3DORBIT** command active, right-click to display the shortcut menu and select **Pan**.

This causes AutoCAD to change to the pan mode within the 3DORBIT command.

2. Click and drag to adjust the view of the model.

3. Right-click and pick **Zoom**.

4. Zoom in and out on the model.

5. Display the shortcut menu and select **Orbit**.

The arcball returns.

Switching from Parallel to Perspective Projection

Consider the two buildings shown in Fig. 36-9. Both are shown from the same point in space. However, the one on the left is shown in *parallel projection*. This is the standard projection that AutoCAD uses when displaying 3D models, and this is how you are currently viewing the 3D model. The lines are parallel and do not converge toward one or more vanishing points. The building on the right is a true *perspective projection*. Parts of the building that are farther away appear smaller, the same as in a real-life photograph.

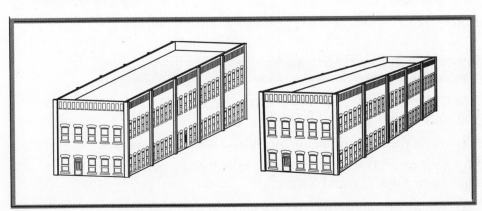

Fig. 36-9

1. With **3DORBIT** still active, right-click to display the shortcut menu.

2. Pick **Projection** and then **Perspective**.

You are now viewing the model in perspective projection.

3. Rotate the model.

4. If necessary, zoom out.

You are now getting a more realistic view of the model.

Shading the Object

1. Display the shortcut menu, pick **Shading Modes**, and pick **Flat Shaded**.

INFOLINK

You will learn more about shading in Chapters 42 and 43.

AutoCAD flat-shades the model. *Flat shading* is a very basic method of shading a model using a single shade of color. Notice that the cylinder has a coarse appearance. You will learn more about flat, or facet, shading later in this book.

2. Rotate the model.

The model remains shaded, which is another benefit of using 3D Orbit to rotate models.

3. From the shortcut menu, select **Shading Modes** and **Gouraud Shaded**.

Gouraud shading is a method of smooth shading that results in a more realistic model. The cylinder now has a smooth appearance.

4. Rotate the model.

The last two shading modes, Flat Shaded, Edge On and Gouraud Shaded, Edge On, shade the model and display the edges that represent the model.

5. Select **Flat Shaded, Edge On** and then select **Gouraud Shaded, Edge On**.

Notice the difference in the appearance of the model.

Controlling Visual Aids

AutoCAD provides several visual aids to help orient you in 3D space as you work with objects in 3D Orbit.

1. From the shortcut menu, pick **Visual Aids** and **Compass**.

This displays a 3D sphere within the arcball composed of three lines representing the X, Y, and Z axes.

2. Rotate the model and pay attention to the 3D sphere.

3. From the shortcut menu, turn off Compass.

The same shortcut menu controls the grid, which represents the 0 elevation baseplane discussed earlier in this chapter. It also controls the display of the UCS icon, which is the red, green, and blue icon that shows the current position of the X, Y, and Z axes.

4. From the shortcut menu, turn off **Grid** and **UCS Icon**, and rotate the model.

5. Turn on **Grid** and **UCS Icon** and rotate the model.

Setting Clipping Planes

A *clipping plane* is a plane that slices through one or more 3D objects, causing the part of the object on one side of the plane to be omitted. It is useful in viewing the interior of a part or assembly. You can use clipping planes to achieve views of an object that are similar to sectional views.

1. Display the shortcut menu, select **More**, and pick **Adjust Clipping Planes.**

2. In the Adjust Clipping Planes window, depress the **Front Clipping On/Off** button. (It may already be depressed.)

3. With the cursor inside this window, click and drag up and down.

This moves the clipping plane up and down and provides a corresponding view of the model in the drawing area.

4. In the Adjust Clipping Planes window, depress the **Back Clipping On/Off** button, as well as the **Adjust Back Clipping** button.

5. With the cursor inside this window, click and slowly drag up and down to adjust the location of the back clipping plane.

6. Pick the **Create Slice** button, click, and move the cursor back and forth.

Now, with the front and back clipping planes on, you can adjust their location at the same time.

7. Close the Adjust Clipping Planes window.

The last two options on the shortcut menu help you control the current view of the object. Reset View resets the view back to how it was when you first started 3D Orbit. The Preset Views option displays a list of predefined views such as Top, Front, and NW Isometric.

8. Display the shortcut menu and pick **Reset View.**

9. Change the projection to parallel and the shading to wireframe, and turn off the grid.

10. Press **ESC** or **ENTER** to exit 3D Orbit.

11. Close the View toolbar, save your work, and exit AutoCAD.

Chapter 36
Review & Activities

Review Questions

1. Describe the purpose of the VPOINT command.

2. The extrusion thickness of an object is specified with which command?

3. Briefly explain the process of creating objects (within the same model) at different elevations and thicknesses.

4. When you are using the VPOINT command and the small crosshairs are in the exact center of the globe, what is the location of the viewpoint in relation to the object?

5. With what command are hidden lines removed from 3D objects?

6. Sketch the VPOINT globe shown in Fig. 36-10 and indicate where you must position the small crosshairs to view an object from the rear and underneath.

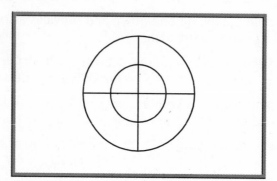

Fig. 36-10

7. Why is the 3D Orbit feature especially useful, compared to other 3D viewing options?

8. What is the difference between parallel projection and perspective projection?

9. What is a clipping plane?

Chapter 36
Review & Activities

Challenge Your Thinking

1. Match the VPOINT globe representations in Fig. 36-11 with the objects. The first one has been completed to give you a starting point. Write your answers on a separate sheet of paper.

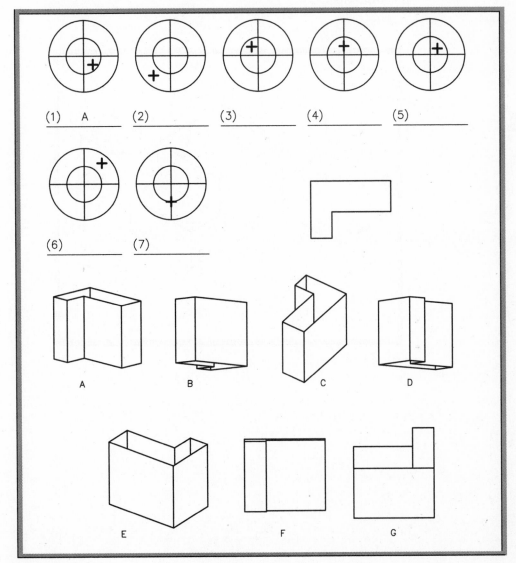

Fig. 36-11

Chapter 36
Review & Activities continued

2. Of what benefit are front and back clipping planes? Describe at least one situation in which clipping planes could save drawing time or reveal important information about a design.

3. Enter the *x*, *y*, and *z* coordinate values for the alphabetically identified points in Fig. 36-12. To get you started, point A has coordinate values of 0,0,0. Write your answers on a separate sheet of paper.

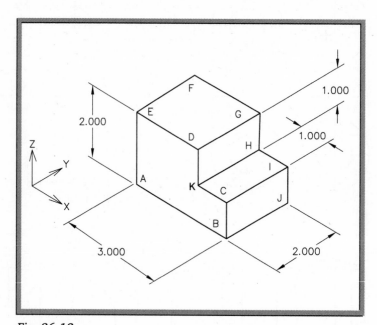

Fig. 36-12

"Challenge Your Thinking question 3 courtesy of Mark Schwendau, Kishwaukee College

Chapter 36
Review & Activities

Applying AutoCAD Skills

Work the following problems to practice the commands and skills you learned in this chapter. Draw the following objects and generate a 3D model of each.

1. Set the elevation at 1″ to create a 3D model of the guide shown in Fig. 36-13.

2. To draw the shaft shown in Fig. 36-14, set the elevation for the inner cylinder at 0. Set the snap resolution prior to picking the center point of the first circle. Do not use object snap to select the center point of the second circle.

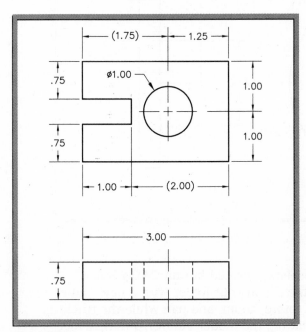

Fig. 36-13

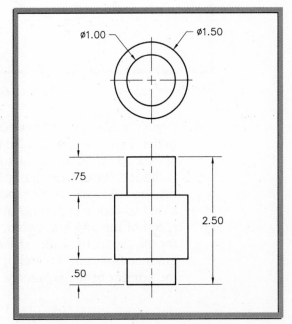

Fig. 36-14

Chapter 36
Review & Activities continued

3. Draw the gear blank and hub shown in Fig. 36-15 as a basic 3D model. View the model using each of the buttons on the View toolbar. Are all the views practical for this model? Which ones are not? Enter the HIDE command and view the model again from each position. Is the HIDE command practical for all views? Why?

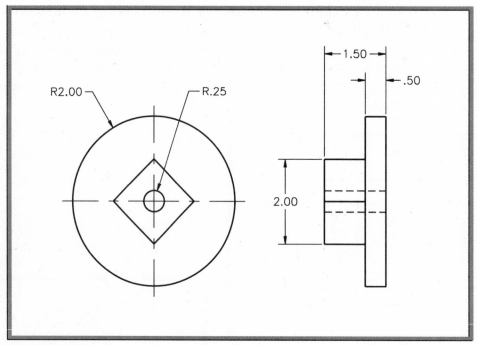

Fig. 36-15

4. Open the file named Truck model.dwg located in AutoCAD's Sample folder. Using Save As..., save the file in your folder with a new name of truck2.dwg. Use 3D Orbit to rotate, zoom, and pan while the truck is shaded. As you are manipulating the model, AutoCAD may reduce the shaded mode of the model to wireframe, depending on the speed of your computer. AutoCAD uses this method to maintain speed of the rotation, zoom, or pan, rather than require you to wait for the model to shade. Use each of the shading modes in 3D Orbit. Produce front and back clipping planes through the truck to view the interior of the geometry.

Chapter 36
Review & Activities

Using Problem-Solving Skills

Complete the following activities using problem-solving skills and your knowledge of AutoCAD.

1. You are teaching a class in AutoCAD. Use the gear blank and hub you created in "Applying AutoCAD Skills" problem 3 to demonstrate the basic principles of 3D Orbit. In your demonstration, show the effects with and without the grid present. Also use the HIDE command at various times. Remember that some actions will discontinue the 3DORBIT command, and you must reenter the command to continue with your demonstration.

2. The gear manufacturer for whom you supply the gear blank (see "Applying AutoCAD Skills" problem 3) wants to see the blank in a realistic form. Provide a three-dimensional view of the gear blank flat-shaded with the edges on. Save and plot the drawing. Did it plot the way you expected?

3. Use your knowledge of AutoCAD to duplicate the shaft, gear, and cam shown in Fig. 36-16 as closely as possible.

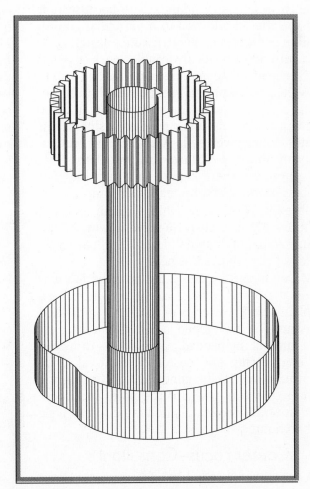

Fig. 36-16

Careers Using AutoCAD

The Stock Market/Tom & Dee McCarthy

Telephone Towers

As soon as telephones became portable, people began talking on the phone everywhere—in the bleachers, in the car, in line at the market. The growth of wireless communications has been so rapid that cellular and digital telephone companies have been hard-pressed to build antenna poles and towers fast enough.

A telecommuncation pole or tower takes time to design and manufacture because it is site-specific. In addition to a particular configuration of antennas, the structure's design must take into account local variables such as wind and ice loads. One manufacturer needed two to six weeks to complete and deliver the components of a pole, ready for construction. To keep up with the accelerating pace of business, this manufacturer invested in AutoCAD and brought in a consultant, R A Maxey and Associates, Inc., to automate their design and engineering processes and speed up manufacturing.

Based in Westerville, Ohio, Randall Maxey is an AutoCAD specialist. He serves clients by tailoring AutoCAD capabilities to the particular needs of the client's business.

Maxey began by analyzing how the job was being done and how it might be automated. Applying his knowledge of AutoCAD and allied software, he developed a solution that now generates all the needed information—drawings for engineering and customer approval, code for the numerically controlled (NC) plasma burner that cuts steel for the pole, and documentation for the installation crew—in as little as two hours.

Career Focus—Consultant

A consultant is an independent businessperson who helps clients in an area of technical or business expertise, often on a short-term or project basis. Typically, a company hires a consultant when it is about to make a change in its way of doing business. For companies implementing AutoCAD, a consultant such as R A Maxey and Associates, Inc. may offer training, finished drawings, research and recommendations, or customer software solutions. In addition to serving clients, a consultant must also devote time to running the business, marketing, and looking for future clients. Randall Maxey maintains contacts with Autodesk dealers and attends industry conferences.

Education and Training

Expertise is the "product" that a consultant offers for sale. To achieve this expertise, most consultants work for several years in a specialized field before leaving regular employment to try consulting.

From the client's point of view, education is an important part of a consultant's credibility. As a Registered Autodesk Developer, Maxey stays up-to-date with in-depth technical and marketing support from Autodesk.

 Career Activity

- Make a design for a telephone pole or tower that rises 60 feet above the ground. Design the tower to optimize two criteria: 1) quick installation, and 2) easy attachment and removal of antennas.

Chapter 37 Point Filters

Objectives:

- Apply X/Y/Z point filters to specify points in 3D space
- Create 3D faces
- Draw lines in 3D space

Key Terms

plan view
point filtering
3D face
X/Y/Z filtering

This chapter continues with AutoCAD's 3D modeling capabilities. X/Y/Z filters enable you to enter coordinates using both the pointing device and the keyboard. With the LINE and 3DFACE commands, you can create lines and surfaces in three-dimensional space using x, y, and z coordinates. For instance, you can create inclined and oblique surfaces, such as a roof on a building.

Creating 3D Faces

The 3DFACE command creates a three-dimensional object similar in many respects to a two-dimensional solid object. The 3DFACE prompt sequence is identical to that of the SOLID command. However, unlike the SOLID command, the 3DFACE command allows you to enter points in a natural clockwise or counterclockwise order to create a normal *3D face,* or planar surface. *Z* coordinates are specified for the corner points of a 3D face, forming a section of a plane in space.

1. Start AutoCAD and create a new drawing as follows.
    - Use the **Quick Setup** wizard.
    - Set the units to **Architectural**.
    - Set the area to **120′ × 90′**
    - **ZOOM All**.

2. Create a **10'** grid and a **5'** snap.

3. Create a new layer named **Objects**, assign the color red to it, and make it the current layer.

4. Enter the **ELEV** command and set the new default elevation at **10'** and the new default thickness at **30'**.

5. Draw the outline of the building shown in Fig. 37-1 using the **LINE** command. The width and length of the building are **40'** and **60'**, respectively. Approximate the remaining dimensions. (These dimensions will later change.)

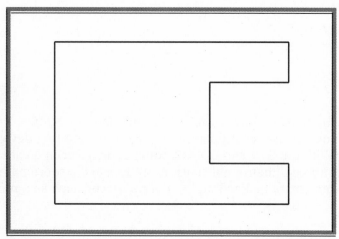

Fig. 37-1

6. Save your work in a file named **3d2.dwg**.

7. With the **VPOINT** command, view the outline of the building from above, front, and left.

8. Remove hidden lines using the **HIDE** command.

9. Undo the last two steps so that you again see the top view.

10. Display the **Surfaces** toolbar.

Surfaces

11. From the Surfaces toolbar, pick the **3D Face** button.

This enters the 3DFACE command.

X/Y/Z Point Filtering

The process of using the pointing device to enter any two of the 3D coordinates that define a point and then entering the third coordinate from the keyboard is known as *point filtering*. This method allows you to specify points in 3D space without constantly changing your point of view to make the points accessible.

1. Type **.xy** and press **ENTER**.

2. In reply to of, approximate the *x,y* position of point 1 in Fig. 37-2 and pick that point. (Point 1 is about 5′ from the corner of the object.)

This is the start of one side of the roof.

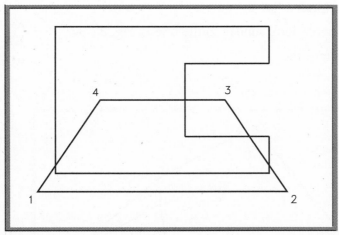

Fig. 37-2

As shown on the Command line, the first point also requires a *z* coordinate.

3. Enter **35′** for the *z* coordinate.

This method of entering 3D points is referred to as *X/Y/Z filtering*.

4. In reply to Second point, enter **.xy** and pick point 2, as shown in Fig. 37-2.

5. Enter **35′** for the *z* coordinate.

6. Repeat these steps for points 3 and 4, but enter **70′** for the *z* coordinate of these points. (Be sure to enter **.xy** before picking the point.)

7. Press **ENTER** to terminate the 3DFACE command.

Mirroring a 3D Face

The MIRROR command is not limited to two dimensions. When you mirror an object such as the 3D face you just created, the object is mirrored in 3D space. Let's use the MIRROR command to create the other half of the roof.

1. Enter the **MIRROR** command, select the 3D face object you just created, and press **ENTER**.

2. Place the mirror line so that the two sections of the roof meet along the shorter horizontal edge of the 3D face. Do not delete the "old object."

Now let's view the object in 3D.

3. View the model from above, front, and right side.

4. Remove hidden lines.

Your model should look similar to the one in Fig. 37-3.

5. Save your work.

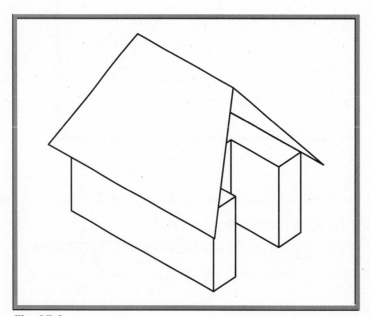

Fig. 37-3

Snapping to Points in 3D Space

A third way to create a 3D face is to snap to existing points at different *x,y,z* coordinates. Let's use this method to finish the roof so that it looks similar to the one in Fig. 37-4.

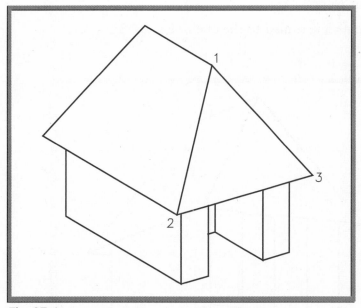

Fig. 37-4

Surfaces

1. With the current 3D view still on the screen, enter the **3DFACE** command and snap to point 1, as shown in Fig. 37-4.

2. Snap to point 2.

3. Snap to point 3.

4. Snap to point 1 to complete the surface boundary.

5. Press **ENTER** to terminate the 3DFACE command.

6. Remove hidden lines.

7. View the object from above, front, and left side.

8. Create the remaining portion of the roof using the **3DFACE** command and object snap, and then remove hidden lines.

HINT:

To complete Step 8, you may choose to return to the plan view and use the MIRROR command to complete the roof.

9. Save your work.

10. Generate a view similar to the one in Fig. 37-5.

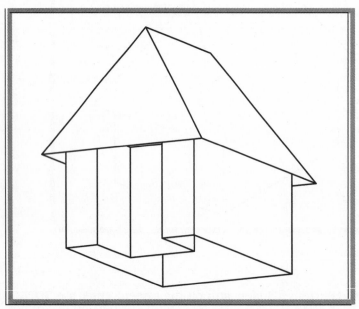

Fig. 37-5

Controlling Visibility of Edges

3D faces can contain visible and invisible edges. The Invisible option is useful for connecting two or more 3D faces to create a 3D model.

1. Generate the plan view.

HINT:

The *plan view* is the top view. You can generate it by entering VPOINT and 0,0,1. These coordinates define the direction from which you view the model.

2. Enter **ZOOM** and **.6x**.

3. Using **3DFACE**, create the following pentagon anywhere on the screen and at any size. Pick the points in the order shown in Fig. 37-6.

Surfaces

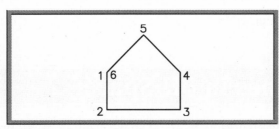

Fig. 37-6

Notice that the object contains two lines that do not appear in Fig. 37-6.

4. Press **ENTER** to terminate the command.

5. Create another pentagon similar to the first one, but create this one according to the following sequence.

Command: **3DFACE**
Specify first point or [Invisible]: (pick point 1)
Specify second point or [Invisible]: (pick point 2)
Specify third point or [Invisible] <exit>: (pick point 3)
Specify fourth point or [Invisible] <create three-sided face>: (enter **i** and pick point 4)
Specify third point or [Invisible] <exit>: (pick point 5)
Specify fourth point or [Invisible] <create three-sided face>: (enter **i** and pick point 6)
Specify third point or [Invisible] <exit>: (press **ENTER**)

The system variable SPLFRAME controls the display of invisible edges in 3D faces. When SPLFRAME is set to a nonzero value, invisible edges are displayed. This allows you to edit them as you would a visible 3D face.

Figure 37-7 shows the same pentagon with SPLFRAME set at a nonzero value.

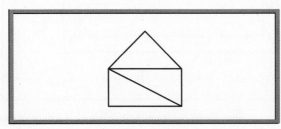

Fig. 37-7

6. Enter **SPLFRAME** and enter **1**.

7. Enter **REGEN**.

The pentagon should now look similar to the one in Fig. 37-7.

8. Set **SPLFRAME** back to **0** and enter **REGEN**.

9. Erase both pentagons.

Creating Lines in 3D Space

The LINE command creates lines in 3D space. Therefore, all lines you have created with AutoCAD are three-dimensional. Their endpoints are made up of *x, y,* and *z* coordinates. In 2D drawings, the *z* coordinate is 0, so it is not usually mentioned.

Let's create a 3D property line around the building using LINE.

1. **ZOOM All**.

2. Enter the **ELEV** command and change the elevation to **10′** and the thickness to **0**.

3. Enter the **LINE** command and create the property line shown in Fig. 37-8. (Approximate its location.)

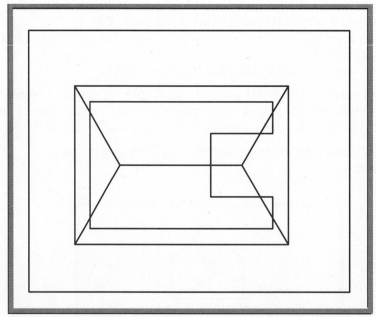

Fig. 37-8

4. Obtain a view similar to the one in Fig. 37-9.

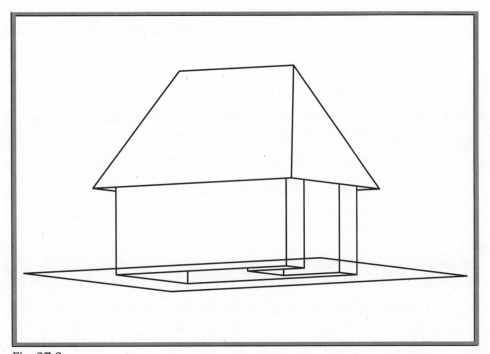

Fig. 37-9

5. Experiment on your own with the **LINE** and **3DFACE** commands.

6. Save your work.

7. Close the Surfaces toolbar and exit AutoCAD.

Chapter 37
Review & Activities

Review Questions

1. Describe a 3D face object.

2. What basic AutoCAD command is the 3DFACE command most like?

3. What can be created with the 3DFACE command that cannot be created with the LINE command? Be specific.

4. Describe the X/Y/Z filtering process of entering 3D points.

5. What is the primary benefit of using X/Y/Z point filters?

6. How can you create lines with thickness in 3D space? Explain.

Challenge Your Thinking

1. Experiment with xlines and mlines in three dimensions. Can you use the X/Y/Z point filters to place these objects? In what ways do they behave differently in 3D space than ordinary AutoCAD lines?

2. Experiment with other combinations of the X/Y/Z point filters, such as .xz and.yz. In what situations might these combinations be useful?

Applying AutoCAD Skills

Work the following problems to practice the commands and skills you learned in this chapter.

1. Open the slide.dwg file from Chapter 22 ("Applying AutoCAD Skills," problem 1), and redraw it in three dimensions. Make the thickness 3 units. Use X/Y/Z point filtering to create the angled surface. Save the drawing as ch37slide.dwg.

Chapter 37
Review & Activities

2. Using AutoCAD's 3D modeling features, add a door and chimney to the two-story building you created earlier. When finished, the model should look similar to the one in Fig. 37-10. Save the drawing as prb37-2.dwg.

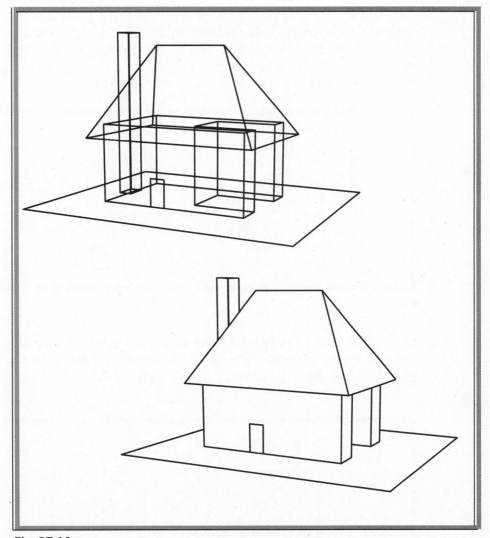

Fig. 37-10

3. Embellish the drawing further by adding other details to the building. Save the finished drawing as prb37-3.dwg.

Chapter 37
Review & Activities continued

 Using Problem-Solving Skills

Complete the following problem using problem-solving skills and your knowledge of AutoCAD.

1. Your lampshade design needs to be drawn in three dimensions before your supervisor will approve the design. Draw a 3D model of the lampshade shown in Fig. 37-11; omit the hanger. Use the appropriate method(s) for creating the 3D faces. Save the drawing as ch37lampshade.dwg. Plot it in the view of your choice.

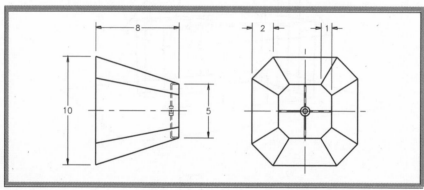

Fig. 37-11

2. Draw the 3D model of the die head shown in Fig. 37-12. Complete the four angled surfaces with the 3DFACE command and X/Y/Z point filtering. Save the drawing as ch37diehead.dwg.

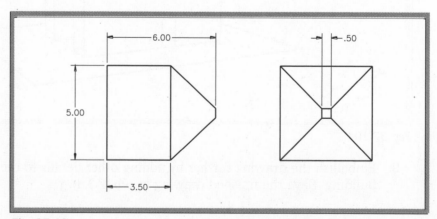

Fig. 37-12

Chapter 38 | User Coordinate Systems

AutoCAD's default coordinate system is the world coordinate system (WCS). In this system, the X axis is horizontal on the screen, the Y axis is vertical, and the Z axis is perpendicular to the XY plane (the plane defined by the X and Y axes). The WCS is indicated in the lower left corner of the screen by the coordinate system icon shown in Fig. 38-1.

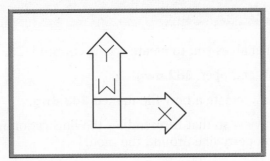

Fig. 38-1

A *coordinate system icon* shows the orientation of the axes in the current coordinate system. You will learn more about variations of this icon as you work through this chapter.

You may create drawings in the world coordinate system, or you may define your own coordinate systems, called *user coordinate systems (UCSs)*. The advantage of a UCS is that its origin is not fixed. You can place it anywhere within the world coordinate system. You can also rotate or tilt the axes of a UCS in relation to the axes of the WCS. This is a useful feature when you're creating a 3D model.

Consider the 3D model shown in Fig. 38-2. A UCS was defined to match the inclined plane of the roof as indicated by the arrows. Once the UCS has been established and made current, all new objects lie in the same plane as the roof.

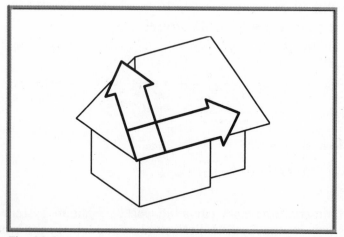

Fig. 38-2

Creating UCSs

The UCS command enables you to create a new current UCS.

1. Start AutoCAD and open **3d2.dwg**.

2. Using **Save As...**, create a new file named **3d3.dwg**.

3. Alter the plan view so that it resembles the illustration in Fig. 38-3. Remove the property line around the model.

 HINT:

Be sure to set the elevation at 10′ and the thickness at 30′ before drawing new lines.

4. Obtain a view similar to the one shown in Fig. 38-2.

5. Display the **UCS** toolbar.

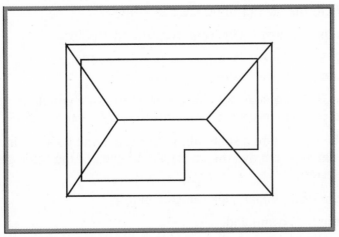

Fig. 38-3

Using Three Points

One of the most common and useful ways of defining a UCS in AutoCAD is to select three points that lie on the axes. The first point becomes the origin, and the other two define the X and Y axes.

1. From the UCS toolbar, pick the **3 Point UCS** button.

AutoCAD enters the UCS command. Because you chose the 3 Point UCS button, AutoCAD entered the 3point option automatically.

2. In reply to Specify origin point, snap to point 1 (shown in Fig. 38-4).

UCS

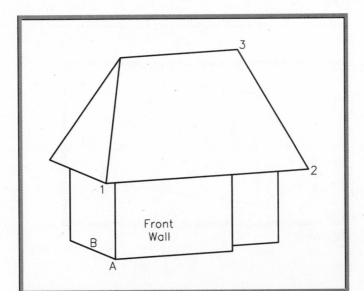

Fig. 38-4

3. In reply to the next prompt, snap to point 2.

This defines the positive X direction from the first point.

4. In reply to the next prompt, snap to point 3. (Point 3 lies in the new XY plane and has a positive *y* coordinate.)

Note that the drawing does not change. However, the crosshairs, grid (if it is on), and coordinate system icon shift to reflect the new UCS.

5. With the coordinate display on, notice the *x* and *y* values in the status bar as you move the crosshairs to each of the three points you selected.

6. From the UCS toolbar, pick the **UCS** button.

UCS

This enters the UCS command.

7. Enter **S** for *Save*, and enter **frtroof** for the UCS name.

This names and saves the current UCS.

Suppose you want to construct the line of roof intersection between the roof and a chimney passing through the roof.

8. Change both the elevation value and the thickness value to **0**.

9. With the **LINE** command, draw the line of intersection where a chimney would pass through (as shown in Fig. 38-5) by creating a rectangle at any location on the roof.

 HINT: ───────────────────────────

Turn on snap.

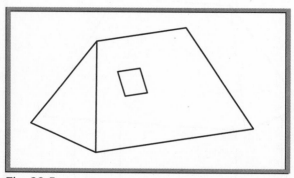

Fig. 38-5

10. Pick the **World UCS** button from the UCS toolbar to return to the WCS.

UCS

11. To prove that the rectangle lies on the same plane as the roof, view the 3D model from different points in space.

HINT:

The viewpoint 4,−.1,1 illustrates it well.

Interpreting the Coordinate System Icon

The coordinate system icon indicates the positive directions of the X and Y axes. Variations of the icon provide additional information, as shown in Fig. 38-6. The letter W appears in the icon if the current UCS is the world coordinate system. If the icon is located at the origin of the current UCS, a + is displayed at the base of the icon. A box forms at the base of the icon if you're viewing the UCS from above. The box is absent if you are viewing the UCS from below.

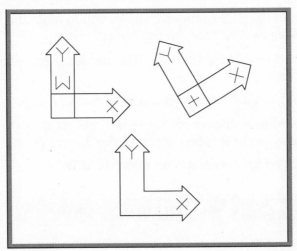

Fig. 38-6

The UCSICON command permits you to control the visibility and position of the coordinate system icon.

1. Enter **UCSICON** and **OR** for ORigin.

The coordinate system icon moves to the origin of the current UCS.

2. Enter **UCSICON** and select the **Noorigin** option.

The coordinate system icon returns to its original location.

3. Enter **UCSICON** and **OFF**.

The coordinate system icon disappears.

4. Enter **UCSICON** and **ON**.

The coordinate system icon reappears.

The UCSICON All option applies changes to the coordinate system icons in all active viewports.

Using an Object

Another convenient way to create a UCS is to align it with a planar object. AutoCAD allows you to specify any object that lies entirely within one plane (such as a 3D face) to define the UCS. Let's create a second UCS, this time using the Object option of the UCS command.

1. View the model from an orientation similar to the one used before (as shown in Fig. 38-2).

2. Pick the **Object UCS** button from the UCS toolbar.

UCS

The Object option of the UCS command allows you to align the UCS with a planar object, such as a 3D face, on the screen.

3. Select the bottom edge of the roof. (Use the same section of the roof as before.)

The crosshairs shift to reflect the orientation of the new UCS.

4. With the coordinate display on, move the crosshairs to the same three corners you chose when applying the 3point option.

Notice that the Object option creates an identical UCS.

Using a Z Axis Vector

You are likely to use the 3point and Object options for most applications, at least at first. However, other UCS creation options are available. In the Z Axis Vector option, you specify the positive direction of the Z axis. AutoCAD then determines the directions of the X and Y axes using an arbitrary but consistent method.

UCS

1. From the UCS toolbar, pick the **Z Axis Vector UCS** icon. (Turn off ortho.)

2. Snap to point A (see Fig. 38-2) in reply to Specify origin point.

3. In reply to the next prompt, pick any point on line B using the **Nearest** object snap.

The new UCS is on the same vertical plane as the front wall of the building. In relation to the wall, notice the positive X, Y, and Z directions that make up the UCS, and notice the coordinate system icon.

Using Other Options

The View UCS button creates a new UCS perpendicular to the current viewing direction; that is, parallel to the screen. This is helpful when you want to annotate (add notes to) the 3D model.

UCS

1. Pick the **View UCS** button.

2. Using the **DTEXT** command, place your name near the model.

The Z Axis Rotate option creates a new UCS by rotating the current UCS a specified number of degrees around the Z axis.

UCS

3. Pick the **Z Axis Rotate UCS** button.

4. Enter **30** for the rotation angle.

The current UCS rotates 30° about the Z axis. The X Axis Rotate UCS and Y Axis Rotate UCS options rotate the current UCS about the X and Y axes.

The Origin option allows you to create a new UCS by moving the origin of the UCS, leaving the orientation of the axes unchanged.

UCS

5. Pick the **Origin UCS** button.

6. Pick the lower right corner of the roof (point 2; see Fig. 38-4).

Notice how the coordinate system icon moves.

Managing UCSs

The purpose of creating user coordinate systems is to make it easier to work with 3D models. When you have created several UCSs, you need to be able to move from one to another quickly. You may also want to delete unused UCSs. The UCS command and toolbars allow you to manage UCSs efficiently.

Changing the Current UCS

AutoCAD provides several methods to change from one UCS to another. Examples include the Previous and World options. The UCS II toolbar also allows you to switch from one UCS to another quickly.

1. From the UCS toolbar, pick the **UCS Previous** button.

This restores the previous UCS. AutoCAD saves the last ten user coordinate systems in a stack, so you can step back through these systems with repeated UCS Previous operations.

2. Pick the **World UCS** button.

This restores the world coordinate system. Notice that W (for World) appears in the coordinate system icon.

NOTE:

> If you want to place a point in the WCS while you are in another UCS, do so by entering an asterisk prior to the coordinates (*e.g.*, *7.5,4). During line construction, you can also place relative coordinates (*e.g.*, @*5,3) and polar coordinates (*e.g.*, @*3.5<90) in the WCS regardless of the current UCS.

3. Display the **UCS II** toolbar and pick the down arrow.

This displays all named UCSs, as well as predefined orthographic UCSs, such as Top, Bottom, and Front. Notice also that the list also contains the Previous and World options.

4. Pick the **frtroof** UCS from the list.

This restores the frtroof UCS.

5. Pick **Back** from the list.

This restores the Back UCS. This UCS would be suitable for editing the back side of the building.

NOTE:

> You can also restore named UCSs from the command line. Enter the UCS command and press R (for Restore). AutoCAD asks for the name of the UCS to restore. The disadvantage of using this method is that you must remember the exact name of the UCS you want to restore.

Generating the Plan View

The PLAN command enables you to generate the plan view of any user coordinate system, including the WCS.

1. Enter the **PLAN** command.

2. Press **ENTER** to generate the plan view of the current UCS.

3. Enter **PLAN** and **W** for World.

A "broken pencil" icon, shown in Fig. 38-7, replaces the coordinate system icon whenever the XY plane of the current coordinate system is perpendicular to the computer screen. This indicates that drawing and selecting objects is limited in this situation.

UCS

4. Pick the **World UCS** button from the UCS toolbar.

5. Erase your name and **ZOOM All**.

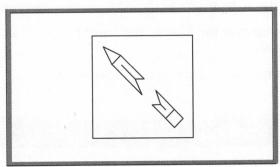

Fig. 38-7

NOTE: _____

If you want to display the plan view of previously saved user coordinate systems, enter the UCS option of the PLAN command.

Deleting a UCS

The UCS Del option, which does not have a corresponding button in the UCS toolbar, allows you to delete one or more saved coordinate systems. You may want to delete a UCS that you have defined incorrectly, or you may need to decrease the file size.

1. Enter the **UCS** command and press **D** (for Del).

2. Type **frtroof** and press **ENTER**.

AutoCAD deletes the definition of the frtroof UCS from the drawing database.

3. Pick the **Undo** button to undo the deletion of the UCS.

4. Close the UCS and UCS II toolbars, save your work, and exit AutoCAD.

Chapter 38
Review & Activities

Review Questions

1. What is the purpose and benefit of using AutoCAD's user coordinate systems?

2. Describe the UCS command's 3point option.

3. What purpose does the coordinate icon serve?

4. What does the W represent in the coordinate system icon?

5. If a box is present at the base of the coordinate system icon, what does this mean?

6. Briefly describe the UCS command's View option.

Challenge Your Thinking

1. Explain the difference between a view and a UCS, and describe how the two can be used together to create complex 3D models.

2. Using AutoCAD's online help, research the UCSFOLLOW system variable. What is the purpose of this variable? Why might it be useful when you are working with several different UCSs?

 ## Applying AutoCAD Skills

Work the following problems to practice the commands and skills you learned in this chapter.

1. Create a new drawing with a drawing area of 24' × 18'. Set the thickness to 8', and draw a 16' × 12' rectangle to represent a room. View the room from the SE Isometric viewpoint. Use the 3Point option of the UCS command to create a new UCS that will allow you to create a door on the right end wall of the room. Using the new UCS, draw a door that is 3' wide and 6'6" high. Center the door on the right wall of the room. Save the drawing as ch38room.dwg.

Chapter 38
Review & Activities

2. With the UCS 3point option, create a UCS using one of the walls that make up the building in this chapter. Save the UCS using a name of your choice. With this as the current UCS, add a door and window to the building. Save this drawing as prb38-2.dwg.

3. Obtain a 3D view of the building from this chapter. With the UCS View option, create a UCS. Create a border and title block for the 3D model. Include your name, the date, the file name, etc., in the title block. Save your work as prb38-3.dwg.

 ## Using Problem-Solving Skills

Complete the following activities using problem-solving skills and your knowledge of AutoCAD.

1. Make a 3D drawing of the track shown in Fig. 38-8. Your supervisor wants further design work on the 60° angular face. Set up a UCS on the face in preparation for the redesign. Save the drawing as ch38track.dwg.

2. Your job as a heating contractor requires that you show the exact location of baseboard hot water heating elements. Open ch38room.dwg, which you created in "Applying AutoCAD Skills" problem 1. Locate two holes, one for input and the other for output, in each of the remaining three walls. Each hole is 2" in diameter, and the center of each hole is 4" up from the floor and 1' in from the adjoining wall. Make a UCS appropriate for each wall before drawing the respective holes. Save the drawing as ch38heating.dwg.

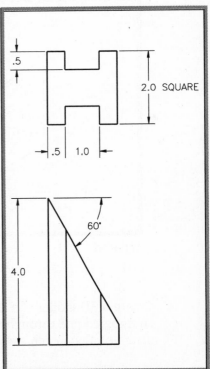

Fig. 38-8

Part 7 Project

3D Modeling

Sometimes, to model an object accurately, you need an alternative to the multiview drawings. Two such alternatives are isometric views and 3D models. AutoCAD provides special options to facilitate such drawings.

Description

This project incorporates two problems:

1. Draw an isometric version of a die (one of two dice), as shown in P7-1. Include the spots, indicating values of one, two, and three on the three faces showing. Center the spots accurately, and make the length of each side 2″ long. Dimension the length, height, and depth of the die.

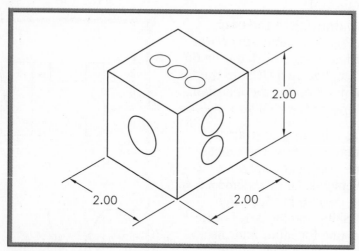

Fig. P7-1

2. Create a 3D model of a die the size of a 3″ cube. Draw the spots (numbered one through six) on each of the six faces, symmetrically arranged upon each surface, as shown in Fig. P7-2.

Part 7 Project

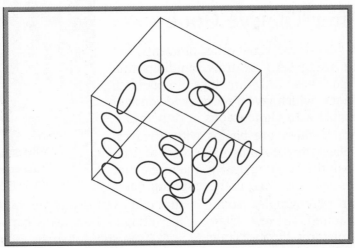

Fig. P7-2

Hints and Suggestions

- Remember that circles appear elliptical in isometric drawings. Use the Isocircle option with the correct isoplane orientation.
- Use the DIVIDE command to place the spots. Remember to set PDMODE to a value that allows you to see the divisions, and use REGEN to make the divisions appear.
- Use the HIDE command before viewing and plotting the final 3D model.
- You may wish to use the BOX primitive to create the 3D model; this command allows you to create a perfect cube of any size.
- Create two viewports of the 3D model to show all six surfaces. You should only need to create one model.

Summary Questions/Self-Evaluation

1. What effect does toggling the isometric plane have upon the use of the ortho feature?
2. Distinguish between thickness and elevation.
3. Of what value is the 3D globe (compass) when you work on 3D models?
4. Describe one possible advantage of reorienting the UCS icon.

Careers
Using
AutoCAD

See the Sites After They've Gone

Draw first, then build—that is the usual order of events. But historical researchers are using CAD to draw buildings that were demolished years ago. These CAD drawings provide data for walk-through animations, which give a more vivid sense of "being there" than is possible with photographs or drawings.

Re-creation of historical architecture begins with archives: design drawings, floor plans, any documents related to construction. With the shape of the building drawn, the researcher looks for the right colors and textures. If the building dates back farther than color photography, some detective work may be required. The researcher might compare other structures by the same builder or look at materials that came from the same source—for example, stonework from the same quarry. Newspaper accounts or diaries may reveal key details.

CAD re-creations have brought back "lost" works of Frank Lloyd Wright. Similar techniques are being used in preservation of standing historic structures, such as the Nebraska State Capitol.

Archaeologists, too, are using CAD to reconstruct settlements from hundreds and even thousands of years ago. A fourteenth-century site in Greece, discovered through satellite imaging, has been identified as construction by Crusaders on their way to the Holy Land. Working from artifacts and other evidence, researchers have built a model of this outpost in CAD.

On the other side of the world, a town in pre-Columbian Central America has come to life again through CAD animation. Researchers obtained much of their fine detail from imprints in volcanic ash.

Career Focus—College Faculty

College and university faculty teach classes within a specialty, such as anthropology or architecture. They also serve on committees

The Stock Market/Jean Miele

that carry out departmental functions, such as recruiting and updating curriculum.

College and university faculty stay abreast of research within their field by reading journals and attending professional gatherings. At many institutions, faculty members are expected to pursue research, adding to the knowledge in their field. Publication of research in scholarly journals and books is often a requirement for tenure (permanent employment).

Education and Training

Most faculty positions require a doctoral degree. In some cases, special experience in art, music, or law may qualify one for a faculty position. A doctoral program typically requires four to seven years of study. The candidate must pass examinations covering the major subjects within the field. The candidate then writes a doctoral dissertation, documenting his or her research into a specialized subject.

Career Activity

- Interview someone who can tell you about faculty work at a four-year college or university. Are the years of extra schooling worth the effort? Explain.

Part 8

Surface Modeling and Rendering

Objectives:

- Create a 3D template
- Control the appearance of surface meshes
- Create revolved and ruled surfaces

Key Terms

path curve
profile
revolved surface
ruled surface
surface
surface mesh

Most 3D models include revolved and ruled surfaces. A *surface* is a mesh of polygons, such as triangles, or a mathematical description using spline information. The REVSURF command creates a revolved surface by rotating a *path curve* (also called a *profile*) around a selected axis. The base of the lamp shown in Fig. 39-1 is an example of a *revolved surface*. The RULESURF command creates a polygon mesh representing a surface between two curves. The lampshade is an example of a *ruled surface*.

Fig. 39-1

Developing a 3D Template File

In preparation for creating a lamp similar to the one shown in Fig. 39-1, let's create a new template file. The template will apply to subsequent 3D exercises in this and the following chapters.

1. Start AutoCAD and use the following information to create a new drawing file.
   * Use the **Quick Setup** wizard.
   * Choose **Decimal** for the units.
   * Create a drawing area of **96" × 72"**.
   * **ZOOM All**.

2. Set the grid at **12** and the snap at **6**.

3. Create the three layers shown in Table 39-1.

Layer Name	Color	Linetype	Lineweight
Box	white	CONTINUOUS	0.15 mm
3dobject	red	CONTINUOUS	0.40 mm
Object2	cyan	CONTINUOUS	0.40 mm

Table 39-1

4. Set the layer named **Box** as the current layer.

5. Enter the **ELEV** command and enter **0** for the elevation and **24** for the thickness.

6. In the center of the screen, construct a rectangular box using a polyline. Make it **36"** on the X axis by **24"** on the Y axis.

7. Specify a viewpoint in front, above, and to the right of the rectangular 3D box.

8. Enter **ZOOM** and **.9x**.

9. Save your work in a template file named **3dtemp.dwt**. For Template Description, enter **3D template file**.

10. Make **3dobject** the current layer.

Controlling the Appearance of Surfaces

Before creating a surface, it's helpful to understand how they are constructed and what controls the quality of their appearance. AutoCAD uses two system variables, SURFTAB1 and SURFTAB2, to control the density (also referred to as the *resolution*) of surface meshes created by commands such as REVSURF and RULESURF. A *surface mesh* is a three-dimensional grid that is defined in terms of $M \times N$ vertices. Envision the vertices as a grid consisting of columns and rows, with M and N specifying the column and row position of any given vertex.

The system variable SURFTAB1 controls the N direction of a surface mesh. An example is the resolution of the lampshade's ruled surface. Both SURFTAB1 and SURFTAB2 affect the base of the lamp because both M and N vertices are used when creating a revolved surface.

If you increase the values of these two system variables, the appearance of models improves, but they require more time to generate on the screen. Removing hidden lines will also consume more time. With small models, you may not notice a difference. If you decrease the values of these system variables excessively, the model will generate more quickly on the screen, but the model may appear rough, and it may not adequately represent the intended design.

1. Enter the **SURFTAB1** system variable.
2. Enter **20** for its value.
3. Enter the **SURFTAB2** system variable.
4. Enter **20** for its value.
5. Enter **ELEV** and set both elevation and thickness to **0**.
6. Save your work.

Producing a Revolved Surface

Let's use the REVSURF command to create a surface to represent the base of the lamp.

Creating the Path Curve

The first part of the process is to create the path curve that will be revolved to produce the surface.

HINT:

When constructing 3D models, consider using multiple viewports. The illustration below provides an example.

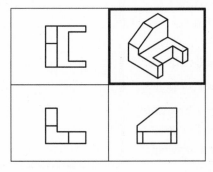

1. Using **Save As...**, begin a new drawing named **lamp.dwg**.

2. Display the **UCS II** toolbar and make **Front** the current UCS.

3. Enter the **HIDE** command to remove hidden lines.

4. Using the bottom of the right front corner of the 3D box as the starting point, approximate the shape shown in Fig. 39-2 with the **PLINE** command's **Arc** and **Line** options.

5. With the **LINE** command, draw a line that passes through the corner of the box and extends upward as shown in Fig. 39-2.

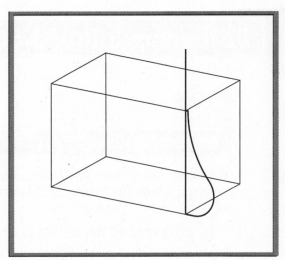

Fig. 39-2

Creating the Surface

Now you can use the REVSURF command to revolve the path curve, creating the surface of the lamp base.

1. Display the **Surfaces** toolbar and pick the **Revolved Surface** button.

This enters the REVSURF command.

Surfaces

2. Pick the polyline in reply to Select object to revolve.

3. Pick the vertical line at the right front corner of the 3D box in reply to Select object that defines the axis of revolution.

4. Enter **0** (the default value) in reply to Specify start angle.

5. Enter **360** (the default value) in reply to Specify included angle.

A 3D model of the lamp base appears.

6. Thaw the layer named Box and remove hidden lines.

Your lamp base should look similar, to the one shown in Fig. 39-3.

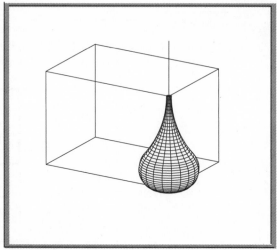

Fig. 39-3

Producing a Ruled Surface

Let's create a surface to represent the lamp shade using the RULESURF command.

Defining the Curves

Before we can create the surface, we need to define the two curves between which the surface will appear. For the lamp shade, we can use circles to define these curves.

1. Make **Object2** the current layer.

2. From the **UCS II** toolbar, make **Top** the current UCS.

3. Enter the **CIRCLE** command and pick the upper right front corner of the box for the circle's center point. (Refer to Fig. 39-4.) Enter **10** for the radius of the circle.

This circle lies in the Top UCS, and it will remain on this plane.

4. Create another circle using the same center, but enter a radius of **6**.

This circle also lies in the Top UCS, but you will move it upward in Step 8.

5. Make **Front** the current UCS.

This will allow you to move the circle upward.

6. Move the smaller circle up **18″**.

HINT:

Use polar tracking and enter 18.

The small circle should now be 18″ above the Top UCS. Your 3D model should now look similar to the one in Fig. 39-4.

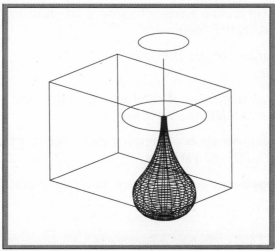

Fig. 39-4

Creating the Surface

Surfaces

1. From the Surfaces toolbar, pick the **Ruled Surface** button.

2. Pick one of the two circles in reply to Select first defining curve.

HINT:

You may need to turn snap off.

3. Pick the other circle in reply to Select second defining curve.

Did the lamp shade appear as you envisioned it?

4. Freeze layer Box and remove hidden lines.

Your 3D model should look similar to the one in Fig. 39-1.

5. View the lamp from various orientations in space.

6. Close the Surfaces and UCS II toolbars, save your work, and exit AutoCAD.

Chapter 39
Review & Activities

Review Questions

1. What is the purpose of the REVSURF command?

2. Describe the use of the RULESURF command.

3. How are user coordinate systems beneficial when using the REVSURF and RULESURF commands?

4. Explain the purpose of the SURFTAB1 and SURFTAB2 system variables.

5. What are the consequences of entering high values for the SURFTAB1 and SURFTAB2 variables?

6. What are the consequences of entering low values for the SURFTAB1 and SURFTAB2 variables?

7. In AutoCAD terms, surface meshes are made up of what?

Challenge Your Thinking

1. Could you have created the lamp in this chapter without using the Top and Front coordinate systems? Explain.

2. It is possible to edit the lamp you created in this chapter using PEDIT, the same command you have used previously to edit 2D polylines. Experiment with the use of PEDIT with objects such as the lamp base and lamp shade. Be sure to save the file with a new file name first. Then write a short essay explaining the use of the PEDIT command with 2D and 3D objects. How do the options differ? What do the 3D options do?

Chapter 39
Review & Activities

Applying AutoCAD Skills

Work the following problems to practice the commands and skills you learned in this chapter.

1-4. Using the 3D features covered in this chapter, create the 3D models shown in Figs. 39-5 through 39-8. Approximate all sizes. Use REVSURF to create the models in Figs. 39-5 and 39-8, and use RULESURF to create the models in Figs. 39-6 and 39-7. Use the 3dtmp.dwt template file.

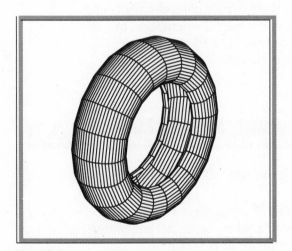

Fig. 39-5

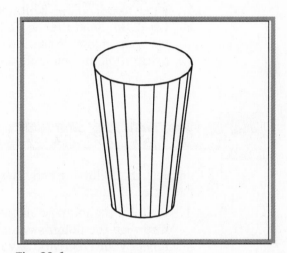

Fig. 39-6

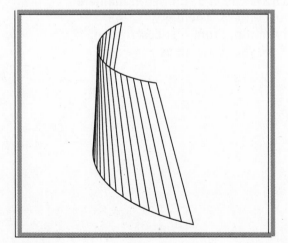

Fig. 39-7

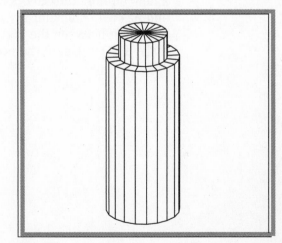

Fig. 39-8

Chapter 39
Review & Activities continued

5-8. Recreate each of the models in problems 1 through 4 using different values for SURFTAB1 and SURFTAB2.

9. Open lamp.dwg and save it as a new file named lampshade.dwg. Erase the current lampshade. Then use the commands you learned in this chapter to design a new, different lampshade. Save your work.

10. Refer to visual.dwt and its description contained on the Instructor's Resource Guide CD-ROM. Load visual.dwt and review its contents. Apply it to the development of a new 3D model.

11. Refer to 3dport.dwg contained on the Instructor's Resource Guide CD-ROM. Load the file and review the model orientation in each viewport. Move from viewport to viewport and change the orientation and magnification of each view.

 Using Problem-Solving Skills

Complete the following activities using problem-solving skills and your knowledge of AutoCAD.

1. Create a preliminary 3D revolution of a clay vase you intend to throw on the potter's wheel in your pottery store. Make the design reflect your creativity. Save the drawing as clay.dwg.

2. The packaging division of your company has designed a bottle for a new holding tank deodorant. The bottle is 6" tall, 2" in diameter, and has a 1" diameter by ³/₄" threaded top. There is a ¹/₂" 45° transition between the body and the top. From this description, make a 3D model of the bottle. Save the drawing as bottle.dwg.

Objectives:

- Create tabulated surfaces
- Produce Coons surface patches
- Define basic surface meshes using coordinate entry
- Edit basic surface meshes
- Change the type and appearance of a surface mesh

Key Terms

Coons surface patch
direction vector
tabulated surface

AutoCAD produces different surface types for different applications. For example, you might use a tabulated surface, created with the TABSURF command, to create an I-beam such as the one in Fig. 40-1. You might use a Coons surface patch to create a freeform shape such as the one at the right in Fig. 40-1. The most basic of the 3D surface commands is 3DMESH. It defines a surface mesh by specifying its size (in terms of M and N) and the location of each vertex in the mesh. M and N equal the number of vertices that you must specify. The third surface shown in Fig. 40-1 is an example.

Let's create each of these 3D objects.

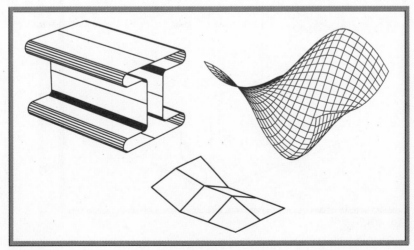

Fig. 40-1

Producing Tabulated Surfaces

A *tabulated surface* is one in which AutoCAD calculates, or tabulates, a surface based on a defined path curve (profile) and a direction vector. A *direction vector* is a 3D line that is used to provide a direction for a path to follow.

Defining the Path and Direction Vector

Let's use the TABSURF command to construct the I-beam shown in Fig. 40-1.

1. Start AutoCAD and begin a new drawing using the template file named **3dtmp.dwt**.

2. Display the **UCS II** toolbar and make **Right** the current UCS.

3. Using the **PLINE** command's **Arc** and **Line** options, approximate the shape shown in Fig. 40-2. It represents one quarter of the I-beam's profile.

HINT:

It's important to use the PLINE command rather than the LINE and ARC commands so that AutoCAD will treat the profile as a single object. It's also helpful to have polar tracking on.

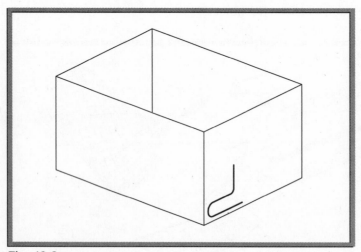

Fig. 40-2

4. Complete the I-beam profile so that it looks like the one in Fig. 40-3.

HINT:

After creating one quarter of the object, mirror it to complete half of the object. Then mirror that to complete the entire object.

5. Save your work in a file named **i-beam.dwg**.

6. Make **Top** the current UCS, and make **Object2** the current layer.

7. Beginning at the right edge of the box, as shown in Fig. 40-3, draw a line approximately **28"** long.

The line provides the direction vector for use with the TABSURF command.

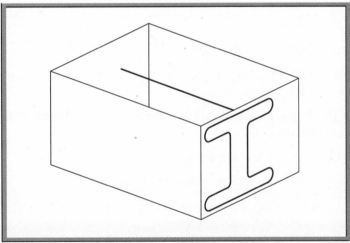

Fig. 40-3

8. Make **3dobject** the current layer.

The SURFTAB1 system variable controls the density of tabulated surfaces.

9. Enter the **SURFTAB1** system variable and enter **8**.

Creating the Surface

1. Display the **Surfaces** toolbar.

2. Pick the **Tabulated Surface** button.

This enters the TABSURF command.

Surfaces

3. In reply to Select object for path curve, pick a point on the lower right quadrant of the I-beam profile.

4. Pick the 28″ line in reply to Select object for direction vector. The point you pick must be closer to the right endpoint of the line than the left endpoint in order for the tabulated surface to extend in the desired direction.

One quarter of the I-beam appears.

5. Complete the remaining parts of the I-beam using the **TABSURF** command. (See the following hint.)

Surfaces

HINT:

Complete the parts in a clockwise direction—lower left quadrant, upper left, then upper right. Otherwise, a previously created polygon mesh may interfere with the selection of new points.

6. Freeze layers **Box** and **Object2**.

7. Using the **3DORBIT** command, create a perspective projection of the 3D model and remove hidden lines.

The model of the I-beam should look similar to the one shown at the beginning of this chapter.

8. Exit 3D Orbit and save your work.

The lines will remain hidden after you exit 3D Orbit.

Creating Coons Surface Patches

The EDGESURF command is used to construct a *Coons surface patch*. A Coons patch is a surface mesh interpolated (approximated) between four adjoining edges. Coons surface patches are used to define the topology of complex, irregular surfaces such as land formations and manufactured products such as car body panels.

Preparing the Edges

Let's apply the EDGESURF command to the creation of a topological figure similar to the one shown at the beginning of this chapter. First, we need to create the four edges that will define the surface patch.

1. Use the **3dtmp.dwt** template file to begin a new drawing.

2. Make **Object2** the current layer.

3. Using the **LINE** command, draw four vertical lines directly on top of the existing four vertical edges of the 3D box.

The lines will permit you to snap to the corners of the box during the construction of the Coons patch.

4. Enter **OSNAP**, make certain that **Endpoint** and **Nearest** are among the checked object snap modes, and pick **OK**.

5. Make **3dobject** the current layer.

6. Make **Front** the current UCS.

7. Using the **Arc** option of the **PLINE** command, approximate the construction of polyline A as shown in Fig. 40-4.

 HINT: ───────────────────────────────

> When picking the first and last points of the polyline, use the Nearest object snap mode. This will snap these points onto the vertical line. If you do not do this, the polyline endpoints will not meet accurately. If they do not meet, the EDGESURF command will not work.

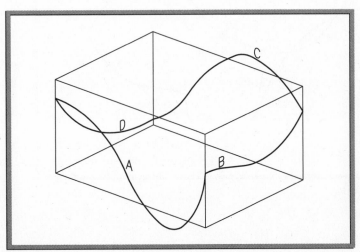

Fig. 40-4

8. Make **Right** the current UCS.

9. Approximate the construction of polyline B as shown in the previous illustration. Use the **Endpoint** and **Nearest** object snap modes for the first and last endpoints of the polyline.

10. Make **Back** the current UCS and construct polyline C using the same procedure described in the preceding step.

11. Make **Left** the current UCS and construct polyline D. Use the **Intersection** object snap mode to connect polyline D to polyline A.

12. Freeze layers **Box** and **Object2**.

13. Save your work in a file named **contour.dwg**.

Creating the Surface

Surfaces

1. Pick the **Edge Surface** button from the Surfaces toolbar.

This enters the EDGESURF command.

2. In reply to Select object 1 for surface edge, pick a point on polyline A near corner 1. (Refer to Fig. 40-5.)

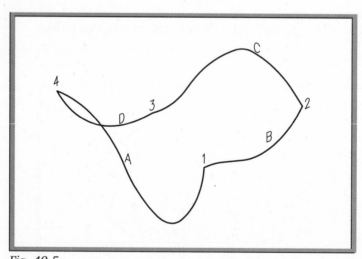

Fig. 40-5

 NOTE: ─────────────────────────────

When you select polyline A, it is important that you pick a point on the polyline that is near corner 1. The same is true when you select the remaining three polylines: pick a point near the corner specified for each polyline.

3. In reply to Select object 2 for surface edge, pick a point on polyline B near corner 2.

4. Select polylines C and D in the same fashion.

A contour similar to the one shown in Fig. 40-1 appears.

5. Using the **3DORBIT** command, shade the surface using **Flat** and **Gouraud** shading and view the contour from various orientations in space.

6. Save your work.

Defining Basic Surface Meshes

The 3DMESH command produces a basic surface mesh.

1. Begin a new drawing using the **3dtmp.dwt** template file.

2. Pick the **3D Mesh** button from the Surfaces toolbar.

Surfaces

This enters the 3DMESH command.

3. Enter **4** in reply to Enter size of mesh in M direction.

4. Enter **3** in reply to Enter size of mesh in N direction.

5. Enter the following coordinates in reply to the series of prompts. Be sure to include the decimal points.

Specify location for vertex (0,0): **5,4,.2**
Specify location for vertex (0,1): **5,4.5,.3**
Specify location for vertex (0,2): **5,5,.3**
Specify location for vertex (1,0): **5.5,4,0**
Specify location for vertex (1,1): **5.5,4.5,.2**
Specify location for vertex (1,2): **5.5,5,0**
Specify location for vertex (2,0): **6,4,0**
Specify location for vertex (2,1): **6,4.5,.2**
Specify location for vertex (2,2): **6,5,0**
Specify location for vertex (3,0): **6.5,4,0**
Specify location for vertex (3,1): **6.5,4.5,0**
Specify location for vertex (3,2): **6.5,5,0**

6. If the mesh does not appear, **ZOOM All** or **Extents** and then zoom in on the mesh.

Your mesh should look similar to the one in Fig. 40-1.

As you can see, specifying even a small mesh is very tedious. The 3DMESH command is meant to be used primarily with AutoLISP and not in the fashion presented above.

NOTE:

The PFACE command is similar to 3DMESH. PFACE produces a polygon mesh of arbitrary topology called a *polyface mesh*.

7. Save your work in a file named **3dmesh.dwg**.

Editing Surface Meshes

The vertices of a mesh can be edited with the PEDIT command in a manner similar to editing a polyline.

1. Enter the **PEDIT** command, pick the mesh, and enter the **Edit vertex** option.

An x appears at one corner of the mesh.

2. Press **ENTER** four times.

This moves the x to four consecutive vertices.

3. Enter the **Move** option and pick a new location for the vertex.

4. Enter **eXit** twice to exit the PEDIT command.

Controlling the Type of Mesh Displayed

Surface meshes can be viewed in any one of the surface types shown in Fig. 40-6.

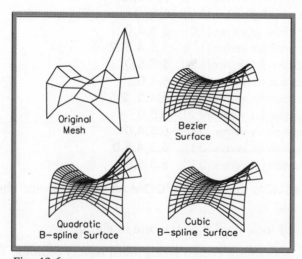

Fig. 40-6

The type of mesh displayed is controlled by the SURFTYPE system variable, as shown in Table 40-1.

Value of SURFTYPE	Surface Type
5	Quadratic B-spline
6	Cubic B-spline
8	Bezier

Table 40-1

NOTE:

The differences among the various types of surface meshes are mathematical in nature and are beyond the scope of this textbook. The type of mesh used by engineers and designers depends on the effects they want to create and the mathematical constraints within which they are working.

Let's create a quadratic B-spline from our original mesh.

1. Enter the **SURFTYPE** system variable.

2. Enter **5** to specify the quadratic B-spline surface.

3. Enter the **PEDIT** command, pick the mesh, and enter the **Smooth surface** option.

The 3D polygon mesh changes to a quadratic B-spline surface.

4. Enter the **eXit** option to exit the PEDIT command.

5. Close the Surfaces and UCS II toolbars.

6. Save your work and exit AutoCAD.

Chapter 40
Review & Activities

Review Questions

1. Describe the steps in using the TABSURF command.

2. In connection with the TABSURF command, what is the purpose of the direction vector?

3. Describe the EDGESURF command.

4. How are user coordinate systems used in conjunction with the EDGESURF command?

5. Why is the use of the 3DMESH command not generally recommended for even the simplest 3D meshes?

Challenge Your Thinking

1. Refer to the I-beam drawing you created in this chapter. Describe a way you could have created the I-beam by using TABSURF only once.

2. Find out more about the differences among a Bezier surface, a quadratic B-spline surface, and a cubic B-spline surface. What are the applications of each surface type?

Applying AutoCAD Skills

Work the following problems to practice the commands and skills you learned in this chapter.

1. Create a diamond-shaped tabulated surface that is 8″ long. Save the drawing as diamond.dwg.

Chapter 40
Review & Activities

2-5. Construct the 3D models shown in Figs. 40-7 through 40-10. Approximate all sizes.

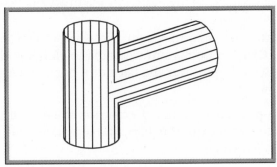

Fig. 40-7

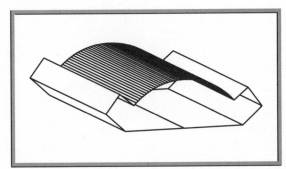

Fig. 40-8

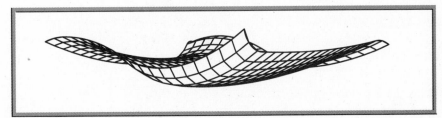

Fig. 40-9

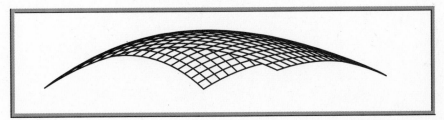

Fig. 40-10

Page 587

Chapter 40
Review & Activities continued

 Using Problem-Solving Skills

Complete the following activities using problem-solving skills and your knowledge of AutoCAD.

1. As a sculptor, you want to see an effect before you start cutting the stone. Create a Coons surface patch to simulate the way a saddle blanket drapes over the back of a horse. Save the drawing as saddle.dwg.

2. Design a 5-pointed star-shaped extrusion to be used to make a die for extruding pasta. Use the 3dtmp.dwt template, and save the drawing as pasta.dwg.

Chapter 41 — 3D Primitives

Objectives:

- **Place predefined 3D surface primitives in a drawing**
- **Edit the placement and orientation of 3D surface primitives**

Key Terms

apex
frustum
primitives
torus

AutoCAD permits you to create several predefined 3D shapes called *primitives*. The name comes from their basic, or primitive, nature. Primitives are usually used as building blocks to create more complex shapes. You can edit these and other 3D shapes using the ALIGN, ROTATE3D, and MIRROR3D commands.

Placing Predefined Primitives

The Surfaces toolbar offers a selection of predefined 3D surface primitives. These primitives include a wedge, box, torus, dish, pyramid, cone, dome, and sphere.

1. Start AutoCAD and begin a new drawing using the **3dtmp.dwt** template file.

2. Make **3dobject** the current layer.

3. If the WCS is not the current UCS, make it the current UCS now.

Wedge

4. Display the **Surfaces** toolbar.

5. Pick the **Wedge** button from the toolbar.

Surfaces

6. In reply to Specify corner point of wedge, pick point 1 as shown in Fig. 41-1. (Snap should be on.)

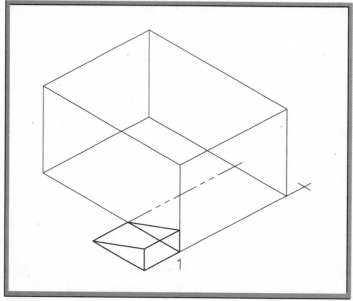

Fig. 41-1

7. Enter **12** for the length, **8** for the width, and **4** for the height, but do not yet enter a rotation angle.

8. Move the crosshairs and notice that you are able to rotate the wedge around the Z axis.

The Z axis is perpendicular to the current UCS.

9. Enter **180** for the rotation angle.

This rotates the wedge 180° counterclockwise. The wedge appears as pictured in the previous illustration.

10. Save your work in a file named **edit3d.dwg**.

Box

1. Freeze layer Box and remove hidden lines.

2. Pick the **Box** button from the Surfaces toolbar.

3. Pick a point (anywhere) in reply to Specify corner point of box.

4. Enter **15** for the length.

Surfaces

NOTE: ─────────────────────────────────

At this point, you could create a cube by entering C for Cube.

5. Enter **8** for the width, **6** for the height, and **15** for the rotation angle.

6. Remove hidden lines.

A box similar to the one in Fig. 41-2 should appear.

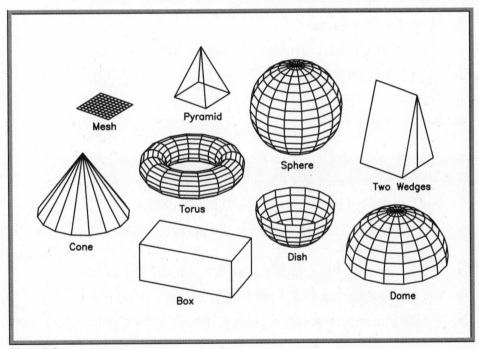

Fig. 41-2

Note, however, that your 3D box may appear different because the viewpoint may be different.

Torus

A *torus* (plural *tori*) is the inner-tube-shaped object shown in Fig. 41-2.

Surfaces

1. Select the **Torus** button from the Surfaces toolbar.

2. Pick a center point (anywhere) and enter **8** for the radius of the torus.

The torus radius is the distance from the center of the torus to the center of the tube.

3. Enter **2** for the tube radius.

4. Accept the next two default values.

These values specify the density of the wireframe mesh used to create and display the torus.

5. Remove hidden lines.

The torus should appear similar to the one in Fig. 41-2.

Dish

Surfaces

1. Select the **Dish** button.

2. Pick a center point (anywhere) and enter **6** for the radius.

3. Accept the next two default values.

A dish appears.

4. Remove hidden lines.

Pyramid

Surfaces

1. Pick the **Pyramid** button from the Surfaces toolbar.

2. Pick a point for the first corner of the pyramid's base anywhere in the drawing area.

3. For the second corner, pick a point **6″** from the first point.

4. For the third corner, pick a point **6″** from the second point.

5. For the fourth corner, pick a point **6″** from the third point.

6. For the apex point, pick a point above the base. (The *apex* defines the top of the pyramid as a point.)

A pyramid appears.

Cone

Surfaces

1. Pick the **Cone** button from the toolbar.

2. For the center point of the cone's base, pick a point anywhere.

3. Enter **7** for the radius of the base.

4. Enter **0** for the radius of the cone's top.

5. Enter **10** for the cone's height.

6. Press **ENTER** to accept the default value for number of segments.

A cone appears.

NOTE:

Entering a value higher than zero for the radius of the cone's top creates a *frustum* of the cone. The cone appears as though the tip has been cut off perpendicular to the base plane of the cone, as shown on the right.

Dome

Surfaces

1. Pick the **Dome** button from the toolbar.

2. For the center point of the dome, pick a point anywhere.

3. Enter **8** for the dome's radius.

4. Accept the next two default values for the number of segments.

A dome appears.

Sphere

Surfaces

1. Pick the **Sphere** button from the toolbar.

2. For the center point of the sphere, pick a point anywhere.

3. Enter **7** for the sphere's radius.

4. Accept the next two default values for the number of segments.

A sphere appears.

Using the Command Line

The 3D command displays the entire list of primitives as command options. This makes it possible to enter B for Box, C for Cone, and so on, to access the 3D primitives.

1. Enter **3D**.

2. Enter **M** for Mesh.

3. Pick four points to form a polygon of any size and shape.

4. Enter **10** for both the *M* size and the *N* size.

5. Remove hidden lines.

Your drawing should now look similar to the one in Fig. 41-2.

INFOLINK

Refer to Chapters 39 and 40 for more about M and N values.

Editing 3D Primitives

You can edit the placement and orientation of 3D primitives by aligning them or by rotating and mirroring them.

Aligning a Primitive

The ALIGN command enables you to move objects in 3D space by specifying three source points and three destination points. Let's change the alignment of the wedge.

1. Thaw layer Box.

2. Enter **ZOOM** and **.7x**, if necessary, to create working space around the drawing.

3. From the **Modify** pull-down menu, select **3D Operation** and **Align**.

This enters the **ALIGN** command.

4. Select the wedge primitive and press **ENTER**.

5. Pick point 1 as shown in Fig. 41-3 for the first source point.

6. Pick point 2 for the first destination point.

7. Pick point 3 for the second source point.

8. Pick point 4 for the second destination point. (Point 4 is located **6"** from point 2 in the positive Y direction.)

9. Pick points 5 and 6 for the third source and destination points, respectively.

The wedge moves and rotates as shown in Fig. 41-4.

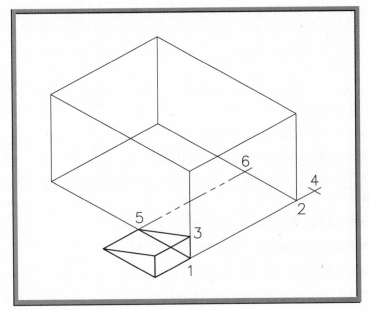

Fig. 41-3

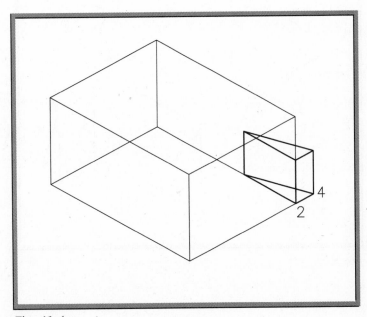

Fig. 41-4

Rotating a Primitive

The ROTATE3D command permits you to rotate an object around an arbitrary 3D axis.

1. From the **Modify** pull-down menu, select **3D Operation** and **Rotate 3D**.

This enters the ROTATE3D command.

2. Select the wedge and press **ENTER**.

AutoCAD presents you with several options for defining the axis of rotation.

3. Enter **2** for the 2points option.

4. Pick points 2 and 4 as shown in Fig. 41-4.

5. Enter **90** for the rotation angle.

The wedge rotates as shown in Fig. 41-5.

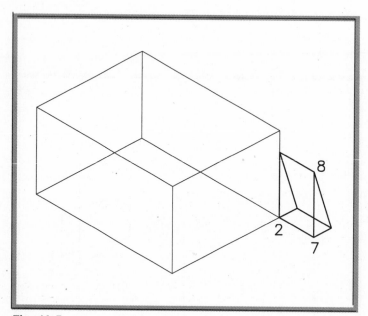

Fig. 41-5

Mirroring a Primitive

You can mirror 3D objects around an arbitrary plane with the MIRROR3D command.

1. From the **Modify** pull-down menu, select **3D Operation** and **Mirror 3D**.

This enters the MIRROR3D command.

2. Select the wedge and press **ENTER**.

AutoCAD presents several options for defining the plane. Notice that 3points is the default setting.

3. For the three points, pick points 2, 7, and 8 as shown in Fig. 41-5 to define the plane, and do not delete the old object.

The mirrored object appears.

4. Freeze layer **Box** and remove hidden lines.

The mirrored wedge should look very similar to the one in Fig. 41-6.

5. Save your work and exit AutoCAD.

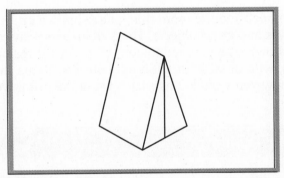

Fig. 41-6

Chapter 41
Review & Activities

Review Questions

1. Name at least five 3D primitives that you can create using the Surfaces toolbar.

2. When creating a torus, you must specify the torus and tube radius values. What is the difference between these two values?

3. After selecting the 3D Mesh button from the Surfaces toolbar, what must you enter to create a 3D mesh?

4. Explain how you can move and rotate a 3D object using the ALIGN command.

5. Briefly state the purpose of the ROTATE3D command.

6. Describe the purpose of the MIRROR3D command.

7. Discuss the use of 3D primitives in AutoCAD. Under what circumstances might these objects save drawing time? When would you use them?

8. Most 3D models require more than one primitive or a combination of primitives and other objects. Think of an everyday object that could be drawn using a combination of at least three of the primitives, with or without additional objects. Name the object and describe how you would construct it using the primitives.

Challenge Your Thinking

1. What is the difference between the MIRROR and MIRROR3D commands? Between the ROTATE and ROTATE3D commands? Experiment as necessary with these commands and then write a paragraph explaining your findings.

2. The Mesh option of the 3D command creates a mesh using M and N vertices. Investigate M and N vertices in greater detail. Then write a paragraph comparing and contrasting M and N vertices with the X and Y axes in the Cartesian coordinate system.

Chapter 41
Review & Activities

Applying AutoCAD Skills

Work the following problems to practice the commands and skills you learned in this chapter.

1. Start a new drawing from scratch and create the 3D object shown in Fig. 41-7. Make it 2.5 units long by 1 unit wide by .5 unit high, and rotate it –90°.

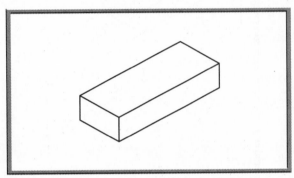

Fig. 41-7

2. Using either the ALIGN or the ROTATE3D command, reorient the previous object so that it looks like the one in Fig. 41-8.

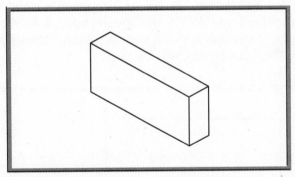

Fig. 41-8

Chapter 41
Review & Activities continued

3. Apply MIRROR3D to the previous object to create the object shown in Fig. 41-9.

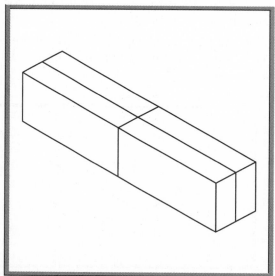

Fig. 41-9

4. The Cub Scouts have an event called the Pinewood Derby; wooden vehicles race down an inclined track. A typical example of these vehicles is shown in Fig. 41-10. Draw a Pinewood Derby vehicle using primitives. Make a box for the body, a smaller box for the "engine," and four tori for wheels. Change the UCS as necessary. Save the drawing as ch41pine.dwg.

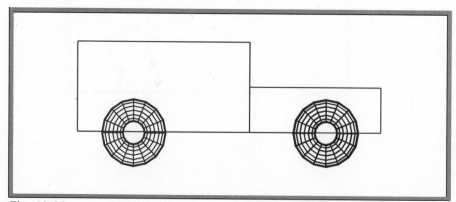

Fig. 41-10

Chapter 41
Review & Activities

 Using Problem-Solving Skills

Complete the following activities using problem-solving skills and your knowledge of AutoCAD.

1. You have been assigned the job of designing a skirt for a pool table. First you must draw the pool table structure (the slate and pockets) in 3D. The slate measures 88″ long by 44″ wide and 1″ thick. The pockets are 4″ in diameter. Draw the slate using the ELEV command and THICKNESS system variable. Insert the six pockets, one in each corner and two side pockets, as primitive dishes from the Surfaces toolbar. Change the UCS as necessary. Make four viewports to show various views of the pool table. Save the drawing as ch41pool.dwg.

2. Nostalgia is popular, and the company for which you design note cards is introducing a new line of "World's Fair" cards. The symbol for the 1939 New York World's Fair was the Trylon and Parisphere (Fig. 41-11). The Trylon was a tetrahedron, or a triangular pyramid, and the Parisphere was a sphere. You are designing a note card that is a caricature of the real thing. However, to achieve the proper relative distances, you need to draw the Trylon and Parisphere at their true sizes in AutoCAD. The Trylon was approximately 1300′ feet tall with a base formed by an equilateral triangle 100′ per side. The Parisphere had an approximate diameter of 400′. The two were separated by approximately 175′ at a distance of 100′ above the ground. Draw the Trylon and Parisphere using primitives. Change the UCS as necessary. Create various views so that you can decide which view will look best in your caricature. Save the drawing as ch41fair.dwg.

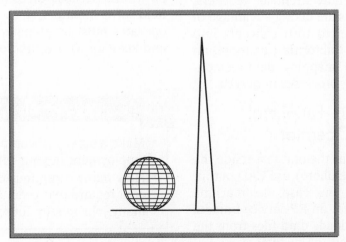

Fig. 41-11

Careers
Using
AutoCAD

The Stock Market/Douglas Whyte

Rapid Prototyping

Demand for prototype models, which are used to test new products and set up manufacturing, is driving the growth of an industry called rapid prototyping (RP). RP encompasses several technologies that have one purpose: to convert CAD data directly into a physical model. Traditionally, models have been created by hand, a time-consuming and expensive process. As the *rapid* in rapid prototyping suggests, the goal of these technologies is to do the job faster, producing models with exact detail in hours or days rather than weeks or months.

Several RP strategies produce a model by building up ultra-thin layers, one on top of another. For example, in laminated object manufacturing (LOM), a sheet of paper is laser-cut to the desired shape and fused by heating to the previous sheet below. In fused deposition modeling (FDM), an extrusion head lays down a succession of exactly shaped layers of thermoplastic. Stereolithography (SL) uses a polymer material that solidifies on contact with a computer-guided laser.

As RP technology becomes more widespread, applications become more varied. Engineers still use prototypes for wind tunnel testing and similar purposes. But now scientists, teachers, salespeople, and others are taking advantage of the power of a model to turn concepts into hands-on reality. In a California classroom, an RP model built from mapping data demonstrates terrain-forming processes in deserts.

Career Focus—Peripheral Equipment Operator

RP machines (such as the one that made the SL model shown in the photo) use CAD output to make a model. In this case, the machine operator might work for an RP service bureau, a company that downloads data files from the Internet to make models for customers.

In general, computer and peripheral equipment operators are responsible for starting up jobs on the machine, monitoring their progress, and diagnosing and correcting problems by making adjustments to the machine.

Education and Training

Experience on a particular system is the most desired qualification for employment as an operator. An associate degree or other two-year curriculum in computer science provides general preparation for computer work. A bachelor's degree is stronger preparation for advancement to supervisory or more analytical positions. Because of continual change in the field of computers, a peripheral equipment operator must be prepared to learn new skills and keep up with technology.

Career Activity

- Make a survey of computer-related employment in your town or region.
- How many computer-related jobs in your area require two years of training? How many jobs require four years?

Chapter 42 Shading and Rendering

Objectives:

- Shade a 3D model using flat and Gouraud shading
- Create a basic rendering of a 3D model
- Save and view a 3D model in various file formats

Key Terms

facet
facet shading
rendering
shading
smooth shading
wireframe

AutoCAD offers several options for shading and rendering a 3D model. The terms *shade* and *render* are often used synonymously. However, a *rendering* usually refers to a model with an appearance that is more realistic than that of basic shading. *Shading* produces a shaded image more quickly than rendering, especially when the model is large and complex. Figure 42-1 shows an example of one type of shaded image.

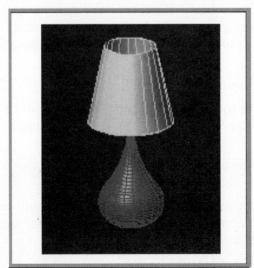

Fig. 42-1

Basic Shading

You can shade a model quickly using the buttons in the Shade toolbar.

1. Start AutoCAD and open the drawing you created in Chapter 39 named **lamp.dwg**.

2. Display the **Shade** toolbar.

3. Rest the pointer on each of the buttons to review the tooltips.

Flat Shading

The Flat Shaded button fills each polygonal facet with a single shade of color. A *facet* is a three- or four-sided polygon that represents a piece of a 3D surface. As discussed in Chapter 39, the facets become smaller and less noticeable as you increase the values of the SURFTAB1 and SURFTAB2 system variables prior to creating the model. The models you have created thus far, such as lamp.dwg shown in Fig. 42-1, include many facets. Flat shading is also called *facet shading*.

Shade

4. Pick the **Flat Shaded** button.

AutoCAD flat-shades the model. Notice that you can see each of the individual rectangular facets.

Gouraud Shading

The Gouraud Shaded button shades the model and smoothes the edges between the facets, giving the model a smooth, realistic appearance. Gouraud (pronounced *ga-ROO*) shading is also referred to as *smooth shading*.

Shade

5. Pick the **Gouraud Shaded** button.

AutoCAD smooth-shades the model. You can no longer see the rectangular facets because AutoCAD has blended their edges.

Displaying the Facets with Shading

The Flat Shaded, Edges On and Gouraud Shaded, Edges On buttons produce shaded views with the wireframe showing through. A *wireframe* is a "stick" representation of a 3D model.

Shade

6. Pick the **Flat Shaded, Edges On** button.

7. Pick the **Gouraud Shaded, Edges On** button.

Shade

This option is helpful when you want to display the wireframe model superimposed on the smooth-shaded model.

Displaying the Wireframe

The 3D Wireframe button displays a model using lines and curves to represent its boundaries. If the model is currently shaded, 3D Wireframe removes the shading. This is the default mode of viewing a 3D model.

The Hidden button also displays a model using the 3D wireframe representation, but it hides the lines representing the back faces. Picking the Hidden button is equivalent to entering the HIDE command.

Shade

8. Pick the **3D Wireframe** button.

9. Pick the **Hidden** button.

Shade

NOTE:

As a reminder, you can also apply these shading options using 3D Orbit. Use it when you need to rotate, zoom, or pan while the model is shaded. If you only need to quickly shade the model, it may be faster to shade it from the Shade toolbar if it is open.

Rendering

The RENDER command enables you to produce high-quality views of a 3D model.

Render

1. Display the **Render** toolbar and pick the **Render** button.

This displays the Render dialog box.

2. In the Rendering Options area, make sure that the **Smooth Shade** check box has been checked.

3. Pick the **Render** button at the bottom of the dialog box.

The rendering process may take a few seconds, depending on the speed of your computer. The rendered image contains soft edges, which improves the overall appearance of the lamp.

You will learn more about the Render **dialog box in Chapter 43.**

Rendering Statistics

The Statistics dialog box displays details about the last rendering. You can save this information in a file, but you cannot change it.

Render

1. Pick the **Statistics** button from the Render toolbar.

This displays the Statistics dialog box, as shown in Fig. 42-2.

2. Review the information, noticing the total rendering time and total faces.

3. Pick the **OK** button.

Fig. 42-2

Saving Rendered Images

Using the Save Image dialog box, you can save a rendering as a BMP, TGA, or TIFF file (see Table 42-1). Many programs are capable of reading, displaying, and printing these file types.

File Type	Description
BMP	Device-independent bitmap format
TGA	32-bit RGBA Truevision
TIFF	32-bit RGBA Tagged Image File Format

Table 42-1

1. Enter the **SAVEIMG** command or pick **Display Image** and **Save...** from the **Tools** pull-down menu.

The Save Image dialog box appears, as shown in Fig. 42-3.

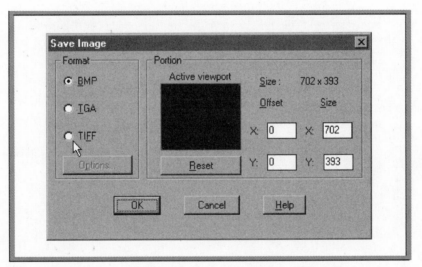

Fig. 42-3

2. Pick the **TIFF** radio button and pick the **OK** button.

3. Name the file **lamp.tif**, select the folder with your name, and pick the **Save** button.

AutoCAD creates a file named lamp.tif.

Displaying a Saved Image

You can display BMP, TGA, and TIFF files using the Replay dialog box.

1. Start a new drawing from scratch.

2. Enter the **REPLAY** command or pick **Display Image** and **View...** from the **Tools** pull-down menu.

The Replay version of the standard Files dialog box appears, as shown in Fig. 42-4.

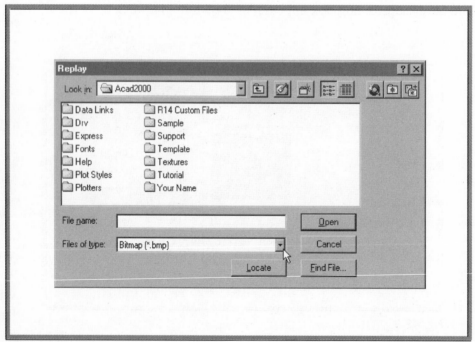

Fig. 42-4

3. To the right of Files of Type, pick the down arrow.

4. Pick **TIFF (*.tif)** from the list.

5. Find and double-click **lamp.tif**.

This displays the Image Specifications dialog box, as shown in Fig. 42-5.

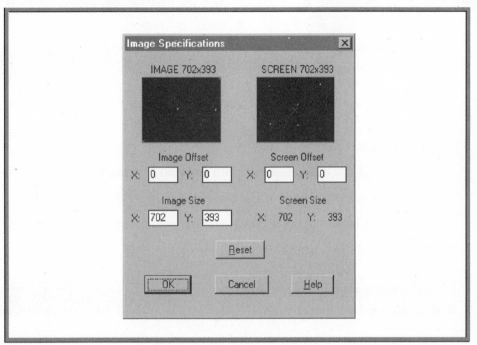

Fig. 42-5

6. Pick the **Help** button if you want to read about this dialog box.

7. Pick the **OK** button to view the file.

AutoCAD reads the lamp.tif file and displays it.

8. Pick the **3D Wireframe** button from the Shade toolbar.

Nothing happens because the lamp is a TIFF file, not an AutoCAD object.

9. Close the Shade and Render toolbars.

10. Exit AutoCAD without saving.

Chapter 42
Review & Activities

Review Questions

1. When would it be practical to use the shading options from the Shade toolbar rather than the shading options in 3D Orbit?

2. What is a facet?

3. What is the difference between flat and Gouraud shading?

4. What is the purpose of the Statistics dialog box?

5. What file formats can you create using the Save Image dialog box?

6. What is the purpose of the Replay dialog box?

Challenge Your Thinking

1. Explore the system variables that control the appearance of shaded objects. What options are available? Explain why people seldom need to change the initial values, and identify a situation in which the options might need to be changed.

2. In this chapter, you saved a rendered image of the lamp.dwg file as a TIFF file. The TIFF format can be read by many graphic and illustration programs other than AutoCAD. Find out what programs available in your school or company can display a TIFF file. If a suitable program is available, load lamp.tif using that program and view the file. How does it compare to the original AutoCAD image? Explain any differences you see.

Chapter 42
Review & Activities

Applying AutoCAD Skills

Work the following problems to practice the commands and skills you learned in this chapter.

1. Shade and render the 3d2.dwg file that you created in previous chapters (Fig. 42-6).

2. Load the Pine Wood Derby vehicle from Chapter 41. Shade the vehicle with Gouraud shading. Save the drawing.

3. Create a BMP file of the rendering of 3d2.dwg. Name it 3d2.bmp.

4. Display the 3d2.bmp rendering in AutoCAD.

5. Shade and render i-beam.dwg (Fig. 42-7).

6. Display statistics on the rendering of the I-beam. How many faces and triangles does it contain?

Fig. 42-6

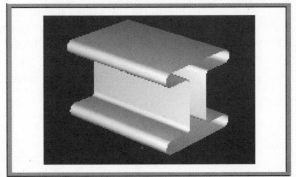

Fig. 42-7

Chapter 42
Review & Activities continued

7. Draw and render the electric motor cover shown in Fig. 42-8.

Fig. 42-8

Begin by drawing a polyline and center line as shown in Fig. 42-9. Use the REVSURF command with a large value for SURFTAB1 to create the cover. To add the lettering, first change the viewpoint so you are looking at the top view; change the UCS to the current view and then insert the lettering. Use the Properties dialog box to give the text the appropriate elevation and thickness. It may be easier to manipulate the views if you use the ROTATE3D command to rotate the object so the plan view is truly the top view of the object. A better approach would be to set up three viewports for top, front, and isometric views, then draw the object in the front viewport. After you have drawn the cover, render it in the isometric view.

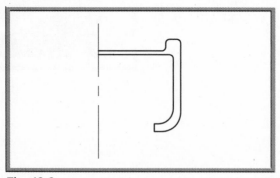

Fig. 42-9

Chapter 42
Review & Activities

 Using Problem-Solving Skills

Complete the following activities using problem-solving skills and your knowledge of AutoCAD.

1. Load the pool table from Chapter 41 ("Using Problem-Solving Skills" problem 1). Shade the slate and pockets with flat shading, edges on. Figure out a way to shade the pool table in colors that look as realistic as possible. Document your procedure, and save the shaded drawing as ch41poolshade.dwg.

2. Open the Trylon and Parisphere drawing from Chapter 41 ("Using Problem-Solving Skills" problem 2). Gouraud-shade the drawing, and then render it. Decide which version is preferable for the note card and explain why. Save both versions of the drawing with different names for later comparison.

Chapter 43 Advanced Rendering

Objectives:

- **Apply Phong shading to a 3D model**
- **Produce a photorealistic rendering of a 3D model**
- **Add material finishes, lights, and backgrounds to photorealistic renderings**
- **Insert and edit the appearance of bitmapped landscape objects in a drawing**

Key Terms

ambient
gradient
interpolation
landscape object
Phong shading
photorealistic
raytracing
scene

The right combination of rendering options can give a model a photorealistic appearance. *Photorealistic* means "resembling a photographic image." AutoCAD offers several options for changing a 3D object's material, lighting, and background to enhance the realistic appearance of a model. For example, adding a shiny metallic finish to a machine part gives it an appearance that you would expect.

Phong Shading

Phong shading is similar to Gouraud shading, except that it uses a more advanced method of interpolation by calculating the light at several points across the model's surface to create a more realistic rendering. *Interpolation* is a method of averaging the edges so that they blend together and provide a smooth appearance.

1. Start AutoCAD and open the drawing you created earlier named **lamp.dwg**.

2. Display the **Render** toolbar and pick the **Render** button.

Render

This displays the Render dialog box, as shown in Fig. 43-1.

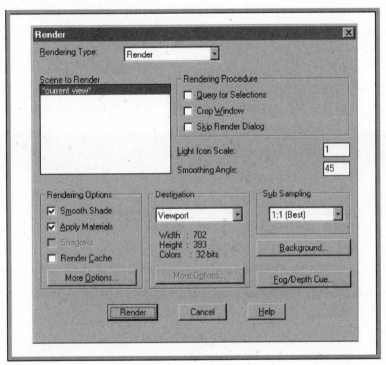

Fig. 43-1

3. Review the information presented in the dialog box.

4. Pick the **More Options...** button in the **Rendering Options** area.

The Render Options dialog box permits you to select either Gouraud or Phong shading. Gouraud shading produces a smooth (non-faceted) rendering by interpolating the adjacent color intensities of a model. As stated earlier, Phong shading uses a similar but more advanced method of interpolation.

5. Pick the **Phong** radio button and then pick **OK**.

6. If **Smooth Shade** is not checked, check it now.

Smooth shading causes AutoCAD to blend the facets and produce a smooth-shaded rendering using either the Gouraud or Phong technique. Note, however, that the smooth shading takes longer than facet shading. As the quality of the rendering increases, so does the time to produce the rendering.

7. Pick the **Help** button and read about the remaining items in this dialog box.

8. After exiting Help, pick the **Render** button.

This causes AutoCAD to render a *scene,* which is a combination of a named view and one or more lights. In this case, since we did not define a scene, AutoCAD uses the current view and a default over-the-shoulder distant light source.

9. Enter **REGEN** to remove the rendering.

Photorealistic Rendering

Render

AutoCAD offers additional rendering options, such as photorealistic and raytrace rendering. *Raytracing* is a calculation-intense process in which lines are drawn from the eye to every point on the screen.

1. Display the **Render** dialog box and pick the **Rendering Type** drop-down box located at the top of the dialog box.

The Photo Real and Photo Raytrace options produce more realistically shaded images of 3D models.

2. Select **Photo Real** and render the lamp.

 NOTE: ───────────────────────────

Depending on your computer's display subsystem, you may not see a significant difference in the rendering quality. Factors include the number of colors available and the resolution of the subsystem.

Editing the Rendered Image

In AutoCAD, you are not limited to rendering objects in the AutoCAD colors assigned to them. You can assign many different colors, materials, textures, and finishes to objects individually. These characteristics do not appear in the standard wireframe view of a model, but when you render them using a photorealistic method, the characteristics appear. You can also define various light sources to illuminate the model appropriately.

Attaching Material Finishes

You can define and attach materials and surface finishes to objects using the Materials dialog box. For example, you can make the surface of an object appear reflective by using a glass finish. AutoCAD also provides dozens of predefined materials that you can assign to objects in your drawing files. Let's assign materials to the lamp to make it look more realistic. We will make the base of the lamp appear to be made of teak, and we'll assign a fabric-like material to the lamp shade.

1. Zoom in so that the lamp fills most of the screen.

Render

2. From the Render toolbar, pick the **Materials** button.

This displays the Materials dialog box.

Objects without a material finish attached to them use the default *GLOBAL* finish, as indicated in the upper left area of the dialog box.

3. Pick the **Materials Library...** button.

The Materials Library dialog box appears, as shown in Fig. 43-2.

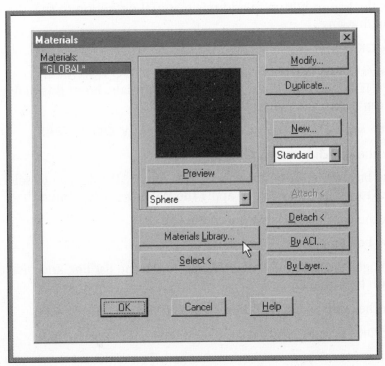

Fig. 43-2

The left side of this dialog box shows the materials that are currently defined in the drawing. The list box on the right side of the dialog box lists the predefined materials AutoCAD provides for use with your drawings.

4. Scroll down the list of predefined materials and notice the various types that are available.

5. Pick **WOOD – TEAK** and then pick the **Preview** button in the middle of the dialog box.

A sphere appears in the preview window with the WOOD – TEAK finish applied. This allows you to see what a material or finish will look like before you apply it to objects in your drawing.

6. Pick the **Import** button.

Notice that WOOD – TEAK is now listed below *GLOBAL* on the left side of the dialog box. We have imported the teak material definition into the current drawing.

7. From the list of available materials, find and click **ZIGZAG PATTERN**.

8. Pick the **Import** button to add this material definition to the lamp drawing.

9. Pick **OK**.

The Materials dialog box reappears. Notice that WOOD – TEAK and ZIGZAG PATTERN are now listed as available materials. From this dialog box, you can assign the materials to the lamp.

10. In the Materials list box on the left, highlight **WOOD – TEAK**.

11. Pick the **Attach** button.

The dialog box temporarily disappears, and AutoCAD asks you to select the objects to which you want to apply the WOOD – TEAK material.

12. Pick the base of the lamp and press **ENTER**.

The dialog box reappears.

13. Highlight the **ZIGZAG PATTERN** material in the list box and pick the **Attach** button again.

14. Pick the lamp shade and then press **ENTER**.

15. Pick **OK** to close the Materials dialog box.

The drawing doesn't appear to have changed. To see the materials you have attached, you have to render the drawing.

Render

16. Pick the **Render** button from the Render toolbar.

17. In the **Rendering Type** drop-down box, select **Photo Raytrace**.

NOTE:

To display material finishes, you must select either Photo Real or Photo Raytrace. The standard Render option does not apply the finishes.

18. Check to be sure the **Apply Materials** check box is checked.

19. Pick the **Render** button in the dialog box.

The rendering process may take several seconds—possibly a minute or longer—depending on the rendering speed of your computer. Note the spinning line and the percentages on the Command line. This indicates that AutoCAD is processing data.

When AutoCAD finishes, a teak lamp with a zigzag-patterned shade appears on the screen.

Defining Light Sources

You can control several types of lighting with the Lights dialog box. First, let's save the current view.

1. Enter the **VIEW** command and save a view named **view1**.

2. Pick **OK**; pick **OK** again.

This will allow us to return to this view later in the exercise.

Render

3. Pick the **Lights** button from the **Render** toolbar.

This displays the Lights dialog box, as shown in Fig. 43-3 (page 620).

Notice that the intensity of the *ambient* light (the general light surrounding the lamp) is at .30. Let's increase the ambient light so that the lamp becomes more visible.

4. Use the slider bar to increase the intensity to **.70** and pick **OK**.

5. After making sure the **Render** dialog box is still set to apply materials and to apply the photo raytrace method, render the lamp.

The lamp should be much more visible.

Using the advanced light control features in AutoCAD, you can make the lamp appear to be lit.

View

6. Display the **View** toolbar and pick the **Front** button.

7. If the drawing is not currently displayed in its wireframe form, change to wireframe now. (You may need to use the **3DORBIT** command to do this.)

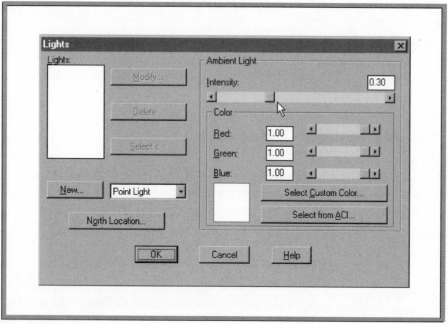

Fig. 43-3

8. Zoom in so that the lamp fills most of the drawing area.

9. From the Render toolbar, pick the **Lights** icon.

Render

10. In the drop-down box next to the New... button, select **Spotlight**.

11. Pick the **New...** button.

The New Spotlight dialog box appears, and AutoCAD requests the name of the new light.

12. Enter **lamp** for the new light name.

13. In the Intensity box, enter a value of **101**.

Now focus on the numbers in the Red, Green, and Blue boxes in the Color area of the dialog box. The value for each color is 1.00 by default, which specifies a white light. Let's change the color of the light source to simulate lamp light.

14. In the box next to Red, enter **.98**.

15. In the box next to Green, enter **.91**.

16. In the box next to Blue, enter **.52**.

The sample color box in the lower left of the dialog box shows a new light color—a light amber shade.

NOTE: ———————————————————————————

You can also specify the color of a light source by moving the slider bars to the right of the color boxes, or by picking the Select from ACI... button, which displays AutoCAD's Select Color dialog box. In addition, you can create new custom colors by picking the Select Custom Color... button.

Now we need to position the light.

1. In the Position area of the dialog box, pick the **Modify** button.

AutoCAD allows you place the light source by picking points in the drawing. In this case, we want to position the spotlight inside the lamp shade to simulate a light bulb.

2. In response to Enter light target, pick a point just above the widest part of the base of the lamp.

3. In response to Enter light location, use the **Nearest** object snap to snap to a point about two-thirds of the way up the red axis line that extends up the middle of the lamp shade.

4. In the New Spotlight dialog box, change the value of Hotspot to **92** and the value of Falloff to **126**.

The falloff angle defines the full cone of light from the spotlight. The hotspot value defines the brightest cone of light within the falloff area.

5. Pick **OK**.

The Lights dialog box returns. Notice that the new LAMP light appears in the list box.

6. Pick **OK** again.

Notice the small object inside the lamp shade where you placed the light. Light sources are AutoCAD objects, so you can edit their position using grips if necessary.

7. Enter the **VIEW** command and return to **VIEW1**, and render the lamp.

The lamp should appear to be lit. Let's define another light source to make the lamp more visible.

1. Enter **REGEN** to return to the wireframe view.

2. Enter **ZOOM** and **.5x**.

3. Pick the **Lights** button from the **Render** toolbar.

Render

4. In the drop-down box next to the New... button, select **Distant Light**.

5. Pick the **New...** button.

The New Distant Light dialog box appears.

6. Name the new light **distant**.

7. In the Light Source Vector area of the dialog box, pick the **Modify** button.

8. In response to Enter light direction TO, pick a point near the bottom of the lamp base.

9. In response to Enter light direction FROM, pick a point near the middle of the left border of the drawing area.

10. Enter an intensity of **.80**.

11. Pick **OK**; pick **OK** again.

12. Render the lamp.

The lamp and lamp shade should now be more visible.

Defining a Scene

A scene is similar to a named view, but a scene can contain one or more light sources. Let's define a scene using the current view and the two light sources you have created.

Render

1. Pick the **Scenes** button from the Render toolbar.

This displays the Scenes dialog box, as shown in Fig. 43-4.

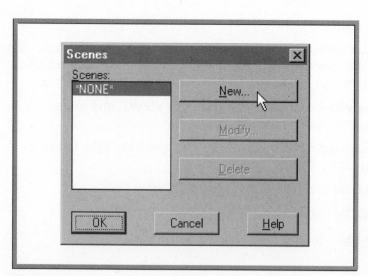

Fig. 43-4

2. Pick the **New...** button.

The New Scene dialog box expects you to name the scene.

3. Enter **scene1**, make sure that the current view and all lights are selected, and pick **OK**.

SCENE1 appears in the Scenes list box, indicating that you have created a new scene. AutoCAD will use the selected scene the next time you render the model.

Render

4. Pick **OK** and pick the **Render** button from the toolbar.

Notice that SCENE1 appears in the Scene to Render list box and that it is selected.

5. To render SCENE1, pick the **Render** button in the dialog box.

6. Save your work.

Adding a Background

In the Render dialog box, you may recall seeing the Background... button. This button displays the Background dialog box.

Render

1. Pick the **Background** button from the Render toolbar.

This also displays the Background dialog box. Notice that many options are grayed out and not available in the dialog box.

2. Pick the **Gradient** radio button.

In this context, *gradient* means to transition gradually from one color to another. When you choose Gradient, many of the unavailable options in the Background dialog box become available, as shown in Fig. 43-5.

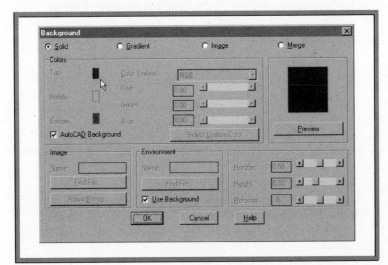

Fig. 43-5

Let's create a blue display background for the lamp.

3. Pick the colored box next to Top in the Colors area.

4. In the Red, Green, and Blue edit boxes, enter **.25**, **.58**, and **.58**, respectively.

This sets the color for the top one-third of the screen.

5. Pick the colored box next to Middle and enter **.44**, **.57**, and **.78**, respectively, in the Red, Green, and Blue edit boxes.

6. Pick the colored box next to Bottom and enter **.30**, **.44**, and **.85**, respectively in the Red, Green, and Blue edit boxes.

7. Pick the **Preview** button to see how the gradient will appear.

8. Pick **OK**.

9. Render the lamp. The lamp now appears against a graduated blue background.

10. Save your work.

You can also use background images saved in various graphic formats. AutoCAD allows you to use files with BMP, PNG, TGA, JPG, TIF, GIF, or PCX extensions as backgrounds.

11. Redisplay the **Background** dialog box and pick the **Image** radio button.

12. Find the file named **acadsig.jpg**, which is located in the **ACAD2000** folder.

HINT:

You may need to change the file type in the Background Image dialog box to *.jpg to find the file.

13. Double-click this file and pick the **OK** button.

14. Render the lamp.

The JPG file fills the background, replacing the blue gradient.

15. Redisplay the **Background** dialog box, pick the **Solid** radio button, and pick **OK**.

Render

16. Render the lamp.

The standard black background reappears.

Other Rendering Options

With fog and depth cueing, you can provide more visual information about the distance of objects from the camera.

Render

1. Display the **Render** dialog box and pick the **Fog/Depth Cue...** button.

This displays the Fog/Depth Cue dialog box. The Fog button in the Render toolbar also displays this dialog box.

2. Check the **Enable Fog** check box located in the upper left corner of the dialog box.

Most of the options in the dialog box become available.

3. Pick the **Help** button and read about the settings in this dialog box.

4. After reading the help information, pick the **Cancel** button in the Fog/Depth Cue dialog box.

5. Pick the **More Options...** button.

The Photo Raytrace Render Options dialog box appears because Photo Raytrace is the rendering option selected in the Render dialog box.

6. Pick the **Help** button and read about the options available in this dialog box.

7. After reading the help information, pick the **Cancel** button to close the Photo Raytrace Render Options dialog box.

8. Pick the **Help** button in the Render dialog box to read about the remaining options. Close it and the dialog box when you are finished.

9. Save your work and close the lamp.dwg file.

Predefined Landscape Objects

A *landscape object* is an object with a bitmap image mapped onto it. Landscape objects permit you to add realistic landscape items such as trees and bushes to a drawing. A bitmapped image mapped to an object gives the object a grain, texture, or other quality that can make it look real.

Placing a Landscape Object

1. Begin a new drawing using the **3dtmp.dwt** template file.

2. Freeze layer **Box** and make **3dobject** the current layer.

3. From the Render toolbar, pick the **Landscape New** button.

This displays the Landscape New dialog box, as shown in Fig. 43-6.

Render

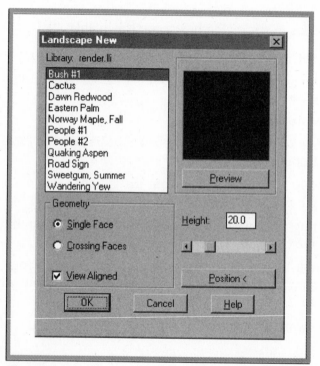

Fig. 43-6

4. Pick **Cactus** from the list box, and then pick the **Preview** button to preview Cactus.

5. Pick the **Position** button and pick a point anywhere near the center of the screen for the base of the landscape object.

6. When the Landscape New dialog box reappears, pick the **OK** button.

A triangle appears.

7. Enter **ZOOM All** and turn off the grid.

Render

8. Display the **Render** dialog box, select **Photo Real** for Rendering Type, and pick the **Render** button.

A landscape object of a cactus appears.

9. View the object from a new viewpoint in space and then render it again using the **Photo Real** option.

10. Enter **REGEN**.

Editing a Landscape Object

Landscape objects, before being rendered, have grips at their base, top, and each corner. You can use the base grip to move the object, the top grip to adjust its height, and the bottom corner grip to scale it. All standard AutoCAD grip editing modes, such as stretch, scale, and rotate, work with landscape objects.

1. Select the landscape object.

2. Using the object's grips, change the shape and position of the cactus and render it again.

Two related commands, LSEDIT and LSLIB, permit you to modify landscape objects and maintain landscape object libraries. You can enter these commands by picking the Landscape Edit and Landscape Library buttons from the Render toolbar.

3. Close the Render toolbar, as well as any others you may have opened.

4. Exit AutoCAD. Do not save your work.

Chapter 43
Review & Activities

Review Questions

1. Why does smooth shading require more computer processing time than faceted shading?

2. Explain how to apply a predefined material to an object in AutoCAD.

3. What does a scene contain that a saved view does not contain?

4. How would you apply a bitmap (BMP) file of a mountain scene to the background of a drawing?

5. Explain the difference between the Render and Photo Real rendering types in the Render dialog box.

6. What is a landscape object?

Challenge Your Thinking

1. Experiment further with the options presented in the Lights dialog box. As you know, AutoCAD allows you to assign colors to the lights. What happens when you shine a light of one color onto an object of another color? When and why might this be useful?

2. Shade several drawings using both Gouraud and Phong shading. Compare the effects. Phong shading uses a more advanced method of interpolation; does this always mean that the shaded image looks better? Does Phong shading have any disadvantages? Explain.

Applying AutoCAD Skills

Work the following problems to practice the commands and skills you learned in this chapter.

1. Create a photorealistic rendering of the edit3d.dwg drawing from Chapter 41. Assign a different color or material finish to each object. After changing several attributes, such as Color, Ambient, Reflection, and Roughness, create a raytrace rendering of edit3d.dwg. Save the drawing as edit3d-a.dwg.

Chapter 43
Review & Activities

2. Reopen edit3d-a.dwg, which you created in problem 1, and create and insert a distant light. After moving the light block in the X, Y, and Z directions, render the drawing. Produce a new scene using the new light. Save the drawing as edit3d-b.dwg.

3. Open i-beam.dwg, which you created in Chapter 40. Apply an appropriate material and render the drawing to resemble an actual I-beam. Save the drawing.

 ## Using Problem-Solving Skills

Complete the following activities using problem-solving skills and your knowledge of AutoCAD.

1. The Trylon and Parisphere at the 1939 World's Fair were in a park-like setting. Create a background for them that will be appropriate for the caricature note card. (Refer to Chapter 41, "Using Problem-Solving Skills" problem 2, if necessary.)

2. The Pinewood Derby vehicle you created in Chapter 41 ("Applying AutoCAD Skills" problem 4) can be painted as you see fit. Apply materials to the various parts to custom-"paint" the vehicle. Render the drawing and save it as paint.dwg.

3. Open lamp.dwg, which you created in Chapter 39. Create several different lights to simulate the lamp off and on each of the three settings of a three-way light bulb. Save each setting as a different scene. Save the drawing as 3-way.dwg.

Part 8 Project

Bakery Brochure

AutoCAD permits you to create realistic drawings by shading and rendering objects in various ways. This capability can be used to create illustrations that are both accurate and artistic. In this project, you will create three versions of a typical bakery product to be used in a bakery's sales brochure.

Description

Your advertising company's newest client is a bakery. To start off the ad campaign, your company will create a one-page "flier" to illustrate some of the bakery's decorating capabilities.

1. Design a surface model of a round, two-tiered cake. Make the cake look as realistic as possible. Follow these guidelines:

    *   The cake should rest on a board about 12" × 24" × 24".
    *   Each tier should be 3" thick.
    *   The lower tier should be about 20" in diameter, and the top tier should be about 12".

2. Make two additional copies of the cake and add lettering and other decorations for three different special occasions (birthday, good luck, etc.)

3. Apply the following treatments to the cakes:

    *   Flat shading
    *   Gouraud shading
    *   Phong shading
    *   Photorealistic rendering

4. Create the flier using all three of the cakes you have created. You may use any programs or materials you wish to create the finished flier.

Part 8 Project

Hints and Suggestions

- Create the cake using REVSURF so that you can simulate the frosting and any decorations that are part of the cake.
- Create the board using the Box primitive.
- Use the ROTATE3D command and change the viewpoint as necessary to create the cake efficiently.
- Experiment with different materials, lighting, and backgrounds to show each cake off to its best advantage.
- If time permits, you may want to experiment with creating a new type of material called "frosting" to apply to the cake.

Summary Questions/Self-Evaluation

1. In what way are 3D primitives considered time-savers?

2. Briefly compare Gouraud and Phong shading.

3. Describe the capabilities or effects of changing the following variables when modifying lights and materials: color, ambient, reflection, roughness, transparency, and refraction.

4. Generally speaking, what influence does altering the lighting have on the RENDER command?

5. What is a gradient background? Why might an advertiser or designer choose to use a gradient for a brochure such as the one you created for this project?

Careers
Using
AutoCAD

Ocean Engineering

It's dark down there, miles below the reach of daylight. The surrounding pressure probes for weakness in every square inch of a diving vessel's shell. If it finds a weak point, deep-sea pressure will crush a submersible like tin foil.

Submersibles are vessels capable of operating under water. High-precision design, as well as careful manufacture and testing, will be critical for a submerisble called *Deep Flight II*. It may soon enable a pilot and scientist to explore the world's deepest ocean floor, the Mariana Trench in the western Pacific.

Deep Flight II already exists as a set of Auto-CAD drawings at Hawkes Ocean Technologies (HOT), a research and development company in northern California. To make the design a reality, HOT president Graham Hawkes must find investors, corporate sponsors, and other sources of financial support.

HOT already has noteworthy successes to its credit. An earlier submersible, *Deep Flight I,* was featured on a National Geographic television special and on the news-magazine show "Dateline." In the movie *Sphere,* the escape sub was another HOT-designed craft, *Wet Flight.* When it's not performing in front of the camera, *Wet Flight* serves as a camera platform for underwater filming.

The next step will be to send digital drawings of *Deep Flight II*'s components to contractors, who will make the parts using computer-aided manufacturing. The components will then be assembled and tested to ensure that all systems work as designed in deep-sea conditions.

Career Focus—Science Technician

Science technicians use the techniques and equipment of science to help scientists and engineers carry out investigations in research

The Stock Market/George Disaro

and development, in industrial production, or in oil and gas exploration. Science technicians who work in research and development specialize according to the focus of their employer. Oceanography is the specialty at a firm like HOT. The specialty might be organic chemistry at an agriculture-related company, or geology at an energy-related company. Within their fields, science technicians construct or maintain experimental equipment, set up and monitor experiments, calculate and record results, and help scientists and engineers in other ways.

Education and Training

Two-year programs in various specialties are available from community colleges and technical institutes. However, in a field such as ocean engineering, a bachelor's degree in one of the sciences or in mathematics is more likely training for a science technician.

 Career Activity

- Write a report about ocean research occupations. What institutions and industries are involved in underwater exploring?
- Talk to someone who can tell you more about the work of science technicians.

Part 9

Solid Modeling

Chapter 44 Solid Regions

Objectives:

- Create regions using Boolean operations
- Perform mass properties calculations on a region
- Create a region from boundaries
- Extrude a solid region to form a 3D model

Key Terms

centroid

composite region

extruding

inner loops

loops

mass properties

outer loop

regions

region primitive

You can produce closed 2D areas known as *regions* using AutoCAD. Regions are the result of combining two or more 2D objects, which involves the Boolean union, subtraction, and intersection operations. Regions are useful when you want to shade 2D objects, easily find an object's area, or calculate mass properties. *Mass properties* are characteristics of an object or material, such as its center of gravity.

INFOLINK

You will learn more about Boolean operations in Chapter 47.

Creating a Solid Region

The REGION command creates objects called regions, which consist of closed 2D shapes called *loops*. Loops can be a combination of lines, polylines, arcs, circles, elliptical arcs, ellipses, and splines, 3D faces, traces, and solids. A loop must be a closed shape.

Let's create a representation of a bicycle sprocket guard manufactured from sheet metal, as shown in Fig. 44-1.

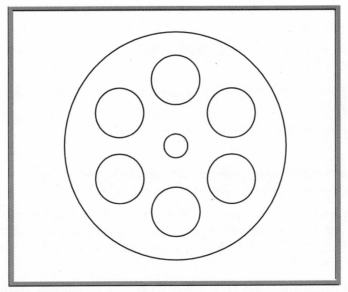

Fig. 44-1

Creating a Region Primitive

1. Start AutoCAD and use the following information to create a new drawing.
   - Use the **Quick Setup** wizard.
   - Use **Decimal** units.
   - Create a drawing area of **17″ × 11″**.
   - **ZOOM All**.

The basic shape from which a more complex 2D region is formed is known as a *region primitive*. In this case, the circle that represents the sprocket guard will be made into a round region primitive.

2. Create a new layer named **Objects**, assign the color red to it, and make it the current layer.

3. Draw a circle **7″** in diameter.

4. From the docked Draw toolbar, pick the **Region** button.

Draw

This enters the REGION command.

5. Select the round object and press **ENTER**.

AutoCAD converts the object to a region primitive.

6. Select the object, right-click, and pick **Properties** from the shortcut menu.

This confirms that the object is indeed a region. Note that the area of the region is 38.4845.

7. Close the Properties window and press **ESC** twice to deselect the region.

8. Save your work in a file named **region.dwg**.

Subtracting to Form a Composite Region

The SUBTRACT command creates a *composite region* by subtracting the area of one set of 2D objects or regions from another set. A composite region is the result of applying one or more Boolean operations, such as a subtraction.

1. Create a **1.5"**-diameter hole. Make its center **2"** upward vertically from the center of the round region.

2. Array the hole **360°** as shown in Fig. 44-1.

3. At the center of the region, create a **.75"**-diameter hole.

Draw

4. Enter the **REGION** command, select the seven new holes, and press **ENTER**.

5. Enter **DBLIST**.

As you can see, AutoCAD created seven individual regions.

6. Close the AutoCAD Text Window.

7. Open the **Solids Editing** toolbar.

Solids Editing

8. From the Solids Editing toolbar, pick the **Subtract** button.

This enters the SUBTRACT command.

9. Pick the large region (the first one you created) and press **ENTER**.

10. Pick the seven small regions and press **ENTER**.

AutoCAD subtracts the areas of the seven smaller region primitives from the large region primitive. The large circle is called the region's *outer loop*. The smaller circles are called the *inner loops*. The result is a single composite region.

11. List the properties of the composite region.

Notice the area of the region. How does it compare with the area before you subtracted the holes?

12. Close the Properties window and enter the **SHADE** command.

AutoCAD shades the region.

13. Save your work.

Adding to Form a Composite Region

The UNION command creates a composite region by combining the area of two or more 2D objects or regions.

1. Set snap at **.25″** and draw a rectangle across the top circle using a polyline, as shown in Fig. 44-2.

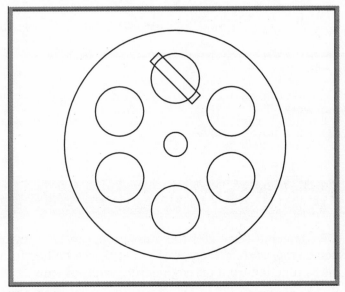

Fig. 44-2

2. Create a region primitive from the polyline.

3. Array the new region primitive so that it appears in the other five holes. Rotate the objects as they are copied.

4. From the Solids Editing toolbar, pick the **Union** button.

Solids Editing

This enters the UNION command.

5. Select the composite region as well as the six new region primitives, and press **ENTER**.

AutoCAD combines the regions to create a new composite region similar to the one shown in Fig. 44-3.

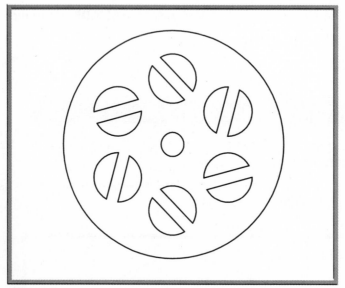

Fig. 44-3

6. Save your work.

The sprocket guard is now complete.

Calculating Mass Properties

The MASSPROP command calculates the mass properties of a region. Mass properties, such as the *centroid* (center of gravity), can be helpful when analyzing a part's function from an engineering point of view.

1. Enter the **MASSPROP** command.

2. Select the composite region and press **ENTER**.

AutoCAD displays information about the region, as shown in Fig. 44-4.

3. Enter **Yes** in reply to Write analysis to a file?

4. When the dialog box appears, choose the directory with your name to store the file, and pick the **Save** button.

This creates a file named region.mpr.

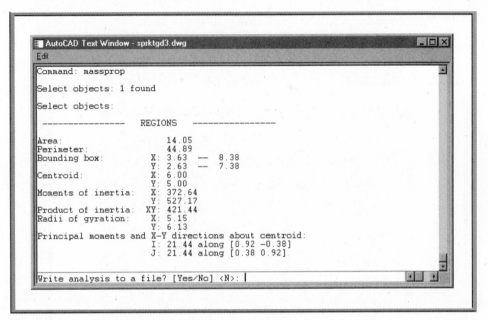

Fig. 44-4

5. Minimize AutoCAD.

6. Open Notepad.

7. Select **Open...** from the **File** pull-down menu.

8. Locate and open the file named **region.mpr**.

The file contains the mass properties information. From here, you could copy and paste this information into a report or AutoCAD drawing.

9. Close Notepad and maximize AutoCAD.

Creating a Region from Boundaries

The BOUNDARY command creates a region or polyline from overlapping objects. These objects must define an enclosed area.

1. Begin a new drawing from scratch.

2. Create two overlapping circles of any size, as shown in Fig. 44-5.

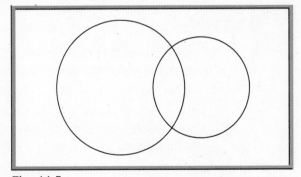

Fig. 44-5

3. Pick **Boundary...** from the **Draw** pull-down menu.

This displays the Boundary Creation dialog box, as shown in Fig. 44-6.

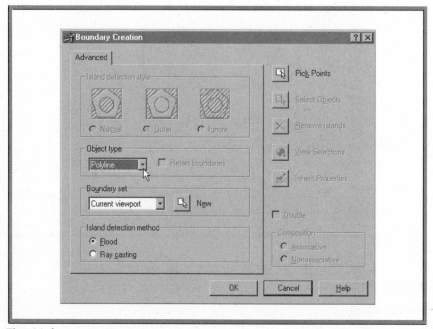

Fig. 44-6

4. In the Object Type drop-down box, choose **Region**.

5. Pick the **Pick Points** button located in the upper right area of the dialog box.

6. Pick a point inside the area defined by the overlapping circles and press **ENTER**.

This creates a region in the shape of the area defined by the overlapping circles.

NOTE:

AutoCAD creates the new region on the current layer, regardless of the layer on which the original objects were created.

7. Enter **MOVE** and **Last** and move the region to a new location on the screen.

8. Close the Solids Editing toolbar and exit AutoCAD. Do not save your work.

Extruding a Solid Region

You can apply most of the commands, Boolean concepts, and mass properties you've used in this chapter to 3D solid modeling. You can also produce 3D solid models from the regions by *extruding* them, or giving them thickness. For example, you can use the EXTRUDE command to produce the 3D representation of the sprocket guard shown in Fig. 44-7.

1. If **region.dwg** is not open, open it now.

2. Use **Save As...** to create a new file named **region2.dwg**.

3. If the region is not currently shown in wireframe form, change the display to wireframe.

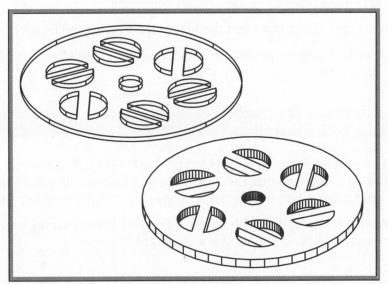

Fig. 44-7

4. Enter the **EXTRUDE** command at the keyboard.

5. Pick the sprocket guard and press **ENTER**.

6. Enter a height of **.25** and press **ENTER** to accept the default taper angle of **0**.

7. From the **View** pull-down menu, select **3D views** and **SW Isometric**.

As you can see, the region is now a 3D model.

8. Shade the model.

The following chapters cover AutoCAD's solid modeling features and commands, such as EXTRUDE, in greater detail.

9. Save your work and exit AutoCAD.

Chapter 44
Review & Activities

Review Questions

1. Explain inner and outer loops.

2. Describe a composite region.

3. With what command can you create a region by subtracting one object from another?

4. How do you determine the area of a region?

5. What is the purpose of the MASSPROP command?

6. Explain the purpose of the UNION command.

7. What two object types can the BOUNDARY command create?

8. Describe a simple method of transforming a 2D solid region into a 3D model.

Challenge Your Thinking

1. Review the information provided by the MASSPROP command. If you do not know what the properties are, find out. In what types of occupations might people need to use this information? Explain.

2. Write a paragraph describing the difference between a region and a closed polyline. What are the advantages of each?

Applying AutoCAD Skills

Work the following problems to practice the commands and skills you learned in this chapter.

1. Using a circle and a rectangular polyline, create the object shown in Fig. 44-8 as a composite region.

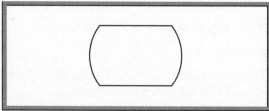

Fig. 44-8

2. From the object you created in problem 1, create the object shown in Fig. 44-9.

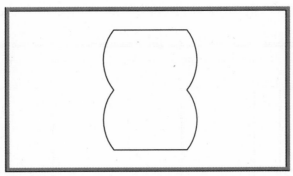

Fig. 44-9

3. From the object you created in problem 2, create the part of an alignment fixture for use in assembling a subsystem for a milling machine (Fig. 44-10).

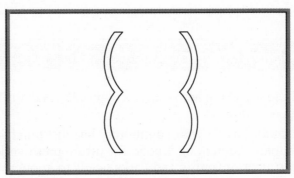

Fig. 44-10

4. Calculate the area (in square inches) of each of the objects you created in problems 1 through 3.

5. Display the mass properties of the object you created in problem 3.

6. Shade the object you created in problem 3.

7. Extrude the two regions you created in problem 3 to a height of .5 unit. Use 3D Orbit to smooth-shade and rotate the solid extrusions.

Chapter 44
Review & Activities continued

8. Draw the front view only of the die head shown in Fig. 44-11. Produce a 3D solid model using the EXTRUDE command. Set a height of 5 units and a taper angle of 0. View the model at various angles. Save the drawing as diehead.dwg.

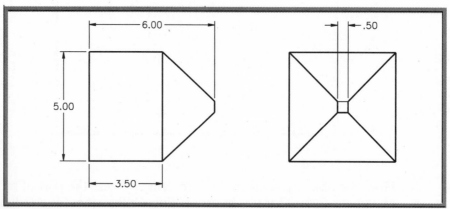

Fig. 44-11

 Using Problem-Solving Skills

Complete the following activities using problem-solving skills and your knowledge of AutoCAD.

1. The Land Trust in your community has just purchased a piece of property containing a pond for future preservation. The State Department of Forests and Parks needs to know the total area of the property (including the pond), as well as the area of the land exclusive of the pond. You are a member of the board of directors of the Land Trust. Use the map in Fig. 44-12 and your knowledge of primitive regions to provide the state with the information it needs. Estimate the location of the pond. Save the drawing as andarea.dwg.

Chapter 44
Review & Activities

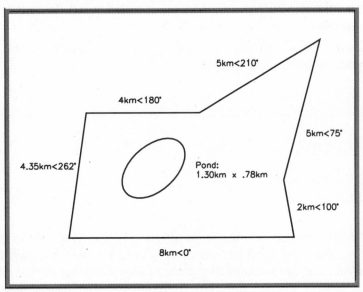

Fig. 44-12

2. The new land purchase abuts the existing Land Trust parcel shown in Fig. 44-13. Determine the total area of the two parcels including the pond. Save the drawing as landarea2.dwg.

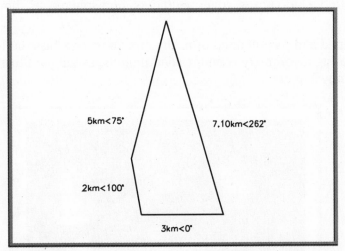

Fig. 44-13

Chapter 45 Solid Primitives

Objectives:

- Explain the difference between solid models and surface models
- Create solid primitives, including a cylinder, torus, cone, wedge, box, and sphere

Key Terms

solid models

solid primitives

Solid modeling is becoming increasingly popular. One reason is that designs created as solid models more accurately reflect a physical prototype. The appearance of a solid model is similar to that of a surface model, but the two are different. *Solid models* are closed volumes like a real physical object, and they contain physical and material properties. Surface models can represent closed volumes too, such as the lamp base you created in an earlier chapter. The lamp shade is not a closed volume because its walls do not have any thickness to them. AutoCAD permits you to generate predefined *solid primitives* (basic shapes you can use to build more complex models) by specifying the necessary dimensions. Examples include boxes, cones, cylinders, and spheres, as shown in Fig. 45-1.

Predefined and user-defined primitives make up the basic building blocks for creating moderately complex solid models, as will be illustrated in the following chapters.

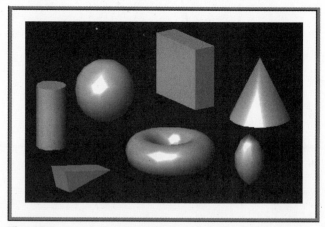

Fig. 45-1

Creating a Cylinder

The CYLINDER command allows you to create a solid cylinder primitive of any size.

1. Start AutoCAD and use the following information to create a new drawing.
   - Use the **Quick Setup** wizard.
   - Use **Decimal** units.
   - Create a drawing area of **11 × 8.5**.
   - **ZOOM All**.

2. Create a new layer named **Objects**, assign the color red to it, and make it the current layer.

3. Set grid to **1** and snap to **.25**.

4. Save your work in a template file named **solid.dwt**. For the template description, enter **For solid models**, and pick **OK**.

5. Close solid.dwt and use it to create a new drawing file named **primit.dwg**.

6. Turn off the grid.

7. Display the **Solids** toolbar and pick the **Cylinder** button from it.

This enters the CYLINDER command.

Solids

NOTE: ─────────────────────────────

The Elliptical option enables you to create an elliptical cylinder using prompts similar to those used in the ELLIPSE command.

8. Pick a point at any location on the screen.

9. Enter **.5** for the radius and **2** for the height of the cylinder.

10. Select a viewpoint that is to the left, in front of, and above the cylinder.

The cylinder should appear similar to the one in Fig. 45-2.

11. From the **View** toolbar, pick **Shade** and **Gouraud Shaded**.

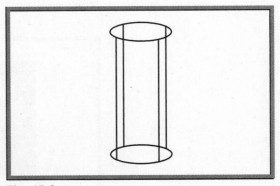

Fig. 45-2

Creating a Torus

The TORUS command lets you create donut-shaped solid primitives.

1. Enter **ZOOM** and **.5x**. Pan if necessary to make space for the torus.

Solids

2. Pick the **Torus** button from the Solids toolbar.

This enters the TORUS command.

3. Pick a center point for the torus (anywhere) and enter **1** for the radius of the torus.

The radius of the torus specifies the distance from the center of the torus to the center of the tube.

4. Enter **.5** for the radius of the tube.

The torus should appear as shown in Fig. 45-3.

5. Save your work.

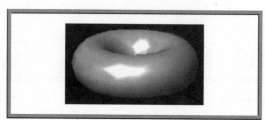

Fig. 45-3

 NOTE:

If you enter a negative number for the radius of the torus, a football-shaped solid primitive appears as shown in Fig. 45-4. Try it.

Fig. 45-4

Creating a Cone

The CONE command allows you to create a cone-shaped solid primitive.

Solids

1. From the Solids toolbar, pick the **Cone** button.

This enters the CONE command.

 NOTE: ———————————————————

The function of the Elliptical option is similar to that of the CYLINDER command.

2. Pick a center point at any location on the screen.

3. Enter **1** for the radius and **2** for the height of the cone.

A cone appears, as shown in Fig. 45-5.

4. Enter **ZOOM** and **.7x** to create more space on the screen.

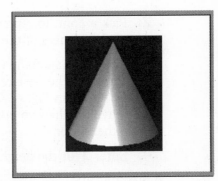

Fig. 45-5

Creating a Wedge

It is possible to create a wedge using the WEDGE command.

Solids

1. From the Solids toolbar, pick the **Wedge** button.

This enters the WEDGE command.

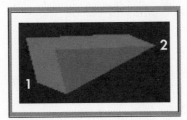

Fig. 45-6

2. Pick two corner points as shown in Fig. 45-6.

3. Enter **.75** for the height.

A wedge-shaped solid primitive appears.

Creating a Box

The BOX command enables you to create a solid box.

1. From the Solids toolbar, pick the **Box** button.

This enters the BOX command.

Solids

 NOTE:

The Center option permits you to create a box by specifying a center point.

2. Pick a point to represent one of the corners of the box.

3. Pick the diagonally opposite corner of the base rectangle as shown in Fig. 45-7. You may pick the two points in either order.

4. Enter **2** for the height.

A box appears.

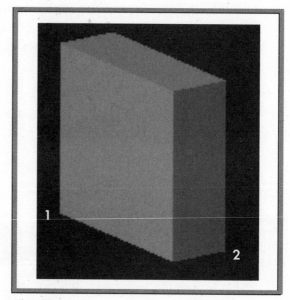

Fig. 45-7

 NOTE:

To create a solid cube, you can select the BOX command's Cube option and enter a length for one of its sides.

Creating a Sphere

The SPHERE command creates a solid sphere primitive.

Solids

1. From the Solids toolbar, pick the **Sphere** button.

This enters the SPHERE command.

2. Pick a point for the center of the sphere.

3. Enter **1** for the radius of the sphere.

The sphere appears as shown in Fig. 45-8.

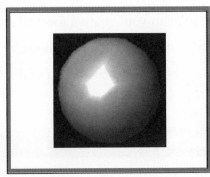

Fig. 45-8

4. Display the **Render** toolbar.

Render

5. Pick the **Lights** button from the Render toolbar, increase the **Ambient Light** intensity, and pick **OK**.

Render

6. Pick the **Render** button from the Render toolbar.

7. Select **Photo Raytrace** for Rendering Type and pick the **Render** button.

8. Close the Solids and Render toolbars.

9. Save your work and exit AutoCAD.

Chapter 45
Review & Activities

Review Questions

1. Describe the function of each of the following commands.
   CYLINDER
   TORUS
   CONE
   WEDGE
   BOX
   SPHERE

2. With what command can you create a football-shaped primitive?

3. With what command can you create a solid cube?

4. What is the purpose of the Elliptical option of the CYLINDER and CONE commands?

Challenge Your Thinking

1. How are the primitive shapes you created in this chapter different from the ones you created in Chapter 41, "3D Primitives"?

2. Experiment with the CYLINDER and CONE commands. Is it possible to create a solid primitive cylinder or cone at other than a 90° (right) angle to its base? Explain.

Applying AutoCAD Skills

Work the following problems to practice the commands and skills you learned in this chapter.

1. Create a solid shaft that measures .5125″ in diameter by 10″ in length. Rotate the shaft so that its length is parallel to the WCS.

Chapter 45
Review & Activities

2. Create a solid ball bearing that measures .3625″ in diameter. Use a polar array to copy the bearing 16 times using a 1″ radius. View the bearings from above, and smooth-shade them as shown in Fig. 45-9.

3. Create an inflated bicycle inner-tube that measures 24″ in diameter. Make the tube diameter 2″.

4. Create a football-shaped primitive of any size on top of a 6″ cube.

5. Open ch41pine.dwg, which you created in Chapter 41 ("Applying AutoCAD Skills" problem 4). Replace the wheels you made using 3D primitive tori with solid wheels. Save the drawing as ch45pine.dwg.

Fig. 45-9

Using Problem-Solving Skills

Complete the following activities using problem-solving skills and your knowledge of AutoCAD.

1. Your company's graphics department is designing ornaments and decorations for various holidays. They need forms for the art decals. Create the solid primitives for them, as shown in Fig. 45-10. Shade the primitives with Gouraud shading. Save the drawing as ornament.dwg.

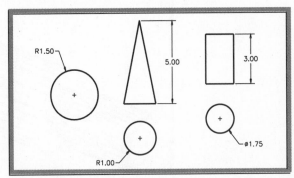

Fig. 45-10

2. You need forms around which to design a laptop computer carrying case. Create two solid boxes: one that measures 13″ × 10″ × 2″ to simulate the computer, and another that measures 5″ × 6″ × 1″ to simulate a plug-in module. Shade the boxes as appropriate. Save the drawing as case.dwg.

Objectives:

* Revolve a 2D polyline to create a 3D solid model
* Adjust the quality, accuracy, and appearance of a solid model
* Extrude a polyline to produce a solid model

Key Terms

revolve

tessellation

Many solid models are produced by first creating a 2D object. For example, it's possible to *revolve* a polyline (turn it around an axis) to create a solid volume. That's how the model of a steel shaft shown in Fig. 46-1 was created. In this chapter, we will do this. We will also extrude a polyline to create a solid.

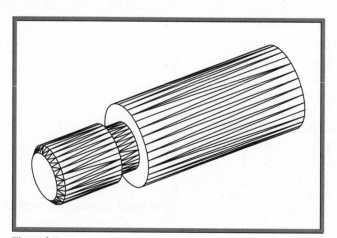

Fig. 46-1

Solid Revolutions

We will use the REVOLVE command to create a solid model of the shaft shown in Fig. 46-1. This command is similar to the REVSURF command, which is used to create a surface of revolution.

Defining the Shape

1. Start AutoCAD and use the **solid.dwt** template file to create a new drawing named **shaft.dwg**.

2. Create the top half of the shaft using a polyline as shown in Fig. 46-2. Make sure the polyline is closed, place it in the center of the screen, and omit the dimensions.

 HINT: _____

Begin at point 1, work counterclockwise, and take advantage of polar tracking. Be sure that the polar tracking increment angle is set at 45°, which will help you create the 45° chamfer.

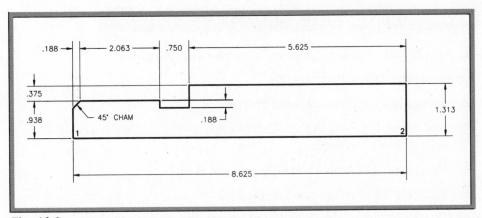

Fig. 46-2

3. Using one of the 3D viewing options, select a viewpoint that is to the left, above, and in front of the polyline.

You should now see a view similar to the one in Fig. 46-3.

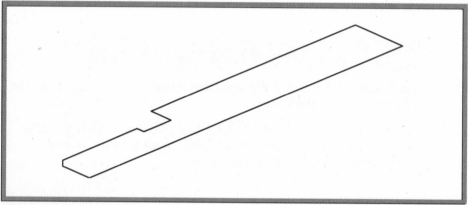

Fig. 46-3

4. Enter **ZOOM** and **.7x**.

5. Save your work.

Revolving the Polyline

Solids

1. Display the **Solids** toolbar and pick the **Revolve** button.

This enters the REVOLVE command.

2. Select the polyline and press **ENTER**.

3. Pick point 1 shown in Fig. 46-2 in reply to the default Specify start point for axis of revolution.

4. Pick point 2 in reply to Specify endpoint of axis.

5. Press **ENTER** to select the default **360**.

The 3D solid model of the shaft generates on the screen, as shown in Fig. 46-4.

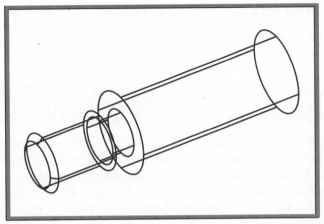

Fig. 46-4

Tessellation

ISOLINES is a system variable that controls the number of tessellation lines used in solid objects. *Tessellation* is the creation of the lines used to describe a curved surface. The default value of 4 often produces a crude-looking object, but the solid model generates more quickly than when ISOLINES is set at a higher value of up to 2047. A higher value, however, improves the quality and accuracy of the curved areas on the solid object.

1. Enter **ISOLINES** and a new value of **20**.

2. Enter **REGEN**.

As you can see, increasing the value of ISOLINES improves the appearance of the solid model. However, it increases the time required by the Boolean subtraction, union, and intersection operations. You will have the opportunity to practice these operations in the following chapters.

Note that the solid model looks like a conventional 3D surface model—not a solid model. However, it is a solid model. Let's prove it.

3. List the properties of the model.

Notice that the object is a 3D solid.

4. Close the window and turn off the grid.

5. Remove hidden lines.

6. Enter **ISOLINES**, enter **4**, and enter **REGEN**.

7. Set **ISOLINES** at **10** and enter **REGEN**.

8. Set **ISOLINES** back to **20** and enter **REGEN**.

9. Save your work and close the drawing file.

Solid Extrusions

The solid model of an aluminum casting shown in Fig. 46-5 was created using the EXTRUDE command. The command extrudes and creates a closed volume from a planar 3D face, closed polyline, polygon, circle, ellipse, closed spline, donut, or region. It is not possible to extrude objects contained within a block or a polyline that has a crossing or self-intersecting segment.

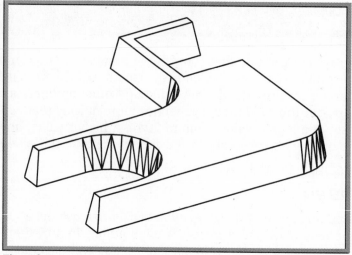

Fig. 46-5

1. Begin a new drawing using the **solid.dwt** template file.

2. Using the **PLINE** command's **Arc** and **Line** options, approximate the size and shape of the casting shown in Fig. 46-6. The casting's overall size is **6.25″ × 6.25″**. Pick the points in the order shown. Do not place the numbers on the drawing.

3. Save your work in a file named **extrude.dwg**.

4. Select a viewpoint that is to the left, above, and in front of the polyline.

5. Pick the **Extrude** button from the Solids toolbar.

Solids

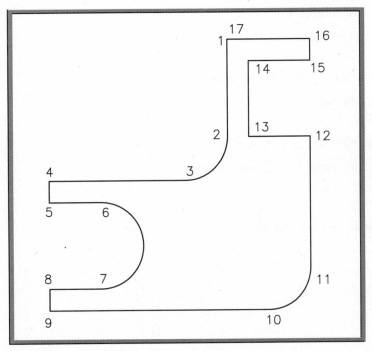

Fig. 46-6

6. Select the polyline and press **ENTER**.

7. Enter **.75** in reply to Specify height of extrusion.

8. Enter **5** (for degrees) in reply to Specify angle of taper for extrusion.

A solid primitive appears, as shown in Fig. 46-7.

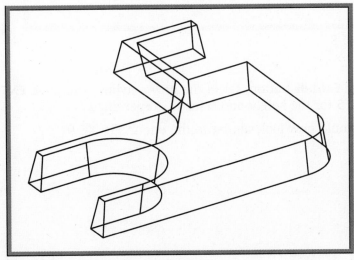

Fig. 46-7

9. Remove hidden lines.

Let's attempt to place a slotted hole in the object.

10. Enter the **PLAN** command to see the plan view of the current UCS; then enter **ZOOM Extents**.

11. Using **PLINE**, place a slotted hole in the object as shown in Fig. 46-8.

12. Next, view the object in 3D space as before.

Note that the "hole" is still two-dimensional.

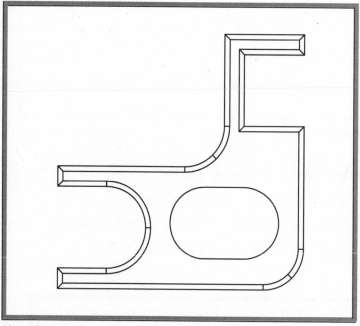

Fig. 46-8

Solids

13. Pick the **Extrude** button, select the new polyline, and press **ENTER**. Enter **.75** for the height and **0** for the taper angle.

The object should now look similar to the one in Fig. 46-9.

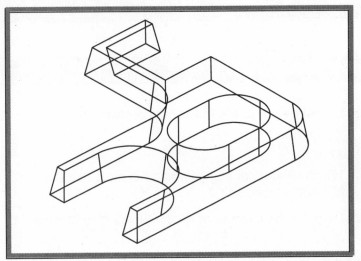

Fig. 46-9

14. Enter **SHADE**.

The slotted hole is filled in because it is solid, not hollow. Creation of a hole requires a Boolean subtraction operation similar to those you performed on solid regions in Chapter 44. In other words, the solid object that represents the slotted hole must be subtracted from the larger solid object.

15. Erase the larger model so that only the slotted hole is present.

As you can see, it is indeed a solid object and not a hole.

16. Undo to restore the larger model.

17. Erase the "slotted hole" by picking its edge.

18. Close the Solids toolbar, as well as any others you may have opened.

19. Save your work and exit AutoCAD.

INFOLINK

You will learn more about Boolean subtraction with solid models in Chapter 47.

Chapter 46
Review & Activities

Review Questions

1. Describe each of the following commands.
   REVOLVE
   EXTRUDE
2. Which solid modeling command is most like the REVSURF command?
3. Describe the purpose of the ISOLINES system variable.

Challenge Your Thinking

1. Discuss the factors you should consider when setting the ISOLINES system variable. Be specific.

2. Is it possible to extrude a wide polyline (one that has a defined width)? Explain.

Applying AutoCAD Skills

Work the following problems to practice the commands and skills you learned in this chapter.

1. Design a frame to fit a picture that measures 11″ × 14″. Create the frame with polylines and extrude it to a thickness of ³/₄″. Save the drawing as ch46frame.dwg.

2. Construct a solid bowling pin as shown on the left in Fig. 46-10. Then increase the number of tessellation lines and enter REGEN to create a higher resolution version of the bowling pin, as shown on the right.

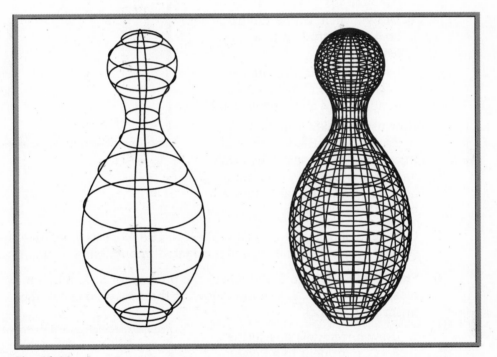

Fig. 46-10

3. Identify a simple object in the room. Create a profile of the object and then extrude it to create a solid model. Note that only certain objects—those that can be extruded—can be used for this problem.

Chapter 46
Review & Activities continued

4. Use the following three ways to create a solid cylinder with a radius of .4 and a height of 2 (Fig. 46-11). Set ISOLINES to 8 before you begin.

    a. Draw a rectangle measuring .2 × 2 and revolve it 360°.

    b. Draw a circle with a radius of .4 and extrude it to a height of 2.

    c. Use the CYLINDER command to create a solid cylinder with a radius of .4 and a height of 2.

    Now use the AREA command to determine the surface area of all three objects. Calculate the exact answer and compare it to the area AutoCAD calculated for each of the three cylinders.

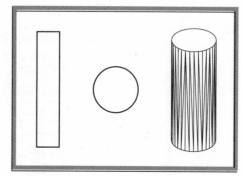

Fig. 46-11

5. Change ISOLINES to 1 and enter REGEN. Use AREA to determine the surface area of the three cylinders created in problem 4 again.

6. Set ISOLINES back to 8 and repeat the process for a right prism measuring .8 × .8 × 2. Use the following three ways to create the prism (Fig. 46-12).

    a. Draw a rectangle measuring .8 × 2. Extrude it to a height of .8.

    b. Draw a square measuring .8 on each side and extrude it to a height of 2.

    c. Use the BOX command to insert a box measuring .8 × .8 × 2.

    Use the AREA command to find and compare the surface area of the three objects.

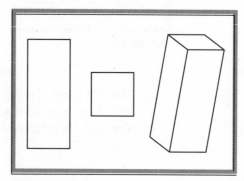

Fig. 46-12

Chapter 46
Review & Activities

 Using Problem-Solving Skills

Complete the following activities using problem-solving skills and your knowledge of AutoCAD.

1. Create a solid model of the boat trailer roller from Chapter 3 ("Using Problem-Solving Skills" problems 1 and 2). Define the shape of the roller with a polyline as shown in Fig. 46-13, revolve it, and view it from the SW Isometric viewpoint. Save the drawing as ch46roller.dwg.

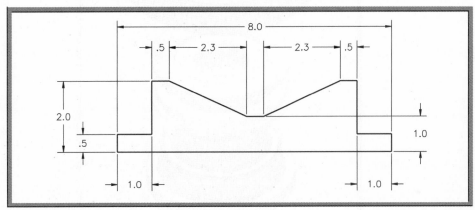

Fig. 46-13

2. The boat trailer manufacturer requires a variation of the boat trailer roller for use on a smaller trailer. Open the drawing and save it as ch46roller2.dwg. Reduce its size by one third. Change the value of the ISOLINES system variable to define the roller more clearly. Shade the roller as appropriate. Save the drawing. Your drawing should look similar to the one in Fig. 46-14.

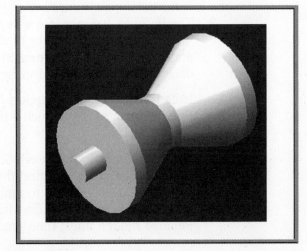

Fig. 46-14

Chapter 46
Review & Activities continued

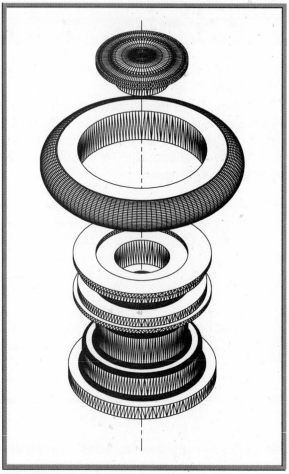

Fig. 46-15

3. The bumper assembly for a bumper pool table is shown in Fig.
46-15. Use the dimensions of the three parts (shown in Fig. 46-16),
and use the REVOLVE command to model the parts as solids. Create
each part as a separate drawing.

Note that the bumper support has a square hole. Later you will
find out how to create such a hole. For now, assume the hole
to be circular.

After you have finished each of the three parts, place it into an
isometric view. You will find this much easier to do if you first use
the ROTATE3D command to make the plan view of the world
coordinate system coincide with the top view of the object.
Remove hidden lines; then produce a smooth-shaded rendering.

Problem 3 courtesy of Gary J. Hordemann, Gonzaga University

Chapter 46
Review & Activities

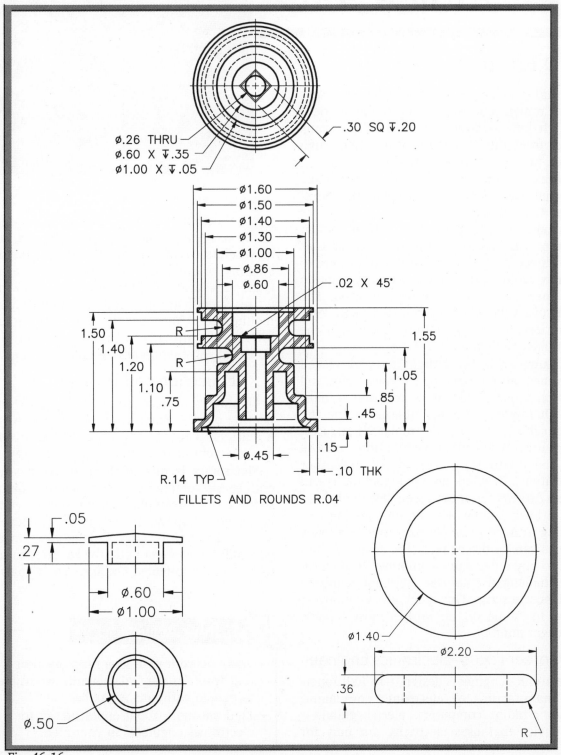

.30 SQ ↧.20

Ø.26 THRU
Ø.60 X ↧.35
Ø1.00 X ↧.05

Ø1.60
Ø1.50
Ø1.40
Ø1.30
Ø1.00
Ø.86
Ø.60

.02 X 45°

R

1.50
1.40
1.20
1.10
.75

R

1.55
1.05
.85
.45
.15
.10 THK

Ø.45

R.14 TYP
FILLETS AND ROUNDS R.04

.05
.27
Ø.60
Ø1.00

Ø.50

Ø1.40
Ø2.20
.36
R

Fig. 46-16

Careers Using AutoCAD

CAD Electronics

"It didn't make me a different person," says Jerry Cromer, reflecting on the skiing accident that left him paralyzed. He wanted the same things from life as before. Just out of high school, he wanted to study CAD.

Access to computers was one of the many new challenges he faced. A sympathetic professor at Central Carolina Technical College let Jerry take the drafting course with a student assistant, who keyed in AutoCAD commands as Jerry directed. Out of curiosity and a desire to do more on his own, Jerry asked permission to take apart the digitizer he used in class. The professor swallowed hard, but said okay. Jerry memorized the configuration of wires in the input plug. He had an idea.

Working with his father, Jerry produced Quadpuck, an interface that converts sips and puffs on a mouthpiece into electronic signals. The Quadpuck system, together with the Quadpuck AirStick (which emulates a mouse), allows users who cannot push buttons to work with full productivity in AutoCAD and other applications. In Windows, Quadpuck users can call up AirKeys for an onscreen keyboard.

As a businessman with an engineering and drafting firm, Jerry emphasizes that Quadpuck systems can be disengaged. Employers do not need to buy a dedicated computer for a physically challenged employee. Without requiring an elaborate setup, Quadpuck gives employers access to "some of the world's most capable and eager minds."

Career Focus—Electronics Engineer

Electronics engineers design, test, and supervise manufacture of electronic equipment, including radios, computers, electric guitars— any device that uses electricity not only for power but for encoding a signal. On the job, electronics engineers develop a specialty, such as wireless communications, aircraft guidance systems, or computer peripherals (such as the Quadpuck AirStick). Engineers make drawings, write specifications, and set maintenance requirements. They also develop cost estimates for manufacturing, test prototypes, and diagnose and solve equipment problems.

Education and Training

Employers usually require a four-year engineering degree. Some engineers go on to earn a master's or other advanced degree to enter management or to learn a new technology.

Electronics is one of those fields in which enthusiasts sometimes follow their own path to success. For example, an innovator with a home workshop might see a need and develop a new product, as Jerry Cromer did. On any nontraditional career path, the key to success is to be highly motivated and have clear goals.

Career Activity

- Write up your own idea for a business that would make CAD benefits accessible to people without CAD skills.
- Find out more about opportunities for electronics engineers in your area.

Objectives:

- **Prepare solid primitives for Boolean operations**
- **Use the Boolean subtraction operation to remove portions of a solid model**
- **Use the Boolean union operation to combine composite models**

Key Terms

Boolean mathematics
Boolean logic
composite solids

Much of AutoCAD solid modeling is based on the principles of *Boolean mathematics.* Boolean math, also called *Boolean logic,* is a system created by mathematician George Boole for use with logic formulas and operations. AutoCAD Boolean operations, such as union, subtraction, and intersection, are used to create composite solids.

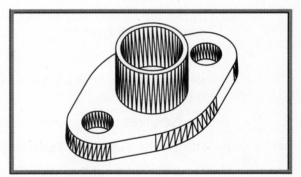

Fig. 47-1

Composite solids are composed of two or more solid primitives. Using Boolean operations, let's create a composite solid of the steel support shown in Fig. 47-1.

Preparing the Base Primitive

1. Start AutoCAD and begin a new drawing using the **solid.dwt** template file.

2. Set the snap to **.25**.

3. Using the **PLINE Arc** and **Line** options, create the shape shown in Fig. 47-2 using the information below.

   - The grid of dots, representing the snap grid, are spaced **.25″** apart. Use the grid to produce the polyline accurately.
   - Use the four points in the order shown. Each point falls on the snap grid.
   - In the PLINE command, use the Arc's **Radius** option.

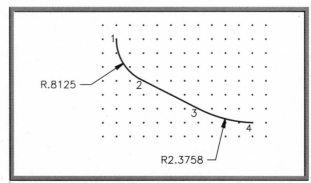

Fig. 47-2

You have just completed one quarter of the object shown in Fig. 47-3.

4. Mirror the shape to create one half of the object, and mirror that to create the entire object.

5. Use the **PEDIT Join** option to join the four polylines into one.

6. Select a viewpoint that is to the left, in front of, and above the object.

7. Set **ISOLINES** to **8** and save your work in a file named **compos.dwg**.

8. Display the **Solids** and **Solids Editing** toolbars.

9. From the Solids toolbar, pick the **Extrude** button.

Solids

This enters the EXTRUDE command.

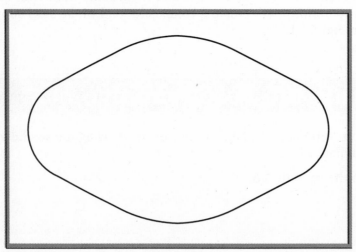

Fig. 47-3

10. Select the polyline and press **ENTER**.

11. Enter **.5** for the height, and specify a taper angle of **3** degrees.

A view similar to the one in Fig. 47-4 appears.

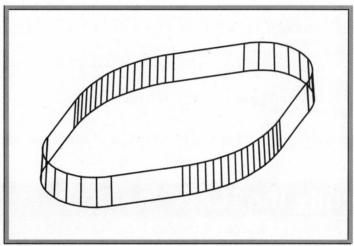

Fig. 47-4

12. Enter the **PLAN** command and press **W** for World; **ZOOM All**.

The plan view allows you to see the taper, as shown in Fig. 47-5, but without the two holes and the dimensions.

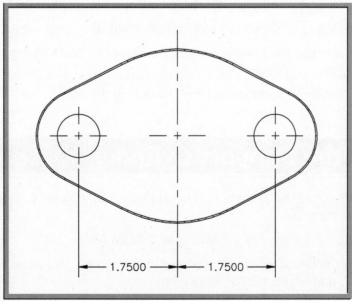

1.7500   1.7500

Fig. 47-5

13. Pick the **Cylinder** button from the Solids toolbar and place two cylinders as shown in Fig. 47-5 to represent the holes. Make their radius **.375** and their height **.5**.

Solids

HINT:

Create one of them and use the COPY or MIRROR command to create the second.

You have just created two solid cylinders within the original extruded solid.

14. Select a viewpoint to the left, in front of, and above the object.

Subtracting the Holes

As you may recall from Chapter 44, the SUBTRACT command performs a Boolean operation that creates a composite solid by subtracting one solid object from another.

Solids Editing

1. From the Solids Editing toolbar, pick the **Subtract** button.

This enters the SUBTRACT command.

2. Select the extruded solid (but not the cylinders) as the solid to subtract from, and press **ENTER**.

3. Select the two cylinders as the solids to subtract, and press **ENTER**.

AutoCAD subtracts the volume of the cylinders from the volume of the extruded solid.

4. The composite solid should look similar to the one in Fig. 47-6.

Adding the Support Cylinder

The UNION command permits you to join two solid objects to form a new composite solid.

1. Create a plan view of the WCS and **ZOOM All**.

2. Place a solid cylinder at the center of the model. Make the diameter **1.75"**, and make it **1.75"** in height.

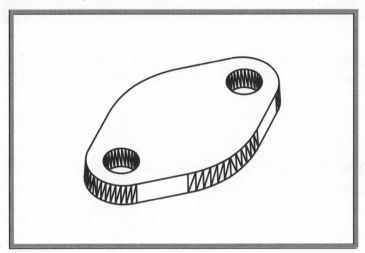

Fig. 47-6

HINT: ─────────────────────────────

Use the snap grid to snap to the center of the model.

---

3. Using the same center point, place a second cylinder inside the first. Specify a diameter of **1.5″** and a height of **1.75″**.

4. Specify a viewpoint in 3D space as you did before.

5. Subtract the smaller cylinder from the larger cylinder.

The model is currently made up of two separate solid objects. Let's use the UNION command to join them to form a single composite solid.

Solids Editing

6. From the Solids Editing toolbar, pick the **Union** button.

This enters the UNION command.

Solids Editing

7. Select the two solid objects and press **ENTER**.

The two objects become a single composite solid.

8. Enter **HIDE**.

Your model should look similar to the one in Fig. 47-1 (page 669).

9. Save your work.

Chapter 47
Review & Activities

Review Questions

1. Describe the procedure used to create a hole in a solid object.

2. What is a composite solid?

3. Explain the purpose of the following solid modeling commands.
   SUBTRACT
   UNION

Challenge Your Thinking

1–2. Describe how you would create the solid models shown in Figs. 47-7 and 47-8. Be specific.

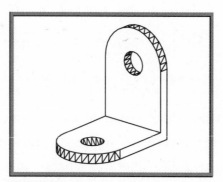

Fig. 47-7

Fig. 47-8

Chapter 47
Review & Activities

Applying AutoCAD Skills

Work the following problems to practice the commands and skills you learned in this chapter.

1–2. Apply the commands you learned in this chapter to create the composite solids shown in Figs. 47-9 and 47-10.

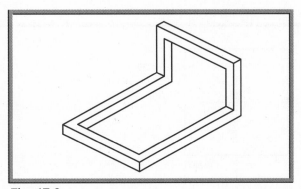

Fig. 47-9

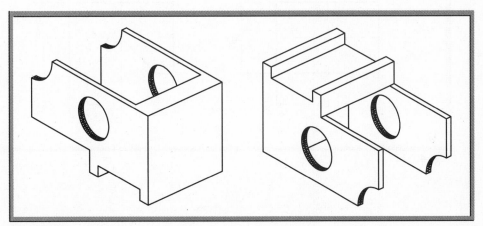

Fig. 47-10

Chapter 47
Review & Activities continued

3. Identify a machine part, such as a gear on a shaft or a bracket. Sketch the part in your mind or on paper. Using AutoCAD's solid primitives and Boolean operations, shape the part by adding and subtracting material until it is complete.

4. Draw a 3D model of the height gage shown in Fig. 47-11 using the EXTRUDE command and Boolean operations. Can you use a box or a wedge to create the V-groove? Why or why not? Save the drawing as ch47htgage.dwg.

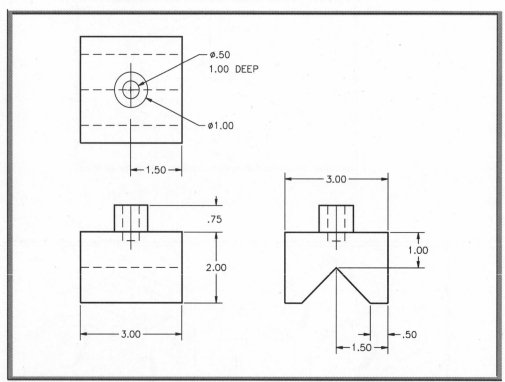

Fig. 47-11

5. Your customer wants to see a 3D view of the base plate shown in Fig. 47-14A. Create the base, extrude it, and then use Boolean operations to create the 3D model. Shade the base plate with flat shading edges on, as shown in Fig. 47-14B. Save the drawing as ch47baseplate.dwg.

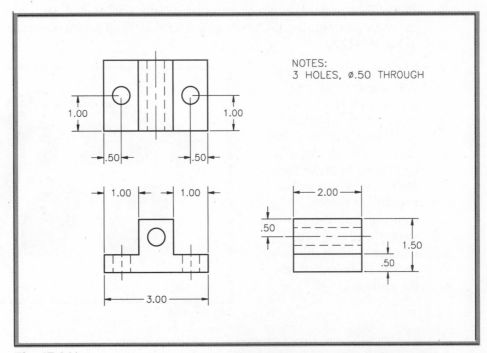

NOTES:
3 HOLES, ⌀.50 THROUGH

Fig. 47-14A

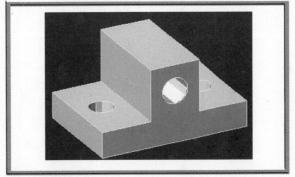

Fig. 47-14B

Chapter 47
Review & Activities continued

Using Problem-Solving Skills

Complete the following activities using problem-solving skills and your knowledge of AutoCAD.

1. Create the fluid coupling end cap shown in Fig. 47-12A using the REVOLVE command. Begin by creating a layer named Endcap, and set the color to cyan. Set ISOLINES to 25. The object's profile is shown in Fig. 47-12B.

After you have finished drawing the end cap, produce an isometric view. You will find this much easier to do if you first use the ROTATE3D command to make the plan view of the world coordinate system coincide with the top view of the end cap. Remove hidden lines and then smooth-shade the end cap.

Fig. 47-12A

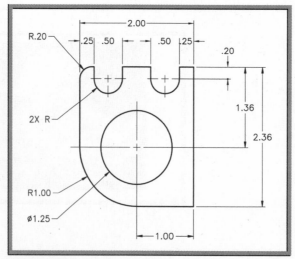

Fig. 47-12B

Problem 1 courtesy of Gary J. Hordemann, Gonzaga University

2. Create the shaft clamp shown in Fig. 47-13 using the SUBTRACT and EXTRUDE commands. Create the .40 diameter hole by inserting and subtracting a cylinder.

After you have finished the object, place it into an isometric view. Remove the hidden lines and then smooth-shade the clamp.

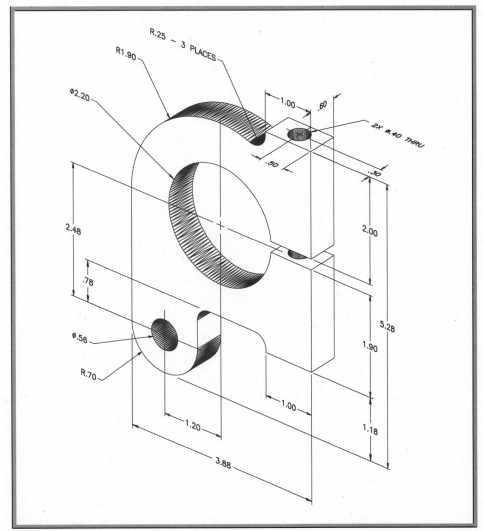

Fig. 47-13

Problem 2 courtesy of Gary J. Hordemann, Gonzaga University

3. Create a solid model of the adjustable pulley half shown in Fig. 47-15. It will help if you first change the view to the right side view and the UCS to the current view. If you allow for the counterbored hole when drawing the profile, you will not have to bring in and subtract cylinders. After revolving the profile, go to the front view and insert and subtract the cylinders for the two smaller holes.

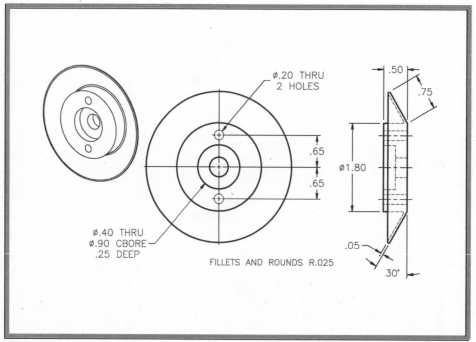

Fig. 47-15

Problem 3 courtesy of Gary J. Hordemann, Gonzaga University

Chapter 48 Tailoring Solid Models

Objectives:

- Apply bevels and rounded edges to a composite solid
- Create a composite solid by intersection
- Shell a solid to create a hollow object
- Check for interference between adjacent objects in 3D space

Key Terms

interference
interference solids
intersection
shelling

AutoCAD provides commands and options for modifying primitives and solid composites. You are already familiar with the CHAMFER and FILLET commands; in this chapter, you will see how they can be used to modify 3D objects.

Using a method called *intersection,* you can create a solid from the overlapping portion of two intersecting solid objects. Another method, called *shelling,* enables you to remove material from an object to create a shell or hollow object. *Interference* occurs when the two objects overlap or collide in 3D space. The INTERFERE command permits you to determine whether two solids or objects overlap.

In this chapter, you will create the model of a table and bowl shown in Fig. 48-1 (page 682). You will create the basic table, round the top edges, bevel the legs, and place the bowl on top of it. Using the INTERFERE command, you will ensure that the bowl is positioned correctly on top of the table and does not collide with it.

> **INFOLINK**
>
> See Chapter 15 for more information about chamfers and fillets in 2D drawings.

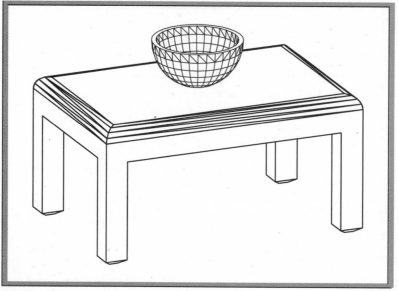

Fig. 48-1

Beveling and Rounding Solid Edges

Let's create the table shown in Fig. 48-1 using the methods we've used in previous chapters. Then we will apply the CHAMFER and FILLET commands to make the table look more like the one in the previous illustration.

1. Start AutoCAD and begin a new drawing using the **solid.dwt** template file.

2. Display the **Solids** and **Solids Editing** toolbars.

Solids

3. Pick the **Box** button from the Solids toolbar and create a solid box **6** units in the X direction and **4** units in the Y direction. Make it **3** units in height and place it in the center of the screen.

NOTE: ────────────────────────────────

In this file, 1 unit = 1'.

4. Select a viewpoint that is to the left, in front of, and above the object.

5. Enter **ZOOM** and **.8x**.

Solids

Solids Editing

6. Using the **BOX** and **SUBTRACT** commands, create and subtract two rectangular boxes from the first one to form the table shown in Fig. 48-2.

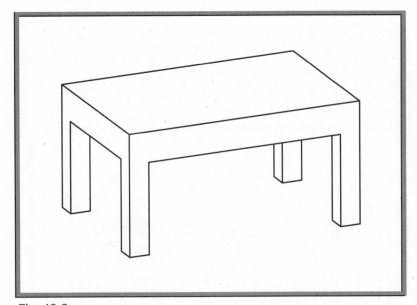

Fig. 48-2

HINT:

Create one box 5 units by 4 units by 2.25 units tall and another box 6 units by 3 units by 2.25 units tall. Subtract both from the first box. It may be helpful to switch to the plan view.

7. Enter **HIDE**.

The solid model should look similar to the one in Fig. 48-2.

8. Save your work in a file named **table.dwg**.

9. Pick the **Undo** button on the docked *Standard* toolbar to return to the wireframe representation.

Beveling the Table Legs

Modify

1. Enter the **CHAMFER** command and select the bottom of one of the four table legs.

2. If AutoCAD selects the side of the table rather than the bottom of the leg, type **N** for Next, and press **ENTER**. When AutoCAD selects the bottom, press **ENTER**.

3. Enter a base surface chamfer distance of **.15**.

4. At the Specify other surface chamfer distance prompt, enter **.1**.

5. At the next prompt, pick all four edges that make up the bottom of the leg and press **ENTER**.

AutoCAD chamfers the edges.

6. Repeat Steps 1 through 5 to chamfer the remaining three legs.

The bottoms of the legs should now look similar to those in Fig. 48-3.

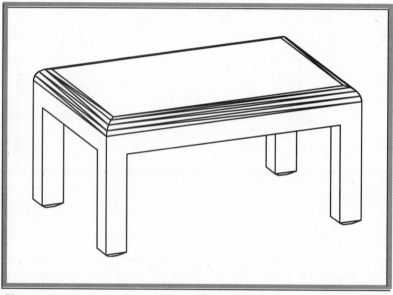

Fig. 48-3

Rounding the Table Top

7. Enter **FILLET**, select any edge of the table top, and enter **.3** for the radius.

Modify

8. Enter **C** for Chain, select the remaining three edges of the table top, and press **ENTER**.

9. Enter **HIDE** and save your work.

The model should look very similar to the one in Fig. 48-3.

10. Pick the **Undo** button to return to the wireframe representation.

Boolean Intersection

You have learned that you can create composite solids by adding and subtracting solid objects. You can also create them by overlapping two or more solid objects and calculating the solid volume that is common (intersecting) to each of the objects. AutoCAD's INTERSECT command performs this Boolean operation.

Solids

1. Enter the **BOX** command and create a 2-unit cube at any location on the floor (WCS) under the table. (The floor and WCS are on the same plane.) Be sure snap is on.

HINT:

You may want to return to the plan view to perform this step. After picking the first corner of the box, select the Cube option and enter 2 in reply to Specify length.

2. View the objects in 3D as you did before. If necessary, move the box away from the table or select a viewpoint so the two objects do not overlap.

3. Enter the **UCS** command, **N** for New, and enter the **3point** option.

4. Using the **Endpoint** object snap, pick any one of the top corners of the cube.

5. To specify the positive X direction, snap to an adjacent top corner of the cube.

6. For the positive Y direction, snap to the other adjacent top corner of the cube.

The UCS and UCS icon change to reflect the new UCS.

7. Enter **PLAN** to review the plan view of the current UCS.

Solids

8. Pick the **Sphere** button from the Solids toolbar.

9. Place the center point of the sphere at the center of the top of the cube and enter **1** for the radius.

A solid sphere appears.

10. Change to the **SW Isometric** viewpoint.

You should see a view that is similar to the one in Fig. 48-4.

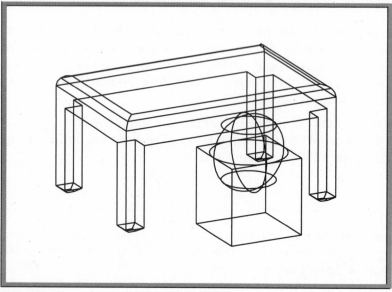

Fig. 48-4

Solids Editing

11. From the Solids Editing toolbar, pick the **Intersect** button.

12. Select the cube and sphere and press **ENTER**.

A half-sphere (hemisphere) appears. AutoCAD has calculated the solid volume that is common (intersecting) with each of the primitives.

13. Save your work.

Shelling a Solid Object

Solids Editing

We want to make the hemisphere into a bowl. The fastest way to do this is to shell it out.

1. Pick the **Shell** button from the Solids Editing toolbar.

This enters the Shell option of the SOLIDEDIT command.

2. Select the hemisphere.

Next, we need to remove the top face from the shelling. If we don't, AutoCAD will hollow out the hemisphere, but it will not have an opening.

3. In reply to Remove faces, pick the edge of the hemisphere.

The top face and the main body face share the same edge, so AutoCAD displays 2 faces found, 2 removed. We don't want to remove the main body face, so we need to add it back.

4. Enter **A** for Add, select the rounded part of the hemisphere (do not select the top edge), and press **ENTER**.

5. For the shell offset distance, enter **.05** and press **ENTER** twice.

AutoCAD produces a shell, as shown in Fig. 48-5.

6. Enter **HIDE**.

7. Pick the **Undo** button and save your work.

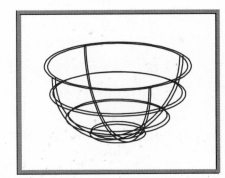

Fig. 48-5

Positioning Adjacent Objects

Let's move the bowl to a more logical position—on top of the table.

1. Produce a front view, as shown in Fig. 48-6.

HINT:

From the View pull-down menu, pick 3D Views and Front.

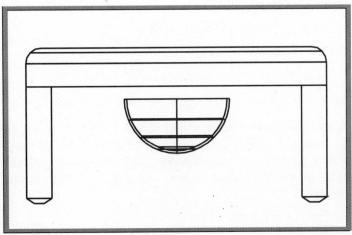

Fig. 48-6

Approximating the Location

2. Enter **ZOOM** and **.5x**.

3. Move the bowl upward so that it appears to rest on the tabletop.

4. Produce a plan view.

5. If the bowl is not resting at the center of the table, move it to the approximate center.

6. Produce a viewpoint to the left, in front of, and above the tabletop.

7. Create a layer named **Bowl**, make it green, and move the bowl to this layer.

8. Turn off the grid, smooth-shade the table and bowl, and save your work.

Testing for Interference

The INTERFERE command enables you to check for interference (overlap) between two or more solid objects. It is particularly useful when fitting together solid objects into an assembly.

1. Zoom in on the bowl.

2. From the Solids toolbar, pick the **Interfere** button.

This enters the INTERFERE command.

Solids

3. Pick either the table or the bowl and press **ENTER**.

4. Pick the other composite solid and press **ENTER**.

The message Solids do not interfere should appear. If you don't receive this message, the two interfere with one another. If the solids don't interfere with each other, proceed to Step 6.

5. If they do interfere, press **ESC** to cancel.

6. Move the bowl downward **.25** unit.

HINT:

From the View pull-down menu, select 3D Views and Front. Enter the MOVE command, select the bowl, and press ENTER. For the base point, pick any point on or near the bowl. For the second point of displacement, move downward a short distance, and with polar tracking on, enter .25.

Solids

7. Enter the **INTERFERE** command.

8. Pick one of the two composite solids and press **ENTER**; pick the other and press **ENTER**.

You should receive the message Interfering pairs: 1. This is because the two interfere with one another.

When solids interfere with one another, you can create *interference solids* (solids that consist of only the interfering part of the two objects). Interference solids make it easier to see which parts of the solids overlap.

9. Enter **Yes** in reply to Create interference solids?

10. Using **MOVE Last**, move the new solid up and away from the bowl and table.

11. Produce a viewpoint similar to before.

The new solid should look similar to the one in Fig. 48-7.

Fig. 48-7

12. Erase the new solid and move the bowl back to its previous location.

13. Close the toolbars you opened in this chapter.

14. Save your work and exit AutoCAD.

Chapter 48
Review & Activities

Review Questions

1. How do the CHAMFER and FILLET commands affect the appearance of solid objects containing outside corners?

2. Explain how the INTERSECT command is used to form a composite solid.

3. Explain why shelling is useful.

4. Describe the purpose of the INTERFERE command.

5. How might the INTERFERE command be useful when designing parts of an assembly?

Challenge Your Thinking

1. How do the CHAMFER and FILLET commands affect the appearance of solid objects containing inside corners?

2. What factors should you consider when you are planning to create a solid model using interference solids? Explain.

Applying AutoCAD Skills

Work the following problems to practice the commands and skills you learned in this chapter.

1–2. Apply the commands you learned in this chapter to create the solid composites of a metal shroud shown, as in Figs. 48-8 and 48-9. Approximate all sizes. Use the INTERSECT command to complete the shroud in Fig. 48-8. Then modify it to complete the one in Fig. 48-9. (Increase the value of ISOLINES before you begin.)

Fig. 48-8

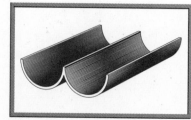

Fig. 48-9

3. Open ch47base plate.dwg. Chamfer the two edges where the upper
   horizontal and vertical surfaces meet. Use a .25″ chamfer at 45°.
   Create fillets with a radius of .15 where the upper vertical surfaces
   meet the plate's horizontal surfaces. Create fillets with a radius of
   .15 at the two outer corners. Refer to Fig. 48-10 for specific locations.

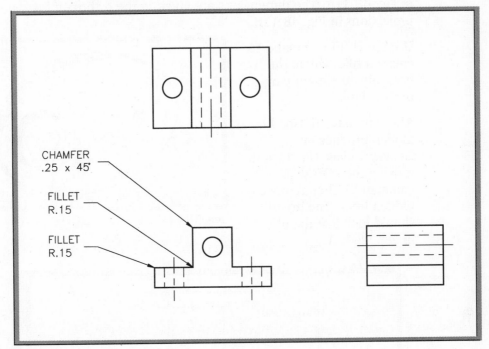

Fig. 48-10

Chapter 48
Review & Activities continued

? Using Problem-Solving Skills

Complete the following activities using problem-solving skills and your knowledge of AutoCAD.

1. Using the SUBTRACT and UNION commands, model the plate shown in Fig. 48-11A. The dimensions are given in the orthographic projections in Fig. 48-11B.

   Use the FILLET command to create a fillet where the boss joins the main part of the plate.

   After you have finished the model, produce an isometric view. (First rotate it using the ROTATE3D command.) Then remove hidden lines. The drawing should look like the one in Fig. 48-11A.

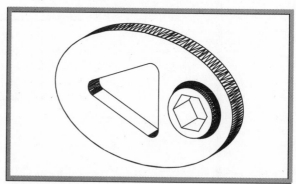

Fig. 48-11A

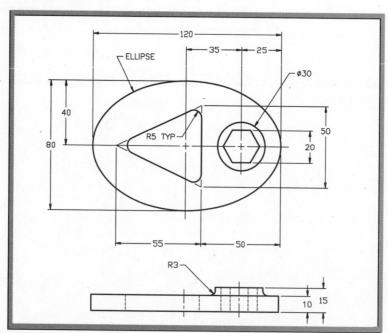

Fig. 48-11B

Problems 4 and 5 courtesy of Gary J. Hordemann, Gonzaga University

2. Using the SUBTRACT and EXTRUDE commands, model the tube bundle support shown in Fig. 48-12A. The dimensions are given in the orthographic projections shown in Fig. 48-12B.

The rounded top can be produced in several ways. One way is to use the REVOLVE command to create a piece to be removed from the original extrusion (using the SUBTRACT command). A second way is to use the REVOLVE command to create a rounded piece (without holes); then use the INTERSECT command to obtain the common geometry. Try both ways. Obtain an isometric view, and then remove the hidden lines.

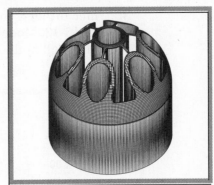

Fig. 48-12A

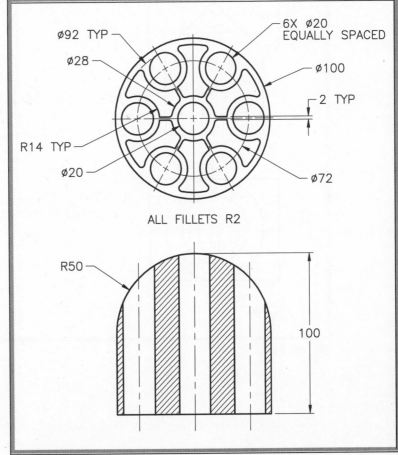

Fig. 48-12B

3. Three orthographic views of a T-swivel support are shown in Fig. 48-13. Model the piece as a solid. Use the FILLET command to fillet all of the indicated edges.

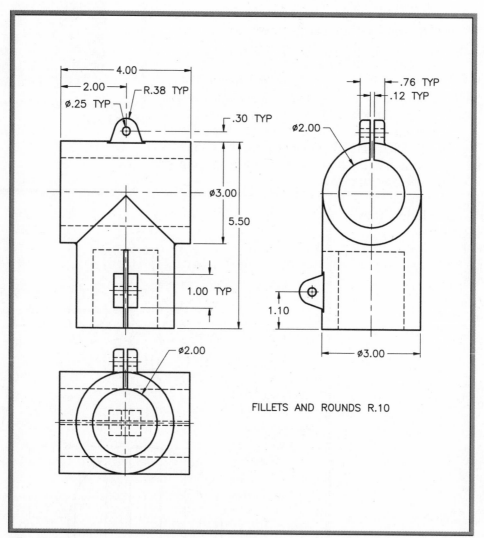

FILLETS AND ROUNDS R.10

Fig. 48-13

Careers Using AutoCAD

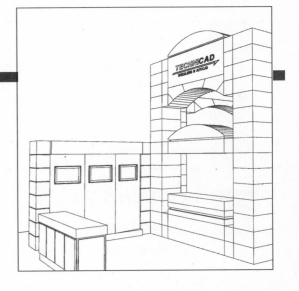

3D Visualization

A two-dimensional overhead view, or floor-plan, has always been the basic planning tool for laying out a trade show exhibit. Yet it takes a three-dimensional view to answer questions such as: "What will our sign look like within the exhibit space?" "What impression will this display make on customers as they arrive?" With 3D AutoCAD drawings, planners combine the precision of a blueprint with the visual quality of an "artist's conception" drawing.

Being able to see the finished product before it is constructed is the heart of business at TechniCAD (Brookside, NJ), where consultant Phil Gauntt specializes in rendering "virtual environments" for customers ranging from civil engineers to interior designers. Some customers are architects who, after creating a 3D AutoCAD drawing (like the one accompanying this story), want output that is photorealistic. Phil scans in samples of a wallpaper pattern, carpet color, or any textures and finishes the customer wants to see in the final image. Using 3D Studio Viz, he applies the textures and finishes to the blocks created in AutoCAD. The result is a visually convincing image of a place that only existed, until that moment, in the mind's eye of the designer. When output to videotape, the image can come to life with animation. Often a customer gets more from ten seconds of video than from hours of poring over a written proposal.

Career Focus: Graphics Consultant

Phil Gauntt got into the consulting business after some years as a drafter, "pushing a pen on mylar." He enrolled in a semester-long CAD course and found new employment as an AutoCAD operator. Eventually Phil set out on his own to found TechniCAD. "Having a background in drafting is a big help," Phil says. Clients often want the finished product in a hurry, so software knowledge and good planning are essential.

Education and Training

In addition to experience as a drafter and later as an AutoCAD operator, Phil Gauntt also took classes to become an Autodesk-certified instructor. He teaches at Pratt Manhattan, an Autodesk Training Center (ATC), where students interested in 3D computer graphics complete a 24-month program of evening classes, gaining experience with AutoCAD and other software, as well as output techniques such as photo-quality printing and videotape.

Career Activity

- Research 3D visualization on the Internet.
- Write a report detailing what types of companies use 3D visualization. What do they use it for?

Objectives:

- **Create a full section from a solid model**
- **Produce a profile view from a solid model**

Key Terms

cutting plane
full section
profile
tangential edges

Solid modeling provides many benefits that are "downstream" from the solid model. Examples include mass properties generation, detail drafting, finite element analysis, and the fabrication of physical parts.

In some cases, solid models can now be used directly to guide production processes. In others, the solid model is used to create production drawings such as sections and profiles. (A *profile* is an outline or contour, such as a side view, of an object.) In this chapter, you will create full section and profile views of a pulley.

1. Start AutoCAD and begin a new drawing using the **solid.dwt** template file.

2. Set snap to **.125** and grid to **.25**.

3. Set **ISOLINES** to **30**, and display the **Solids** toolbar.

4. Create the polyline shown in Fig. 49-1 using the following information.

    - The dots of the grid are spaced **.125** apart. The bold dots are spaced **.25** apart.
    - All polyline endpoints fall on the grid, so use it to produce the object accurately.
    - Produce the polyline in the top half of the drawing area.
    - The axis of revolution is an imaginary line of arbitrary length.

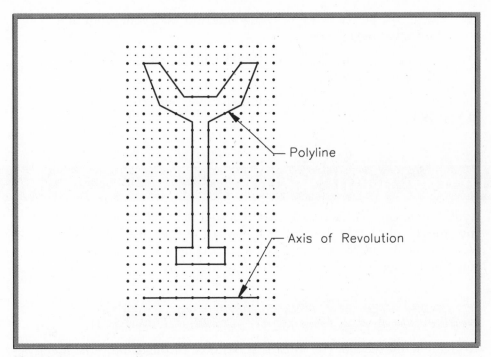

Fig. 49-1

5. Using the **REVOLVE** command, revolve the polyline to produce the pulley.

The pulley should appear, as shown on the left in Fig. 49-2.

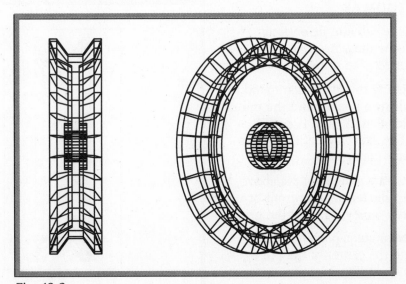

Fig. 49-2

6. Center the pulley in the drawing area and save your work in a file named **pulley.dwg**.

7. Select a viewpoint that is to the left and in front of the model, and enter **ZOOM .8x**.

Your view of the model should be similar to the one on the right in Fig. 49-2.

Creating a Full Section

A *full section* of a part is a view that slices all the way through the part, showing the interior of the part as it appears at that slice, or cross section. The SECTION command helps you create a full cross section of a solid model, such as the pulley shown on page 701.

Defining the Cutting Plane

To create a full section, you must first define a *cutting plane*—the plane through the part from which the sectional view will be taken.

1. Enter the **PLAN** command and press **ENTER** to accept the Current UCS default setting.

A plan view of the pulley should appear, as shown in Fig. 49-3. You will not see the box around the pulley.

2. **ZOOM All**.

The pulley should lie entirely within the drawing area. (The grid reflects the drawing area.)

3. Using the **PLINE** command, draw a box around the pulley as shown in Fig. 49-3. The exact size of the box is not important.

4. Create a view that is above, to the left, and in front of the pulley; turn off the grid.

The rectangular polyline passes through the center of the pulley, as shown in Fig. 49-4.

5. Enter **HIDE**.

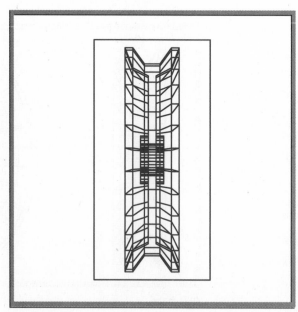

Fig. 49-3

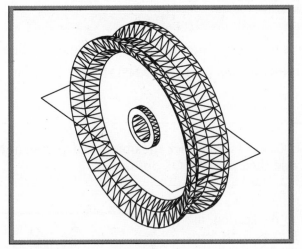

Fig. 49-4

Extracting the Section

1. Enter **PLAN** to obtain the plan view of the current UCS and enter **ZOOM All**.

2. Make layer **0** the current layer.

Solids
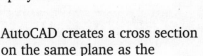

3. Pick the **Section** button from the Solids toolbar.

This enters the SECTION command.

4. Pick the pulley and press **ENTER**.

AutoCAD asks you to define a plane.

5. Enter **O** (for object) and pick the rectangular polyline.

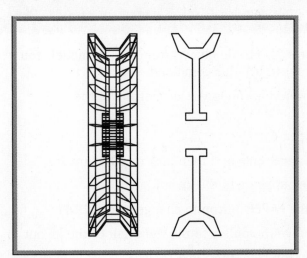

AutoCAD creates a cross section on the same plane as the rectangular polyline.

6. Enter **MOVE Last** and move the cross section to the right of the pulley.

7. Erase the rectangular polyline object and save your work.

The drawing should look similar to the one in Fig. 49-5.

Fig. 49-5

Completing the Section

The full section is not complete until you hatch it and add lines.

1. Hatch the cross section using the **ANSI31** hatch pattern.

HINT:

Pick the Select Objects button, pick the object that you want to hatch, press ENTER, and then pick the OK button.

2. Add lines to the section, as shown in Fig. 49-6, to complete the sectional view.

3. Make **Objects** the current layer, and freeze layer **0**.

4. Produce a left view of the pulley, **ZOOM .7x**, and save your work.

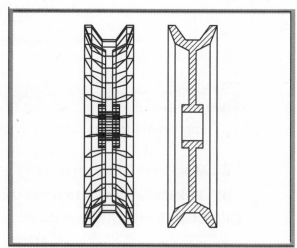

Fig. 49-6

Creating a Profile

The SOLPROF command creates 2D profile objects from a solid model. You must be in a paper space layout to use this command.

1. Pick the **Layout 1** tab located in the lower left portion of the drawing area.

The Page Setup dialog box appears.

2. For the layout name, enter **Profile** and then pick the **OK** button.

A single viewport of the pulley appears in the layout.

3. In the status bar, pick the **PAPER** button to change it to **MODEL**.

4. Enter **ZOOM .5x** and move the pulley to the right half of the layout.

5. Pick the **Setup Profile** button from the Solids toolbar.

Solids

This enters the SOLPROF command.

6. Select the pulley and press **ENTER**.

7. Enter **Yes** in reply to Display hidden profile lines on separate layer?

This means that AutoCAD will place hidden lines on a separate layer, allowing you to control the display of these lines.

8. Enter **Yes** in reply to Project profile lines onto a plane?

This projects the visible and hidden profile lines onto a single plane.

9. Enter **Yes** in reply to Delete tangential edges?

This deletes transition lines, called *tangential edges,* that occur when a curved face meets a flat face.

AutoCAD creates profile objects on top of the pulley.

10. With polar tracking or ortho on, move the pulley away and to the left of the profile objects. (The pulley is red, and the profile objects are black.)

11. Using the **LIST** command, select the profile objects.

AutoCAD created a block for the visible profile lines and a block for the hidden profile lines. Also, AutoCAD created new layers to hold the profile objects.

12. Assign the **Hidden** linetype to the new layer beginning with PH.

AutoCAD stores hidden profile lines on this layer and visible lines on the layer that begins with PV.

13. Pick the **Model** tab and **ZOOM All**.

The new profile should look similar to the one in Fig. 49-7.

14. Close all floating toolbars, save your work, and exit AutoCAD.

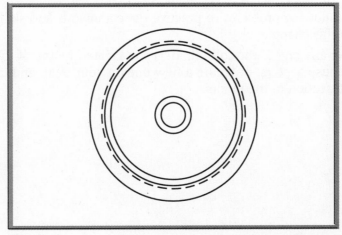

Fig. 49-7

Chapter 49
Review & Activities

Review Questions

1. Name two downstream benefits of solid modeling.

2. Explain the purpose of the SECTION command.

3. Explain how to define the cutting plane when using the SECTION command.

4. What kind of objects does the SOLPROF command produce?

5. What is the purpose of the layers that SOLPROF creates?

Challenge Your Thinking

1. Discuss how you would orient and print together the full section and profile views that you created in this chapter.

2. Investigate current manufacturing practices. Then explain why solid models can sometimes be used directly to guide production processes, but at other times, the solid model is used merely to create production drawings such as sections and profiles.

 ## Applying AutoCAD Skills

Work the following problems to practice the commands and skills you learned in this chapter.

1. Select any one of the solid models you created in any of the previous chapters, or create a new one on your own, and create a full section of the model.

2. Draw a solid model of the bushing and insert shown in Fig. 49-8. Then create a full sectional view to completely describe them. Remember to change your hatch for the insert. Save the drawing as ch49bushing.dwg.

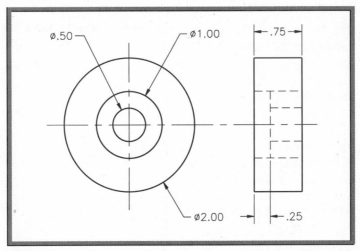

Fig. 49-8

3. You are given a file containing the solid model and sectional view of the bushing and insert (problem 2 above). The manufacturing department needs a set of 2D working drawings. Create a 2D profile, and add any other views you think are necessary to describe the bushing and insert fully. Save the drawing as ch49bushing2.dwg.

Chapter 49
Review & Activities continued

Using Problem-Solving Skills

Complete the following problem using problem-solving skills and your knowledge of AutoCAD.

1. Fig. 49-9 shows a bumper assembly for a bumper pool table. The dimensions for the bumper support, bumper ring, and cap are given in Chapter 46. Model these three objects as solids. Note that the bumper support has a square hole with a chamfer of .02 × .02. Create a box and use the SUBTRACT command to create the hole.

   After you have created the three solids, put them together as they would fit if assembled. Create a full section view using the SECTION command. Use the INTERFERE command to check for interferences among the three pieces.

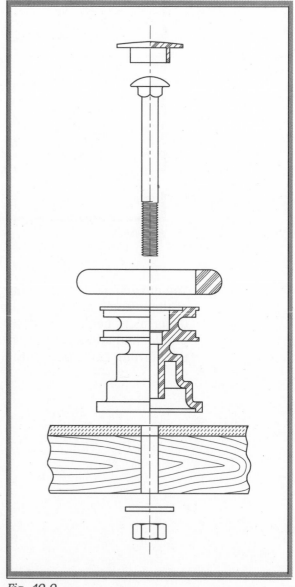

Fig. 49-9

Problem 1 courtesy of Gary J. Hordemann, Gonzaga University

Chapter 49
Review & Activities

2. A solid model of the rocker arm shown in Fig. 49-10 is needed, along with a full section and a complete set of 2D drawings, for the manufacturing and design departments. Provide the necessary drawings. Save the drawings using appropriate names.

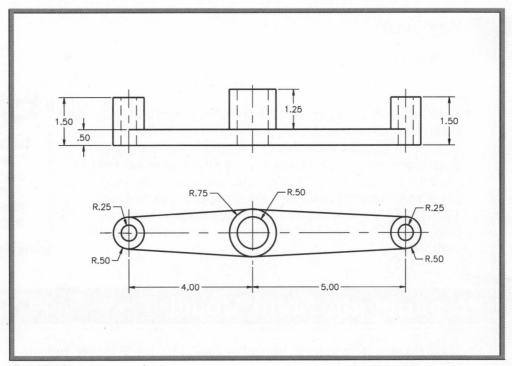

Fig. 49-10

Chapter 50 — Documenting Solid Models

Objectives:

- Prepare a solid model for use in creating orthographic and section views
- Produce orthographic and section views from a solid model
- Position and plot orthographic, sectional, and isometric views

Key Term

auxiliary view

AutoCAD enables you to use orthographic projection to lay out multiview and section drawings of solid models. As you may recall, orthographic projection is a method of creating views of an object that are projected at right angles onto adjacent projection planes. The SOLVIEW command uses AutoCAD's paper space layout to establish the viewports and new layers. The SOLDRAW command uses the viewports and layers to place the visible lines, hidden lines, and section hatching for each view.

INFOLINK

See Chapter 11 for more information about <u>orthographic projection</u>.

Preparing the Solid Model

The following steps prepare the pulley model for the SOLVIEW and SOLDRAW commands.

1. Open the drawing file named **pulley.dwg**.

2. Using **Save As...**, create a new drawing file named **pulley2.dwg**.

3. Freeze the layers that begin with PH and PV, set the linetype scale to **.5**, and open the **Solids** toolbar.

PH and PV are the layers that AutoCAD used to store the hidden and visible profile lines.

4. Pick the layout tab named **Profile** to switch to the layout.

5. Click **MODEL** on the status bar so that it now reads **PAPER**.

The paper space icon replaces the coordinate system icon.

6. Erase the single viewport.

 HINT: ———————————————————————————

Pick the red viewport outline.

This deletes the viewport and its contents.

7. Create a new layer named **Vports**, assign the color cyan to it, and make it the current layer.

8. From the **View** pull-down menu, select **Viewports** and **4 Viewports**, and enter **F** (for Fit) at the Command line.

AutoCAD fits four new views of the pulley in the drawing area, as shown in Fig. 50-1.

9. Save your work.

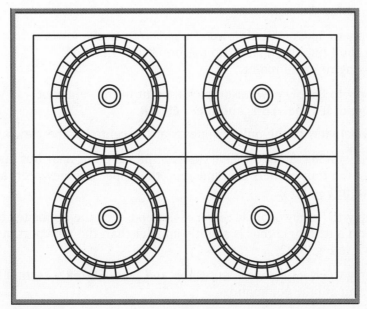

Fig. 50-1

Creating the Views

AutoCAD provides commands that can calculate and display orthographic, auxiliary, and sectional views. The SOLVIEW command performs the calculations and stores the necessary view-specific information. The SOLDRAW command uses that information to produce the final drawing views.

Producing the Top View

Let's produce a top view of the drawing in the upper left viewport.

1. Pick the **Setup View** button from the Solids toolbar.

Solids

This enters the SOLVIEW command and displays four options at the Command line. The Ucs option creates a profile view relative to a user coordinate system. The Ortho option creates an orthographic view from an existing view. Auxiliary creates an auxiliary view from an existing view. An *auxiliary view* is one that is projected onto a plane perpendicular to one of the orthographic views and inclined in the adjacent view. The Section option creates a sectional view, complete with crosshatching.

2. Enter **O** for **Ortho**.

3. In reply to Specify side of viewport to project, pick the lower horizontal line that makes up the upper left viewport, snapping to its midpoint.

4. In reply to Specify view center, pick a point in the center of the upper left viewport, and press **ENTER**.

An orthogonal view of the pulley appears at the center of the viewport.

5. In reply to Specify first corner of viewport, pick a point near but inside one of the four corners that make up the upper left viewport, as shown in Fig. 50-2.

6. In reply to Specify opposite corner of viewport, pick the opposite corner to form a rectangle that is slightly smaller than the viewport, as shown in Fig. 50-2.

7. In reply to Enter view name, enter **top** and press **ENTER** a second time to terminate the command.

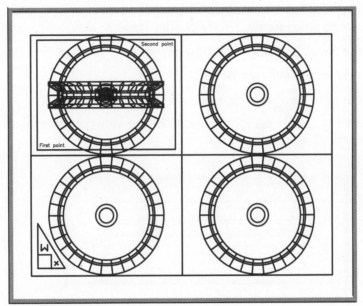

Fig. 50-2

8. Display the list of layers, as shown in Fig. 50-3.

Note the three layers that SOLVIEW created for visible lines, hidden lines, and dimensions.

9. Erase the old upper left viewport. Do not erase the new viewport containing the new orthographic view.

10. Save your work.

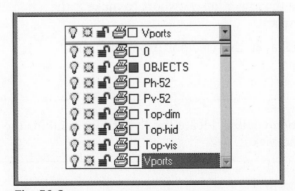

Fig. 50-3

Now that SOLVIEW has accumulated the required information, you can use SOLDRAW to generate the final drawing view.

11. Pick the **Setup Drawing** button from the Solids toolbar.

This enters the SOLDRAW command.

12. Pick the new viewport that you created and press **ENTER**.

AutoCAD creates an orthographic view from the viewport, complete with hidden lines, as shown in Fig. 50-4.

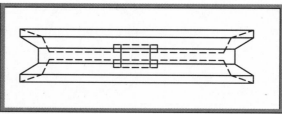

Fig. 50-4

13. On the status bar, change **PAPER** to **MODEL**.

14. Enter **ZOOM** and **.7x**.

15. Save your work.

Creating the Front View

Let's create the front view of the pulley in the bottom left area of the screen.

1. Click inside the lower left viewport to make it the current viewport.

2. View the pulley in this viewport from the front.

HINT: ────────────

From the Standard toolbar, locate the flyout that contains the Front View button and pick it.

3. Enter **ZOOM** and **.7x**.

4. Pick the **Setup View** button from the Solids toolbar to enter the SOLVIEW command, and enter the **Ortho** option.

5. Pick the left vertical line that makes up the lower left viewport, snapping to its midpoint.

6. In reply to Specify view center, pick a point in the center of the viewport and press **ENTER**.

7. In reply to Specify first corner of viewport, pick a point inside one of the four corners of the viewport, as you did for the top left viewport.

8. Pick the opposite corner to form a rectangle that is slightly smaller than the viewport.

9. Enter **front** for the view name and press **ENTER** a second time to terminate the command.

10. Save your work.

11. In the status bar, change **MODEL** to **PAPER** and erase the old lower left viewport. Do not erase the new viewport containing the orthographic view.

Solids

12. Pick the **Setup Drawing** button from the Solids toolbar to enter the SOLDRAW command.

13. Pick the newest viewport and press **ENTER**.

AutoCAD creates an orthographic view from the viewport drawing, complete with hidden lines, as shown in Fig. 50-5.

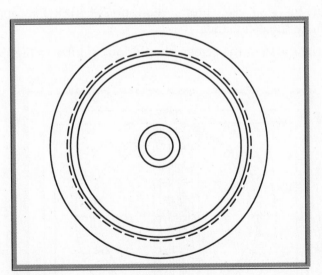

Fig. 50-5

14. Save your work.

Creating a Full Section View

Let's use the SOLVIEW command's Section option to create a full section.

1. In the status bar, change **PAPER** to **MODEL**.

2. Click in the lower right viewport to make it active.

3. Pick the **Setup View** button from the **Solids** toolbar and enter **S** for Section.

Solids

4. In reply to Specify first point of cutting plane, snap to the center of the pulley in the lower right viewport.

5. In reply to Specify second point of cutting plane, pick a point a short distance directly above or below the center point.

This defines the cutting plane.

6. In reply to Side to view from, pick a point on either side of the center point. (The resulting section is the same either way.)

7. In reply to Enter view scale, press **ENTER** to accept the default value.

8. In reply to Specify view center, pick a point near the center of the viewport and press **ENTER**.

9. For the first and second corners, define a viewport outline as you did before.

10. In reply to View name, enter **section**; press **ENTER** a second time to terminate the command.

The lower right area of the screen should look similar to Fig. 50-6.

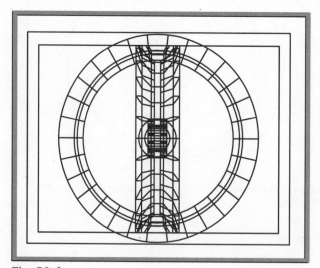

Fig. 50-6

Fig. 50-7

11. Review the list of layers and save your work.

12. Change **MODEL** to **PAPER** in the status bar.

13. Erase the old lower right viewport. Do not erase the new viewport containing the new view.

14. Pick the Setup Drawing button from the Solids toolbar.

Solids

15. Pick the newest viewport and press **ENTER**.

AutoCAD creates a full section from the viewport drawing, complete with section lines, as shown in Fig. 50-7.

16. Change **PAPER** to **MODEL** in the status bar, enter **ZOOM** and **.7x**, and save your work.

Creating the Isometric View

Finally, let's add an isometric view of the pulley in the upper right viewport.

1. Click the upper right viewport and enter **ZOOM** and **.7x**.

2. Produce an isometric view of the pulley in this viewport.

 HINT:

From the Standard toolbar, locate the flyout that contains the SW Isometric View button and pick it.

Standard

The pulley in the upper right viewport should look similar to the one in Fig. 50-8 (page 714).

Positioning and Plotting the Views

The drawing is a few steps from being complete. All you need to do is position the viewports and turn on hidden line removal in the upper right viewport.

Positioning the Viewports

You can reposition the viewports using the MOVE command. Note that you do not have to resize the viewports. It is okay for the border lines to overlap as shown in the following illustration.

1. Change **MODEL** to **PAPER** in the status bar.

2. Position the four viewports as you would on a drawing sheet, as shown in Fig. 50-8. (See the following hint.)

HINT:

When you move the left and bottom viewports, move them in pairs with polar tracking or ortho on. For example, move the bottom two viewports upward at the same time. This keeps the views perfectly aligned with one another.

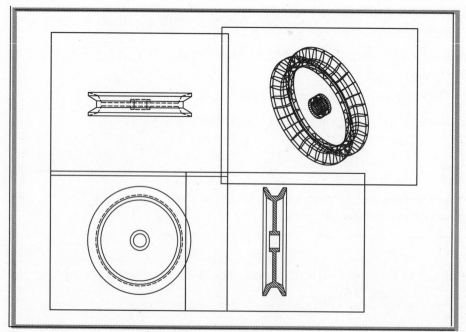

Fig. 50-8

Removing Hidden Lines

1. Select the upper right viewport and then right-click.

2. From the shortcut menu, pick **Hide Plot** and **Yes**.

This causes AutoCAD to remove hidden lines from this view when plotting.

3. Make **Objects** the current layer and freeze layer Vports.

Plotting the Drawing

Standard

1. Pick the **Plot** button from the Standard toolbar.

2. In Plot area, pick the **Display** radio button.

3. In the Plot scale area, select **Scaled to Fit**.

4. Do a full preview, make corrections if necessary, and print the drawing.

The printed sheet should look similar to Fig. 50-9.

5. Save your work, close the toolbars you opened in this chapter, and exit AutoCAD.

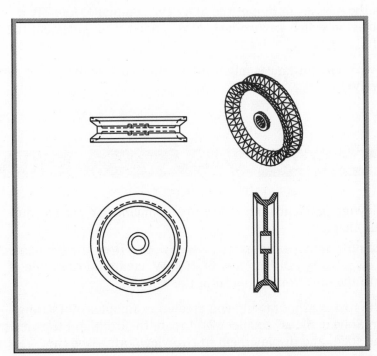

Fig. 50-9

Chapter 50
Review & Activities

Review Questions

1. What is the purpose of the SOLVIEW command?

2. Explain the purpose of each of the following SOLVIEW options.
   a. UCS
   b. Ortho
   c. Auxiliary
   d. Section

3. What layers does the SOLVIEW command create?

4. How is the SOLDRAW command used in conjunction with the SOLVIEW command?

5. Is it possible to use SOLDRAW without SOLVIEW? Explain.

Challenge Your Thinking

1. Discuss the Auxiliary option of the SOLVIEW command. Identify a solid object in which an auxiliary view is needed to document the object fully.

2. Under what circumstances might it be helpful to use the Ucs option of SOLVIEW?

Applying AutoCAD Skills

Work the following problems to practice the commands and skills you learned in this chapter.

1. Create a new drawing file named compos2.dwg from the file named compos.dwg. Using SOLVIEW and SOLDRAW, create four new views similar to the ones you produced in this chapter.

2. Open the rocker arm drawing you created in Chapter 49 ("Using Problem-Solving Skills" problem 2). Create the front and top views from the solid, and display them in two viewports. Save the drawing as ch50arm.dwg.

Chapter 50
Review & Activities

Using Problem-Solving Skills

Complete the following activities using problem-solving skills and your knowledge of AutoCAD.

1. Your supervisor has asked you to create production drawings for the base plate you drew in Chapter 47 ("Using Problem-Solving Skills" problem 1). She wants you to use viewports to display the front view, top view, longitudinal full section, and an isometric view. Open the drawing (or create it, if you haven't already). Save it as ch50base.dwg, and create the required views.

2. Your company's publications department needs some illustrations for its advertising copy. Open the drawing of the height gage you drew in Chapter 47 ("Applying AutoCAD Skills" problem 4). Save the drawing as ch50gage.dwg, and display viewports showing the front view, right-side view, a full section through the V-groove, and an isometric view.

Careers

Using AutoCAD

Collaboration: How to Share Drawings

Gigantic projects, such as building an airliner, making a Hollywood movie, or designing a major sports arena, depend on collaboration. Now, with AutoCAD drawings exchanged via the Internet, keeping "everybody on the same page" is more than just a figure of speech.

Cox Richardson Architects used the Internet to coordinate work among more than 30 consulting companies in designing the Multi-Use Arena in Sydney, Australia. All the necessary drawings—mechanical, civil, structural, and electrical—were done in AutoCAD and integrated into construction drawings. As various participants made additions or modifications, the updated information was posted for the other design partners. This strategy made for a seamless integration of systems as construction got under way.

The Multi-Use Arena is a distinctive landmark. The Cox Richardson design makes a regal impression with a coronet-like roof made of an exterior, cable-suspended system. Inside, the roof system creates a series of trusses from which hang the flags of Olympic nations, giving the arena an international identity.

The Multi-Use Arena reduces cooling costs for the glassed-in lobby by means of louvers and sun screens that project from the facade to block direct sun. AutoCAD-based animation made quick work of modeling the sun's position at different times of year.

Career Focus—Architect

An architect designs buildings. The design may attract the eye with its aesthetic qualities, but it must also meet requirements for function, safety, and budget. After a client approves the design, the architect prepares drawings for construction, including coordination of heating/cooling, plumbing, and electrical systems. The architect bears legal responsibility for the fitness of the building.

Education and Training

To obtain the necessary license in the United States, an architect must earn a five-year bachelor of architecture degree (three- and four-year programs are available for candidates who already have a degree in another field) and complete a three-year internship, in addition to passing the Architect Registration Examination. Professionals who have not completed all these requirements may work under the supervision of a licensed architect.

Career Activity

- Interview someone who can tell you about the daily work of people in an architectural firm. What kinds of jobs do they do, besides designing and drafting?
- With what other types of businesses do architects typically do business?

Objectives:

- **Split a solid model in half**
- **Calculate mass properties of a solid model**
- **Create an STL file of a solid model for rapid prototyping**

Key Terms

rapid prototyping (RP)

XYZ octant

Solid modeling offers many benefits in addition to those discussed in previous chapters. One of the benefits is the ease at which a solid allows you to calculate mass properties. Another benefit is the natural link to *rapid prototyping (RP)*. RP is a method of quickly producing physical models and prototype parts from solid model data. Once you have a solid model, it becomes very easy to output the data needed to drive an RP system.

In this chapter, you will split the pulley you created in Chapter 49, as shown in Fig. 51-1. You will use the pulley half to calculate mass properties and create data that can be used to produce an RP part.

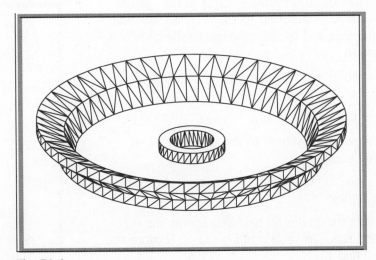

Fig. 51-1

Slicing the Pulley

The SLICE command cuts through a solid and retains either or both parts of the solid. The command is especially useful if you want to create a mold for half of a symmetrical part, such as a pulley.

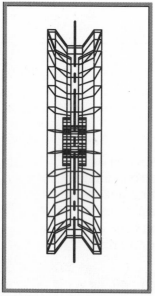

1. Start AutoCAD and use the **pulley.dwg** drawing file to create a new file named **half.dwg**.

2. Freeze all layers except for layer Objects.

3. With snap on, move the pulley to the center of the screen.

4. Enter **UCS** and enter **W** for World.

5. Enter **PLAN** and enter **W** for World.

The view of the pulley should be similar to the one in Fig. 51-2, without the center line.

6. Display the **Solids** toolbar.

7. From the Solids toolbar, pick the **Slice** button.

This enters the **SLICE** command.

8. Select the pulley, press **ENTER**, and enter the **YZ** option.

Solids

Fig. 51-2

AutoCAD asks you to pick a point on the YZ plane. Try to visualize the orientation of this plane, which is vertical and perpendicular to the screen. "Vertical" is the Y direction and "perpendicular to the screen" is the Z direction.

9. Pick a point anywhere on the center line as shown in Fig. 51-2.

This defines the cutting plane.

10. Pick a point anywhere to the right of the cutting plane.

This tells AutoCAD that you want to keep the right half of the pulley.

AutoCAD cuts the pulley in half and leaves the right half on the screen, as shown in Fig. 51-3. (Ignore the leaders and text.)

11. Save your work.

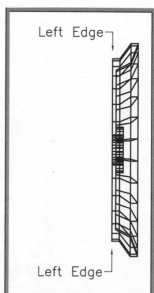

Fig. 51-3

Calculating Mass Properties

Because solid models contain volume information, it is possible to obtain meaningful details about the model's physical properties. For example, you can generate mass properties, including the center of gravity, mass, surface area, and moments of inertia.

1. From the **Tools** pull-down menu, pick **Inquiry** and **Mass Properties**.

This enters the MASSPROP command.

2. Select the solid model and press **ENTER**.

AutoCAD generates a report similar to the one shown in Fig. 51-4. Properties such as mass and volume permit engineers to analyze the part for structural integrity.

3. Press **ENTER** to see the entire report.

```
---------------------------- SOLIDS ----------------------------
Mass:                   18.1501
Volume:                 18.1501
Bounding box:      X:   4.5000       --      6.2500
                   Y:   0.7500       --      7.7500
                   Z:  -3.5000       --      3.5000
Centroid:          X:   5.3750
                   Y:   4.2500
                   Z:   0.0000
Moments of inertia:  X: 467.1366
                     Y: 596.7763
                     Z: 924.6118
Products of inertia:  XY: 414.6154
                      YZ: 0.0000
                      ZX: 0.0000
Radii of gyration:   X:   5.0732
                     Y:   5.7341
                     Z:   7.1374
Principal moments and X-Y-Z directions about centroid:
             I:   139.3011 along [1.0000 0.0000 0.0000]
             J:    72.4097 along [0.0000 1.0000 0.0000]
             K:    72.4097 along [0.0000 0.0000 1.0000]
Write to a file? <N>:
```

Fig. 51-4

4. Enter **Yes** in reply to Write to a file? and press **ENTER** or pick the **Save** button to use the default half.mpr file name.

AutoCAD creates an ASCII (text) file, assigning the MPR file extension to it automatically. The MPR file extension stands for "Mass Properties Report." Let's locate and review the file.

5. Close the AutoCAD Text Window if you haven't already, and minimize AutoCAD.

6. Start Notepad.

7. Select **Open...** from the **File** pull-down menu.

8. Locate and open the file named **half.mpr**.

HINT:

Change the file name extension from txt to mpr to list only those files that have the MPR extension.

The contents of the file should be identical to the information displayed by the MASSPROP command.

9. Exit Notepad and maximize AutoCAD.

Outputting Data for Rapid Prototyping

STLOUT creates an STL file from a solid model. STL is the file type required by most rapid prototyping systems, such as stereolithography and Fused Deposition Modeling (FDM). RP systems enable you to create a physical prototype part from a solid model. The part can serve many purposes, such as a pattern for prototype tooling.

Positioning the Solid Model

Before creating the STL file, you must orient the part in the position that is best for the RP system. In most cases, the part should be lying flat on the WCS. You can accomplish this in the current drawing by rotating the part around the Y axis.

1. Enter the **ROTATE3D** command, select the solid, and press **ENTER**.

2. Enter the **Yaxis** option because you want to rotate the part around the Y axis.

3. Pick a point anywhere on the left edge of the part (shown in Fig. 51-2 on page 720) to define the Y axis.

4. Enter **–90** for the rotation angle.

NOTE: _____

It is important that you rotate the part in the negative direction. A positive rotation positions the part upside-down.

5. **ZOOM All** and, if necessary, move the part so that it lies entirely within the drawing area, as reflected by the grid.

If the part lies outside the positive XYZ octant, the STLOUT command will not work. The *XYZ octant* is the area in 3D space where the x, y, and z coordinates are greater than 0.

6. Pick a viewpoint that is slightly above the part to achieve a view similar to the one in Fig. 51-5.

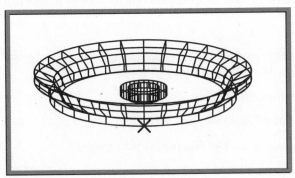

Fig. 51-5

The part should appear to lie on the WCS. Let's check to be sure.

7. Using the **ID** command, pick a point at the bottom of the part (indicated by the x in Fig. 51-5).

HINT: _____

Use the Nearest object snap to snap to a point at the bottom of the part.

The value of the *z* coordinate should be 0.0000. If it is 0.0000 or a value less than 0, you must move the solid upward into the positive XYZ octant.

View

8. Select the **Front** preset view and save your work.

9. With the **MOVE** command, move the solid upward **.1** unit.

Creating the STL File

Now that the pulley is properly positioned, you can create the STL file.

1. Enter the **PLAN** command to view the plan view of the WCS and **ZOOM All**.

2. Enter the **STLOUT** command, pick the solid, and press **ENTER**.

AutoCAD wants to know whether you want to create a binary STL file. Normally, you would create a binary file because they are much smaller than ASCII STL files. However, you cannot view the contents of a binary STL file, and we want to view its contents.

3. Enter **No** to create an ASCII STL file.

4. Pick the **Save** button or press **ENTER** to accept half.stl for the name of the file.

Reviewing the STL File Data

Let's review the contents of the STL file.

1. Minimize AutoCAD and start WordPad.

2. Select **Open...** from the **File** pull-down menu.

3. In the File name box, enter ***.stl**. Be sure to press **ENTER**.

4. Locate the file named **half.stl**, select it, and pick the **Open** button.

The contents of the file appear. The text listing in Fig. 51-6 shows the first part of the file.

STL files consist mainly of groups of *x, y,* and *z* coordinates. Each group of three defines a triangle.

5. After you've reviewed the file, pick **Exit** from the **File** pull-down menu, and maximize AutoCAD.

6. Close the **Solids** toolbar, as well as any other toolbars you may have opened.

7. Save your work and exit AutoCAD.

```
solid AutoCAD
      facet normal 0.0000000e+000  0.0000000e+000  -1.0000000e+000
            outer loop
                  vertex 8.4399540e+000  4.8596573e+000  1.0000000e-001
                  vertex 8.5000000e+000  4.2500000e+000  1.0000000e-001
                  vertex 5.8750000e+000  4.2500000e+000  1.0000000e-001
            endloop
      endfacet
      facet normal 0.0000000e+000  0.0000000e+000  -1.0000000e+000
            outer loop
                  vertex 4.8750000e+000  4.2500000e+000  1.0000000e-001
                  vertex 2.2500000e+000  4.2500000e+000  1.0000000e-001
                  vertex 2.3100460e+000  4.8596573e+000  1.0000000e-001
            endloop
      endfacet
      facet normal 0.0000000e+000  0.0000000e+000  -1.0000000e+000
            outer loop
                  vertex 4.9130602e+000  4.4413417e+000  1.0000000e-001
                  vertex 4.8750000e+000  4.2500000e+000  1.0000000e-001
                  vertex 2.3100460e+000  4.8596573e+000  1.0000000e-001
            endloop
      endfacet
      facet normal 0.0000000e+000  0.0000000e+000  -1.0000000e+000
```

Fig. 51-6

The view shown in Fig. 51-7 (page 726) was created using a special utility called stlview.lsp. This utility reads and displays the contents of an ASCII STL file.

NOTE:

See the Instructor's Resource Guide CD-ROM for a copy of stlview.lsp and the instructions needed to run it.

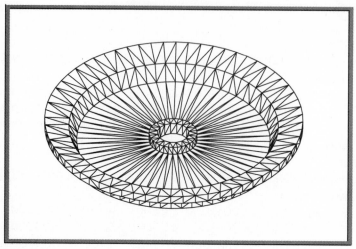

Fig. 51-7

As you can see, an STL file is indeed made up of triangular facets. You can adjust the size of the triangles using the FACETRES (short for "facet resolution") system variable by changing its value before you create the STL file. A high value creates a more accurate part with a smoother surface finish, but the file size and the time required to process the file increase. You should set it as high as necessary to produce an accurate part. Knowing what this value should be for a given part comes with experience and experimentation.

Prototype Part Creation

More than 300 RP service bureaus around the world can create plastic prototype parts from STL files. If you were to send the half.stl file to one of them, they could create a part for you for a fee.

Figure 51-8 shows pictures of a completed plastic prototype part. It was created from the half.stl file using the SLA-250 stereolithography system from 3D Systems, Inc. (Valencia, CA). The SLA-250 is one of the most widely used RP systems in the world.

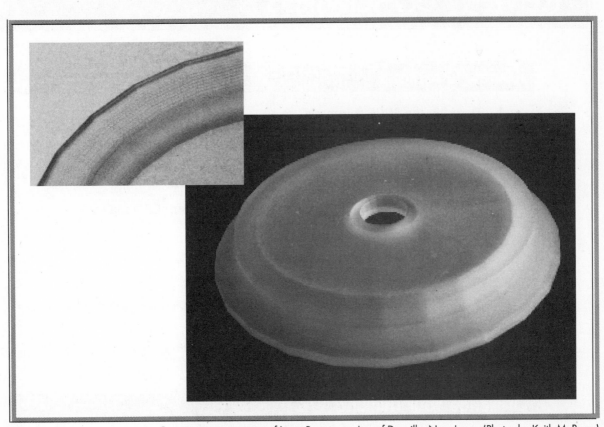

Fig. 51-8   Prototype part courtesy of Laser Prototypes, Inc. of Denville, New Jersey (Photos by Keith M. Berry)

Notice the flat segments that make up the large diameter of the pulley. They are caused by the triangular facets. You could reduce their size by increasing the value of FACETRES before creating the STL file. You could also reduce their visibility by sanding the part, but this could change the accuracy of the part and mold. The part pictured above was built from a version of half.stl with FACETRES set at a relatively low value to show the undesirable effect of large facets.

Chapter 51
Review & Activities

Review Questions

1. Describe the SLICE command.

2. When would the SLICE command be useful?

3. AutoCAD can create files with an MPR file extension. How are these files created, and what information do they provide?

4. Name at least three mass properties for which AutoCAD makes information available about a solid model.

5. Explain the purpose of the STLOUT command.

6. What is the purpose of creating an STL file?

7. What AutoCAD system variable affects the size of the triangles in an STL file?

8. What is the primary advantage of creating a binary STL file? An ASCII STL file?

Challenge Your Thinking

1. Write a paragraph comparing and contrasting the function and purpose of the SECTION and SLICE commands.

2. Describe a situation in which you might need to slice an object using the 3points option.

3. Explore the different rapid prototyping systems that are commercially available. Write an essay that compares each of them.

Applying AutoCAD Skills

Work the following problems to practice the commands and skills you learned in this chapter.

1. Create ASCII and binary STL files from the solid model stored in compos.dwg. Compare the sizes of the two files and then view the contents of the ASCII STL file.

2. Make two slices of the height gage you created in Chapter 47 ("Applying AutoCAD Skills" problem 4). Create one slice through the V groove and the second slice perpendicular to the first and in the same plane.

3. If you know of a rapid prototyping service bureau in your area, request a demonstration of their RP system(s). Explore the possibility of having them create a prototype part for you from an STL file.

4. Open the file named shaft.dwg. Use the SLICE command to split the model down the middle to create a pattern. Use STLOUT to create a binary STL file.

5. Produce the mass properties for shaft.dwg.

 ## Using Problem-Solving Skills

Complete the following activities using problem-solving skills and your knowledge of AutoCAD.

1. The bushing and insert shown in Fig. 51-9 are part of an aerospace project. Weight is critical! Slice the bushing and insert. Then determine the mass of each of the parts. Write each of the reports to a file and print the files.

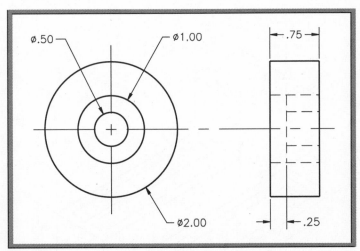

Fig. 51-9

2. Position the bushing and insert from problem 1. Then make an ASCII STL file of the bushing only. View the STL file, but do not print it.

Part 9 Project

Designing and Modeling a Desk Set

AutoCAD is capable of creating 2D drawings, 3D surface models, and 3D solid models. The differences are meaningful. Although all three types of drawings yield information, 3D models allow more realistic views. Solid modeling, in addition, allows determination of mass properties and the generation of production data such as that used in automated machine tool processes. Familiarity with solid modeling enhances the ability of the drafter/designer to create solutions; it is well worth the effort to learn.

Description

Your company has sold the desk set shown in P9-1 for several years. Your supervisor has determined that it is time to modernize and streamline the desk set to meet current market demand.

1. Create a solid model of the desk set shown in P9-1. Include the round space to store pens, pencils, and the like; a large, flat, shallow area for items such as paper clips and rubber bands; and a medium-height section for a note pad, business cards, and similar papers. Wall thickness should range between 3/32″ and 3/16″. Assign a plastic material to the model. Save it as desk1.dwg.

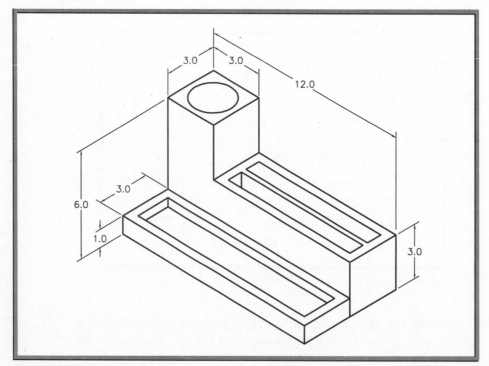

Fig. P9-1

Part 9 Project

2. Save desk1.dwg with a new name of desk2.dwg. Alter the design to modernize it or to make it more useful. For example, you may decide to provide an inclined surface on the two medium-height holes so that the one in back rises above the one in front of it. You may also want to round the edges to give it a streamlined look, or even add or subdivide compartments.

3. Prepare a presentation for your supervisor. In your presentation, compare and contrast the new and old designs and explain why the new design is superior. Prepare rendered images of both desk sets to support your presentation.

Hints and Suggestions

- There is often more than one way to construct a solid model. If your approach does not seem to be working, try another.
- Remember to increase the value of ISOLINES and FACETRES to improve appearance. The tradeoff is the increased time required for operations such as regenerations and rendering.
- Create the recesses for the compartments (such as the pencil holder) using the SUBTRACT command.
- Use the UNION command as necessary so that the final product is a single composite solid.

Summary Questions/Self-Evaluation

1. In your desk set design, how could the area actually being used to hold items be calculated?

2. Could the entire desk set design be created via extrusion? Explain how it could or why it could not.

3. What are some advantages of using preloaded solid primitives in the creation of more complex solid models?

4. Explain how to use the SUBTRACT command to create a thin-walled cylinder.

5. What is meant by "downstream benefits" of solid modeling?

6. Describe the types of information that are provided by the MASSPROP command.

Careers
Using
AutoCAD

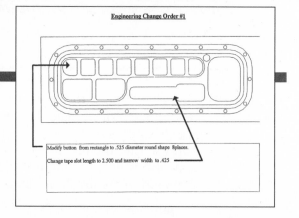

Engineering Change Order #1

Modify button from rectangle to .525 diameter round shape 8places.

Change tape slot length to 2.500 and narrow width to .425

Test Your Real-World Skills

A team of three specialists enters the test area. The CAD specialist goes to work immediately, drawing just enough part geometry for the CAM specialist to determine a tool path. The CAM specialist creates NC code to implement the tool path and passes it to the CNC specialist, who has the milling machine ready to turn out a prototype cover plate (see illustration). Judges inspect the prototype, then ask for another with two last-minute changes.

The specialists are students at the SkillsUSA Championships, getting a taste of how teamwork combines with technology in real-world manufacturing. The annual event draws nearly 3,500 students to compete in more than 50 occupational and leadership skill areas.

The competition is put on by SkillsUSA-VICA, formerly known as VICA (Vocational Industrial Clubs of America). With more than 13,000 local chapters, SkillsUSA-VICA is an alliance of educators, businesses, unions, and others with an interest in providing students with skills that are in demand in the workplace.

SkillsUSA-VICA programs offer quality experiences for students in leadership, teamwork, citizenship, and character development, as well as practical lessons in goal setting and job interviewing. SkillsUSA-VICA builds self-confidence, work attitudes, and communication skills. It emphasizes total quality at work, including high ethical standards, superior work skills, lifelong education, and pride in the dignity of work. SkillsUSA-VICA also promotes understanding of the free enterprise system and involvement in community service.

Career Focus: Vocational Teacher

Vocational education teachers, also called *career and technical education teachers,* prepare students for occupations that require job-specific skills, such as automobile repair, medical and dental services, or construction technology. A large part of the instruction consists of hands-on practice in a shop or laboratory setting.

In addition to their classroom work, vocational education teachers prepare lessons and assignments, evaluate student performance, and serve on faculty committees. Vocational education teachers keep up with developments in their field by reading journals and attending professional meetings.

Education and Training

Preparation for vocational education teachers varies by subject and by region. Typically, these teachers combine work experience in their field with a bachelor's, master's, or doctoral degree. Teachers who work in secondary schools are required to complete a one- or two-year program in education theory and practice to obtain a teaching certificate.

 Career Activity

- Find out more by visiting the SkillsUSA-VICA site—http://www.skillsusa.org—on the World Wide Web.
- Write a report on how SkillsUSA-VICA would be helpful to you if you were a vocational education teacher.

Part 10

Menus

Chapter 52 An Internal Peek at AutoCAD's Menus

Objectives:

- Examine the contents of AutoCAD's acad.mnu file
- Study the individual parts that make up acad.mnu and associate them with the menus, menu items, and toolbars in AutoCAD
- Locate each of AutoCAD's menu files

Key Terms

binary file

bitmap

macro

This chapter focuses on the individual parts that make up the acad.mnu file. This is the file that contains the menus, toolbars, and buttons you see and use in AutoCAD. In the steps that follow, you will review the file in its raw form—the parts you don't see in AutoCAD. As you learn more about the contents of this file, you will begin to associate its individual parts with the menus, menu items, and toolbars in AutoCAD. In subsequent chapters, you will learn how to modify and create your own menu.

Reviewing the Contents of Acad.mnu

Let's use Microsoft WordPad to review acad.mnu. First, let's make a backup copy of acad.mnu.

1. Produce a copy of acad.mnu. Name the copy **acadback.mnu**.

 HINT:

This file is located in the Acad2000/Support folder.

2. Open WordPad by picking the Windows **Start** button and selecting **Programs**, **Accessories**, and then **WordPad**.

3. Pick the **Open** button in WordPad and open **acadback.mnu**.

HINT:

In WordPad's Open dialog box, enter *.mnu to see only the MNU files. Be sure to press ENTER.

The file contains the source code for AutoCAD's menus.

4. Review the first part of the file.

You should see information similar to that shown in Table 52-1 on pages 736-737.

The ASCII file contains more than 5,000 lines of code, including the blank lines. Much of the code consists of AutoCAD's pull-down menu items. Examples, such as ***POP2, are presented in this chapter.

Individual Menu Elements

AutoCAD uses several menu section labels and special menu characters in acad.mnu. Each of them serves a specific purpose.

1. Locate each of the following menu sections in the menu file and read the brief descriptions located at the right.

HINT:

Pick Find... from WordPad's Edit pull-down menu, or use the Find button.

Menu Sections

***BUTTONS1	— specifies a buttons menu 1 for the buttons on a mouse or digitizer cursor control
***AUX1	— specifies auxiliary device menu 1
***TABLET1	— specifies tablet menu area 1
***POP2	— pull-down menu sections are defined using ***POP1, ***POP2, etc.
***MENUGROUP=	— specifies a menu file group name
***TOOLBARS	— specifies a toolbar
***HELPSTRINGS	— specifies help comments that appear in place of the status bar when pointing at a button.
***ACCELERATORS	— specifies accelerator keys

```
//
//     AutoCAD 2000 Menu
//     28 April 1997
//
//     Copyright (C) 1986, 1987, 1988, 1989, 1990, 1991, 1992, 1994, 1996, 1997, 1998
//     by Autodesk, Inc.
//
//     Permission to use, copy, modify, and distribute this software
//     for any purpose and without fee is hereby granted, provided that
//     the above copyright notice appears in all copies and that both
//     the copyright notice and the limited warranty and restricted rights
//     notice below appear in all supporting documentation.
//
//     AUTODESK, INC. PROVIDES THIS PROGRAM "AS IS" AND WITH ALL FAULTS.
//     AUTODESK, INC. SPECIFICALLY DISCLAIMS ANY IMPLIED WARRANTY OF
//     MERCHANTABILITY OR FITNESS FOR A PARTICULAR USE.  AUTODESK, INC.
//     DOES NOT WARRANT THAT THE OPERATION OF THE PROGRAM WILL BE
//     UNINTERRUPTED OR ERROR FREE.
//
//     Use, duplication, or disclosure by the U.S. Government is subject to
//     restrictions set forth in FAR 52.227-19 (Commercial Computer
//     Software - Restricted Rights) and DFAR 252.227-7013(c)(1)(ii)
//     (Rights in Technical Data and Computer Software), as applicable.
//
//
//     NOTE:  AutoCAD looks for an ".mnl" (Menu Lisp) file whose name is
//          the same as that of the menu file, and loads it if
//          found.  If you modify this menu and change its name, you
//          should copy acad.mnl to <yourname>.mnl, since the menu
//          relies on AutoLISP routines found there.
//


//
//     Default AutoCAD NAMESPACE declaration:
//
***MENUGROUP=ACAD


//
//   Begin AutoCAD Digitizer Button Menus
//
```

Table 52-1

```
***BUTTONS1
// Simple + button
// if a grip is hot bring up the Grips Cursor Menu (POP 500), else send a carriage return
// If the SHORTCUTMENU sysvar is not 0 the first item (for button 1) is NOT USED.
$M=$(if,$(eq,$(substr,$(getvar,cmdnames),1,5),GRIP_),$P0=ACAD.GRIPS $P0=*);
$P0=SNAP $p0=*
^C^C
^B
^O
^G
^D
^E
^T


***BUTTONS2
// Shift + button
$P0=SNAP $p0=*


***BUTTONS3
// Control + button


***BUTTONS4
// Control + shift + button


//
//    Begin System Pointing Device Menus
//
***AUX1
// Simple button
// if a grip is hot bring up the Grips Cursor Menu (POP 500), else send a carriage return
// If the SHORTCUTMENU sysvar is not 0 the first item (for button 1, the "right button")
// is NOT USED.
$M=$(if,$(eq,$(substr,$(getvar,cmdnames),1,5),GRIP_),$P0=ACAD.GRIPS $P0=*);
$P0=SNAP $p0=*
^C^C
^B
^O
^G
^D
^E
^T
```

Table 52-1 (continued)

2. Locate each of the following pull-down menus in the menu file and read the brief descriptions located at the right.

Pull-Down Menus

[&Style...]	– displays Style... in the menu; the ampersand (&) character denotes the mnemonic key (i.e., underlines the S in the word Style, indicating that you can enter S at the keyboard to activate this menu item)
->	– as a prefix, indicates that the pull-down or cursor menu has a submenu
[->Inquir&y]	– indicates that Inquiry has a submenu; the ampersand (&) denotes the mnemonic key
<-	– as a prefix, indicates that the pull-down or cursor menu item is the last in the submenu
[<-&Save...]	– indicates that Save... is the last item in a submenu
~	– "grays out" a prompt (displays it in halftone), disabling the menu item
[–]	– specifies a separator line

3. Locate each of the following items (toolbar and image tile menus) in the menu file and read the brief descriptions located at the right.

Toolbars

**TB	– specifies a toolbar

Image Tile Menus

***image	– specifies an image tile menu
**image_poly	– specifies an image tile submenu named image poly
$I=	– *addresses an image tile menu*
[acad(Cone,Cone)]	– addresses the cone slide image contained in the acad slide library for display as an image tile; also displays Cone in the list box

4. Locate each of the following special characters in the menu file and read the brief descriptions located at the right.

Special Characters

;	– issues ENTER
'	– an apostrophe specifies transparent command entry
\	– the backslash stops the computer, and the computer expects input from the user

(a space)	— an empty space is the same as pressing the spacebar
+	— menu item continues on the next line
=*	— displays the current top-level image, pull-down, or cursor menu
$	— special character code that loads a menu section
^	— this character (called a *caret*) automatically presses the CTRL key
^B	— toggles snap on or off
^C	— issues a cancel
*^C^C	— prefix for a repeating item
^D	— toggles coordinates on or off
^E	— sets the next isometric plane
^G	— toggles grid on or off
^H	— issues a backspace
^O	— issues the CTRL and O keys to toggle the ortho mode
^P	— toggles MENUECHO on or off
^T	— toggles tablet on or off
^V	— changes the current viewport
^Z	— null character that suppresses the automatic addition of SPACEBAR at the end of a menu item
ID_	— AutoCAD uses this prefix to begin all name tags in acad.mnu.
//	— specifies a comment (as opposed to program code)

Other menu elements exist, although the previous lists provide enough to get you started with menu development and customization.

Menu Items

You can combine the individual elements described in the previous section to create menu items (also referred to as *macros*) that perform specific AutoCAD functions. For example, [Redo]^C^C_redo displays Redo in the menu, issues cancel twice, and enters the REDO command.

1. Study each of the following menu items from the menu file.

   ***POP5

   — specifies the fifth (Format) pull-down menu

ID_ZoomWindo [&Window]'zoom _w

— ID_Zoomwindo is the name tag; Window appears in the menu; the W is underlined; the apostrophe allows for transparent command entry; AutoCAD ignores the underscore and then enters the ZOOM command; then enters W for Window

ID_CircleDia [Center, &Diameter]^C^C_circle _d

— ID_CircleDia is the name tag; Center, Diameter appears in the menu; the D is underlined; issues cancel twice; enters the CIRCLE command; presses the spacebar, pauses for user input; enters D for Diameter

2. Locate each of the previous menu items in the menu file.

3. Start AutoCAD. Begin a new drawing from scratch.

4. Locate each of the previous items in AutoCAD's menus and pick each of them.

HINT:

It can be very difficult to find the second and third menu items without knowing in which pull-down menus they are located. After locating each item in WordPad, scroll up a short distance and find the first ***POP menu item. The number corresponds with the position of the menu. For example, ***POP1 specifies the File pull-down menu.

AutoCAD's Menu Files

AutoCAD's main menu file is acad.mnu, but other menu files also exist. As you create and customize menus in the chapters that follow, it is important that you be familiar with the files listed in Table 52-2 and understand their purpose.

File Name	Description
acad.mns	An ASCII file that looks very similar to acad.mnu; this file stores changes to the toolbars or menus made from within AutoCAD.
acad.mnc	Compiled version of the acad.mns file; AutoCAD loads this binary version of the menu file. (A *binary file* is one that uses the binary numbering system, which consists entirely of 0s and 1s.)
acad.mnr	Menu resource file containing the bitmaps for the buttons (a *bitmap* is an image in which bits are referenced to pixels); this binary file updates as you modify a button using the button editor in AutoCAD.
acad.mnl	Menu LISP file, which contains AutoLISP expressions that are used by acad.mnu; you will learn more about AutoLISP later in the book.
acad.mnu	AutoCAD's template menu file; this file has not been compiled, so you can open and edit it.

Table 52-2

1. Locate each of the files listed in Table 52-2 in AutoCAD's Support folder.
2. Close WordPad (do not save any changes) and exit AutoCAD.

Chapter 52
Review & Activities

Review Questions

Describe what each of the following menu items will do.

1. [&Properties]^C^C_properties

2. [Dimstyle]_d

3. [->&Zoom]

4. **LINE 3

5. [New...]^C^C_new

6. [Window]_w

7. [Text &Style...]'_style

8. [&Start, Center, End]^C^C_arc _c

Challenge Your Thinking

Note: To answer all questions and problems that require opening acad.mnu, use the acadback.mnu file you created instead. This will help prevent any unintentional changes to acad.mnu.

1. In AutoCAD, identify a menu item and then execute it. Try to envision what the menu item would look like in the acad.mnu file. What individual menu elements would AutoCAD use?

2. In acad.mnu, locate the menu item that you identified in the previous question. Is it similar to what you had envisioned?

 ## Applying AutoCAD Skills

Work the following problems to practice the commands and skills you learned in this chapter.

1. Print a portion of the acad.mnu file and then start AutoCAD. Experiment with different command and submenu sequences and attempt to locate the sequences on the printout.

Chapter 52
Review & Activities

2. Locate items in the printout that you cannot fully visualize. Attempt to find the corresponding items in the menus and execute them.

3. Open the acadback.mnu file you produced on page 734, Step 1. Where in the file does the following comment appear?

   // Long command names are truncated (not abbreviated) to fit.

   If you do not know what *truncated* means, find out. How is it different from *abbreviated*?

Using Problem-Solving Skills

Complete the following activities using problem-solving skills and your knowledge of AutoCAD. *Note:* For the these exercises, work from the acadback.mnu file you created on page 734. *Do not* work directly from AutoCAD's acad.mnu file.

1. As a third-party developer, you need to modify some of AutoCAD's menus. Specifically, you want to eliminate the grips, add toolbars, and change some of the command options. Before you can do so, you must familiarize yourself with the location of these items in the menu file. What ***POP number is assigned to grips? Where does the Toolbars section begin? Are commands with no options, such as RAY, included with command option menus?

2. Before you can market your AutoCAD add-on product, you need to comply with software copyright laws. Find out the following information about acad.mnu (remember to use the acadback.mnu file). Who can use the acad.mnu file? For what purposes? What is the fee, and what provisions apply? What is the latest copyright date?

Chapter 53 Custom Menus and Toolbars

Objectives:
- Develop a pull-down menu
- Load and use a pull-down menu file
- Create and load a partial menu
- Create and delete a custom toolbar

Key Terms
base menu
partial menus

AutoCAD users can create pull-down menus and toolbars that can include a wide range of AutoCAD commands. Users can develop custom macros that automatically execute any series of inputs. For example, a simple two-item macro can enter ZOOM Previous in one step. Sophisticated macros can execute numerous AutoCAD commands and functions in a single step.

Every command, option, or combination that you can enter at the keyboard can be entered automatically using macros. Thus, you have the flexibility to develop a menu at any level of sophistication.

Developing a Pull-Down Menu

Let's create a simple pull-down menu named pull.mnu.

1. Using Notepad, store the code shown in Fig. 53-1 in a file named **pull.mnu**. Skip the first line so that the line above ***POP1 is blank.

NOTE:
See also the Instructor's Resource Guide CD-ROM. It includes the pull.mnu file, which contains the code shown in Fig. 53-1.

```
***POP1
[Construct]
[Line]^C^CLINE
[Circle]^C^CCircle
[Arc]^C^CARC
[--]
[->Display]
        [Pan]'PAN
        [->Zoom]
        [Window]^C^CZOOM W
        [Previous]^C^CZOOM P
        [All]^C^CZOOM A
        [Extents]^C^CZOOM E
```

Fig. 53-1

2. Save the file and exit Notepad.

3. Make a backup copy of the file.

Loading and Using a Menu File

The MENU command permits you to load menus. Loading a menu is required in order to use it.

1. Start AutoCAD and begin a new drawing from scratch.

2. Enter the **MENU** command.

This displays the Select Menu File dialog box.

3. In Files of type, select **Menu Template (*.mnu)**.

4. Locate and select **pull.mnu**.

AutoCAD displays a warning message indicating that loading a template menu (MNU) file overwrites the menu source (MNS) file if one exists. In this case, pull.mns does not exist.

NOTE:

If the computer you are using is also used by others, check with your supervisor or instructor to make sure you are not overwriting someone else's file.

5. Pick the **Yes** button to continue loading the file.

The new Construct pull-down menu appears, replacing all of the others, including the docked toolbars, as shown in Fig. 53-2.

6. Try each of the options in the Construct menu.

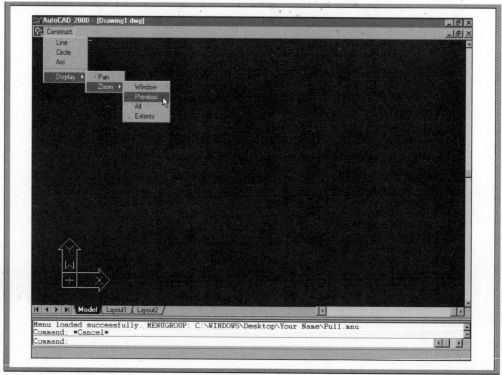

Fig. 53-2

NOTE:

Even though the menus and toolbars are gone, you still have full access to all AutoCAD commands. Just enter them at the keyboard. Notice, however, that the shortcut menus (those that appear when you right-click), are no longer available.

AutoCAD automatically compiles the pull.mnu file into pull.mnc and pull.mnr files. As mentioned in the previous chapter, the MNC file is the compiled version, and the MNR file contains the bitmaps, such as the buttons if they exist. Both are binary files.

When AutoCAD creates the MNC file, it also creates an MNS file. Initially, the MNS file is identical to the MNU file, without comments or special formatting. AutoCAD changes it each time you change the contents of the MNU file through the standard AutoCAD interface, such as changing the contents of a toolbar. You will create and edit a toolbar later in this chapter. See the on-line *AutoCAD Customization Guide* for more information on the MNU, MNC, MNR, and MNS files.

7. Try each of the buttons on the mouse or cursor control.

They do not function the same as before because acad.mnc is not loaded.

8. Enter the **MENU** command and select **acad.mnc**, which is located in the AutoCAD Support directory.

All the standard menus and toolbars reappear.

Loading and Using a Partial Menu

AutoCAD offers base and partial menus. A *base menu* is one that is loaded when you first start AutoCAD, or when you use the MENU command. The MENULOAD command loads *partial menus* that work with, not instead of, the current base menu. Let's convert pull.mnu to a partial menu that we can use with the standard acad menu.

Converting a Menu to a Partial Menu

1. Minimize AutoCAD.

2. Make a copy of pull.mnu and name it **pullpart.mnu**.

Note that the pullpart.mnu file is also available on the Instructor's Resource CD-ROM.

3. Using Notepad, open **pullpart.mnu**.

4. At the top of the file, add *****MENUGROUP=PULLPART**, with a blank line after it.

NOTE:

The menu group name can be different than the file name. We used PULLPART for both to make it easier to remember the names.

5. Select **Save** from the **File** pull-down menu and exit Notepad.

6. Maximize AutoCAD.

Loading a Partial Menu

1. From the **Tools** pull-down menu, select **Customize Menus...** or enter the **MENULOAD** command.

The Menu Customization dialog box appears.

2. Pick the **Menu Groups** tab to display it unless it is already displayed.

3. Pick the **Browse...** button and find and select the new **pullpart.mnu** file; then pick the **Open** button.

4. Pick the **Load** button.

5. In reply to the warning message, pick the **Yes** button.

PULLPART appears in the Menu Groups list box.

6. Select **PULLPART** in this box.

Adding a Menu to the Menu Bar

1. Pick the **Menu Bar** tab to change the contents of the dialog box.

2. Pick the **Menu Group** down arrow and select **PULLPART**.

3. In the Menu Bar list box, select **Tools**.

4. Pick the **Insert** button.

This inserts Construct between Format and Tools, as shown in Fig. 53-3.

5. Pick the **Close** button.

AutoCAD adds the Construct menu to the menu bar.

6. Try the selections on the Construct menu.

Removing a Partial Menu

The Menu Customization dialog box allows you to remove partial menus.

1. Select **Customize Menus...** from the **Tools** pull-down menu.

2. Pick the **Menu Bar** tab.

3. In the **Menu Bar** list box, select **Construct**, and pick the **Remove** button.

This removes the Construct pull-down menu from AutoCAD's menu bar.

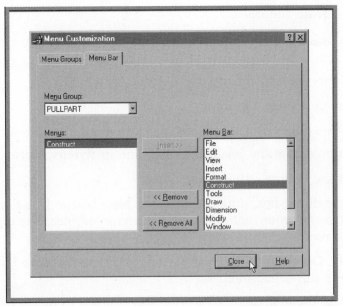

Fig. 53-3

4. Pick the Menu Groups tab.

5. If **PULLPART** is not selected in the Menu Groups list box, select it.

6. Pick the **Unload** button to remove PULLPART from the Menu Groups list box.

7. Pick the **Close** button.

Working with Toolbars

AutoCAD permits you to add new toolbars based on your needs to increase your productivity. You can also delete toolbars from within AutoCAD.

Creating a Toolbar

Let's create a new toolbar to contain some commonly used buttons.

1. Pick **Toolbars...** from the **View** pull-down menu.

This enters the TOOLBAR command, which displays the Toolbars dialog box, which you may recall from Chapter 2.

2. Pick the **New...** button.

3. Enter **Custom** for the toolbar name, and pick the **OK** button.

This adds Custom to the Toolbars list box. Notice also that the beginning of a toolbar appears near the top of the AutoCAD window.

4. Pick the **Customize...** button.

5. In the **Categories** drop-down box, pick **Standard** from the list.

All of the buttons from the Standard toolbar and its flyouts appear.

6. Review the buttons using the scrollbar.

7. From the dialog box, drag and drop the **New** button (the first button in the group) into the new Custom toolbar near the top of the screen.

8. Drag the **Open** button (second button) and then the **Save** button (third) to the Custom toolbar.

As you can see, the toolbar grows as you drag buttons to it. You can also copy and drag buttons from AutoCAD's toolbars.

9. From the **Draw** toolbar, drag and drop the **Line** button onto the Custom toolbar.

10. From the **Modify** toolbar, drag and drop the **Erase** button onto the Custom toolbar.

11. From the **Object Properties** toolbar, drag and drop the **Layers** button onto the Custom toolbar.

The Custom toolbar should look similar to one in Fig. 53-4.

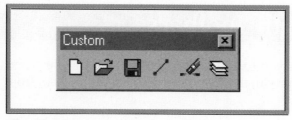

Fig. 53-4

 NOTE: ─────────────────────────────

Your toolbar may look different than the one shown above, depending on where you drop the buttons. After you have finished the toolbar, you can change its shape in the same way you do other toolbars.

12. Delete the Erase button from the Custom toolbar by dragging and dropping it to an open area in the drawing area.

13. Pick the **Close** button to close the Customize Toolbars dialog box; pick **Close** again to close the Toolbars dialog box.

AutoCAD updates its acad.mns, acad.mnc, and acad.mnr files.

14. Drag the Custom toolbar to a location of your choice and try each of the buttons.

15. Reshape the toolbar and then dock it along the right edge.

Whenever you change the position or visibility of a toolbar, AutoCAD stores the change in the computer's registry. This is why the screen may look different after someone else uses AutoCAD, even if this person did not create or save any drawing files.

Deleting a Toolbar

AutoCAD also makes it easy to delete toolbars. Be careful not to delete toolbars that you want to keep.

1. Pick **Toolbars...** from the **View** pull-down menu.

2. In the **Toolbars** list box, highlight **Custom**. Be sure that this is the one you've selected.

3. Pick the **Delete** button. If you're certain that you have selected Custom, pick the **Yes** button.

 NOTE: _____

The Properties... button displays the Toolbar Properties dialog box. It provides information about the toolbar and permits you to change the toolbar name and help text.

4. Pick the **Close** button.

AutoCAD updates its acad.mns, acad.mnc, and acad.mnr files once again.

5. Exit AutoCAD. Do not save.

Chapter 53
Review & Activities

Review Questions

1. Briefly define an AutoCAD macro.

2. Why are custom macros useful?

3. Explain the concept of base and partial menus.

4. Which AutoCAD command allows you to load a base menu?

5. AutoCAD compiles MNU files into what two binary file types?

6. What must you do to convert a simple base menu, such as pull.mnu, to a partial menu?

7. Explain how you would add the Polyline button from the Draw toolbar to a new custom toolbar.

8. Where does AutoCAD store changes to the position and visibility of toolbars?

Challenge Your Thinking

1. The INSERT command can be included in a macro like any other command. In conjunction with a drawing file name, how could this be useful?

2. When creating a custom toolbar, is it possible to copy a button from a flyout to the custom toolbar? Explain.

Applying AutoCAD Skills

Work the following problems to practice the commands and skills you learned in this chapter.

1. Open the pullpart.mnu file in Notepad and save it as pull2part.mnu. Modify the file to remove all the Zoom commands. Make sure it is still a partial menu. Add the modified menu to the AutoCAD menu bar and try the options. Does it work correctly? Remove the menu from the menu bar.

Chapter 53
Review & Activities

2. Create the pull-down menu shown in Fig. 53-5. Name it mine.mnu. Position the new partial menu between the Modify and Window pull-down menus. Try each of the menu selections. If available, refer to the Instructor's Resource CD-ROM. It contains the following code stored in a file named mine.mnu.

```
***MENUGROUP=MINE

***POP1
[MINE]
[LINE]^C^Cline
[ERASE W]^C^Cerase w
[Flip Snap] ^b
[--]
[->Display]
        [Pan]'pan
        [->Zoom]
                [ZOOM W]^C^Czoom w
                [ZOOM P]^C^Czoom p
                [<-ZOOM All]^C^Czoom a
        [<-Viewpoint]^C^Cvpoint;;
--]
[->Text]
        [COMP S]^C^Cstyle comp complex;;;;;;;
        [His Name]^C^Ctext 6,2 .2 0 John Doe;;;Mechanical Engineer;
        [<-My Name]^C^Ctext s comp 6,3 .2 0;
[--]
[Units]^C^Cunits
[--]
[->Object Snap]
        [Mid Point]mid
        [Cen Point]cen
        [<-Nearest]near
[--]
[*Cancel*]^C^C
```

Fig. 53-5

Chapter 53
Review & Activities continued

3. Remove and unload the partial menu that you created in Problem 1.

4. Create the custom toolbar shown in Fig. 53-6 and name it Frequent. Be sure to press the CTRL key when copying buttons from the toolbars.

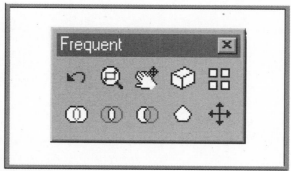

Fig. 53-6

5. Delete the Frequent toolbar that you created in Problem 4.

 ## Using Problem-Solving Skills

Complete the following activities using problem-solving skills and your knowledge of AutoCAD.

1. As you are working with AutoCAD, you discover that you use the LINE, PLINE, CIRCLE, OFFSET, TRIM, and EXTEND commands frequently. Create a custom toolbar named Common that contains these commands.

2. You now realize you need a few more buttons on the Common toolbar. Add ERASE, ZOOM Window, ZOOM All, and EXPLODE to the toolbar.

3. In your opinion, what buttons would make up the "ideal" toolbar? Create the toolbar.

Careers
Using
AutoCAD

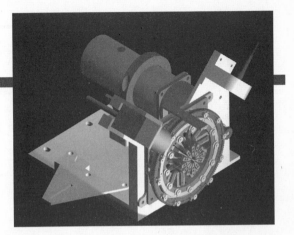

Manufacturing in Microgravity

In a microgravity environment, steam doesn't rise and water droplets do not fall. In microgravity, materials behave differently—which is fun for astronauts. For manufacturers it's an opportunity that may lead to amazing new materials: lighter, stronger, and more durable than those made on Earth.

Leading the way toward this manufacturing future, NASA Lewis Research Center and a team of private companies are developing technology for materials processing in microgravity. One example is the reducing diameter device (see illustration), which creates a flow of liquid polymer that is free of shearing forces. The challenge is that liquids in microgravity don't behave as they do on Earth. Once perfected, this technology will have applications in fiber spinning, blow molding, sheet drawing, extrusion, and injection molding.

In developing the reducing diameter device, David Haydu of ADF, Inc. (Cleveland, Ohio) used Autodesk Mechanical Desktop and related software to preview how a ride aboard a rocket would affect his design. Mechanical Desktop is a 3D modeling tool, a next step for AutoCAD users in the fields of industry and engineering.

Career Focus—Aerospace Engineer

Aerospace engineers design, test, and supervise the production of aircraft, missiles, and spacecraft. They develop new technologies in aviation, defense, and space exploration. Often, they specialize in an area such as structure, navigation and control, or production methods.

Aerospace engineers use computers to solve mathematical equations that describe how a component will operate within a system. Many use PC-based CAD to produce and analyze designs. In the case of the reducing diameter device, David Haydu used this approach to ensure that his design would meet a requirement that the center of gravity and center of pressure be equal.

Education and Training

A bachelor's degree is required for most engineering jobs in aerospace. For some jobs, such as those related to production, two years of technical education may be enough. An engineer may pursue a master's or other post-graduate degree to develop expertise within a specialty or new technology. Training at a school that emphasizes aerospace is an advantage when seeking a first job.

An aerospace engineer needs teamwork skills because of the collaborative efforts for large products. For the same reason, oral and written presentation skills are useful.

 Career Activity

- On the World Wide Web, visit and explore www.nasa.gov/index.
- Search the index for "internship." Are any summer jobs in the aerospace field available near you?

Chapter 54 Tablet Menus

Objectives:

- **Develop a tablet menu**
- **Configure a digitizing tablet to work with a custom tablet menu in AutoCAD**
- **Load and use a custom tablet menu**

Key Terms

cell
digitize
digitizing tablet
puck
tablet menu overlay
tablet menus

A *digitizing tablet* is an electronic input device that consists of a board or tablet and a pointing device called a *puck* or a stylus that resembles a pen. The puck is similar to a mouse, but it contains more buttons and can be programmed to perform several different functions.

AutoCAD permits you to designate areas of a digitizing tablet for *tablet menus*. Tablet menus allow you to enter a wide variety of AutoCAD commands and functions as an alternative to or in conjunction with using pull-down menus, toolbars, and the keyboard. Tablet menus can make command selection easier and increase the speed with which you work in some applications. They also help you customize AutoCAD to meet your specific needs.

In order to complete this chapter, you must have a digitizing tablet connected to and working properly with your computer. AutoCAD requires that you use a Wintab-compatible digitizing tablet, which functions as a Windows pointing device. Any digitizing tablet that works in Windows will also work in AutoCAD. Use the System tab in the Options dialog box to change the current pointing device to a Wintab-compatible device.

Developing a Tablet Menu

The first step in designing a new tablet menu is to ask yourself what you would like to include in the menu. The best way to answer this question is to sketch the tablet menu overlay on paper so that you gain some sense of the placement of each menu item. (The *tablet menu overlay* is a template that you develop to use with a tablet menu. It contains icons or words that show you where to pick on the tablet to enter a certain command or sequence.) After you refine the sketch to your liking, you can use it to develop the actual menu file. In this chapter, an example tablet menu overlay has been provided. Later, after you've learned the process, you'll be able to design your own.

Preparing the Menu Overlay

The example in Fig. 54-1 (page 758) is a relatively simple, but functional, tablet menu overlay. Let's use it as the model for the following steps.

1. Make an enlarged photocopy of the tablet menu overlay so that it comes close to fitting your digitizing tablet. If you do not have access to a copier that has enlargement capability, use the overlay at its existing size.

NOTE: ——————————————————————————————

The drawing in Fig. 54-1 is also available as overlay.dwg on the Instructor's Resource Guide CD-ROM.

The overlay must not extend outside the active area of the digitizer. For instance, if the active area is 11″ × 11″, then the overlay must not be larger than 11″ × 11″.

Creating the Menu File

Next, let's create the menu file that holds the menu items. The menu file will contain three menus: Buttons, Tablet2, and Tablet3. Notice how the items in these menus correspond to the items on the tablet menu overlay, starting with tablet menu 1.

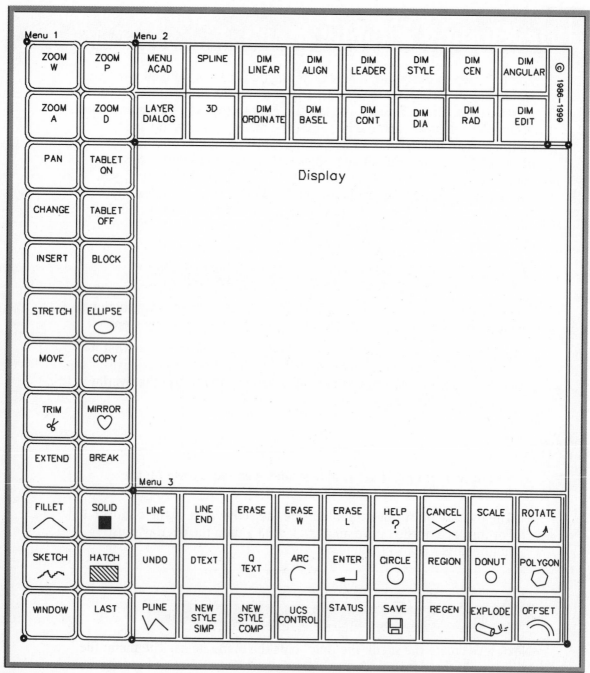

Fig. 54-1

NOTE: ————————————————————————————————

The Buttons menu is for a three-button pointing device.

2. Start AutoCAD and begin a new drawing from scratch.

3. Using Notepad, enter the items in Fig. 54-2 on pages 760-761. Name the file **tab.mnu**. Be sure to enter the items exactly as shown, and press **ENTER** after each entry.

NOTE: ————————————————————————————————

The Instructor's Resource CD-ROM contains a copy of tab.mnu.

4. Be sure to save the menu contents.

5. Make a backup copy of the file.

Configuring the Menu Areas

To make the menu file work with the digitizing tablet and AutoCAD, you must configure the tablet so that it recognizes the areas specified in the menu file. This is done by placing the menu overlay on the digitizing tablet and using the TABLET command to enter key points on the overlay. You must define the area of each menu on the overlay and specify the number of rows and columns of cells each menu contains. (A *cell* is an area on the tablet menu that, when picked, enters a command or function in AutoCAD.)

1. Secure the menu overlay to the digitizer tablet with tape.

2. Enter the **TABLET** command, and then enter the **Cfg** option.

AutoCAD prompts for the number of tablet menus you want to create.

3. Enter **3**.

4. If you receive the message Do you want to align tablet menu areas? enter **Y** for Yes.

AutoCAD prompts you to digitize the upper left corner of the menu area.

```
***BUTTONS
;
^C^CREDRAW

***TABLET1
^C^Czoom w
^C^Czoom p
^C^Czoom a
^C^Czoom d
^C^Cpan
^C^Ctablet on
^C^Cchange
^C^Ctablet off
^C^Cinsert
^C^Cblock
^C^Cstretch
^C^Cellipse
^C^Cmove
^C^Ccopy
^C^Ctrim
^C^Cmirror
^C^Cextend
^C^Cbreak
^C^Cfillet
^C^Csolid
^C^Csketch
^C^Chatch
window
last

***TABLET2
^C^Cmenu acad.mnc
^C^Cspline
^C^Cdimlinear
^C^Cdimaligned
^C^Cleader
^C^Cdimstyle
^C^Cdimcenter
^C^Cdimangular
^C^Cddlmodes
^C^C3d
^C^Cdimordinate
^C^Cdimbaseline
```

Fig. 54-2

```
^C^Cdimcontinue
^C^Cdimdiameter
^C^Cdimradius
^C^Cdimedit

***TABLET3
^C^Cline
^C^Cline end
^C^Cerase
^C^Cerase w
^C^Cerase l
^C^Chelp
^C^C
^C^Cscale
^C^Crotate
^C^Cundo
^C^Cdtext
^C^Cqtext
^C^Carc
;
^C^Ccircle
^C^Cregion
^C^Cdonut
^C^Cpolygon
^C^Cpline
^C^Cstyle simp simplex;;;;;;;
^C^Cstyle comp complex;;;;;;;
^C^Cdducs
^C^Cstatus
^C^Cqsave
^C^Cregen
^C^Cexplode
^C^Coffset
```

Fig. 54-2 (continued)

To digitize a point, center the crosshairs of the puck over the point to be digitized and press the pick button. *Digitize* means to convert 2D drawings or 3D objects into a digital format. In this overlay, menu 1 consists of the first two columns.

5. Locate the upper left corner of menu 1, shown in Fig. 54-3, and pick that point. (The point is indicated on the overlay by a small donut.)

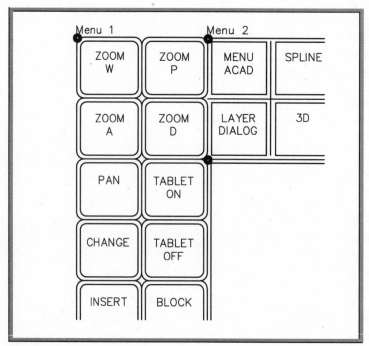

Fig. 54-3

6. Pick the lower left corner of menu 1 (also indicated by a small donut).

7. Pick the lower right corner of menu 1 (the small donut two cells to the right of the preceding point).

You have just defined the boundaries of menu 1.

8. Enter **2** for the number of columns in menu 1.

9. Enter **12** for the number of rows in menu 1.

Now AutoCAD knows the exact size and location of all twenty-four cells in tablet menu 1. Because you have specified three menus, AutoCAD now prompts you for the upper left corner of menu 2.

10. Locate menu 2 and its upper left corner, and pick that point.

Menu 2 consists of the two upper rows, beginning with the cell called MENU ACAD. This menu contains eight columns and two rows.

11. Proceed exactly as you did with menu 1 until you are finished with menu 2. Be sure you select the leftmost donut when you pick the lower right corner of menu 2.

12. Proceed with menu 3. It contains nine columns and three rows.

After you are finished defining menu 3, AutoCAD asks if you want to respecify the fixed screen pointing area.

13. Enter **Y** for Yes.

14. Digitize the lower left and the upper right corners of the display pointing area (the square bounded by the three menus). For the upper right corner, be sure to pick the rightmost donut.

15. Enter **No** in reply to the last question.

You are finished with the tablet configuration.

Loading and Using the Menu File

Before you can use the menu overlay, you must load it in AutoCAD. By default, AutoCAD loads the standard acad.mnu file. Among other things, this file defines the toolbars that appear each time you open AutoCAD. To change the menu file, you can use the MENU command. Let's load the tablet menu called tab.mnu.

1. Enter the **MENU** command and locate and double-click **tab.mnu**.

2. Pick the **Yes** button.

If you correctly completed the above steps, you should now have full access to the new tablet menu. Notice that the docked toolbars disappeared. That's because acad.mnu is no longer the current menu file.

3. Experiment with the tablet menu by picking each of the cells on the overlay, but do not yet pick the item named MENU ACAD located in the upper left corner.

4. To bring back the standard acad.mnu file, pick the tablet menu item called **MENU ACAD**. If this doesn't work, enter **MENU** and select **acad.mnc** from the **Support** directory.

Acad.mnc is a compiled menu file, whereas acad.mnu is not. Selecting either one will load the standard AutoCAD menu.

You no longer have access to the tablet menu.

5. Exit AutoCAD without saving.

Chapter 54
Review & Activities

Review Questions

1. What AutoCAD command and option are used to configure a digitizing tablet menu?

2. Explain the purpose of tablet configuration.

3. What are the minimum and maximum number of tablet menus that can be included on a digitizing tablet?

4. What command is used to load a tablet menu file?

Challenge Your Thinking

1. Discuss the advantages and disadvantages of using a digitizing tablet instead of or in addition to the pull-down menus and the keyboard.

2. Explore the various options available for digitizing tablets. What should you consider when buying one? Describe the distinguishing features of the various tablets. Tell which features would be most important to you and why.

Applying AutoCAD Skills

Work the following problems to practice the commands and skills you learned in this chapter.

1. Plan and sketch a tablet menu overlay for a plumbing supply house to include a symbol library for bathroom plumbing fixtures.

Chapter 54
Review & Activities

2. Develop a new tablet menu and include a symbol library in one section of the menu. Use the previously created library called lib1.dwg and use it to create a new file named lib2.dwg. Then redefine the blocks using single-word names as shown below.

tablesaw
drillpress
jointer
surfaceplaner
workbench

If you do not give the blocks single-word names, they will not work in the menu items. The example in Fig. 54-4 should help you get started.

Name the menu file tab2.mnu. The tab2.mnu and lib2.dwg files are available on the Instructor's Resource CD-ROM.

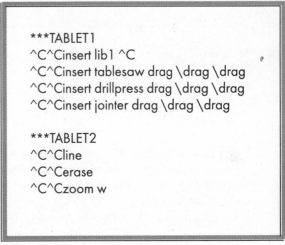

```
***TABLET1
^C^Cinsert lib1 ^C
^C^Cinsert tablesaw drag \drag \drag
^C^Cinsert drillpress drag \drag \drag
^C^Cinsert jointer drag \drag \drag

***TABLET2
^C^Cline
^C^Cerase
^C^Czoom w
```

Fig. 54-4

Note that the first line after ***TABLET1 above inserts the block definitions into the current drawing.

 ## Using Problem-Solving Skills

1. Your job in the PC support department of a plumbing supply house requires that you create a tablet menu for plumbing fixtures. Make a tablet menu with a symbol library for bathroom fixtures. Use the sketch you made in "Applying AutoCAD Skills" problem 1.

2. Expand the symbol library of the tablet menu you created in problem 1 to include kitchen fixtures.

Chapter 55   AutoCAD's Tablet Menu

Objectives:

- Configure the standard AutoCAD tablet menu
- Customize the top area of the standard AutoCAD tablet menu

Key Terms

configuration
menu area 1

This unit steps you through the process of configuring AutoCAD's tablet menu. *Configuration* allows AutoCAD to recognize the commands and functions associated with each portion of the digitizing tablet. Also, you will discover how to develop tablet menu area 1, located at the top of the tablet menu. Techniques are included for positioning new AutoCAD command macros and symbol libraries in this area of the tablet menu. The AutoCAD tablet menu template (overlay) is shown in Fig. 55-1.

The top area has been reserved for you to customize. A total of 225 cells are available. This area is referred to as *menu area 1*.

There is more than one way to develop menu area 1. We will use a basic method that allows you to follow each step of the development easily. This method also shows you how the other portions of the tablet menu were designed.

Configuring the Tablet Menu

To configure AutoCAD's tablet menu, we will use a procedure similar to the one you used in Chapter 54 for the tab.mnu file.

1. Fasten the tablet menu overlay securely to the digitizing tablet.

2. Start AutoCAD and begin a new drawing from scratch.

3. Enter the **TABLET** command and the **Cfg** option.

4. Enter **4** for the number of desired tablet menus.

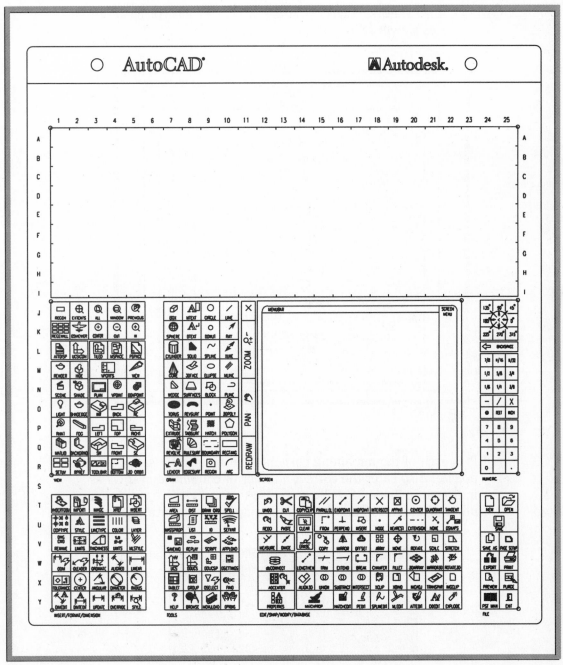

Fig. 55-1

5. As AutoCAD now requests, digitize the upper left corner of menu area 1 as shown in Fig. 55-2.

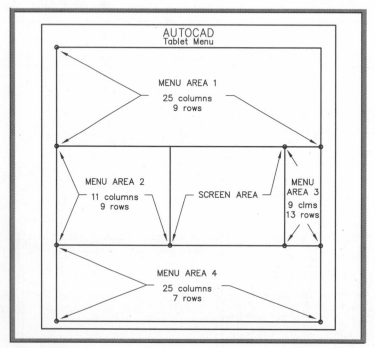

Fig. 55-2

6. Pick each of the remaining corners as instructed by AutoCAD. Be precise. Also enter the correct number of columns and rows as indicated in Table 55-1. Be sure also to specify the fixed screen pointing area. Do not specify a floating screen pointing area.

Menu Area	Columns	Rows
1	25	9
2	11	9
3	9	13
4	25	7

Table 55-1

When you are finished, AutoCAD stores this information and the Command prompt becomes available.

7. Pick several items on the tablet menu to make sure that it is configured properly.

Customizing the Tablet Menu

Now let's customize area 1 of the tablet menu to include the symbol library you created in Chapter 32. To do this, we will need to create a drawing file to use as the tablet menu overlay for this area. Then we will write the code needed to associate the appropriate commands and functions with the tablet area we are designing.

Creating the Overlay

We will begin by creating the rectangular shape of area 1.

NOTE: _____

> The drawing (as printed on page 770) is also available as frame.dwg on the Instructor's Resource CD-ROM.

1. Open the AutoCAD tablet menu overlay drawing file named **Tablet 2000.dwg**, which is located in AutoCAD's **Sample** folder.

2. Using **Save As...**, create a file named tablet **2000bk.dwg**. Store it in the folder with your name.

3. Create and make current a new layer named **Menu** and assign the color red to it.

4. Zoom in on menu area 1.

5. Using the **PLINE** command, trace the outline of menu area 1, as shown in Fig. 55-3.

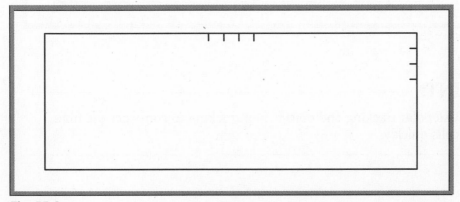

Fig. 55-3

6. Add four short vertical lines beginning at column 12, as shown, and add three short horizontal lines beginning at row A.

7. Create a block named **frame** and enter **0,0** for the insertion base point. Select each of the lines and the polyline outline you created.

8. Enter the **WBLOCK** command, enter **frame** for the file name, and enter **frame** for the block name. Save it in the folder with your name.

9. Save your work.

Designing Area 1

You now have a drawing file, frame.dwg, that matches the size of tablet area 1. Let's open this file and add command macros and symbols. When it is finished, you will be able to plot frame.dwg and position it under the template's transparent area 1.

1. Close Tablet 2000bk.dwg and open the **frame.dwg** file.

2. Enter **ZOOM Extents**, and then enter **ZOOM** and **.95x**.

3. Use horizontal and vertical lines to create the nine cells shown in Fig. 55-4.

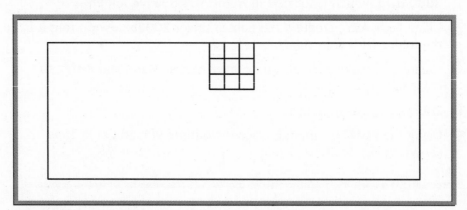

Fig. 55-4

 HINT: ────────────────────────────────

Use polar tracking and object snap tracking to construct the nine cells quickly.

4. Erase the three short horizontal lines and save your work.

Let's insert the symbol library named lib1.dwg that you created in Chapter 32.

5. Using **AutoCAD DesignCenter**, insert each of the five blocks from **lib1.dwg** into the drawing, as shown in Fig. 55-5.

HINT:

You will need to reduce their size. Try .005 for the scale. Checking Uniform Scale in the Insert dialog box will save time. Uniform scale specifies a single scale factor for *x, y,* and *z* coordinates.

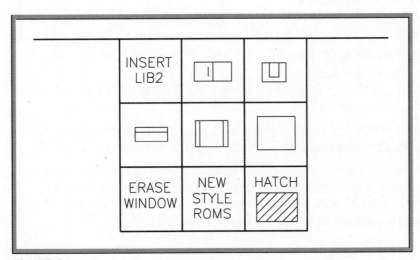

Fig. 55-5

6. Save your work.

7. Place the text in the cells as shown in the previous illustration, and include a small hatch pattern under the word HATCH.

8. **ZOOM All** and save your work.

9. Plot the drawing extents at a scale of **1 = 1**.

10. After plotting, trim around the menu area with scissors, and secure the menu under the template.

11. Exit AutoCAD.

Writing the Code

Now we have to let AutoCAD know what we want to happen when we pick each of the cells we have defined in menu area 1. To do this, we will add macros to the code that controls AutoCAD's menus. As you may recall, a macro is a code-based command that instructs the computer to carry out a set of instructions.

1. Locate the acad.mnu file, which is located in AutoCAD's support folder. Make a copy of it, and name it **acadnew.mnu**.

2. Open WordPad, and open **acadnew.mnu**.

3. Using the Find capability, locate the item *****TABLET1** in the file.

```
***TABLET1
**TABLET1STD
[A-1]\
[A-2]\
[A-3]\
[A-4]\
[A-5]\
```

Fig. 55-6

After you locate ***TABLET1 in the file, notice the similar items (A-1 through A-25, B-1 through B-25, and so on) that follow it. The first few lines of this section are shown in Fig. 55-6.

Each of these numbered items corresponds to a cell in menu area 1. To illustrate this, the upper middle portion of area 1 is shown in Fig. 55-7.

The numbering sequence of the cells begins at the upper left corner of the tablet menu and proceeds to the right. Tablet area 1 contains 225 cells arranged into 25 columns and 9 rows. This is where you enter the macros that correspond to the items in menu area 1.

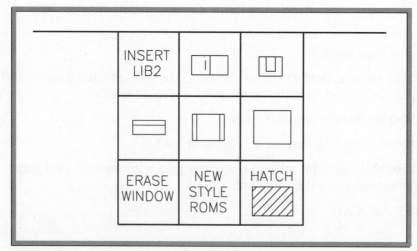

Fig. 55-7

4. Type the menu items over the numbers as shown in Fig. 55-8 (pages 774-775). Notice that their placement corresponds directly to the cells in the upper middle portion of tablet area 1.

 NOTE:

In the code shown in Fig. 55-8, text was omitted where you see blank lines. This was done to conserve space on the page.

5. Save your changes in the acadnew.mnu file, being sure to save the file as a text file (not as a Word document) and exit WordPad.

If you did not create lib2.dwg in Chapter 54, you will do so in the following steps.

1. If you have already created lib2.dwg, skip to Step 4.

2. Open **lib1.dwg** and use **Save As...** to create **lib2.dwg**.

3. Redefine the five blocks using the following single-word names.
   tablesaw
   drillpress
   jointer
   surfaceplaner
   workbench

4. Copy **lib2.dwg** to the **Acad2000** folder.

5. Restart AutoCAD and begin a new drawing from scratch.

6. Using the **MENU** command, load **acadnew.mnu**.

7. Pick the new tablet menu item named **LIB2**.

This item, as you may have noticed when you typed the macro, inserts the small symbol library named lib2.dwg (including its block definitions) into the current drawing.

8. Pick each of the tool symbols and place them one at a time. Use a scale factor of **.05**.

9. Pick the remaining three menu items and notice what each of them does.

This gives you a taste of what can be developed in the top area of AutoCAD's standard tablet menu.

10. Load the **acad.mnc** file, and exit AutoCAD without saving.

```
[***TABLET1
**TABLET1STD
[A-1]\
[A-2]\
[A-3]\
[A-4]\
[A-5]\
[A-6]\
[A-7]\
[A-8]\
[A-9]\
[A-10]\
[A-11]\
^C^Cinsert lib2;^C
^C^Cinsert tablesaw drag \drag \drag
^C^Cinsert drillpress drag \drag \drag
[A-15]\
[A-16]\
[A-17]\
[A-18]\
[A-19]\
[A-20]\
[A-21]\
[A-22]\
[A-23]\
[A-24]\
[A-25]\
[B-1]\
[B-2]\
[B-3]\
[B-4]\
[B-5]\
[B-6]\
[B-7]\
[B-8]\
[B-9]\
[B-10]\
[B-11]\
^C^Cinsert jointer drag \drag \drag
^C^Cinsert surfaceplaner drag \drag \drag
^C^Cinsert workbench drag \drag \drag
```

Fig. 55-8

```
[B-15]\
[B-16]\
[B-17]\
[B-18]\
[B-19]\
[B-20]\
[B-21]\
[B-22]\
[B-23]\
[B-24]\
[B-25]\
[C-1]\
[C-2]\
[C-3]\
[C-4]\
[C-5]\
[C-6]\
[C-7]\
[C-8]\
[C-9]\
[C-10]\
[C-11]\
^C^Cerase w
^C^Cstyle roms romans;;;;;;;
^C^Cbhatch
[C-15]\
[C-16]\
[C-17]\
[C18]\
[C-19]\
[C-20]\
[C-21]\
[C-22]\
[C-23]\
[C-24]\
[C-25]\
```

Fig. 55-8 (continued)

Chapter 55
Review & Activities

Review Questions

1. How do you define the four standard tablet areas when configuring the tablet menu?

2. What is the purpose of the AutoCAD tablet menu area 1?

3. What is the benefit of customizing the tablet menu?

4. Explain the number sequence of cells contained in tablet menus.

Challenge Your Thinking

1. Consider how a custom tablet menu could help you use AutoCAD. Discuss your ideas with others. Then design a custom tablet menu to help you do your work more easily.

2. Review AutoCAD's standard tablet menu overlay. Plan a new design for it that includes all of the functionality but in a different format. Place the commands and functions in locations that would be useful to you.

Applying AutoCAD Skills

1. Experiment with different portions of the AutoCAD tablet menu areas 2, 3, and 4 to discover how these menus were designed.

2. In the acadnew.mnu file, place additional menu items in menu area 1. Categorize the items in area 1 so that related items are grouped together. You may wish to add headings above each group of related items.

3. Create the custom menu you designed in the "Challenge Your Thinking" question and try it. Does it work the way you expected? If not, revise it until it becomes a helpful tool.

4. Configure the tablet menu you designed in Chapter 54 ("Applying AutoCAD Skills" problem 1).

Chapter 55
Review & Activities

 Using Problem-Solving Skills

1. Customize area 1 of AutoCAD's tablet menu to include the symbol library of electronic devices you created in Chapter 32 ("Using Problem-Solving Skills" problem 2).

2. Create the redesigned tablet menu overlay that you planned in the second "Challenge Your Thinking" question.

Chapter 56 Slides and Scripts

Objectives:

- Create and view slide files
- Develop a slide show as a script and run the script
- Create and apply a slide library

Key Terms

script file
slide
slide library
slide show

AutoCAD permits you to develop slides from drawings and include them in a slide show. A *slide* is an image of a drawing that can be displayed and controlled within AutoCAD using a script file. Though it may sound complicated, it is really very simple.

A *script file* is nothing more than an ASCII text file with an SCR file extension. You can execute a script when you start AutoCAD or from within AutoCAD using the SCRIPT command. Figure 56-1 shows an example of a script file.

```
limits 0,0 30,20
zoom a
grid 1
snap on
```

Fig. 56-1

With earlier versions of AutoCAD, people used script files such as the one above to store drawing parameters and settings to expedite the setup process. The use of drawing template files has largely replaced this practice.

Producing a Slide Show

Script files can be used for other purposes, too, such as showing a continuous sequence of drawings in a sort of electronic flipchart. AutoCAD calls this a *slide show*.

Creating the Slides

The first step in creating a slide show is to create slides. You can create slides from existing drawings using the MSLIDE command. Let's create slides from a couple of drawings.

1. Begin a new drawing from scratch, and name it **show.dwg**. Save it in the folder with your name.

2. Pick the **Insert Block** button from the docked Draw toolbar, and insert the drawing named **deck.dwg**. Use **0,0,0** for the insertion point, and accept the remaining default settings.

Draw

3. **ZOOM All**.

4. Enter the **MSLIDE** command, enter **deck.sld** for the file name, and pick the **Save** button.

AutoCAD creates a slide of the deck drawing. The new slide file cannot be edited. The drawing files used to create the slides remain untouched and can be edited.

5. Erase all objects on the screen so that it is blank.

6. Insert **staird.dwg** using the same method of insertion.

7. **ZOOM All** and center the drawing.

8. With the **MSLIDE** command, create another slide. Name it **staird.sld**.

9. Erase staird.dwg so that the screen is blank.

10. If you'd like to make additional slides from other drawings, create them now.

Viewing the Slides

Let's use the VSLIDE command to look at the first slide we created.

1. Enter **VSLIDE**, select **deck.sld**, and pick **Open**.

The deck.sld slide appears on the screen.

2. View the **staird.sld** slide.

3. To restore the original screen, enter **R** (for REDRAW).

Creating a Script

Now let's create a script file (slide show). It's going to be a short one!

1. Minimize AutoCAD and start Notepad.

2. Enter the text shown in Fig. 56-2 using upper- or lowercase letters. In the last entry, press the spacebar once after typing redraw and then press **ENTER**. (Without the space, AutoCAD will not enter the command.)

```
vslide deck
delay 2000
vslide staird
delay 2500
redraw
```

Fig. 56-2

 NOTE:

If you have created additional slides, you can include them also.

3. Name the file **show.scr**, and be sure to store it in the folder that contains the slides you created.

You have just created a simple slide show stored as a script file. In this script, the VSLIDE command displays the slides you specified. The DELAY command tells AutoCAD to hold the slide on the screen for *x* number of milliseconds. Although computer clocks run at different speeds, 1000 milliseconds is approximately a one-second delay. The REDRAW command blanks the screen at the end of the slide show.

Running the Script

Now let's try the slide show.

1. Select **Run Script...** from the **Tools** pull-down menu to enter the SCRIPT command, select **show.scr**, and pick **Open**.

AutoCAD displays each slide on the screen.

2. To repeat the slide show, enter the **RSCRIPT** command.

You can include this command at the end of a script file to repeat the slide show automatically. The backspace key will interrupt a running script. This allows you to issue other AutoCAD commands. If you wish to return to the script, enter the RESUME command.

3. Create new slides from other drawings and include them in the **show.scr file**. Include the **RSCRIPT** command at the end of the script.

4. Run the revised slide show.

Producing a Slide Library

A *slide library* in AutoCAD is a single file in which you can store slides, similar to filling a carousel tray of 35-mm slides. The individual files from which the slides were created need not be present to use a slide library.

Creating the Library

1. Minimize AutoCAD.

2. Locate the AutoCAD file named **slidelib.exe** (located in AutoCAD's **Support** folder) and copy it to the folder with your name, which contains the slide files.

3. From the Windows **Start** menu, pick **Programs** and open a DOS window. (If you are using Windows NT 4.0, pick **Command Prompt**.)

4. At the DOS prompt, change to the directory with your name. An example is shown below.

   cd \Program Files\acad2000\Jody Wozniak

NOTE:

> If you receive the message Too many parameters – Files\acad2000, enter the following text instead, exactly as you see it here:
>
>   cd \Progra~1\acad2000\Jodywo~1
>
> (Substitute the name of your folder for Jodywo~1, making sure that the last two characters are ~1).

5. At the DOS prompt, enter **slidelib tools** to begin a slide library file named tools.slb.

This starts the slidelib.exe program and displays a copyright statement on the screen.

6. Enter the following text exactly as you see it here.

   deck
   staird (press **ENTER** twice)
   Press the **F6** function key and **ENTER**.

You should now have a slide library file named tools.slb in the folder with your name.

7. Close the DOS window.

Applying the Slide Library

1. Move the slide files out of the folder with your name. You can temporarily place them on the Windows desktop.

2. Open the script file **show.scr** and revise it so its contents are identical to the text in Fig. 56-3.

```
vslide tools(deck)
delay 2000
vslide tools(staird)
delay 2500
redraw
```

Fig. 56-3

Note that the slide library name, tools, precedes the slide name, which is enclosed in parentheses.

3. Maximize AutoCAD.

4. Enter the **SCRIPT** command and find and double-click **show.scr**.

AutoCAD runs the script the same as before. The difference is that the slide files themselves are not available to AutoCAD. (Recall that you removed them from the directory in Step 1.) Instead, AutoCAD uses the slide definitions contained in the tools.slb slide library. The benefit of this is two-fold: First, you can store a large number of slides in a single file—a slide library—and you don't have to keep the slide files themselves. The second benefit is that you can easily give a collection of slides to someone in a single file. This makes it easy to send a set of slides to someone over the Internet.

5. Move the slide files from the Windows desktop to the folder with your name.

6. Exit AutoCAD without saving.

Chapter 56
Review & Activities

Review Questions

1. Briefly describe the purpose of each of the following commands.

   MSLIDE
   VSLIDE
   SCRIPT
   RSCRIPT
   DELAY
   RESUME

2. Describe the purpose of an AutoCAD script file.

3. What is the file extension of a script file?

4. What does the number following the DELAY command indicate?

5. From what type of files does the slide library utility create a slide library file?

Challenge Your Thinking

1. Is it more practical to store a drawing setup in a script file or in a drawing template file? Explain why.

2. Brainstorm possible uses for an AutoCAD slide show. Be creative. Then choose one possibility and develop a plan for the slide show.

Applying AutoCAD Skills

Work the following problems to practice the commands and skills you learned in this chapter.

1. Create a dozen or so slides of previously created drawings. Include them in a slide show stored as a script file. Run the show.

2. Develop a script file that includes several AutoCAD commands. Make it elaborate. When you're finished, print the file so that you can work out the bugs as you run it.

3. Create the slide show for which you developed a plan in the second "Challenge Your Thinking" question.

Chapter 56
Review & Activities

Using Problem-Solving Skills

Complete the following activities using problem-solving skills and your knowledge of AutoCAD.

1. In addition to your duties as a professor of mechanical engineering, you provide corporate training in drafting and CAD. You want to create a slide show to demonstrate the concept of layers and their importance in a complex drawing design. Create an orthographic drawing of the base plate shown in Fig. 56-4. Make sure each type of element (object lines, hidden lines, center lines, dimensions, and text) is on a separate layer. Starting off with the object lines layer thawed and all others frozen, make slides with successive layers thawed. Run the slide show. Can you use it to demonstrate the concept of layers?

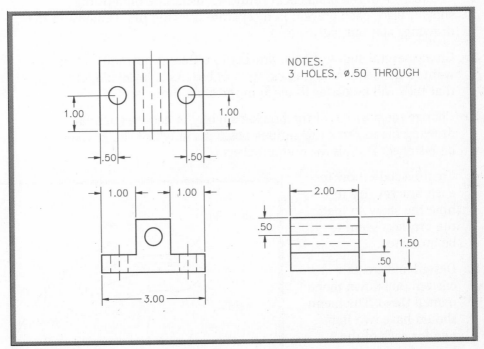

Fig. 56-4

2. Create a slide library of all the drawings you have created in this course. Keep the slide library so that you can develop slide shows for future assignments.

Part 10 Project

Creating a Trade Show Demo

Slide shows created in AutoCAD can be used for many different purposes. For example, you could use a slide show to demonstrate how to perform a specific task in AutoCAD or to demonstrate a new product design at a trade show. In this project, you will create a "demo" for use at an exhibit at a mechanical engineering trade show. To make the demo easily accessible during the exhibit, you will create a custom menu item that allows you to run the demo from a pull-down menu.

Description

Design a slide show that demonstrates the use of the two adjustable brackets shown in Figs. P10-1 and P10-2. Include one slide of each that shows the dimensions; then show each bracket, complete with a bolt or other fastener, in several different positions to show how they adjust.

1. Create both of the brackets in AutoCAD using the dimensions shown. Place each bracket in a separate drawing file. Dimension the drawings appropriately.

2. Create several slides of each bracket in different positions. (You may want to use sequential names, such as brkt1a.sld, brktab.sld, etc., so that they will be easier to place in the slide show later.)

3. Change the position of the bracket slightly in each sequential drawing file to show the various positions in which the bracket can be fastened. Do this for both brackets.

4. Create a slide show for each bracket. Try to time the show so that the brackets seem to be animated.

5. Design and create a custom pull-down menu named Demo. The menu should have two items: one for each of the bracket slide shows you have created. Name the items appropriately so that you know which bracket will be demonstrated when you select each menu item.

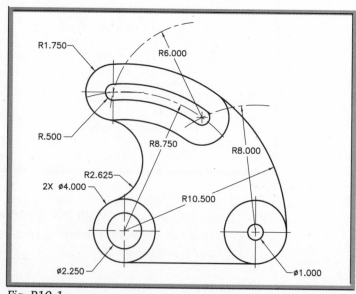

Fig. P10-1

Part 10 Project

Hints and Suggestions

- Timing a slide show to make a drawing appear to be animated takes planning and a certain amount of trial-and-error. You may need to experiment to find the best DELAY times.

- The animation will turn out best if there is only a minimal difference in the

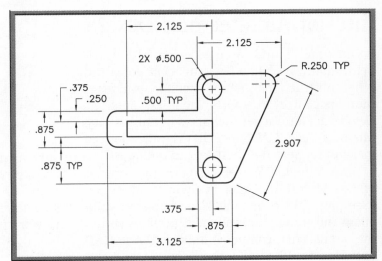

Fig. P10-2

position of the bracket in any two sequential slides. To achieve this, you may want to consider positioning the bracket, saving the slide, then rotating or moving the bracket a small amount and then saving the next slide.

- Consider using layers to manage dimensions, the bolt or other fastener, and any text you may have included in the drawings.

- Review the information in Chapters 52 and 53 before creating the custom menu. Be sure to work with a backup copy of acad.mnu.

Summary Questions/Self-Evaluation

1. How many actual drawing files did you create? Discuss the advantages and disadvantages of using a single drawing file for each slide show.

2. From a commercial standpoint, what is the advantage of creating two separate slide shows, one for each fastener? Are there any advantages to creating a single slide show for both fasteners?

3. Could you alter menu area 1 of AutoCAD's tablet menu so that the slide shows could be run from the tablet menu? Why might you want to do so? Explain.

Careers
Using
AutoCAD

Interior Architectural Design

Old-time sports arenas, even the most beloved, always had a few seats near poles and other obstructions. Today, architectural firms that specialize in sports facilities, such as Devine deFlon Yaeger Architects of Kansas City, Missouri, use CAD and 3D visualization to make every seat the "best seat in the house."

Devine deFlon Yaeger designed the interior of the Multi-Use Arena in Sydney, Australia. (See page 718 of this book for a related story about the arena.) To evaluate sight lines within the arena, the company imported AutoCAD drawings to 3D Studio Max. The 3D software gave perspective views that revealed, for example, whether rail height would pose problems for spectators in wheelchairs.

Visualization from AutoCAD drawings nailed a conflict between the roof design and plans for the lighting system. When lighting was reconfigured from one sport to another, certain spotlights ran afoul of roof trusses. Identifying the problem before construction made the solution easy and inexpensive.

Career Focus—Architectural Drafter

A team of architectural drafters, supervised by a project architect, generated the drawings used in the design and visualization of Sydney's Multi-Use Arena. An architectural drafter prepares technical drawings based on rough sketches and/or calculations by an architect. The drafter uses a knowledge of standard building techniques to fill in details of the structure. The drawings show multiple views of the structure and its elements and include specifications for construction, such as materials, measurements, and construction methods.

With work experience, an architectural drafter may specialize in a type of structure, such as schools, hospitals, or sports arenas. An architectural drafter works in an office environment, often at a computer workstation alongside others. Drafting requires hours of attention to work in fine detail.

Education and Training

For an entry-level architectural drafter, employers generally require training at a two-year college or technical institute. Courses in a typical program include drafting and mechanical drawing fundamentals, CAD techniques, methods of manufacturing and construction, and mathematics, science, and basic engineering techniques.

A beginning drafter usually does routine work under close supervision. After gaining experience, the drafter takes on more challenging tasks and may advance to other positions such as senior drafter or supervisor.

Career Activity

- Think of any sports facility completed recently, and do research on the firm that built it. What other projects have they done?
- If you wanted a job with the firm, what kinds of experience would be important to them? How could you get the appropriate experience?

Part 11

AutoLISP

Chapter 57 Exploring AutoLISP

AutoLISP is AutoCAD's version of the LISP programming language. LISP stands for "LISt Processing." LISP is well suited to graphics applications, which is a reason Autodesk chose to embed it into AutoCAD. For example, you can use AutoLISP to write a program that automates the creation of a staircase in a building. The program asks you for the distance between the upper and lower floors and for the size or number of steps (risers and treads). With this information, it will automatically draw a staircase for you.

AutoCAD has a built-in LISP interpreter that you can use to enter AutoLISP code at the Command line. Alternatively, you can load and run AutoLISP code from a menu item or file. AutoCAD also includes an enhanced version of AutoLISP called Visual LISP (VLISP), which offers a development environment that includes a compiler, debugger, and other tools. VLISP extends the language to interact with objects using ActiveX, a technology developed by Microsoft based on the component object model (COM) architecture. This chapter and the two that follow concentrate on the interpreted implementation of AutoLISP.

Reviewing AutoLISP Examples

Many AutoLISP files that come with AutoCAD are located in AutoCAD's Support folder. They provide examples of good programming practice, and they illustrate the power of AutoLISP, so you are encouraged to review them.

1. Start Notepad and pick **Open...** from the **File** pull-down menu.

2. Change the directory to AutoCAD's **Support** folder.

3. In the File name edit box, enter ***.lsp** to list only the AutoLISP files.

Table 57-1 provides brief descriptions of some of these files. Let's take a look one of these programs.

AutoLISP File	Purpose
3d.lsp	creates various three-dimensional objects, including a pyramid, box, cone, dome/dish, wedge, torus, and 3D mesh
3darray.lsp	creates 3D rectangular arrays by specifying rows, columns, and levels; also creates polar arrays around a specified axis
mvsetup.lsp	sets the drawing units and limits of a new drawing based on the paper size and drawing scale
edge.lsp	lets you interactively change the visibility of the edges of a 3D face
ddview.lsp	provides a quick and easy interface to the VIEW command

Table 57-1

4. Open the file named **3darray.lsp** in AutoCAD's **Support** folder.

5. Review the contents of the file.

The first portion of 3darray.lsp is shown in Fig. 57-1 (pages 792-793).

```
; Next available MSG number is   29
; MODULE_ID LSP_3DARRAY_LSP_
;;;
;;;   3darray.lsp
;;;
;;;   Copyright 1987-1988,1990,1992,1994,1996-1998 by Autodesk, Inc.
;;;
;;;   Permission to use, copy, modify, and distribute this software
;;;   for any purpose and without fee is hereby granted, provided
;;;   that the above copyright notice appears in all copies and
;;;   that both that copyright notice and the limited warranty and
;;;   restricted rights notice below appear in all supporting
;;;   documentation.
;;;
;;;   AUTODESK PROVIDES THIS PROGRAM "AS IS" AND WITH ALL FAULTS.
;;;   AUTODESK SPECIFICALLY DISCLAIMS ANY IMPLIED WARRANTY OF
;;;   MERCHANTABILITY OR FITNESS FOR A PARTICULAR USE.  AUTODESK, INC.
;;;   DOES NOT WARRANT THAT THE OPERATION OF THE PROGRAM WILL BE
;;;   UNINTERRUPTED OR ERROR FREE.
;;;
;;;   Use, duplication, or disclosure by the U.S. Government is subject to
;;;   restrictions set forth in FAR 52.227-19 (Commercial Computer
;;;   Software - Restricted Rights) and DFAR 252.227-7013(c)(1)(ii)
;;;   (Rights in Technical Data and Computer Software), as applicable.
;;;
;;;==========================================================
;;; Functions included:
;;;     1) Rectangular ARRAYS (rows, columns & levels)
;;;     2) Circular ARRAYS around any axis
;;;
;;; All are loaded by: (load "3darray")
;;;
;;; And run by:
;;;     Command: 3darray
;;;        Select objects:
;;;           Rectangular or Polar array (R/P): (select type of array)


;;; ================= load-time error checking =================

(defun ai_abort (app msg)
  (defun *error* (s)
    (if old_error (setq *error* old_error))
    (princ)
```

Fig. 57-1

```
        )
      (if msg
        (alert (strcat " Application error: "
                    app
                    " \n\n "
                    msg
                    " \n"
             )
        )
      )
      (exit)
    )

;;; Check to see if AI_UTILS is loaded, If not, try to find it,
;;; and then try to load it.
;;;
;;; If it can't be found or it can't be loaded, then abort the
;;; loading of this file immediately, preserving the (autoload)
;;; stub function.

    (cond
      ( (and ai_dcl (listp ai_dcl)))        ; it's already loaded.

      ( (not (findfile "ai_utils.lsp"))              ; find it
        (ai_abort "3DARRAY"
              (strcat "Can't locate file AI_UTILS.LSP."
                  "\n Check support directory.")))

      ( (eq "failed" (load "ai_utils" "failed"))         ; load it
        (ai_abort "3DARRAY" "Can't load file AI_UTILS.LSP"))
    )

    (if (not (ai_acadapp))          ; defined in AI_UTILS.LSP
       (ai_abort "3DARRAY" nil)        ; a Nil <msg> supresses
    )                  ; ai_abort's alert box dialog.

;;; ================= end load-time operations =========
;;;
;;;
;;;************************** MODES ***************************
;;;
;;;
;;; System variable save

(defun MODES (a)
  (setq MLST '())
```

Fig. 57-1 (continued)

The beginning of many AutoLISP programs includes copyright information and statements explaining the purpose of the program. Notice that this program allows you to create rectangular and circular three-dimensional arrays.

Loading AutoLISP Files

Now that you have seen the contents of 3darray.lsp, let's load it into AutoCAD.

Using the Command Line

AutoLISP files such as 3darray.lsp can be loaded and invoked by entering the AutoLISP load program at the Command prompt.

1. Exit Notepad. Do not save changes to the file.

2. Start AutoCAD and begin a new drawing from scratch.

3. At the Command prompt, enter the following text exactly as you see it, including the parentheses. (You may use upper- or lowercase letters.) Be sure to press **ENTER**.

   (load "3darray")

If the program loads properly, you will receive the message 3DARRAY loaded, and you will have access to a new command called 3DARRAY. If you receive an error message, refer to the section titled "AutoLISP Error Messages" later in this chapter.

4. Create a 3D cube **1** unit in size.

5. Select a viewpoint that is above, in front, and to the right of the object.

6. Enter **ZOOM** and **1** to reduce the current zoom magnification.

7. Enter **3DARRAY**, select the cube and press **ENTER**, and choose to perform a rectangular array.

8. Enter **4** for the number of rows, **3** for the number of columns, and **2** for the number of levels.

9. Specify a distance of **1.5** between rows, columns, and levels.

Solids

The 3D array generates on the screen.

10. Enter **ZOOM Extents**.

11. Enter the **HIDE** command.

The drawing should look similar to the one in Fig. 57-2.

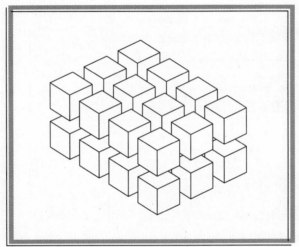

Fig. 57-2

Using a Dialog Box

In addition to loading AutoLISP files at the Command line, you can load them using the Load/Unload Applications dialog box.

1. Select **Load Application...** from the **Tools** pull-down menu.

This displays the Load/Unload Applications dialog box, as shown in Fig. 57-3 (page 796).

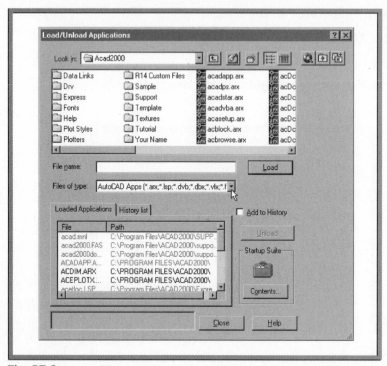

Fig. 57-3

Loaded programs, also called applications, are listed under the tab named Loaded Applications. In addition to loading applications, you can unload them by selecting the file you want to unload and picking the Unload button. Unloading applications frees system memory.

2. At the right of Files of type, pick the down arrow.

This displays the types of applications that AutoCAD can load. ObjectARX is a programming environment in AutoCAD. With ObjectARX, for example, you can use the C++ programming language to customize AutoCAD. VBA, which stands for "Visual Basic for Applications," is an object-based programming environment implemented in AutoCAD.

3. Pick **AutoLISP Files (*.lsp)**.

4. Find and double-click AutoCAD's **Support** folder.

5. Select **3darray.lsp** and pick the **Load** button.

This loads 3darray.lsp, the same as if you had used the AutoLISP load command.

6. Pick the **Close** button.

AutoLISP Error Messages

At one time or another, you may receive one of many AutoLISP error messages. Here are three examples.

Insufficient node space

Insufficient string space

LOAD failed.

The error messages insufficient node space and insufficient string space appear when AutoCAD runs out of "heap" space. The *heap* is an area of memory set aside for storage of all AutoLISP programs and symbols (also called *nodes*). Elaborate AutoLISP programs require greater amounts of heap and stack space.

The *stack* is also an area of memory set aside by AutoLISP. Stack holds programming arguments and partial results; the deeper you "nest" these items, the more stack space is used.

The error message LOAD failed appears when the file named in the load program cannot be found, or when the user does not have read access to the file. This message also appears when the file name is misspelled or the incorrect file location (folder) is specified.

Chapter 57
Review & Activities

Review Questions

1. What is AutoLISP?

2. Name two applications for AutoLISP.

3. What file extension do AutoLISP files use?

4. What purpose does the 3darray.lsp program serve?

5. How would you load a file named project.lsp at the Command line?

6. How would you load a file named project.lsp using a dialog box?

7. Why is the amount of heap space important when you load AutoLISP programs?

Challenge Your Thinking

1. AutoCAD is a complex CAD program that allows you to do most drafting tasks easily. In addition, you can customize the program by creating your own menu items. With these facts in mind, explain why separate AutoLISP programs may be desirable for some AutoCAD applications.

2. Investigate the difference between LISP and Visual LISP. Which would you prefer to use if you were required to do a large amount of AutoCAD customization? Why?

Applying AutoCAD Skills

Work the following problems to practice the commands and skills you learned in this chapter.

1. Load and invoke the AutoLISP programs supplied with AutoCAD.

2. Open and review the contents of the edge.lsp file located in AutoCAD's Support folder.

Chapter 57
Review & Activities

3. Open and review the xplode.lsp file located in AutoCAD's Support folder. What is the purpose of this program?

4. Use Notepad or WordPad to browse through the other LSP files in AutoCAD's Support folder. Which files are loaded automatically by AutoCAD every time a drawing is opened?

 Using Problem-Solving Skills

Complete the following problem using problem-solving skills and your knowledge of AutoCAD.

1. Choose one of the AutoLISP programs provided with AutoCAD, such as xplode.lsp. Prepare a demonstration of the program for a corporate review board.

2. Load the 3darray.lsp file. Create a slide show that shows how this AutoLISP program works. Run the slide show.

Chapter 58 · Basic AutoLISP Programming

Objectives:

* Demonstrate a basic knowledge of the arithmetic format used in AutoLISP functions
* Use basic AutoLISP functions to create an AutoLISP routine
* Store AutoLISP routines as files and as menu items

Key Terms

cadr
car
defun
function
list
routines
setq

The significance of AutoLISP embedded into AutoCAD has captured the interest of many. This chapter introduces AutoLISP programming so that you can decide whether or not it's for you. Like many others, you may remain satisfied with applying the power of already-developed AutoLISP programs. Few AutoCAD users will write sophisticated AutoLISP programs. Typically, drafting, design, and engineering professionals have neither the time nor the interest to learn AutoLISP fully. They are likely to seek ready-made AutoLISP programs, such as those that are included in AutoCAD's Support folder.

On the other hand, you may find AutoLISP programming very intriguing. If you enjoy programming, you should continue with the following chapter and then explore and learn other AutoLISP commands and functions on your own.

AutoLISP Arithmetic

As illustrated in Chapter 57, you can enter AutoLISP programs, such as load, directly at the Command prompt. Let's enter an AutoLISP arithmetic expression.

1. Start AutoCAD and begin a new drawing from scratch.

2. At the Command prompt, type (* 5 6) and press **ENTER**.

The number 30 appears at the bottom of the screen because $5 \times 6 = 30$. Notice the sequence in which the arithmetic expression was entered. The operation (in this case, multiplication) is entered first, followed by the numbers. The entire expression is enclosed in parentheses.

3. Enter (/ 15 3).

The number 5 appears because $15 \div 3 = 5$.

4. Try (+ 25 4).

The number 29 appears.

AutoLISP Functions

In AutoLISP, a *function* is a computer routine or subroutine that performs a calculation with variables you provide and gives a single result. This result is either displayed at the Command line or, if the function is part of a larger routine, is passed to another part of the program.

Assigning Values to Variables

The *setq* function is one of the most basic functions in AutoLISP. It is used to assign values to a variable.

1. Type (setq A 10) and press **ENTER**. (You can type A in upper- or lowercase letters.)

The value of variable A is now 10.

You can also set a variable equal to an arithmetic expression.

2. Enter (setq B (– 20 4)).

Notice the parentheses. The variable B now holds the value of 20 – 4, or 16.

3. To list the value of A, enter **!A**.

The number 10 appears.

4. Enter **!B**.

Let's try something a bit more complex.

5. Enter **(setq STEP1 (– B A))**.

Now POINT1 is equal to the value of B minus the value of A.

Let's try one more.

6. Enter **(setq STEP2 (* (* 2 STEP1)(/ A 5)))**.

NOTE:

The total number of left parentheses must equal the total number of right parentheses. If they are not equal, you will receive a message such as 1>. This means you lack one right parenthesis. Add one by typing another right parenthesis and pressing ENTER.

Listing Multiple Values

Setq by itself can assign only one value to a variable. The *list* function can be used to string together multiple values, such as *x* and *y* coordinates, to form a point. Setq, used with list, can then assign a list of coordinates to a single variable. Let's step through an example.

1. Type **(setq base1 (list 7 6))** and press **ENTER**.

The result (7 6) appears. This is now the value of base1.

2. Enter **!base1**.

The value (7 6) appears again.

3. Enter the **LINE** command and pick a point near the lower left corner of the screen.

4. In reply to Specify next point, enter **!base1**.

A line appears.

5. Press **ENTER** to terminate the LINE command.

6. List the properties of the line.

Notice that the coordinates of the second point are 7,6—the value of base1.

This function is especially helpful when you need to reach a point off the screen. For instance, if the upper right corner of the drawing area is 15,10 and the LINE command has been issued, you can reach point 25,20 (or any point, for that matter) if a variable is assigned to that point. You just enter the variable preceded by the exclamation mark (!) as you did in the base1 example. Let's try it.

1. Enter **(setq base2 (list 35 28))**.

2. Enter the **LINE** command and pick a point anywhere on the screen.

3. In reply to Specify next point, enter !base2.

A line appears that runs off the screen.

4. Press **ENTER** and enter **ZOOM Extents**.

You should now see the endpoint 35,28 of the line.

Obtaining the First Item in a List

The *car* function is used to obtain the first item in a list. For example, you could use car to specify the *x* coordinate in a coordinate pair. So, what would be the car of base1? Let's enter it.

1. Type **(car base1)** and press **ENTER**.

The value 7 appears.

Obtaining the Second Item in a List

Cadr is like car, only cadr gives you the second item of a list—the *y* coordinate, in this case.

2. Enter **(cadr base1)**.

The value 6 appears.

As you can see, we can obtain either the *x* coordinate or the *y* coordinate from a list.

Combining Functions

We can also assign a new variable to a set of coordinates that contains the cadr of base1 (the *y* coordinate) and 0 as the *x* coordinate.

First, we must create a new list containing 0 and the cadr of base1. Then we need to use setq to assign the new list to a variable (we'll call it hole1).

3. Enter **(setq hole1 (list 0 (cadr base1)))**.

AutoCAD displays (0 6) on the Command line.

4. Try a similar function using **car** and **0** (for the *y* coordinate). Use **hole2** for the variable name.

AutoCAD displays (7 0).

HINT:

Enter (setq hole2 (list (car base1) 0)).

5. Enter **!hole1** and then enter **!hole2**.

The values (0 6) and (7 0) return.

The preceding steps gave you a taste of AutoLISP programming. Chapter 58 will pick up from here and will apply most of the preceding AutoLISP programming techniques.

Storing AutoLISP Routines

AutoLISP functions and groups of functions (often called *routines*) can be stored and used in at least two different formats:

*   as files containing an LSP file extension
*   as pull-down or tablet menu items

Lengthy and sophisticated AutoLISP routines are most often stored as AutoLISP files (with the LSP extension), while shorter and simpler routines are typically stored as menu items.

Saving Routines as Files

AutoLISP code that has been stored in an LSP file can subsequently be loaded and invoked at any time. The programs executed in Chapter 57 are examples. Let's store the above functions to illustrate this capability.

1. Minimize AutoCAD.

2. Start Notepad and store the text in Fig. 58-1 exactly as you see it. Name the file **first.lsp**, save it, exit Notepad, and return to AutoCAD.

NOTE:

This file is also available on the Instructor's Resource CD-ROM as first.lsp.

```
(defun c:FIRST ()
(setq a 10)
(setq b (- 20 4))
(setq step1 (- b a))
(setq step2 (* (* 2 base1)(/ a 5)))
)
```

Fig. 58-1

Notice the AutoLISP function defun. The *defun* function allows you to define a new function or AutoCAD command and to invoke the AutoLISP file. Once loaded, the above program can be invoked by entering FIRST at the Command prompt.

NOTE:

Notice the inclusion of the right parenthesis at the program's last line. This right parenthesis evens the number of left and right parentheses. And, in conjunction with the left parenthesis in front of defun, the right parenthesis encloses the defun function.

3. At the Command prompt, enter **(load "first")** or use the **Load/Unload Applications** dialog box to load first.lsp.

HINT:

If the first.lsp file is located in a folder other than the default AutoCAD folder, specify the path when entering the load function. For example, enter (load "user/first") where user is the name of a folder. Notice the slash mark (not a backslash). An alternative to this approach is to open a drawing from the directory that contains first.lsp. AutoCAD will then find the file.

4. Enter **FIRST**, now a new AutoCAD command, to invoke the program. The number 24 appears.

NOTE: ────────────────────────────────

The preceding code can also be stored without the defun function. Also, setq can be entered just once. It would look like this:

```
(setq
a 10
b (− 20 4)
step1 (− b a)
step2 (* (* 2 step1)(/ a 5))
)
```

Entering (load "first") would then load and invoke the routine in one step.

Storing Routines as Menu Items

Let's store a similar AutoLISP routine as an item in a pull-down menu.

1. Minimize AutoCAD.

2. Make a copy of pullpart.mnu. Name it **pullprt2.mnu**.

NOTE: ────────────────────────────────

The pullprt2.mnu file is available on the Instructor's Resource CD-ROM.

3. Using Notepad, open **pullprt2.mnu**.

4. Edit the file so that it matches the text in Fig. 58-2 exactly.

5. Save the **pullprt2.mnu** file and exit Notepad.

6. Maximize AutoCAD.

7. Load the partial menu file named pullprt2.mnu. Place it between the Insert and Format pull-down menus.

```
***MENUGROUP=PULLPRT2

***POP1
[Construct]
[Line]^C^CLINE
[Circle]^C^CCIRCLE
[Arc]^C^CARC
[--]
[Pick This]^C^C(setq a 10) (setq b (- 20 4)) (- b a)
[--]
[->Display]
        [Pan]'PAN
        [->Zoom]
                [Window]^C^CZOOM W
                [Previous]^C^CZOOM P
                [All]^C^CZOOM A
                [<-Extents]^C^CZOOM E
```

Fig. 58-2

HINT:

Pick Customize Menu... from the Tools pull-down menu. Follow the directions on pages 747-748 for loading and using a partial menu.

8. After the Construct pull-down menu appears, select **Pick This** from it.

The AutoLISP code should return 6.

9. From the **Menu Customization** dialog box, unload **PULLPRT2**.

The Construct pull-down menu disappears from the menu bar.

10. Exit AutoCAD without saving.

Chapter 58
Review & Activities

Review Questions

1. What will be returned if you enter (* 4 5) at the Command prompt?

2. Explain the purpose of the setq function.

3. If the value of variable XYZ is 129.5, what will be returned when you enter !XYZ?

4. What purpose do the car and cadr functions serve?

5. What is the purpose of the list function?

6. How do you load and invoke an AutoLISP file named red.lsp that contains (defun c:RED ()?

7. Explain the relative benefits of saving an AutoLISP function as a file and as a menu item.

Challenge Your Thinking

1. You may have noticed that you can use many of the AutoLISP programs provided with AutoCAD without first loading them. AutoCAD provides a way to load AutoLISP functions automatically. Find out how to do this. Then back up the necessary AutoCAD files and experiment. See if you can get first.lsp, which you created in this chapter, to load automatically.

2. Describe a way to add first.lsp permanently to the menu so that it appears automatically each time you start AutoCAD.

Applying AutoCAD Skills

Work the following problems to practice the commands and skills you learned in this chapter.

1. Create an AutoLISP function that assigns a specific coordinate point to variable centerpoint and a specific value to variable radius. Then use these variables to create a circle. Change the values of centerpoint and radius and create another circle. Does the second circle reflect the changes you made?

Chapter 58
Review & Activities

2. Assign the values indicated to the variables listed in Table 58-1.

   Perform the following operations in AutoCAD using AutoLISP. Record the keystrokes you entered to perform the operations as well as your answers.

   a. Subtract bolt from screw.

   b. Add bolt, screw, nut, and washer.

   c. Subtract washer from the product of screw and bolt.

   d. Divide washer by nut.

Variable	Value
bolt	52
screw	174
nut	89
washer	248

Table 58-1

3. Using the same variables you assigned in the previous problem, create a menu item that sets the variables to the values indicated. Record your keystrokes. Add the menu item to pullprt2.mnu and save the file as test.mnu. Load test.mnu and confirm that it works as expected.

4. Since AutoLISP can return the immediate results of calculations, it can be used as a calculator while creating an AutoCAD drawing. You are making the framing drawings of a house. Use AutoLISP to determine how many floor joists you need for a 51' ranch house if each joist is 16" on center. The 2" × 6" studs for the outside walls of your 51' × 26' 6" ranch house are 2' on center. Each corner uses three. Neglecting doors and windows and extra studs for joining interior walls, how many studs do you need for the outside walls of the entire house?

 ## Using Problem-Solving Skills

Complete the following activities using problem-solving skills and your knowledge of AutoCAD.

1. You are designing a triangular space of an office building. The area, A, of the space is 138 square feet and the base, B, of the triangular area is 12'. Use AutoLISP to determine the height, H, of the triangle. The formula is A = 1/2 B * H.

2. The building from problem 1 has numerous triangular areas of different sizes. Create an AutoLISP routine that performs this calculation, and save the routine for future use.

Advanced AutoLISP
Programming

Objectives:

- **Create a routine to implement a collection of AutoLISP functions**
- **Develop an AutoLISP program that creates a drawing border**
- **Use a parametric program to create doors and windows for architectural elevation drawings**

Key Terms

command function

documentation

parametrics

This chapter picks up where the last one left off. You will apply methods of AutoLISP programming that you've already learned, as well as new methods. As part of this work, you will learn how to use AutoLISP to create a border for a drawing. Also, you will use a parametric program created with AutoLISP to create doors and windows for architectural elevation drawings.

Creating a Border

Let's apply several AutoLISP functions to create a border for a drawing. Initially, we'll step through the process. Later, we'll store the function as a routine and enter it at the Command prompt.

Using AutoLISP Functions to Define the Border

We'll use the setq, list, car, and cadr functions to define a rectangular border for a 36″ × 24″ sheet. In order to place the border 1″ from the outer edge of the sheet, we'll define a drawing area of 34″ × 22″.

1. Start AutoCAD and begin a new drawing from scratch.

Let's begin by assigning a variable to each corner of the border. We'll call the lower left corner LL, the lower right corner variable LR, and so on.

2. Enter **(setq LL (list 0 0))**.

This takes care of the lower left corner.

3. For the upper right corner, enter (**setq UR (list 34 22)**).

Now let's use car and cadr for the remaining two corners.

4. Enter (**setq LR (list (car UR)(cadr LL)))**.

5. Enter (**setq UL (list (car LL)(cadr UR)))**.

Let's try out the new variables.

6. One at a time, enter **!LL**, then **!UR**, then **!LR**, and last **!UL**.

The correct coordinates for each corner appear.

Using the LIMITS and LINE commands, and the above variables, let's establish the new border format.

7. Enter the **LIMITS** command.

8. In reply to Specify lower left corner, enter **!LL**.

9. In reply to Specify upper right corner, enter **!UR**.

10. **ZOOM All**.

11. Set the grid to **1** unit.

12. Enter the **LINE** command and enter **!LL** for the first point.

13. For the second point, enter **!LR**; for the third point, **!UR**; for the fourth point, **!UL**; and then close by entering **C**.

14. Enter **ZOOM** and **.9x** so that you can more easily see the border.

Creating a Border Routine

This is all very interesting, but it took a lot of steps. Let's combine all the steps into a single routine and store it.

1. Minimize AutoCAD.

2. Using Notepad, create and store the text shown in Fig. 59-1 in a file named **bord.lsp**.

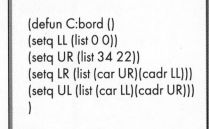

```
(defun C:bord ()
(setq LL (list 0 0))
(setq UR (list 34 22))
(setq LR (list (car UR)(cadr LL)))
(setq UL (list (car LL)(cadr UR)))
)
```

Fig. 59-1

NOTE: ─────────────────────────────

The bord.lsp file is also available on the Instructor's Resource Guide CD-ROM.

3. After saving the file and exiting Notepad, maximize AutoCAD.

4. Begin a new drawing from scratch.

5. Pick **Load Application...** from the **Tools** pull-down menu.

6. Locate and load the file named **bord.lsp** and close the dialog box.

7. At the Command prompt, enter **bord** in upper- or lowercase letters.

This executes the file, although it doesn't do much, at least not yet. The AutoLISP command function is needed to make the file useful.

Using the Command Function

The AutoLISP *command function* executes standard AutoCAD commands from within AutoLISP.

1. Using Notepad, open **bord.lsp**.

2. Using **Save As...**, create a new file named **border.lsp** and change its contents so they are identical to the text shown in Fig. 59-2.

 NOTE: —————————————————————————————

The border.lsp file is also available on the Instructor's Resource Guide CD-ROM.

```
(defun C:border ()
(setq LL (list 0 0))
(setq UR (list 34 22))
(setq LR (list (car UR)(cadr LL)))
(setq UL (list (car LL)(cadr UR)))

(command "LIMITS" LL UR)
(command "ZOOM" "A")
(command "GRID" "1")
(command "LINE" LL LR UR UL LL "")
)
```

Fig. 59-2

AutoCAD commands and command options are enclosed by double quotes ("). The two consecutive double quotes ("") are equivalent to pressing the spacebar.

3. In AutoCAD, load the **border.lsp** file and enter **BORDER** at the Command prompt.

AutoCAD automatically establishes a new drawing area, executes a ZOOM All to show the entire drawing area, sets the grid at 1, and creates a border.

4. Enter **ZOOM** and **.9x** so that you can more easily see the border.

Documenting the Routine

As in all programming, it's good practice to provide explanatory remarks in AutoLISP files. The explanation can be helpful to you and others that use the program. Possibly even more importantly, a well-documented file is easier to change and update later if necessary.

You can provide remarks, or *documentation,* in an AutoLISP file by preceding your comments with a semicolon (;). AutoLISP ignores the text on each line that occurs after a semicolon, so you can include documentation on a line-by-line basis.

Good programming practice also includes making the program as easy to read as possible. One way to achieve this is to use indentation to show which lines of code belong together. In the following routine, for example, you can see at a glance which lines are part of the 34x22 function. The indentation is invisible to AutoLISP, so it makes no difference to the routine.

1. Using Notepad, store the contents of border.lsp in a new file named **34x22.lsp**.

2. Edit the file to include the changes shown in Fig. 59-3. (page 814).

 NOTE: ————————————————

The 34x22.lsp file is also available on the Instructor's Resource CD-ROM.

```
;  This routine establishes a drawing area
;  for a 34" x 22" format (36" x 24" sheet size)
;  and draws a border line.
;

(defun C:34x22 ()
(setq LL (list 0 0))
(setq UR (list 34 22))
(setq LR (list (car UR)(cadr LL)))
(setq UL (list (car LL)(cadr UR)))

        (command "LIMITS" LL UR)         ; sets drawing limits
        (command "ZOOM" "A"); zooms all
        (command "GRID" "1")    ; sets grid
        (command "LINE" LL LR UR UL LL "")
        )
```

Fig. 59-3

3. In AutoCAD, begin a new drawing from scratch.

4. Load the **34x22.lsp** file and enter **34x22** at the Command prompt.

Parametric Programming

You can use AutoLISP to parameterize objects and drawings. This concept, called *parametrics,* enables you to produce unlimited variations of frequently created objects that share the same basic geometry. An example is a program that produces door and window variations, as shown in Fig. 59-4.

With a parametric program such as this, you are not required to insert shapes from a library of drawing files or blocks. Instead, you let the program do the work for you. The doors and windows in Fig. 59-4 were created using a program named dwelev.lsp, written by Bruce Chase. The program code is printed in Fig. 59-5 (pages 816-817).

Let's create and use the dwelev.lsp routine.

1. Using Notepad, accurately enter this AutoLISP routine. Name it **dwelev.lsp**.

Fig. 59-4

NOTE:

The dwelev.lsp file is also available on the Instructor's Resource CD-ROM.

2. Close the open drawings without saving any of them.

3. Begin a new drawing named **dw.dwg** and pick **Use a Wizard**, **Quick Setup**, and **OK**.

4. Prepare an architectural working environment based on a scale of ¼″ = 1′ and a C-size (**24″ × 18″**) sheet.

HINT:

Determine the drawing area by multiplying 4 by 24 and 4 by 18. The drawing area, therefore, is 96′ × 72′.

5. **ZOOM All** and set snap at **2′**.

6. Check the status bar. If the ORTHO, POLAR, and OSNAP buttons are depressed, pick them now to turn them off.

```
;Simple parametric DOOR/WINDOW ELEVATION drawing program

; Copywritten by Bruce R Chase, Chase Systems.
; May be copied for non-commercial use.


(setq hpi (* pi 0.5))
(defun d_we1 (p1 x y off / tp)    ;draws the rectang & offsets
  (command "pline" p1 "w" 0.0 0.0
    (setq tp (polar p1 angl x))
    (setq tp (polar tp (+ angl hpi) y))
         (polar tp (- angl pi)  x)
    "cl"
  )
  (setq e (entlast))
  (if off (command "offset" "t"
        (cons (entlast)(list p1))
        (polar (polar p1 angl off)(+ angl hpi) off) ""))
  (setq ee (entlast))
)


(defun d_we2 (spt offbase offside offtop x y numx numy sx sy offin trim /
        p1 p2 p3 p4 xx yy e ee)
  (d_we1 spt x y (if trim (* -1 trim) nil))      ; base d/w w\trim
  (if (and numx numy)(progn                 ; set base of panels
    (setq p1 (polar
            (if offside (polar spt angl offside) spt)
            (+ angl hpi)
            (if offbase offbase 0.0)))
   (d_we1 p1                           ; build panels
    (setq xx (if numy (/ (- x offside offside (* (- numy 1) sy)) numy) x))
    (setq yy (if numy (/ (- y offtop  offbase (* (- numx 1) sx)) numx) y))
    (if offin offin nil)               ; raised panel or glass trim
  )
  (command "array" e ee "" "R" numx numy p1    ; array the base panel
     (polar (polar p1 angl (+ sy xx))
            (+ angl hpi)(+ sx yy)))
))
)


(defun drwdr2 (spt xx x / tp)                ; getdist or default program
  (terpri)(terpri)
```

Fig. 59-5

Page 816

```
    (prompt (strcat xx " <"))
    (princ (rtos x (getvar "lunits")(getvar "luprec")))
    (if (Null (setq tp (if spt (getdist spt ">: ")(getdist ">: ")))) x tp)
)
(defun d_we3 ()
    (while (null (setq spt (getpoint "\n \nLower left corner of door/window: "))))
    (setq angl (if (null
        (setq tp (getorient spt "\nBase angle of door/window <0.0>: ")))
        0.0 tp))
)

(defun c:dwelev ( / spt offbase offside offtop        ; gather all the info
            x y numx numy sx sy offin trim tp angl)
    (d_we3)
    (setq x (drwdr2 spt "Width of door/window" 36.0))
    (setq y (drwdr2 spt "Height of door/window" 80.0))
    (setq trim (if (zerop (setq tp (drwdr2 spt "Trim width" 0.0))) nil tp))

    (if (setq numy (getint "\nNumber of panel rows <none>: "))(progn
        (setq numx (if (null (setq numx
            (getint "\nNumber of panel columns <1>: "))) 1 numx))
        (setq offbase (drwdr2 spt "Bottom rail width" 10.0))
        (setq offtop  (drwdr2 spt "Top rail width"   6.0))
        (setq offside (drwdr2 spt "Side rail width"   4.0))
        (if (> numx 1)(setq sx
            (drwdr2 nil "Spacing between panel rows" 1.0)))
        (if (> numy 1)(setq sy
            (drwdr2 nil "Spacing between panel columns"  1.0)))
        (setq offin   (if (zerop
                (setq tp (drwdr2 nil "Offset distance for raised panel" 0)))
                nil tp))
    ))
    (d_we2 spt offbase offside offtop x y numx numy sx sy offin trim)
(princ)
)
(prompt "\nCommand: DWELEV \n")


(c:dwelev)       ;call up program with this command
            ;or use within another procedure with actual sizes:

;(progn (d_we3)
; ——— spt offbase offside offtop x   y  numx numy sx  sy  offin trim—
;    (d_we2 spt 10.0   4.0   6.0   40.0 84.0 2   4  2.0 3.0 1.5   4.0)
;)
```

Fig. 59-5 (continued)

7. Load the **dwelev.lsp** file.

Loading the file automatically enters the new DWELEV command created by the program.

NOTE: ─────────────────────────────────

> If the routine does not appear to load, compare your program code to the dwelev.lsp code printed in this chapter. They must be identical.

The program is mostly self-explanatory because it employs easy-to-understand prompts.

8. Use the following as a guide as you enter your responses.

   Lower left corner of door/window: (Pick a point at any location.)
   Base angle of door/window <0.0>: (Press **ENTER**)
   Width of door/window <3'>: **4'**
   Height of door/window <6'8">: (Press **ENTER**)
   Trim width <0">: **1.5"**
   Number of panel rows <none>: **4**
   Number of panel columns <1>: **2**
   Bottom rail width <10">: **8"**
   Top rail width <6">: (Press **ENTER**)
   Side rail width <4">: **5"**
   Spacing between panel rows <1">: **2"**
   Spacing between panel columns <1">: **3"**
   Offset distance for raised panel <0">: **1"**

9. Zoom in on the door and examine it.

The door should look identical to the one in Fig. 59-6.

You can also create windows using dwelev.lsp. Just specify **0** in reply to Number of panel rows and Number of panel columns. To create the casement style windows shown earlier, link two or more windows together (using COPY or ARRAY).

10. Create additional doors and windows by entering **dwelev** at the Command prompt

11. Save your work.

Fig. 59-6

The dwelev.lsp routine is intentionally basic. This makes it easier for you to understand its operation. If you are an accomplished programmer, you may choose to embellish the routine by including doorknobs, window molding, and other details normally included in door and window symbology.

Other AutoLISP Functions

Table 59-1 provides a list of other commonly used AutoLISP functions.

1. Experiment with the functions shown in Table 59-1. Try to discover ways of including them into 34x22.lsp, dwelev.lsp, and other programs.

2. When you are finished, exit AutoCAD.

Function	Description
setvar	Sets an AutoCAD system variable to a given value and returns that value. The variable name must be enclosed in double quotes. *Example:* (setvar "CHAMFERA" 1.5) sets the first chamfer distance to 1.5.
getvar	Retrieves the value of an AutoCAD system variable. The variable name must be enclosed in double quotes. *Example:* (getvar "CHAMFERA") returns 1.5, assuming the first chamfer distance specified most recently was 1.5.
getpoint	Pauses for user input of a point. You may specify a point by pointing or by typing a coordinate in the current units format. *Example:* (setq xyz (getpoint "Where? "))
getreal	Pauses for user input of a real number. *Example:* (setq sf (getreal "Scale factor: "))
getdist	Pauses for user input of a distance. You may specify a distance by typing a number in AutoCAD's current units format, or you may enter the distance by pointing to two locations on the screen. *Example:* (setq dist (getdist "How far? "))
getstring	Pauses for user input of a string. *Example:* (setq str (getstring "Your name? "))

Table 59-1

Chapter 59
Review & Activities

Review Questions

1. Why is it often more efficient to store AutoLISP routines than to use a collection of AutoLISP functions directly at the Command prompt?

2. Explain the purpose of the AutoLISP command function.

3. For what reason are semicolons used in AutoLISP routines?

4. Briefly explain the purpose of the following functions:

   getvar

   getpoint

   getdist

5. Explain the benefits of applying parametric programming techniques.

Challenge Your Thinking

1. Explain what makes an application a good choice for parametric programming. Give examples of parametric applications you think could or should be accomplished within AutoCAD using AutoLISP.

2. In this chapter, you created an AutoLISP routine that automatically places a border on a 36" × 24" sheet. Explain how you could revise this routine to create a parametric AutoLISP program that allows you to draw a border for any size sheet of paper.

 ## Applying AutoCAD Skills

Work the following problems to practice the commands and skills you learned in this chapter.

1. Write the AutoLISP code you would use to create a 4 × 6 unit rectangle on the screen with corners at points pt1, pt2, pt3, and pt4. Enter the code and confirm that it works as expected.

2. Using dwelev.lsp and AutoCAD commands, create an architectural elevation drawing similar to the one shown in Fig. 59-7. This elevation, minus doors and windows, is available in the elev.dwg file on the Instructor's Resource CD-ROM.

Fig. 59-7

3. Modify the border you created for a 34″ × 24″ sheet to work with a 12″ × 9″ A-size sheet.

4. The AutoLISP program shown in Fig. 59-8 (page 822), is titled v1. The v1.lsp program does the following:

   - opens two viewports with horizontal orientation
   - makes the upper one the top view and the lower one the front view
   - zooms both to 80% of the current view

   The program shown in Fig. 59-9, titled v2, resets to a single viewport with an 80% zoom.

   Write a program named v3 that will create three viewports, with top, front, and isometric orientations. Write another program, named v4, that creates four viewports with top, front, right-side, and isometric view orientations. Create a three-dimensional object to check your programs. They should generate viewports like the ones shown in Figs. 59-10 and 59-11 (also on page 822). The v1.lsp and v2.lsp files are available on the Instructor's Resource CD-ROM.

Chapter 59
Review & Activities continued

```
(setvar "CMDECHO" 0)
(defun c:V1 ()
        (command "VPORTS" "2" "H")
        (setq cvpt (getvar "CVPORT"))
        (IF (= (getvar "CVPORT") 2) (setvar "CVPORT"
3) (setvar "CVPORT" 2))
        (command "VPOINT" "0,-1,0")
        (command "ZOOM" ".8x")
        (setvar "CVPORT" cvpt)
        (command "PLAN" "W")
        (command "ZOOM" ".8x")
        (princ)

)
                                    Program v1
```

Fig. 59-8

```
(setvar "CMDECHO" 0)
(defun c:V2()
        (command "VPORTS" "SI")
        (command "ZOOM" "E")
        (command "ZOOM" ".8x")
        (princ)
)
                                    Program v2
```

Fig. 59-9

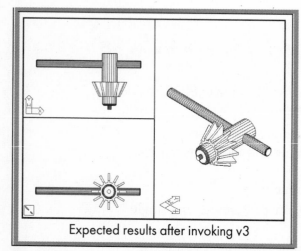

Expected results after invoking v3

Fig. 59-10

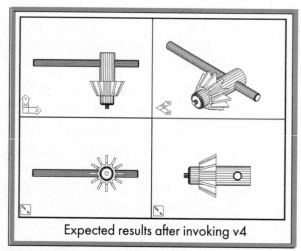

Expected results after invoking v4

Fig. 59-11

Problem 4 courtesy of Gary J. Hordemann, Gonzaga University

Chapter 59
Review & Activities

 Using Problem-Solving Skills

Complete the following activities using problem-solving skills and your knowledge of AutoCAD.

1. On your own, learn other AutoLISP functions, such as getreal and getdist, and develop them into new AutoLISP routines.

2. Try to create the parametric program discussed in the second "Challenge Your Thinking" question. Refer to the dwelev.lsp program and the documentation provided with AutoCAD if you need help.

   Do not be discouraged if your program does not work the first time you try to run it. Most programs, even those written by professional programmers, have to be "debugged" before they will run properly.

Part 11 Project

Parametric Border Utility

A programming language describes the terms, abbreviations, and syntax that programmers use to create, modify, improve, and debug software. There are many such languages (and more are being developed as need and opportunity arise), and they differ quite a bit in function and appearance. Programmers often specialize in one or more languages to become more proficient at their craft. In a similar vein, being well versed in one language is no guarantee of being able to use another.

Just as one need not be a mechanic to drive a car, a software user has no obligation to view source code or programming routines in order to use the program. There is, however, a decided advantage for the drafter who is capable of some programming expertise. It allows the user to customize existing functions and create new ones. It permits the basic software program to be extended into a more versatile and useful one.

Description

The border.lsp file you created in Chapter 59 is handy for making a 34" × 22" border around your drawings. Not all of your drawings are likely to be that size, however, and it can be a chore to change the routine.

1. In this project, your task is to create an AutoLISP program that provides user input so that you can enter any *x,y* coordinate locations as drawing limits.

2. Modify the program you have created to add your name and the current date.

3. Fully document the program.

Hints and Suggestions

- Use the getreal function, and draw the border with the RECTANGLE command.

- Refer to the programs in Chapters 57 through 59, as well as those provided with AutoCAD, for reference if you need help.

Part 11 Project

Summary Questions/Self-Evaluation

1. Why might someone using AutoCAD benefit from learning to program in AutoLISP?

2. What factors influence the number of programming languages that are available?

3. Why might it benefit a programmer to also be a user of the software he or she has created?

4. What happens if the number of left and right parentheses does not match in an AutoLISP routine? If the numbers don't match, can you simply add an additional parenthesis anywhere in the program, or must you place it in a precise location? (Experiment if necessary to answer this question.)

5. Why do you suppose that very few programming routines work perfectly the first time?

Careers
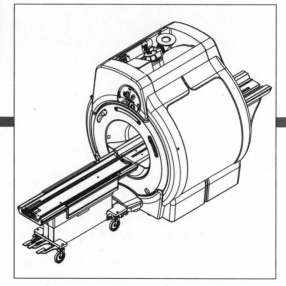
Using
AutoCAD

Planning Medical Installations

For a homeowner buying a washer and dryer, it's wise to take measurements *before* the machines are delivered. The same principle applies when hospitals purchase state-of-the-art diagnostic machines. A magnetic resonance imaging (MRI) system, such as the one shown in the illustration, puts a multi-ton load on floor joints and could fill a small room wall to wall. GE Medical Systems, manufacturer of cardiological and vascular diagnostic equipment, uses AutoCAD to put together all the information needed for installation.

AutoCAD, as customized by GE, has completely replaced hand-drawn field sketches as the way to communicate site information to the home office. This technology increases drawing accuracy and quality with significant speed and cycle-time improvements. An installation specialist with a laptop computer creates a site-specific floor plan in AutoCAD, showing the desired locations of equipment pieces. The AutoCAD file, attached to an e-mail message, goes to a designer at the home office.

Designer Timothy Quinn examines the drawing to determine whether the equipment can be installed without modifications to the room. If the answer is yes, Quinn generates a floor plan for installers, showing positions for machine components and electrical hookups. In some cases, Quinn is able to find a good fit by moving components around within the drawing.

When the site requires structural or electrical modifications, Quinn adds information needed for construction drawings to the AutoCAD file and sends it to the contractor who will do the work. Along with the file, he sends a time-saving LISP routine developed at GE Medical Systems. For example, the LISP routine for the contractor selects drawing layers with structure and electrical information but ignores the layer on equipment positioning.

Career Focus—Architectural Designer

In this application, the designer develops plans for the installation of medical equipment. The designer receives information about the floor plan, wiring, and structure of an installation site and generates drawings for installers and contractors.

Most of the designer's work is done in an office at a computer workstation. However, the designer may travel to an installation site or to meetings with sales and installation staff or with contractors.

Education and Training

Timothy Quinn has a degree in architecture. Others working in his office in related jobs have two-year or four-year degrees in architecture or in structural or electrical engineering.

 ## Career Activity

- Visit a Web site for a major corporation for information about employment.
- What kinds of jobs are most in demand?
- What education and experience are they looking for?
- What salaries do they list for different jobs?

Part 12

Importing and Exporting

Chapter 60 Digitizing Hard-Copy Drawings

Objectives:

- **Calibrate a drawing tablet in preparation for digitizing a drawing**
- **Digitize a hard-copy drawing into AutoCAD**

Key Terms

digital
fixed screen pointing area
tablet mode

In chapters 54 and 55, you digitized points to define and configure tablet menu overlays. This chapter steps through the process of digitizing an entire drawing, thereby converting it from hard copy into an electronic, or digital, format. *Digital* is a data storage format that uses a two-digit binary system. Note that you must have a digitizing tablet connected to your AutoCAD system in order to complete this chapter.

There will be times, especially in a business environment, when you'll wish your hand-completed drawings were stored in AutoCAD. Suppose your firm has recently implemented CAD. All of your previous drawings were completed by hand, and you need to revise one or more of them. As you know, it can be time-consuming to redraw them by hand. Fortunately, most CAD systems, including AutoCAD, offer a method of transferring those drawings to digital form. It is not always practical to digitize drawings, but it is often faster than recreating the drawings.

Since you may not have easy access to a simple drawing not yet in AutoCAD, let's digitize the deck.dwg drawing that you created in Chapter 26. This drawing is shown in Fig. 60-1.

1. Start AutoCAD and open **deck.dwg**.

2. Plot the drawing limits at **1 = 24**.

3. Using **Save As...**, create a new drawing file named **digit.dwg**. Erase the entire drawing so that the screen is blank.

 NOTE: ─────────────────────────────

Be sure snap is set at 6″ and is turned on. Also be sure to display the entire drawing area by entering ZOOM All.

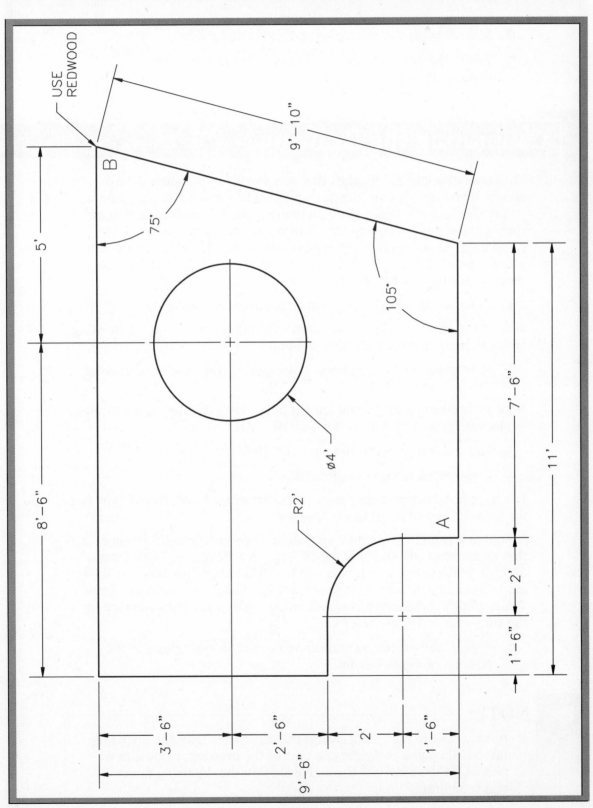

Fig. 60-1

4. Make **Objects** the current layer, if it is not already.

5. Fasten the hard copy of deck.dwg onto the center of the digitizing tablet.

Calibrating the Drawing

The calibration process requires that you specify a minimum of two points. If you specify two points, they should form a rectangle, and the points should not be close to one another. Also, it is important that you specify the lower left coordinates before you specify the upper right coordinates. If you do it in the reverse order, the digitizing process can become unreliable. For large, complex drawings, consider as many as three to five calibration points.

1. Enter the **TABLET** command and the **CALibrate** option.

We need to identify at least two known (absolute) points on the drawing. Let's call point B of the deck absolute point 7'4'.

2. In response to Digitize point #1, pick point A (be precise) and enter the coordinates **7',4'**.

Now we need to pick a second known point. Let's choose point B located at absolute point 17',13'6" to the right of the first point.

3. Pick point B precisely and enter **17',13'6"** for the coordinates.

4. Press **ENTER** to end the calibration.

You have just calibrated the tablet to the drawing. AutoCAD can now size the drawing according to this calibration.

Note that a TABLET button now appears in the status bar. This means that *tablet mode* (the mode in which AutoCAD recognizes input from a digitizing tablet) is on, as it needs to be for the digitizing process. Tablet mode should be off whenever you select buttons from toolbars or menus. Tablet mode can be toggled on and off by clicking the TABLET button in the status bar or by pressing F4.

5. Toggle tablet mode on and off and notice the difference in the position of the crosshairs.

 NOTE:

It is possible to digitize drawings that are larger than the digitizing tablet. When you complete one part of the drawing, move a new part of the drawing into the active area on the digitizing tablet and recalibrate.

Digitizing the Drawing

Let's begin to digitize the drawing, starting at the upper left corner of the drawing.

1. Enter the **LINE** command, and make certain that snap is on.

2. Turn on the tablet mode and digitize (pick) the upper left corner of the deck. (Read the following hint.)

 HINT:

> If you encounter difficulty in reaching the upper left corner or any other portion of the screen with the pointing device, it may be that the *fixed screen pointing area* (the area in which AutoCAD recognizes digitized points) is too small. If this is the case, enter the TABLET command and the Cfg option, specify 0 tablet menus, and enlarge the screen pointing area.

Whenever you digitize points from the drawing, tablet mode must be turned on. Whenever you select buttons or pull-down menus, tablet mode must be turned off.

3. With tablet mode on, digitize the next corner point, working clockwise.

This completes the first line segment.

4. Digitize the next two points.

5. When digitizing the next point, ignore the fillet and pick the approximate location of the corner. (With snap on, your selection of the corner will be accurate. You will insert the fillet later.)

6. Continue around the object until you close the polygon, and save your work.

7. Using the **FILLET** command, specify a fillet radius of **2'** and place the fillet at the proper location. Remember to turn tablet mode off if you want to use the buttons or pull-down menus.

8. Use the **CIRCLE** command to place the hole. Be sure tablet mode is on and digitize the center and radius from the hard-copy drawing.

9. Make **DIM** the current layer.

10. Using AutoCAD's dimensioning commands, dimension the object.

11. Save your work and exit AutoCAD.

Chapter 60
Review & Activities

Review Questions

1. What command and option are used to calibrate a drawing to be digitized?

2. Why is the calibration process necessary?

3. Briefly explain the process of calibrating a drawing to be digitized.

4. Why is the snap resolution important when digitizing?

5. Explain when the tablet mode should be turned on and when it should be turned off.

Challenge Your Thinking

1. If you have a mouse, explain whether you can digitize a drawing using only a conventional mouse.

2. Digitizing can make the task of entering drawings into AutoCAD a much simpler process. However, the digitizing process is not perfect. Discuss any problems you encountered as you digitized deck.dwg in this chapter. What other problems do you think might occur? List the problems and describe ways to minimize each.

Applying AutoCAD Skills

Work the following problems to practice the commands and skills you learned in this chapter.

1. Obtain two hand-completed drawings and digitize each. Save the drawing files as ch60-1A.dwg and ch60-1B.dwg.

2. Your architectural firm has finally moved to computers. All of the vellum drawings need to be placed into the computer. Calibrate the floor plan shown in Fig. 60-2 and digitize it. Save the drawing as ch60floor.dwg.

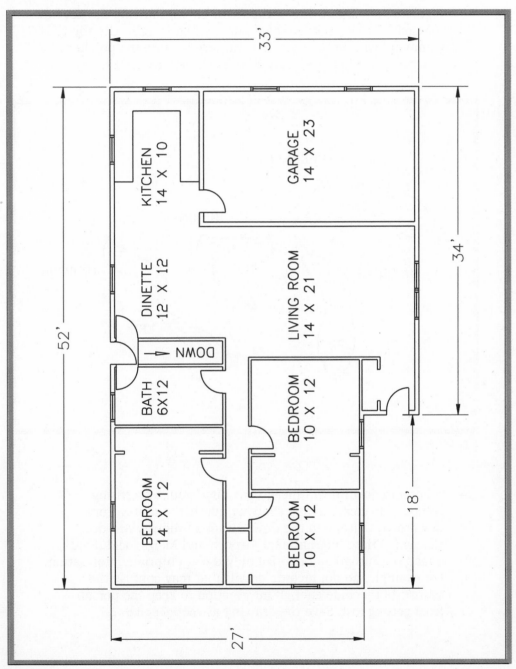

Fig. 60-2

3. Figure 60-3 is a drawing of the Great Lakes, showing the net flow for each lake. Reproduce this figure by digitizing it. Use the SKETCH command with SKPOLY set to 1. Be sure to close the polylines in order to make hatching easy. The hatch pattern is FLEX.

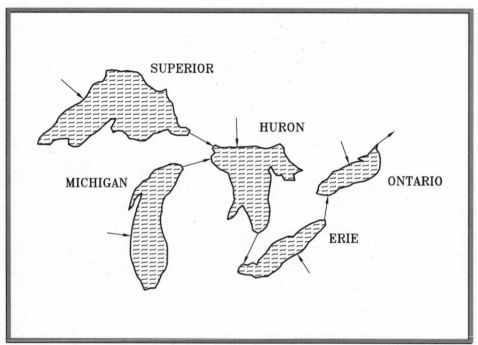

SUPERIOR

HURON

MICHIGAN

ONTARIO

ERIE

Fig. 60-3

4. A hotel in downtown Chicago has hired your advertising company to create a map showing guests the routes from downtown Chicago to the three airports: Midway Airport, Chicago-O'Hare International Airport, and Meigs Field. Find a map of Chicago and digitize only the appropriate information. For example, do not include details that may confuse the tourist, but include enough information to keep the person from getting lost. Save the drawing as ch60airport.dwg.

Problem 3 courtesy of Gary J. Hordemann, Gonzaga University

Chapter 60
Review & Activities

Using Problem-Solving Skills

Complete the following activities using problem-solving skills and your knowledge of AutoCAD.

1. An assembly fixture was damaged, and the manufacturer sent you an isometric drawing (Fig. 60-4). You need to enter it into your computer system before you create an orthographic drawing. Calibrate the drawing and digitize it. Save the drawing.

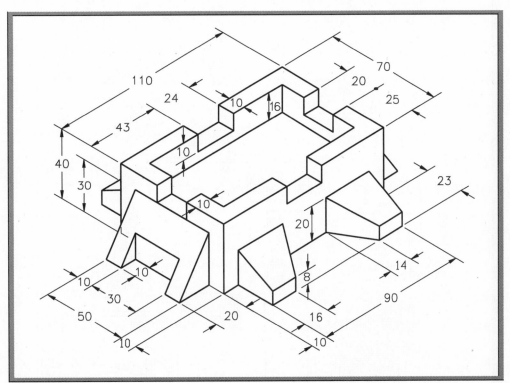

Fig. 60-4

2. Use an enlarged copy of the map in Fig. 60-5 (page 836) to practice digitizing. Use the PLINE command to digitize the streets. Use the LINE command to digitize the heavy dashed lines, which represent buried conduit. Use the SKETCH command to digitize the lake and the river. Determine the compass bearing of one of the streets by changing to surveyor's units and digitizing the north arrow. Use the DONUT command to digitize the small dots (street lights).

Problem 2 courtesy of Gary J. Hordemann, Gonzaga University

Chapter 60
Review & Activities continued

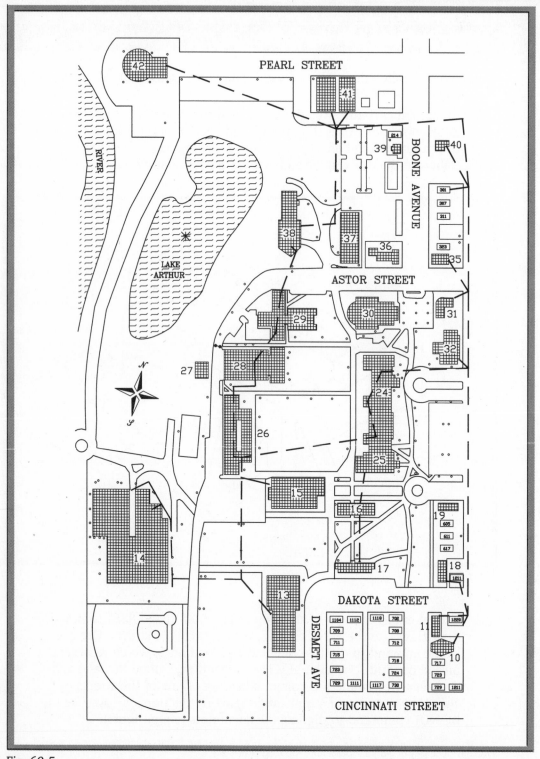

Fig. 60-5

Chapter 61 Standard File Formats

Objectives:

- Import and export **DXF, EPS, 3DS,** and **SAT (ACIS)** files
- Describe the differences between ASCII and binary **DXF** files
- Explain the purpose of **IGES** files

Key Terms

ACIS
drawing interchange format (DXF)
Encapsulated PostScript (EPS)
Initial Graphics Exchange Specification
SAT files
3D Studio

Standard file formats are important for moving graphics and other information from one system or program to another. This chapter steps through the translation process using several industry standard formats.

DXF Files

AutoCAD's *drawing interchange format (DXF)* is a *de facto* standard for translating files from one CAD program, such as AutoCAD, to another, such as CADKEY. You can translate DXF files to other DXF-compatible CAD programs or to programs for specialized applications. For example, certain manufacturing software uses DXF files to generate tool path code for computer numerical control (CNC) mills and lathes.

Creating an ASCII DXF File

You can easily generate a DXF file from an existing AutoCAD drawing.

1. Start AutoCAD and open **dimen.dwg**.

2. Select **Save As...** from the **File** pull-down menu.

3. Display the **Save as type** drop-down box and review the list of options.

Note that AutoCAD offers four variations of DXF.

4. Select **AutoCAD 2000 DXF (*.dxf)**.

5. Pick the **Options...** button.

This displays the Saveas Options dialog box, as shown in Fig. 61-1.

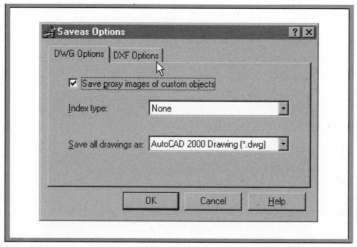

Fig. 61-1

6. Pick the **DXF Options** tab.

AutoCAD permits you to produce either an ASCII or binary file. You can read an ASCII DXF file with a text editor, such as Notepad. Also, ASCII DXF files are compatible with a wider range of applications. Binary DXF files contain all of the information of an ASCII DXF file but in a more compact form.

The Select Objects check box lets you select one or more objects to translate instead of the entire drawing. The Save thumbnail preview image check box specifies whether the Preview area of the Select File dialog box displays an image of the drawing.

AutoCAD gives you the option of entering decimal places of accuracy. The default value is adequate in most cases, although you may need to increase this value for certain applications. The disadvantage of a high number is increased file size.

7. Pick the **ASCII** radio button, the **Save thumbnail preview image** check box, and the **OK** button.

8. In the File name box, accept the suggested **dimen.dxf** name and pick the **Save** button.

The translation can take a while for complex drawings. When the Command prompt returns, the translation is complete.

9. Minimize AutoCAD.

10. Find the file named **dimen.dxf** and compare its size to that of its DWG counterpart.

As you can see, the DXF file is about twice as large.

11. Using WordPad, review the contents of the ASCII DXF file.

The beginning of the DXF text file should look similar to the text in Fig. 61-2.

12. Exit WordPad and maximize AutoCAD.

```
             0
         SECTION
             2
         HEADER
             9
        $ACADVER
             1
         AC1015
             9
     $ACADMAIN TVER
            70
             6
             9
       $DWGCODEPAGE
            31
         ANSI_1252
             9
         $INSBASE
            10
```

Fig. 61-2

Exporting a Binary DXF File

Binary DXF files are approximately 25% to 40% smaller than ASCII DXF files. Also, they can be written and read by AutoCAD much faster.

1. Select **Save As...** from the **File** pull-down menu.

2. Select **AutoCAD 2000 DXF (*.dxf)** from the **Save as type** drop-down box.

3. Pick the **Options...** button and the **DXF Options** tab.

4. Pick the **BINARY** radio button and pick **OK**.

5. In the File name box, change the name to **dim.dxf** and pick the **Save** button.

6. Compare the size of the ASCII DXF file with that of the binary DXF file.

The binary DXF file should be at least 25% smaller.

7. Return to AutoCAD.

Importing a DXF File

Both ASCII and binary DXF files can be imported into AutoCAD.

1. Pick the **Open** button from the docked **Standard** toolbar.

2. From the **Files of type** drop-down box, select **DXF (*.dxf)**.

3. Select a DXF file currently on disk. If one is not available, select one of the two DXF files you just created and pick the **Open** button.

 NOTE: _____

> The Instructor's Resource CD-ROM includes example DXF files. If you use these files, be sure to ZOOM All after importing each file.

The DXF file generates on the screen. At this point, you could save the drawing as a drawing (DWG) file, but don't.

4. Close both open files without saving.

PostScript Files

AutoCAD allows you to export and insert an *Encapsulated PostScript (EPS)* file type that is compatible with many graphics programs for desktop publishing and presentation applications.

Exporting an EPS File

1. Open the drawing named **lamp.dwg** and fill the screen with the lamp.

2. Select **Export...** from the **File** pull-down menu.

3. Display the **Save as type** drop-down box and select **Encapsulated PS (*.eps)**.

4. Pick the **Options...** button.

5. Pick the **Help** button and read about the options available in the dialog box.

6. After exiting help, pick the **OK** button to accept all of the default settings.

7. Pick the **Save** button to accept the lamp.eps name.

AutoCAD creates a file named lamp.eps.

Importing an EPS File

1. Close lamp.dwg and begin a new drawing from scratch.

2. Select **Encapsulated PostScript** from the **Insert** pull-down menu.

3. Unless you have another EPS file, select **lamp.eps** and pick **Open**.

4. Press **ENTER** to accept the **0,0,0** insertion point.

5. In reply to Scale factor, drag the crosshairs to the right until the drawing fills most of the drawing area and pick a point.

The PostScript image appears.

6. View it from different angles in space.

It is now a flat image instead of a 3D model because PostScript does not recognize 3D objects.

 NOTE:

AutoCAD offers the PSDRAG command to control the appearance of a PostScript image as you are dragging it into position. The PSQUALITY system variable is available if you want to adjust the rendering quality of PostScript images. The PSFILL command allows you to fill 2D polyline outlines using a PostScript fill pattern.

7. Close the current drawing file without saving.

3D Studio Files

AutoCAD creates a 3D Studio (3DS) file from selected objects. *3D Studio* is a 3D modeling and animation program produced by an Autodesk company named Kinetix.

Exporting a 3DS File

1. Open the **shaft.dwg** drawing file.

2. Select **Export...** from the **File** pull-down menu.

3. Display the **Save as type** drop-down box and select **3D Studio (*.3ds)**.

4. Pick the **Save** button to accept the **shaft.3ds** name.

5. Select the shaft model and press **ENTER**.

AutoCAD displays the 3D Studio Export Options dialog box, as shown in Fig. 61-3.

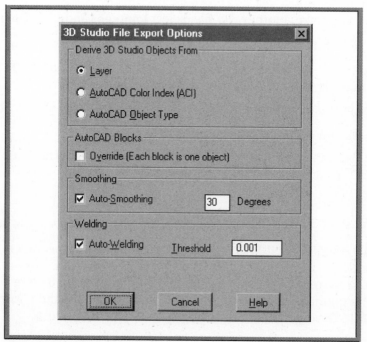

Fig. 61-3

6. Review the options and then pick the **Help** button to read about them.

7. Pick **OK** to accept the default settings.

AutoCAD creates a file named shaft.3ds. You can read this file into the 3D Studio software, as well as other programs that import 3DS files.

Importing a 3DS File

1. Close shaft.dwg without saving changes and begin a new drawing from scratch.

2. Select **3D Studio...** from the **Insert** pull-down menu.

3. Select **shaft.3ds** and pick **Open**.

NOTE: ─────────────────────────

The Instructor's Resource CD-ROM includes example 3DS files.

This displays the 3D Studio File Import Options dialog box, as shown in Fig. 61-4.

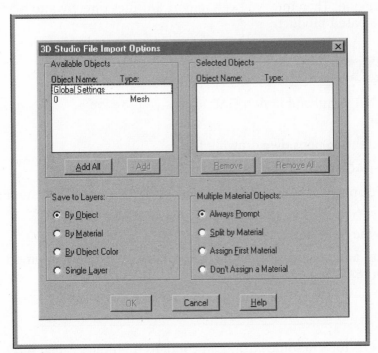

Fig. 61-4

4. Review the options in the dialog box and pick the **Help** button to read about them.

5. After exiting help, pick the **Add All** button to select all available objects, and pick the **OK** button.

AutoCAD imports the 3DS file. At this point, you could save this as a DWG file, but don't.

6. View the file from any angle in space and shade the drawing.

7. Close the file without saving.

ACIS Files

AutoCAD permits you to export objects representing NURBS surfaces, regions, and solids to an ACIS file in ASCII (SAT) format. *ACIS* is the 3D geometry engine that AutoCAD uses to create solid models.

SAT files are machine-independent geometry files. These files contain topological and surface information about models that were created using the ACIS engine. Therefore, you can import SAT files directly into many other ACIS-based products.

Exporting an SAT File

The ACISOUT command in AutoCAD allows you to save an ACIS file in SAT format.

1. Open the **compos.dwg** drawing file.

2. Enter **ACISOUT**, select the solid model, and press **ENTER**.

AutoCAD displays the Create ACIS File dialog box. Notice that the default file type is ACIS (*.sat).

3. Pick the **Save** button.

This creates a file named compos.sat. You can read this file into products that import SAT files.

 NOTE:

The Instructor's Resource CD-ROM contains a product called ACIS Viewer. The product enables you to view SAT files without the expense and learning curve of a CAD product. This free viewing software is particularly useful to companies that want to share proposed designs with people inside and outside the company.

Importing an SAT File

1. Close the compos.dwg file and start a new drawing from scratch.

2. Select **ACIS File...** from the **Insert** pull-down menu.

AutoCAD displays the Select ACIS File dialog box. Notice that the default file type is ACIS (*.sat).

3. Double-click **compos.sat**.

AutoCAD reads and displays the contents of compos.sat.

4. Produce an isometric view of the solid and smooth-shade it.

5. Close this file without saving and exit AutoCAD.

Converting AME Models

In some previous releases, AutoCAD used Advanced Modeling Extension (AME) to create solid models. People who have used older releases of AutoCAD may need to convert these AME files into ACIS files. AutoCAD provides the AMECONVERT command to convert AME Release 2 or 2.1 solid models to AutoCAD 2000's ACIS-based solids or regions.

NOTE:

No conversion is necessary to open an AME file using AutoCAD 2000. The file opens automatically, but it saves the AME solid model as a series of polyface meshes. To recreate the solid model, you must then use the AMECONVERT command.

The Instructor's Resource CD-ROM contains an AME solid model named t-conn.dwg. If you have access to this file, you may want to open it in AutoCAD 2000 and convert it using the AMECONVERT command.

IGES Files

IGES stands for *Initial Graphics Exchange Specification*. IGES is an industry standard approved by the American National Standards Institute (ANSI) for interchange of graphic files between CAD systems.

Translating from one CAD program to another using IGES is useful. However, each CAD program is unique. Consequently, certain characteristics, such as layers, blocks, linetypes, colors, text, and dimensions, are potential problem areas.

For example, some CAD systems use numbers for layer names and do not accept names such as Object or Dimension. If an AutoCAD drawing file is translated to a system using layer numbers, all of the AutoCAD layer names are changed to numbers. These types of problems are also present when you translate files using the DXF file format.

Translator software for importing and exporting IGES is available as an option from Autodesk, Inc.

Chapter 61
Review & Activities

Review Questions

1. What is a DXF file, and what is its purpose?

2. List 10 file formats that AutoCAD can export.

3. Explain the advantages of using binary DXF files over ASCII DXF files.

4. Why might you need to save a drawing in EPS format?

5. Can you store 3D data in a 3DS file? In an EPS file?

6. What file extension does AutoCAD use for a file you export using the ACISOUT command?

7. What is the purpose of the AMECONVERT command?

8. What is IGES, and what is the purpose of an IGES file?

9. What are the potential problem areas associated with translating DXF and IGES files from one CAD system to another?

Challenge Your Thinking

1. Experiment further with importing and exporting DXF files. What happens if you try to import a DXF file into a drawing that already has objects in it?

2. Experiment further with importing and exporting 3DS files. Try each of the options under Save to Layers in the 3D Studio File Import Options dialog box. What, if any, information is lost when you translate an AutoCAD file to a 3DS format and then translate it back to an AutoCAD (DWG) format?

Chapter 61
Review & Activities

Applying AutoCAD Skills

Work the following problems to practice the commands and skills you learned in this chapter.

1. Using deck.dwg, create an ASCII DXF file using the capabilities provided by AutoCAD. After the translation is complete, review the contents of the DXF file.

2. If you have access to another CAD program, load the DXF files from problem 1 into that system. (*Note:* The CAD system must be able to import a DXF file.) If you are successful, review characteristics of the drawings, such as layers, colors, linetypes, blocks, text, and dimensions. Note the differences between the CAD programs.

3. Create EPS and 3DS files from table.dwg.

4. Open ch46roller.dwg from Chapter 46 ("Using Problem-Solving Skills" problems 1 and 2). Export it as an Encapsulated PostScript file. Import the EPS file into AutoCAD. View it in the top view so that it appears as you originally imported it.

5. Create or obtain DXF files from another CAD program and import them into AutoCAD. Note the differences between AutoCAD and the programs used to create the files you imported into AutoCAD. If available, refer to the Instructor's Resource CD-ROM for examples.

6. Obtain or create EPS and 3DS files and import them into AutoCAD. If available, refer to the Instructor's Resource CD-ROM for examples.

Using Problem-Solving Skills

Complete the following problem using problem-solving skills and your knowledge of AutoCAD.

1. You are using 3D Studio to create an animated display of your products for a trade show. Export the solid model of the boat trailer roller you created in Chapter 46 as a 3D Studio file. Import it into AutoCAD, shade it, and view it from various angles. Save the model as an AutoCAD 2000 drawing file.

Chapter 61
Review & Activities continued

2. Create a solid model of the dovetail drawing you created as an isometric drawing in Chapter 35. (The multiview drawing is shown in
   Fig. 61-5 for reference.) Save the file as a DXF file in ASCII format
   for use with the CNC milling machine in your shop. Check it in
   Wordpad to make sure it translated. Import it into AutoCAD 2000.

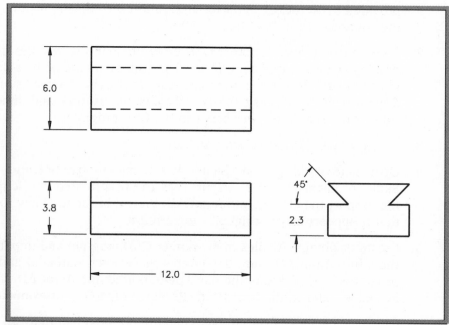

Fig. 61-5

Careers Using AutoCAD

Mountain Bike Design

Replacing steel crankset bolts with titanium bolts can take half an ounce off the weight of a mountain bike. Designers of today's sport and competition bicycles aim for that kind of micro-efficiency for one reason: although bicycles are a mature technology (the fundamentals are well understood), the pressure to keep improving is intense.

Jason Clark is an industrial designer for Syncros Applied Technology Inc., specializing in mountian-bike components such as crank arms and seat posts. For example, shown here is a titanium crankset, attached to gears for a mountain bike.

Jason's job is to think about the rider and the bike and the geometries that connect them, looking for new component shapes that might give the rider even a little extra push in a key situation. Clark experiments with design ideas using 3D drafting software: specifically, Autodesk Mechanical Desktop. As a drafting tool for industrial design, Mechanical Desktop is a next step for many AutoCAD users. CAD makes the design process more efficient, and efficiency is as important to Clark, designing bikes, as it is to riders, always looking for the slightest competitive edge.

Career Focus: Industrial Designer

Industrial designers work for manufacturers of all kinds of products, from cars to cosmetics to cash registers. A designer develops a general idea for a product into specific form that is represented in drawings or in a physical model. In addition to creative talents, designers rely on their knowledge of available materials, production methods, and ways that a product is likely to be used. Design goals vary with the product. For a sports car, the primary goals might be high performance and safety. For other products, the goal might be low cost or eye-catching style.

Industrial designers work mostly in offices with regular hours, but they may come under deadline pressure. Designers sometimes supervise craft workers, such as model makers. Designers may also be involved in work on product prototypes, helping to define cost-effective procedures for mass manufacture of the product.

Education and Training

A few institutions, such as The University of the Arts in Philadelphia, and the Art Center College of Design in Pasadena, California, offer degrees in industrial design. Industrial designers often have their bachelor's or master's degree in manufacturing, engineering, or another related discipline. Different combinations of education and experience may qualify a candidate for a particular employer. Within the field of mountain biking, for example, a few designers began their careers in their own garage, making custom components for friends.

Career Activity

- Write a report about the kinds of jobs that are created when a sport such as mountain biking becomes a worldwide phenomenon.
- For research, look in a mountain biking magazine for specialty products, from bike accessories to clothes to vacation packages.

Objectives:

- Import, display, edit, and manage raster images
- Mix raster and vector objects
- Adjust the quality of a raster image
- Export a raster image file

Key Terms

bitmapped images

bit plane

clip

crop

raster images

vector-based object

AutoCAD allows you to insert raster image files, which are also referred to as *bitmapped images*. *Raster images* use a method of describing a line as a series of dots or pixels. When these lines are spaced closely together, raster images can represent solid-filled objects.

You can combine raster images in a variety of formats with AutoCAD drawings. Some government agencies and contractors use this capability to insert aerial photography into AutoCAD. Using AutoCAD's drawing and editing commands, they trace features such as buildings and roads to create AutoCAD drawings.

Raster images can be 8-bit grayscale, 8-bit color, or 24-bit color. The term "8-bit" refers to the depth in bit planes. A *bit plane* is a portion in memory allocated to deliver graphic information. Knowing the number of bit planes allows you to calculate the number of colors. An image with 8-bit grayscale contains up to 256 shades of gray, and 8-bit color images contain up to 256 shades of color. Images with 24-bit color contain as many as 16,777,216 shades of color.

Attaching a Raster Image File

AutoCAD permits you to import many raster image formats, including (among others) BMP, FLC, JPG, PCX, PNG, TGA, and TIF.

1. Start AutoCAD and begin a new drawing from scratch.

2. Display the **Reference** toolbar.

Reference

3. Pick the **Image Attach** button from the Reference toolbar.

This displays the Select Image File dialog box.

4. Display the **Files of type** drop-down box and review the list of raster file types supported by AutoCAD.

5. Select the **All image files** default.

6. Find and open AutoCAD's **Sample** folder.

As you can see, AutoCAD's Sample folder contains two raster image files in JPG format.

7. Pick the **Details** button located in the upper area of the dialog box to display a detailed listing of the files.

Notice the size of each file.

8. Single-click the first file in the list.

A preview of the raster image appears in the Preview area of the dialog box.

HINT:

If a preview does not appear, pick the Show preview button above the Preview area. When the preview appears, a Hide preview button becomes available.

9. Preview the second raster file.

10. Select the file named **r300-20.jpg** and pick the **Open** button.

The Image dialog box appears, as shown in Fig. 62-1.

Image   [?] [X]

Name: [R300-20 ▼] [Browse...]   ☑ Retain Path

Path: C:\Program Files\ACAD2000\SAMPLE\R300-...

Insertion point	Scale	Rotation
☑ Specify on-screen	☑ Specify on-screen	☐ Specify on-screen
X: [0.0000]	[1.0000]	Angle: [0]
Y: [0.0000]		
Z: [0.0000]		

[OK]   [Cancel]   [Help]   [Details >>]

Fig. 62-1

11. Pick the **Details** button to display more information.

12. Review the options in the dialog box and then pick the **Help** button to read about them.

13. After exiting help, pick the **OK** button to accept the default settings.

Notice that a small box is attached to the crosshairs.

14. In reply to Insertion point, pick a point in the lower left area of the screen.

15. In reply to Scale factor, drag the box up and to the right, filling about half the screen, and pick a point.

16. The raster image is of an air cylinder assembly. Note that AutoCAD references the image instead of importing it into the drawing. The image appears in the current drawing, but it does not actually become a part of the drawing file. Consequently, the size of the file only increases slightly.

NOTE:

You can also use AutoCAD DesignCenter to attach raster image files by dragging them into the current drawing, similar to the way you drag blocks, dimension styles, and other content.

17. Save your work in a file named **raster.dwg**.

Mixing Raster and Vector Objects

A *vector-based object* is one that is stored as one or more line segments, which are described by the position of their endpoints. AutoCAD is a vector-based program, so most of the objects that you create in AutoCAD are vector-based. Examples include lines, circles, splines, and text. AutoCAD attaches raster images to a drawing as a specific object type. Because AutoCAD treats raster images as objects, you can mix raster images with the vector objects in a drawing.

1. Using AutoCAD's realtime zoom and pan capabilities, review the detail in the image.

2. Fill the screen with the image.

3. Create a new layer named text, assign the color yellow to it, and make it the current layer.

4. Create a new text style named tech using the **TechnicBold** font. Accept the default settings for the text style.

5. Near the bottom and center of the image, enter **Air Cylinder Assembly** using text that is **.25** in height.

As you can see, you can add vector objects into a file that contains raster objects.

Editing a Raster Image

You can copy, move, rotate, resize, and clip imported raster images. When you *clip*, or *crop*, an image, you change its boundary.

1. **ZOOM All** and pick the border of the image.

2. Pick one of the four grips.

3. Drag to increase the size of the image, pick a point, and press **ESC** twice.

4. Pick the **Image Clip** button from the Reference toolbar.

5. Select the outline of the image.

6. Read the Command line and press **ENTER** to accept the New default.

7. Form a rectangle of any size by picking two points inside the image.

Reference

AutoCAD crops the image.

8. Pick the **Image Frame** button from the Reference toolbar.

9. Enter **Off**.

Reference

This removes the border line from the raster image.

10. Make the border reappear.

11. Increase the size of the image using its grips.

12. **Cancel** twice to remove the grips.

Adjusting the Image Quality

You can also adjust the image color, contrast, brightness, and transparency of raster images.

1. Pick the **Image Quality** button from the Reference toolbar.

Reference

AutoCAD presents High/Draft options on the Command line. These quality settings do not affect the actual image; they only affect display performance. High-quality images take longer to display.

2. Press **ESC** to cancel.

3. Pick the **Image Transparency** button from the Reference toolbar.

Reference

4. Pick the image and press **ENTER**.

AutoCAD presents ON/OFF on the Command line. Several raster file formats allow images with transparent pixels. If you turn on image transparency, AutoCAD recognizes the transparent pixels so that graphics on the screen show through those pixels.

5. Enter **On**.

6. Pick the **Image Adjust** button from the Reference toolbar, pick the image, and press **ENTER**.

Reference

This displays the Image Adjust dialog box, as shown in Fig. 62-2.

Managing Images

The Image Manager dialog box allows you to unload, reload, attach, and detach raster image files.

1. Pick the **Image** button from the Reference toolbar.

Reference

This displays the Image Manager dialog box, as shown in Fig. 62-3.

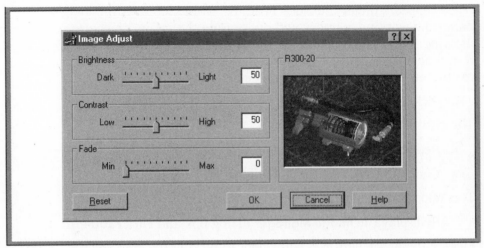

Fig. 62-2

Notice that several buttons are not available because you have not yet selected a raster image.

2. Click **r300-20**.

These buttons become available.

3. Pick the **Unload** button.

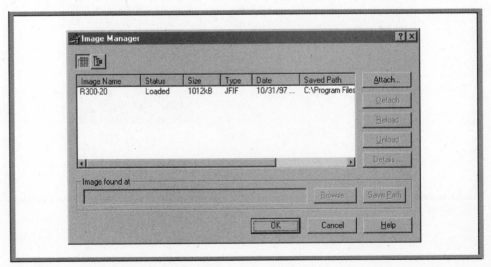

Fig. 62-3

Unload appears under the Status heading. Unload removes the image data from working memory without erasing the image object from the drawing. When the image is no longer needed for editing, Unload is recommended because it increases performance by reducing the memory requirement for AutoCAD without detaching the raster file.

4. Pick the **OK** button.

The outline of the image remains.

5. Redisplay the **Image Manager** dialog box.

Reference

6. Click **r300-20**, pick the **Reload** button, and pick **OK**.

This reloads the image.

7. Redisplay the **Image Manager** dialog box and click **r300-20**.

Picking the Attach button is equivalent to picking the Image Attach button from the Reference toolbar. The Detach button removes the selected image definitions from the drawing database and erases all the associated image objects from the drawing and display.

8. Pick the **Details...** button.

This displays details on the image, as shown in Fig. 62-4.

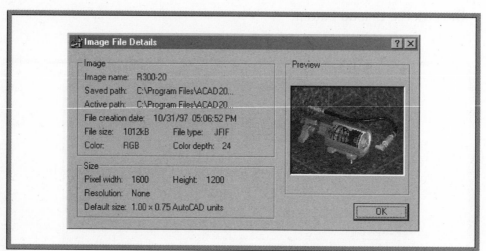

Fig. 62-4

9. Pick the **OK** button.

10. In the upper left area of the Image Manager dialog box, pick the **Tree View** button.

This illustrates the images that are attached to the current drawing.

11. Pick the **OK** button.

12. Minimize AutoCAD and review the size of the raster.dwg file.

The file is very small (probably around 30 Kb) compared to the 1,013 Kb for r300-20.jpg. Raster.dwg remains small because AutoCAD attaches (externally references) the raster file instead of importing its contents.

13. Maximize AutoCAD and save your work.

Exporting Raster Image Files

In addition to attaching raster images to AutoCAD drawings, you can export the contents of drawings as raster files. AutoCAD allows you to export drawings in Microsoft Windows Bitmap (BMP), Windows Metafile (WMF), and Encapsulated PostScript (EPS) formats.

1. Open the file named **hatch.dwg**.

2. Select **Export...** from the **File** pull-down menu.

3. From the **Save as type** drop-down box, select **Bitmap (*.bmp)**.

4. Pick the **Save** button to accept the **hatch.bmp** name.

5. Select the entire drawing and press **ENTER**.

AutoCAD creates a raster image file named hatch.bmp.

6. Close the Reference toolbar and exit AutoCAD without saving.

Chapter 62
Review & Activities

Review Questions

1. List six raster image file formats that AutoCAD can import.

2. How do you resize an attached raster image in AutoCAD?

3. What is the purpose of the Unload and Reload buttons in the Image dialog box?

4. Why do AutoCAD drawing files with "imported" raster files remain small?

5. What raster image file types can AutoCAD export?

6. Discuss potential applications of mixing raster and vector objects. When might this be useful to a company or government agency?

Challenge Your Thinking

1. An Image Quality button is available on the Reference toolbar. When might you need to use it? How does it affect the quality of a printed image?

2. When would you choose to use the Detach button in the Image Manager dialog box?

Applying AutoCAD Skills

Work the following problems to practice the commands and skills you learned in this chapter.

1. Begin a new drawing from scratch. Attach the hatch.bmp file you created in this chapter to the new drawing file. Insert the image to fill most of the screen. Save your work in a file named prb62-1.dwg.

2. Attach the watch.jpg file from AutoCAD's Sample folder to a new drawing file named. ch62chrono.dwg. Enter the title Chronometer on a separate layer. Specify a color and text style. Remove the frame from the watch. Save the drawing.

Chapter 62
Review & Activities

3. Begin a new drawing from scratch. Attach the file named watch.jpg located in AutoCAD's Sample folder. Insert the image to fill most of the screen. Increase the brightness and crop the image. On a new layer named text, add the phrase WATCH DESIGN #477 in the lower right area of the image. Save your work as prb62-2.dwg.

4. Open lamp.dwg and create a BMP raster image file from it.

Using Problem-Solving Skills

Complete the following activities using problem-solving skills and your knowledge of AutoCAD.

1. Your new job as advertising editor for an Autodesk distributor requires creative thinking. You want to demonstrate the capabilities of AutoCAD in the pictorial graphics area, specifically raster images. Since clip art is available in most word processing software packages, you have decided to import some into AutoCAD. Attach a raster image file in one of the compatible formats. Size the image, edit it, enhance its quality, and add a title. Save the file as ch62ad1.dwg.

2. Find a second piece of clip art that you can use with your original file to "make a statement." Superimpose the second image on the first. Size the images, edit them, and enhance their quality so that they clearly show the point you are making with your advertising "statement." Add a title to the completed graphic. Save the file as ch62ad2.dwg.

Chapter 63

External References

Objectives:

- **Attach drawing files to a base drawing as external references (xrefs)**
- **Unload and reload xrefs that have been attached to a drawing**
- **Bind all or part of an xref permanently to a drawing**

Key Terms

base drawing
bind
dependent symbols
external references
xrefs

External references (also called *xrefs*) are ordinary drawing files that have been attached to the current drawing. The process of inserting an xref is similar to inserting a drawing as a block. However, xrefs do not become part of the drawing. Instead, the xrefs are loaded automatically each time the drawing file is loaded. In addition to viewing the xrefs, you can make use of the xref objects by, for example, snapping to them.

External references are helpful if you want to view an assembly of individual components as a master drawing. Xrefs are particularly useful when you are working on a project with other AutoCAD users in a network environment.

Preparing the Drawings

To explore the concept of using xrefs, let's use an architectural drawing project.

1. Start AutoCAD and pick **Use a Wizard**, **Quick Setup**, and **OK**.

2. Pick the **Architectural** radio button and the **Next** button.

3. Enter **88′** for the width and **68′** for the length, and pick the **Finish** button.

The drawing area is based on a scale of $1/8'' = 1'$ on a standard A-size sheet.

4. **ZOOM All**.

5. Set the grid at **10'** and snap at **2'**.

6. Save your work in a template file named **xref.dwt**. Enter **Base drawing** for the template description.

7. Close xref.dwt and use it to create a new drawing named **property.dwg**.

8. Create and make current a new layer named **Property**. Assign the color red to it.

Most property lines do not form a perfect rectangle. However, to keep the process simple, we will use a rectangular shape.

9. Draw a rectangular property line using the **PLINE** command. Make the property line **84' × 64'**, being sure to stay inside the drawing area.

10. Save your work and close this drawing file.

11. Use the **xref.dwt** template file to begin a new drawing.

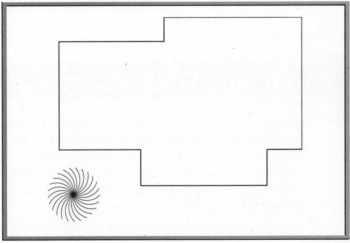

12. Create and make current a new layer named **Building**. Assign the color magenta to it.

13. Create and center the building outline shown in Fig. 63-1 using a polyline. Do not draw the tree symbol at this time.

Fig. 63-1

14. Save your work in a file named **bldg.dwg** and minimize this drawing.

A drawing into which xrefs are inserted is often called the *base drawing*. In this case, the base drawing will be the drawing that is used by the landscape architect.

15. Create another drawing using the **xref.dwt** template file.

16. Create and make current a new layer named **Trees**, and assign the color green to it.

17. Create a simple tree symbol, as shown in the previous illustration. Approximate the size and shape of the tree.

HINT: ————————————————————————————————————

Use the ARC and ARRAY commands to create the tree.

18. Create a block of the tree. Name it **Tree**, and use the tree's center as the insertion base point. Do not retain the tree on the screen.

19. Save your work in a file named **ldscape.dwg**

You should now have the drawing files shown in Table 63-1, containing the layers and layer colors shown.

File Name	Layer	Layer Color
property.dwg	property	red
bldg.dwg	building	magenta
ldscape.dwg	trees	green

Table 63-1

Attaching Files As Xrefs

Suppose you are the landscape architect for this project. You are responsible for completing a landscape design for the building in bldg.dwg. Because you are working on a tight schedule and the customer would like to see a proposed landscape design, you must deliver a preliminary design even though the design of the building is not yet complete.

You can obtain a copy of the bldg.dwg drawing from the building architect and attach it as an xref to your landscape drawing. Even though the building is not yet finished, including it in the drawing will give you an idea of what the architect is planning.

Now let's attach the bldg.dwg file as an external reference.

Reference

1. Open the **Reference** toolbar and pick the **External Reference** button.

The Xref Manager dialog box appears, as shown in Fig. 63-2.

2. Pick the **Attach...** button to display the Select Reference File dialog box.

3. Find and double-click the **bldg.dwg** drawing file.

This displays the External Reference dialog box, as shown in Fig. 63-3.

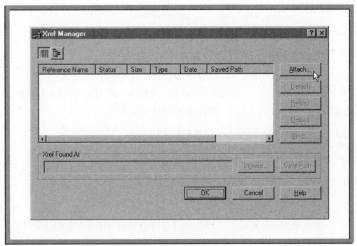

Fig. 63-2

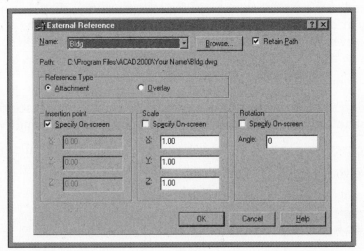

Fig. 63-3

Study the information in the dialog box. Much of it is self-explanatory.

4. Pick the **Help** button to read about the unfamiliar parts of the dialog box, and then exit help.

5. Pick the **OK** button in the **External Reference** dialog box; pick **OK** in reply to the AutoCAD message.

Move the crosshairs and notice that the bldg.dwg drawing appears to be inserting, as if you were using the INSERT command.

6. Enter **0.0** for the insertion point.

The bldg drawing is now present.

7. Select the building and list its properties.

Listing information about an object tells whether it belongs to an xref. Notice that External reference appears in the listing.

We are going to place several trees around the house. Therefore, it is important to see the property line so that none of the trees extend outside the property line. Let's attach property.dwg.

Reference

8. Pick the **External Reference Attach** button.

The Select Reference File dialog box appears.

9. Find and double-click the **property.dwg** file.

10. Pick **OK** in the **External Reference** dialog box, pick **OK** in reply to the message, and position the red property line by entering **0,0** for the insertion point.

The property line is now present.

As a reminder, these new objects are for reference only; they do not become part of the drawing. However, they stay attached to the drawing—even between editing sessions—until they are detached. The xrefs do not cause the drawing file to increase in size.

11. Save your work.

12. Insert the block named **Tree** into the current drawing. Place the tree close to the east side of the building, and accept the default values.

13. Place several other trees (of various sizes) around the building.

14. Save your work and close **ldscape.dwg**.

Changing an Xref

Suppose the customer asks the building architect to expand the east side of the building.

1. Maximize the **bldg.dwg** drawing.

2. Stretch the east part of the building **10'** to the east.

3. Save your work and minimize the drawing.

The bldg drawing has changed, so the building architect should notify the landscape architect of the change. Suppose the landscape architect loads ldscape.dwg to continue working on it.

4. Open **ldscape.dwg** and pick the **OK** button.

The latest change in bldg.dwg is reflected in the ldscape.dwg drawing. As you can see in Fig. 63-4, the trees may now be too close or interfere with the building. Being linked to related drawings that may change is a beneficial feature of an xref.

5. Move the tree(s) away from the building.

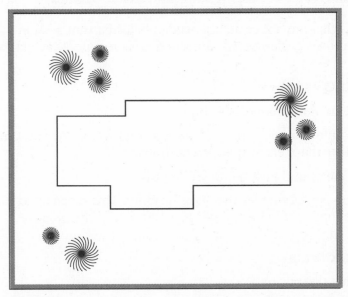

Fig. 63-4

Xref Layers

1. Display the **Layer Properties Manager** dialog box.

Notice that ldscape.dwg contains two additional layers as a result of attaching the two xrefs. Each of the new layer names is preceded by its parent xref drawing file name and is separated by the | (vertical bar) character. You can control the visibility, color, and linetype of the xref layers.

NOTE: ─────────────────────────

You may need to expand the Name column in the dialog box to see the full name of each layer.

2. Close the dialog box and save your work.

Managing Xrefs

Like raster images, xrefs can be unloaded and reloaded as necessary while you are working on a drawing.

Unloading and Reloading Xrefs

Unloading xrefs that are not currently needed helps current session editing and improves performance. Unloaded xrefs remain attached to the drawing.

1. Enter **UNDO** and **Mark**.

2. Pick the **External Reference** button.

Reference

Notice the new information in the Xref Manager dialog box. As you can see, AutoCAD lists the xrefs in the current drawing.

3. Select property and pick the **Unload** button.

Note that Unload appears under the Status heading. This means that AutoCAD will suppress the display and regeneration of the property xref definition.

4. Pick the **OK** button.

The property line disappears.

5. Press the spacebar or **ENTER** to redisplay the **Xref Manager** dialog box, select **property**, and pick the **Reload** button.

The Reload button re-reads and displays the most recently saved version of the drawing.

6. Pick the **OK** button.

The property line reappears.

7. Redisplay the **Xref Manager** dialog box.

Above the Reload button is the Detach button. Use this button only if you want to detach an xref permanently.

Binding an Xref

If you want to make an xref a permanent part of the base drawing, you can *bind* it to the drawing. Binding incorporates the xref into the base file. Note that once you bind an xref, you can no longer unload or detach it.

1. Pick the **Help** button and read about the **Bind...** button.

2. After exiting help, select **property** and pick the **Bind...** button.

3. In the Bind Xrefs dialog box, pick the **Bind** radio button, unless it is already picked, and pick **OK**.

Property disappears from the list of external references in the Xref Manager dialog box because it is now a permanent part of the drawing.

4. Pick the **OK** button.

5. Enter **UNDO** and **Back** to return **property.dwg** to its xref status.

Binding Parts of an Xref

In the future, you may not want to bind an entire xref to a drawing. To bind only part of an xref, you can use the XBIND command. This command permanently binds a subset of an xref's dependent symbols to the current drawing. *Dependent symbols* are named objects in an xref, such as blocks, layers, and text styles.

Reference

1. Pick the **External Reference Bind** button from the Reference toolbar.

The Xbind dialog box appears.

2. Pick the white box with a **plus (+) sign** in it located to the left of **bldg**.

This displays the xref's dependent symbols.

3. Pick the white box with a **plus (+) sign** in it located to the left of **Layer**.

This displays the layer bldg|building.

4. Pick **bldg|building** and pick the **Add** button.

AutoCAD adds bldg|building under Definitions to Bind.

5. Pick the **OK** button.

When binding a dependent layer to the current drawing, AutoCAD renames the layer by replacing the | (vertical bar) with $#$.

6. Display the **Layer Properties Manager** dialog box.

In this case, AutoCAD renamed bldg|building to bldg0Building.

7. Close the dialog box and the Reference toolbar, as well as any other toolbars you may have opened.

8. Save your work and exit AutoCAD.

Chapter 63
Review & Activities

Review Questions

1. What is the primary benefit of using external references (xrefs)?

2. What is the function of each of the following buttons in the Xref Manager dialog box?
   Attach
   Detach
   Reload
   Unload
   Bind

3. Attaching an xref is similar to using what other popular AutoCAD command?

4. What command tells whether a given object belongs to an xref?

5. Describe the purpose of the XBIND command.

Challenge Your Thinking

1. Experiment with the Attachment and Overlay options in the External Reference dialog box. What are the differences between the two? In what ways are they similar?

2. Describe a situation in which you might want to bind an external reference permanently to a drawing. Why might you need to use an external reference initially if you eventually want the reference to become a permanent part of the drawing?

Chapter 63
Review & Activities

Applying AutoCAD Skills

Work the following problems to practice the commands and skills you learned in this chapter.

1. Draw the front and side elevations of the ranch style house shown in Fig. 63-5A and B as separate drawings. Create a master drawing of appropriate size and bring the two drawings together on the master drawing as external references.

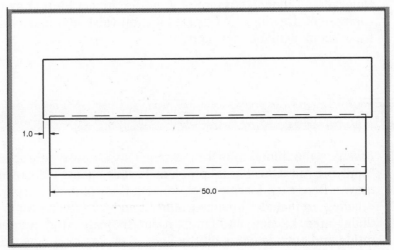

Fig. 63-5A

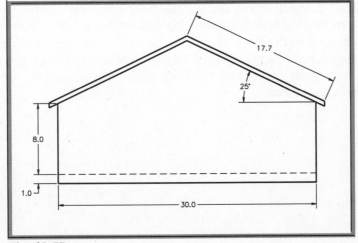

Fig. 63-5B

Chapter 63
Review & Activities continued

2. This problem works best as a small group activity, involving two to four individuals. The group should identify a project. Make it simple; otherwise it may be difficult to organize. The project should involve several components that fit together. Each person on the project should be responsible for completing one or more different components. The project leader must coordinate the effort and be responsible for completing the final assembly made up of the individual drawings.

   While a local area network would aid greatly in the completion of this project, it can be done using individual AutoCAD stations. However, all individuals working on the project must copy their component drawings to a single location (disk and folder) and make them available to others.

 ## Using Problem-Solving Skills

1. Based on the dimensions for the ranch style house described in "Applying AutoCAD Skills" problem 1, your clients wants you to draw a floor plan for approval. Make a floor plan as a separate drawing, to include three bedrooms, two baths, living room, dining area, kitchen, and family room. Indicate windows and both inside and outside doors on the floor plan. Create a new master drawing of appropriate size and bring all three drawings together as external references.

2. Your client wants to change the floor plan you created in problem 1 to include four bedrooms instead of three. Make the necessary changes to the floor plan drawing. Add a 12′ × 20′ deck in the appropriate location at the back of the house. Check the master drawing to ensure that the floor plan changes and the deck are reflected there.

Careers
Using
AutoCAD

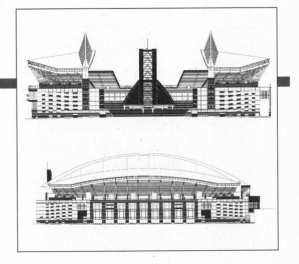

Seattle Seahawks Stadium

As seen by Superman (or anyone else with x-ray vision), the new Seattle Seahawks stadium might resemble a gigantic web of electrical cable and wire. The stadium will have miles of wire, providing the electricity needed to keep cold drinks cold, microphones hot, and hundreds of lights turned on.

Scheduled to open in 2002, the Seahawks stadium is the creation of Ellerbe Becket, an international architecture/engineering/construction (A/E/C) firm. "We maintain staff in a lot of different disciplines," explains Mike Hnastchenko, Director of Technology at Ellerbe Becket. "Given the size of so many of our projects, it is essential that we maintain a high level of coordination through our CAD tools."

Ellerbe Becket uses AutoCAD and related Autodesk software to generate and maintain master drawings. Specialists, such as electrical engineers, select layers within master drawings that relate to their discipline. Any changes they make within their layer are reflected in the master drawings. This unified CAD approach not only saves time (no need to redraw for different disciplines), but greatly facilitates coordination of team members.

When fans enter the stadium, their attention will be drawn to the top of its 12-story tower where a gigantic video screen will show replays and crowd-rousing graphics. Only a few people—such as Superman, electrical engineers, and perhaps a handful of AutoCAD users—will consider all the well-planned work hidden within the tower.

Career Focus—Electrical Engineer

Electrical engineers design, develop, test, and supervise manufacture of electrical equipment. This equipment includes power supply components, such as generating plants and transformer stations. They also include components powered by electricity, such as motors, controls, and lighting.

Within their specialties, electrical engineers set specifications for electrical systems, develop maintenance procedures, and diagnose and solve problems in equipment operation.

Education and Training

A bachelor's degree is required for entry-level jobs in electrical engineering. Coursework emphasizes mathematics, physical sciences, and theory and methods in the profession.

A new engineer typically works under close supervision, developing expertise in the employer's applications. Electrical engineers generally perform as part of a team with professionals in other disciplines.

Career Activity

- Write a report about employment prospects for electrical engineers.
- Identify companies that recruit new graduates and the range of starting salaries for electrical engineers.

Object Linking and Embedding (OLE)

Objectives:

* **Link AutoCAD to a WordPad document using object linking and embedding (OLE)**
* **Edit an AutoCAD server document and view the changes in the client document**
* **Link a Word document to AutoCAD using object linking and embedding**

Key Terms

client document

object linking and embedding (OLE)

server document

As you know, you can copy an AutoCAD drawing and paste it into another Windows program. Using *object linking and embedding (OLE)*, you can maintain a link between AutoCAD and the other program, if the program supports OLE. When you edit the AutoCAD drawing, the change occurs automatically in the second program. This can save time and ensure accuracy if you want to maintain the same version of a drawing in another document.

You can also link information from another program to AutoCAD. For example, you can paste a table from Microsoft Word® or a spreadsheet from Microsoft Excel™ into AutoCAD. When you change the table in Word, or the spreadsheet in Excel, AutoCAD updates automatically to reflect the change.

Linking to Another Document

When you link an AutoCAD view to a document or file in another program, such as Microsoft WordPad, the AutoCAD drawing becomes the *server document*. The document in WordPad becomes the *client document*.

1. Start AutoCAD and open the drawing named **region.dwg**.

2. Using **Save As...**, create a new file named **ole.dwg**.

3. Display the **Shade** toolbar and pick the **3D Wireframe** button to remove the shading.

Shade

4. Pick **Copy Link** from the **Edit** pull-down menu. This enters the COPYLINK command, which copies the current view to the Windows Clipboard. You can paste the contents of the Clipboard into a document as an OLE object.

5. Exit AutoCAD without saving changes and start Microsoft WordPad.

6. In WordPad, select **Object...** from the Insert pull-down menu.

The Insert Object dialog box appears, as shown in Fig. 64-1.

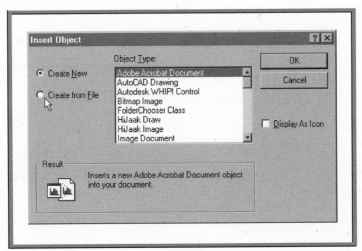

Fig. 64-1

7. Pick the **Create from File** radio button and check the **Link** check box.

8. Pick the **Browse...** button and find and double-click the **ole.dwg** file.

9. Pick the **OK** button.

This opens the file in both AutoCAD and WordPad.

10. Minimize AutoCAD.

11. In **WordPad**, pick **Save** or **Save As...** from the **File** pull-down menu and enter **link.doc** for the file name. Be sure to select the folder with your name.

12. Minimize WordPad.

Editing the Server Document

Let's change the drawing in AutoCAD—the server document.

1. Maximize AutoCAD.

2. Add 10 small holes, as shown in Fig. 64-2.

3. Save your work and minimize AutoCAD.

4. Maximize WordPad.

The change is reflected automatically in the WordPad file.

5. Pick **Links...** from the **Edit** pull-down menu.

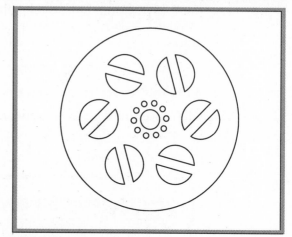

Fig. 64-2

The Links dialog box appears, as shown in Fig. 64-3.

The buttons in this dialog box permit you to update a link, open or change a source, and break a link. Update is currently set to Automatic, which means that the link updates automatically, as it did. Also, the link is listed as Automatic in the Links list box. If Update were set to Manual, you would need to pick the Update Now button to update the link.

6. Pick the **Cancel** button.

7. Save your work and exit WordPad.

8. Maximize AutoCAD, close the **Shade** toolbar, and exit AutoCAD.

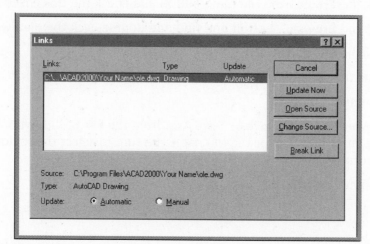

Fig. 64-3

Linking a Document to AutoCAD

The beginning of this chapter mentioned that you can link a document to AutoCAD if you have a product, such as Microsoft Word or Excel, that is OLE-capable. The steps are similar to the previous ones, except Word or Excel becomes the server and AutoCAD becomes the client.

NOTE:

The following steps assume that you are using Microsoft Word. If you are using another OLE-capable application, use these steps as a guide.

1. Start Microsoft Word, begin a new document, and enter the following text:

   Sprocket Guard Design #3179

2. Highlight the text and select **Copy** from the **Edit** pull-down menu.

3. Save your work in a file named **link2.doc** and minimize (do not exit) Word.

4. Start AutoCAD and begin a new drawing from scratch.

5. Select **Paste Special...** from AutoCAD's **Edit** pull-down menu.

The Paste Special dialog box appears. Notice that Source is the Word document you just copied to the Windows Clipboard.

6. Pick the **Paste Link** radio button, and pick **OK**.

The contents of the Word document appear in the AutoCAD drawing area, bounded by a box.

7. Save your work in a file named **link.dwg**.

8. Minimize AutoCAD and maximize Microsoft Word.

9. Edit the sentence to read:

   Sprocket Guard Design #3179 for Prototype B

10. Save your changes, minimize Microsoft Word, and maximize AutoCAD.

The changes you made in Microsoft Word update automatically in AutoCAD. This is because both the client and the server are currently open.

11. Save the change and exit AutoCAD.

12. Maximize Word and exit it also.

Chapter 64
Review & Activities

Review Questions

1. What is the overall purpose of object linking and embedding (OLE)?

2. When you link an AutoCAD drawing to a WordPad document, is AutoCAD the server or the client? Why?

3. If WordPad is a client, describe what you would do in WordPad to update, change, or cancel the link between the server and client.

4. Suppose you want to link a spreadsheet from Microsoft Excel to AutoCAD. Is the spreadsheet document the server or the client?

Challenge Your Thinking

1. The procedures described in this chapter focus on linking an AutoCAD drawing to files from other applications and vice versa. Find out how linking is different from embedding. Write a paragraph explaining the difference and describing at least one application in which each would be useful.

2. The city council is working on plans for a major downtown renovation effort. The architects and development companies are working simultaneously on different parts of the project, but they are all using AutoCAD. The council wants to see the latest revisions at the meeting tonight. Should the project manager use OLE or external references to show the council the latest work efforts? Explain your answer.

Chapter 64
Review & Activities

Applying AutoCAD Skills

Work the following problems to practice the commands and skills you learned in this chapter.

1. Link information from a program such as Microsoft Word or Excel to AutoCAD. Change the server document and then update the client document.

2. Use Microsoft WordPad to create a document named shop.doc. In the document, write two or three paragraphs describing the best way to set up a workshop. Link workshop.dwg (from Chapter 32) to the shop.doc file to illustrate the paragraphs you have written.

3. Make changes to workshop.dwg in AutoCAD and then update the link you created in problem 2.

4. In AutoCAD, save workshop.dwg with a new file name of traffic.dwg. Add arrows showing the projected traffic patterns around the tools in the workshop. Save the changes to traffic.dwg. In the shop.doc file (created in Problem 2), add a paragraph or two discussing the effect of foot traffic through the workshop on worker efficiency. Link traffic.dwg to the shop.doc file to illustrate the point. View and update all the links to the shop.doc file.

5. Make the necessary changes to embed traffic.dwg into the shop.doc file.

Chapter 64
Review & Activities continued

 Using Problem-Solving Skills

Complete the following activities using problem-solving skills and your knowledge of AutoCAD.

1. An advertising brochure that you are helping to create needs the preliminary drawing of the lamination for an electromagnet shown in Fig. 64-4. Create the drawing and dimension it. Link the drawing to WordPad or to the word processor you currently use. In the word processor, type the title "Electromagnetic Lamination" in 20 pt, bold Times New Roman type centered over the lamination drawing.

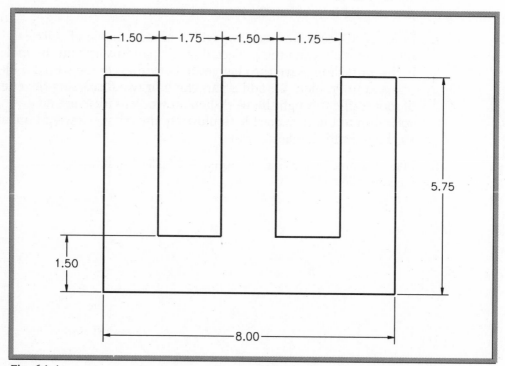

Fig. 64-4

2. Change the AutoCAD drawing from problem 1 to show the five holes as shown in Fig. 64-5; redimension the drawing. Check the advertising brochure to make sure it has been updated.

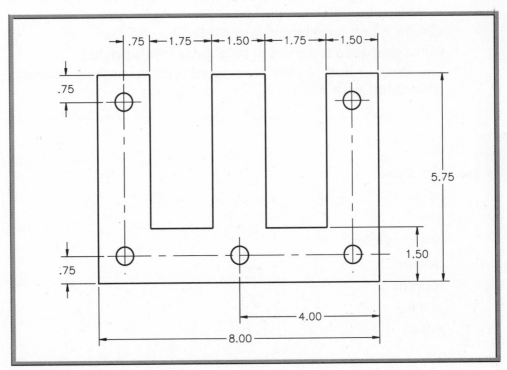

Fig. 64-5

Chapter 65 Database Connectivity

Objectives:

- **Connect to an external database**
- **Configure a data source connection**
- **View and edit records in a database**
- **Perform a query to obtain a specific subset of the records in a table**
- **Create a link template in AutoCAD**
- **Link records in a database to objects in a drawing**
- **Use links to find objects associated with records or records associated with objects**

Key Terms

database
fields
key
link templates
queries
records
sorting
tables

AutoCAD allows you to connect to an external database and associate objects in a drawing file with records in the database. Combined with AutoCAD's Internet connectivity, this capability allows you to use databases to which you have Internet access all over the world. A *database* is a file or group of files that holds a large amount of tabular data. Databases contain one or more *tables*, which consist of any number of *fields* (categories of information) and *records* (groups of related fields).

If you are familiar with Microsoft Excel or a similar spreadsheet program, you can think of a table as a worksheet, the columns as fields, and the rows as records. Each field has a unique name by which you can reference it. You can identify each record individually using a *key*. The purpose of the key is to provide each record with a unique identifier. A table's key may consist of one or more fields that uniquely identify the record. For example, a phone list table might have the following fields:

First Name
Last Name
Work Number
Home Number
Mobile Number

First Name and Last Name can be treated as the key fields in the table if no two people in the list have the same first and last name. To avoid the possibility of duplicate items, it is common to use a numerical ID field to act as the key. In most database programs, you can set the software to assign a unique number to each new record automatically when you create the record.

This chapter uses AutoCAD's dbConnect capabilities to connect to a database, link data in the database to objects in the drawing, and work with the data. For example, you could use dbConnect to associate external data from a corporate or individually developed database with drawing objects. You could then use the drawing to find associated data, or you could use the data to manipulate the drawing.

Connecting to a Data Source

AutoCAD's dbConnect uses a data source as an alias that refers to a specific database. An easy-to-use tree interface in dbConnect allows you to connect to data sources such as external databases, view their connection status, and access their tables. The tree displays links, labels, and queries that have been created and saved in the drawing. It also shows tables available in currently connected databases. Right-clicking on any entry in the tree provides a context-sensitive menu with options that are currently available for that entry.

NOTE:

The queries referenced above are stored in the Microsoft Access database and should not be confused with queries that we will create later in this chapter, which will be stored with the drawing.

Standard

1. Copy the files **db_samp.dwg** and **db_samples.mdb** from AutoCAD's **Sample** folder to the folder with your name.

2. Rename the files **db.dwg** and **db.mdb**, respectively.

3. Start AutoCAD and open the **db.dwg** file in the folder with your name.

4. From the docked **Standard** toolbar, pick the **dbConnect** button.

The dbConnect Manager dialog box appears, as well as a dbConnect pull-down menu in the menu bar. (The dialog box may be docked on the left side of the drawing area, depending on how the last person left it.)

5. In the dbConnect Manager, right-click the **jet_dbsamples** entry under **Data Sources**.

6. If the Connect option is grayed out, pick **Disconnect**. If it is not grayed out, pick **Connect**, pick **Configure**, and skip Step 7.

7. Right-click **jet_dbsamples** and pick **Configure...**

8. Pick the button (...) next to **C:\ACAD2000\SAMPLE\db_samples.mdb** to browse AutoCAD's Sample folder.

9. Double-click the folder with your name.

10. Double-click the **db.mdb** file to finish the database selection process.

Notice that when you connect to the database, the tables in the database are listed in the tree and the link icons change to indicate that the links are now connected. You will learn more about links later in this chapter.

Configuring a Data Source Connection

Before you can attach to one of your own databases as a data source, you must configure it. After a database has been configured, it will be available to you from dbConnect whenever you wish to connect to it. You can link a single drawing to records from multiple databases, and you can link multiple drawings to a single database.

We will create a second data source with a new alias name that points to the same database as jet_dbsamples. This will help you understand the data source configuration process.

1. Right-click **Data Sources** in the **dbConnect** tree.

2. Pick the **Configure Data Source...** item.

The Configure a Data Source dialog box appears.

3. Enter **Facility Data** as the new data source name and pick the **OK** button.

The Data Link Properties dialog box appears, as shown in Fig. 65–1.

You must make changes to the Provider and Connection tabs to establish a link to the data source.

4. Pick the **Provider** tab.

5. Pick **Microsoft Jet 3.51 OLE DB Provider** in the **OLE DB Provider(s)** list.

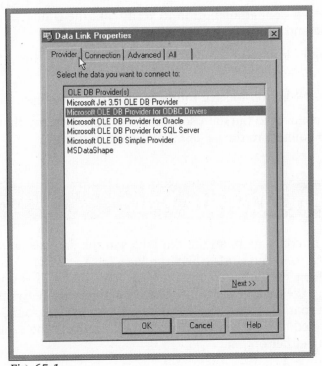

Fig. 65-1

NOTE: ———————————————————

If you see Microsoft Jet 4.0 OLE DB Provider instead, pick it. If you don't see either Jet driver, you may need to install the driver.

6. Pick the **Next** button.

This takes you to the Connection tab.

7. Pick the button (...) located at the right of the **Select or enter a database name** edit box.

8. Find and double-click the folder with your name.

9. Double-click the **db.mdb** file to finish the database selection process.

NOTE: ———————————————————

When you connect to most corporate databases, you will probably need to specify a user name and password. For most Microsoft Access databases, you should be able to connect to the database using the default Admin user name, unless you have configured users.

10. Pick the **Test Connection** button to verify that the connection is available and configured properly.

11. Pick the **OK** button in the box that notifies you that the connection succeeded.

12. Pick the **OK** button to save the new data source configuration.

You can now connect to the data source in the same manner that you used above to connect to the jet_dbsamples data source.

Viewing and Editing Data

You can view or edit data in any of the tables in the database after you have connected to it. AutoCAD offers the View Table and Edit Table options. Use the View Table option rather than the Edit Table option if you want to avoid changing the data in the table. The only difference between the two options is that View Table displays data on a gray background and the data cannot be edited. Other than that, you can perform the same functions using either option.

Editing Records

The Edit Table option allows you to edit fields in existing records, delete records, and add new records.

1. Right-click the **Employee** table in the tree under the **jet_dbsamples** data source.

2. Pick the **Edit Table** option to open the table for editing.

AutoCAD displays the table in a window called Data View, as shown in Fig. 65-2.

Note that you can also double-click on a table to open it. The Data View looks very similar to Microsoft Excel and works almost identically to the table editor in Microsoft Access.

The arrow to the left of the first record indicates the currently selected record. The record counter in the lower left corner of the window also provides an indication of which record is currently selected. The arrows on either side of the record counter are used to step up and down through the records and move to the top and bottom of the table. You may also set the current record by clicking anywhere on the record.

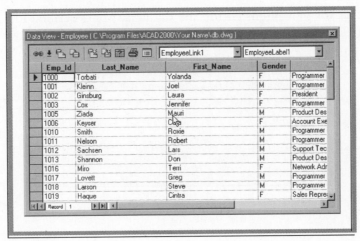

Fig. 65-2

Records are edited by first clicking on a field to make it and its record current and then either typing something new or clicking a second time to highlight the record's contents and editing it. Let's edit a few records.

3. Click in the left margin (gray box) of record **1013** (Don Shannon) to select and highlight the record.

4. Scroll to the right until you see the Department column.

Notice that highlighting the record makes it easy to identify the proper record as you scroll to the right. You may also click the record and use the Current Record arrow for orientation.

5. Under the **Department** field, click **Engineering** in the current record.

6. Enter **IS**.

Notice that the word Engineering is replaced with what you just typed. Also notice that the current record indicator in the gray box of the left margin has been replaced with a pencil, indicating that the record has been changed. Changes to a record are not permanent until you make another record current. While you are editing a record, you can use the ESC key to discard unwanted changes.

7. Click on any other record in the table to store the change.

AutoCAD also commits (stores) a changed record to the database when you close the Data View.

8. Press **CTRL** and the left arrow key at the same time to jump to the first column.

Shortcut keys for movement within the database are described in Table 65-1.

Shortcut Key	Action
CTRL + Left Arrow	Beginning of record
CTRL + Right Arrow	End of record
CTRL + Home	Beginning of table
CTRL + End	End of table

Table 65-1

9. Click the **First_Name** field of record **1018** (Steve).

10. Click again to highlight the name **Steve** for editing.

11. Change it to **Stephen**.

12. Press the down arrow key to move off the edited record.

13. Close the Data View by clicking the **x** in the upper right corner of the window.

14. Double-click the **Employee** table and verify that your changes have been saved.

15. Scroll to the bottom of the table.

16. Select records 1060 through 1063 by clicking in the left margin of boxes next to **1060** and dragging the cursor down until the four records are highlighted.

17. Right-click in the left margin and pick the **Delete record** option.

18. Pick the **Yes** button to confirm the deletion.

The records disappear.

19. Click the **Emp_Id** field of the record that has an asterisk (*) in the left margin.

20. Fill in the new record by typing the following in each of the fields. Be sure to press **ENTER** at the end of the record to move down to the next record.

   2000   Sanborn   Don   M   Consultant   IS   9999

Suppose the entire record was entered by mistake. We can discard the changes rather than committing them.

21. Right-click on the **delta (Δ) symbol** in the upper left corner of the table.

22. Pick the **Restore** item.

AutoCAD closes the Data View window and throws away any changes that were made. Reopen the Employee table and verify that the changes were not committed.

Sorting Data

Sorting (rearranging the order of) tables that contain large amounts of data can make it easier to find the data in which you are interested. Tables are generally sorted by default on their key fields (columns). In the case of the Employee table, you can see that it is sorted on the Emp_Id field.

1. Right-click one of the column headings, such as **Last_Name**.

2. Pick the **Sort** option.

This displays the Sort dialog box, as shown in Fig. 65-3.

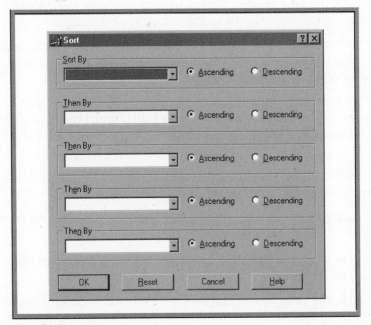

Fig. 65-3

The Sort dialog box permits you to select a number of columns to sort by. Also, it allows you to sort either by ascending or descending order.

3. Select **Department** from the **Sort By** drop-down list.

The Ascending radio button next to it should be checked.

4. Pick the **OK** button to sort the table.

Scroll around and view the sorted results. The records are now listed in alphabetical order according to department.

Another method of sorting involves double-clicking a specific column heading. The column automatically sorts in ascending order. Each subsequent double-click on the same column reverses the sort order.

5. Double-click the **Last_Name** column heading.

6. Double-click it again.

Formatting Columns

Other options in the column shortcut menu allow you to hide and unhide, freeze and unfreeze, and align columns.

1. Right-click the **First_Name** column header and pick the **Hide** option to hide the column.

2. Right-click the **Last_Name** column header and pick the **Freeze** option to "glue" the column to the left side of the table.

3. Scroll left and right and notice how the frozen column does not move with the other data. (You may need to reduce the size of the Data View window so that scrolling is available.)

You can freeze only one column at a time but you can hide as many columns as you wish. You can highlight several columns by clicking and dragging the cursor across the column headings and then hide them all at once.

4. Right-click the **Emp_Id** column header and pick **Align** and then **Center** to center the numbers in the column.

5. Right-click any of the column headers and select **Unhide All** from the shortcut menu to remove the hide state.

6. Right-click one of the column headers and select **Unfreeze All** to remove the freeze state.

7. Right-click the **Emp_Id** column header and pick **Align** and **Standard** to reset the column alignment to its standard format.

Formatting changes are temporary and are limited to the current session in Data View. Each time you exit Data View, the formatting is lost.

8. Close the Data View window.

Querying Data

Database *queries* allow you to view only those records in a database that meet specific criteria. For example, in a facility drawing such as db.dwg, you may wish to identify all rooms that have an area greater than 200 square feet, or you may wish to see a list of all laser printers in the facility.

The result of a query looks, for all practical purposes, like another table. If you store a query, you can view it in Data View like any standard table. Also, you can edit the queried data.

1. Right-click the stored query named **ROOMQUERY1** in the **dbConnect** tree, and pick **Edit...** to edit the query.

This displays the Query Editor, as shown in Fig. 65-4.

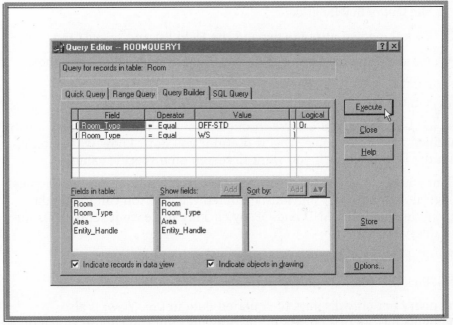

Fig. 65-4

Notice that the Query Editor displays the Query Builder tab. This is the preferred method of building complex queries. If you know how and prefer to write Structured Query Language (SQL) queries, you may use the SQL Query tab. You may also use the SQL Query tab to view the SQL statement corresponding to the query that you build with Query Builder. For simple single-field queries, you may prefer to use the Quick Query or Range Query tabs.

The Query Editor currently shows records from table Room. (The table name is listed in the note above the query tabs.) The two rows of the query specify Room_Type for the field and specify values of OFF-STD and WS. Therefore, when you run the query, the result will only show those records that have a room type of OFF-STD or WS.

2. Pick the **Execute** button to view the query results.

AutoCAD displays the query results in the Data View like a table. As you can see, the query displays only the rooms of type OFF-STD or WS.

3. Pick the **Return to Query** button at the top of the **Data View** window to return to the Query Editor.

To build a complex query, it is often useful to begin with a simple query and then refine it one step at a time until you get the results you want. We will now expand on the query and refine our search to rooms of type OFF-STD and WS that have an area greater than 150 square feet.

4. Click in the open cell directly under the second **Room_Type** entry in the **Query Builder** grid.

5. Pick the down arrow to expand the **Field** list and select **Area**.

6. On the same row in the grid, click the cell under **Operator** and select **> Greater than** from the list.

7. On the same row, click the cell under **Value** and enter **150**.

The (...) button lets you browse through a list of all possible values for the specified field in the table. This is useful if you don't know what kind of data the field holds or if you want to search on a specific value.

When you begin a new entry in the Query grid, it automatically assumes a logical AND with any previously entered criteria. You can toggle the logical field between AND and OR by clicking it. Leave it set to AND for this query.

8. Pick the **Execute** button to view the query results.

9. Pick the **Return to Query** button to return to the Query Editor.

The query not only displays the queried data in Data View; it also highlights the queried objects in the drawing. You can choose which of these methods you prefer by using the check boxes at the bottom of the Query Builder tab. By default, both are checked. The Store button allows you to store your query changes in the Query Editor.

10. Pick the **Store** and **Close** buttons.

11. Close the Data View window.

12. Right-click **ROOMQUERY1** in the tree and read the options in the shortcut menu.

As you can see, right-clicking a query permits you to execute, delete, duplicate, or rename a query in the tree. Double-clicking the query executes it.

13. Click outside the shortcut menu to remove it from the screen.

Working with Link Templates

Link templates provide the key information that AutoCAD needs to link objects in a drawing to unique records in a database table. You may choose to link records from multiple tables to a single object. Likewise, you may link multiple objects to a single record. After a link template has been created, AutoCAD displays it under the drawing file name in the dbConnect tree. The link template can then be used as an alias for the table that it references and serves as another entry point to the Data View window.

Let's create a link template for Computer table in jet_dbsamples.

1. Right-click the **Computer** table in the dbConnect tree and pick the **New Link Template...** option.

This displays the New Link Template dialog box, as shown in Fig. 65-5.

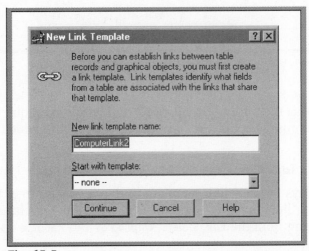

Fig. 65-5

The name dbConnect suggests for the template consists of the table name followed by the word Link and a number. You can keep the suggested name or enter a new name.

2. Change the name to **NewComputerLink**.

3. Under **Start with template**, pick the down arrow and select **ComputerLink1** as a starting point.

4. Pick the **Continue** button.

This displays the Link Template dialog box, as shown in Fig. 65-6.

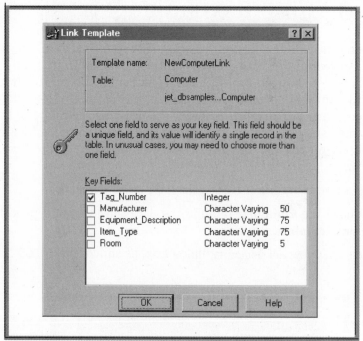

Fig. 65-6

To give dbConnect the information it needs to identify a unique record in a table, you must select the key or keys in the table to which you are creating a link.

5. Check the **Tag_Number** check box and press the **OK** button.

AutoCAD displays the new link template in the dbConnect tree. You can manage existing link templates by right-clicking a link template in the tree and selecting from the Edit, Delete, Duplicate, and Rename options.

NOTE:

You can only edit, delete, or rename a link template when no links exist between the table and the drawing.

Links

Once you have created a link template, you will want to establish links between the objects in the drawing and records in the database.

Working with Links

Before creating new links, let's work with some existing ones to see how they are used to view objects associated with records. We will also see how objects can be used to locate associated records in a table.

1. Double-click the **InventoryLink1** link template.

This displays the Data View window.

2. Pick the **Data View and Query Options** button in the Data View dialog box to display the dialog box shown in Fig. 65-7.

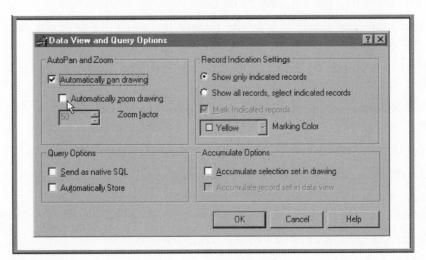

Fig. 65-7

The dialog box permits you to customize the Data View window and how it displays links.

3. Check the **Automatically zoom drawing** check box.

This causes AutoCAD to zoom in automatically on the objects in the drawing that are linked to records that you select in the Data View.

4. Set **Zoom factor** to **20**.

This sets the zoom factor for when AutoCAD zooms in on the linked objects in the drawing. You can experiment with different numbers to find a zoom factor that works the best.

5. Pick **OK** to return to Data View.

6. Pick the **AutoView Linked Objects in Drawing** button in the Data View dialog box.

AutoCAD automatically displays objects associated with any records that you select in Data View. The method that AutoCAD uses to display objects depends on the settings in the Data View and Query Options dialog box.

7. Click in the left margin (gray box) of record **1004** to select the record, and focus your attention on the AutoCAD drawing area.

You may need to move the dbConnect Manager and Data View window out of the way of the AutoCAD window. Notice that AutoCAD zoomed in on and highlighted the drawing objects linked to the selected record.

8. Click the small gray box in the left margin of record 1003.

AutoCAD zooms in on and highlights the objects linked to record 1003.

9. Pick the **AutoView Linked Records in Data View** button in the Data View dialog box.

10. In the AutoCAD drawing, select the desk in room 6020.

Data View refreshes and displays only the records linked to the desk object in the drawing.

11. Select more objects in the drawing and see how the records they are linked to are displayed.

12. Pick the **Data View and Query Options** button in the Data View dialog box.

13. Select the radio button for **Show all records, select indicated records**.

14. Make sure that the **Mark indicated records** check box is checked.

Marking Color should be yellow. With these options selected, Data View always shows the entire table and highlights only the linked records.

15. Pick the **OK** button to return to the Data View window.

16. In the AutoCAD drawing, clear the current selection.

17. Select the desk in room 6020 again.

18. Select the telephone on the desk.

Notice that Data View continues to display all records, but highlights only the selected records.

19. Select more objects in the drawing and see how the records to which they are linked are highlighted in Data View.

AutoViewing data is good for instant feedback. You can also use the corresponding View options to the left of the AutoView buttons in the Data View dialog box to view objects manually. With these buttons, selecting an object or record does not automatically update its linked counterpart. You must pick the View button first to force the display to update.

Creating Links

Now let's look at how to create a link between a record and objects in a drawing. AutoCAD's dbConnect uses the link template configuration to store the record's key fields with selected objects in the drawing so that it can find objects linked to a given record and/or records associated with an object.

1. Close the Data View window.

2. Double-click on **ROOMLINK1** in the dbConnect tree.

3. Pick the **AutoView Linked Objects in Drawing** button in the Data View dialog box.

4. Double-click the gray box in the left margin next to room **6028**.

This zooms in on room 6028 in the drawing because the record has already been linked to the room label.

5. Click in the drawing area to activate it, and then press the **ESC** key twice to clear the selection of the room 6028 label in the drawing.

Now that we have cleared the current selection, we can start fresh with a new selection set to link to a record.

6. Click anywhere in record **6028** in the Data View window.

7. Pick the **Link!** button in the upper left corner of the Data View dialog box.

The Data View window disappears momentarily to allow you to select the objects to link to the currently selected record.

8. Select the internal (magenta) lines that make up the room's perimeter and press **ENTER**.

AutoCAD establishes a link between the selected objects and the current record. It then moves down one record so you can continue linking.

9. Click the gray box in the left margin next to room **6028**.

As you can see, AutoCAD highlights all of the internal lines you selected.

 NOTE:

Links are stored with the objects in the drawing. If you create or delete links, and you do not save the drawing, you will lose your changes.

10. Close the Data View window and dbConnect Manager.

11. Save the changes in the drawing file and exit AutoCAD.

Chapter 65
Review & Activities

Review Questions

1. What is the purpose of connecting to an external database?

2. What does it mean to "configure" a data source connection?

3. Briefly describe how to view and edit database records from within AutoCAD.

4. What is a record?

5. What is a table?

6. What is a key field?

7. What does SQL stand for?

8. How can you narrow down to a specific set of data?

9. What is the purpose of a link template?

Challenge Your Thinking

1. It is possible for the links between the drawing and database to get out of sync when one end of the link is deleted outside of the control of dbConnect. Give some specific examples of when this might occur.

2. Investigate SQL. Why is it considered better, for simple queries, to use dbConnect's Query Builder instead of writing SQL statements from scratch? Why might writing SQL statements be preferable for more complex queries?

 ## Applying AutoCAD Skills

Work the following problems to practice the commands and skills you learned in this chapter.

1. AutoCAD's dbConnect provides the means of labeling linked objects with data from a database. Using the EmployeeLabel1 label definition as a sample, create a new label for computers and apply it to the db.dwg drawing.

Chapter 65
Review & Activities

2. Open the db.dwg file you copied earlier in this chapter. Connect with the data source named Facility Data. View the data in the Employee database. Sort the database by department. How many departments are there? How many employees are there in each department?

3. Create a new query that displays all computers with attached laser printers in db.dwg. Use the query to select all laser printers in the drawing. Then move them to a new layer named Laser. Make the new layer green and turn off all layers except for the layers CPU and Laser to view the results.

4. Use Data View to select all rooms that are linked to the db.mdb database. From the selection set of all linked rooms, and using temporary layers, identify the rooms in the drawing that have no reference in the Room table by hiding the linked room numbers. Turn on only the following layers to view the results: E-B-CORE, E-B-FURR, E-B-GLAZ, E-F-CASE, E-F-SILL, E-F-STAIR, IWALLE, PANELS_201, and RMNUM.

 ## Using Problem-Solving Skills

Complete the following activities using problem-solving skills and your knowledge of AutoCAD. Be sure to use the db.dwg and db.mdb files that you copied earlier in this chapter.

1. A new office building is nearing completion. The telephones will be installed next week, and the installers need a list of the rooms that will have telephones. Prepare a query from the Inventory database to show the inventory ID numbers, rooms, description, and model of the telephones. Sort the query by room in ascending order. Print the query.

2. One of your duties as an administrative assistant is to ensure that the corporate offices (those of the president and the vice-presidents) are ready for occupancy. Open the Data View for EmployeeLink1 and use the AutoView Linked Objects in Drawing button to determine what inventory each of the offices should contain. Sort the Data View by Title to make sure you have checked all the corporate offices. Write down the names of the officers and their room numbers so you can personally check the offices to ensure compliance with drawing specifications.

Part 12 Project

Electronic Art Database

Art for use in technical fields comes in many different forms. Examples include technical illustrations such as engineering drawings, as well as pictorial representations created as conceptual illustrations for prospective clients. Some companies now maintain a database of art created or used by employees. Such a database makes it easy for employees to locate and use the illustrations they need for their current projects.

Description

The purpose of this project is to create a collection of art that would be useful to a specific type of company. The type of company is up to you, but it must have a need for various types of illustrations, including CAD art. You will use AutoCAD to maintain a database of this collection.

1. Determine the type of company for which your collection will be used. Document your choice and list the types of art that the company will need. For example, if you choose an architectural company, you will need to include floor plans, site plans, and other dimensioned drawings. You will also need to include concept drawings and elevations for residential and commercial properties. Fig. P12-1 shows a few examples.

2. Collect the drawings. Include raster and vector drawings created on a computer. Some of your drawings should be done in AutoCAD; others may be collected from other sources. Consider using royalty-free clip art for some of the pictorial representations. Also include at least one hand-drawn drawing that can be digitized. Your collection should have a minimum of 15 illustrations to be useful, but you may add as many more as you think are necessary.

3. Create electronic versions of all the drawings. Digitize any hand-drawn illustrations for which digitizing is practical. For hand-drawn concept illustrations, you may need to consult someone about scanning the images into electronic form.

4. Design a database, or adapt one from another source, to maintain your collection of illustrations. (Many database programs offer example databases that you can adapt to suit your needs.) Be sure each record includes a copy of the illustration to which it refers.

5. Create a hard-copy catalog for use by people in the company who are looking for specific illustrations. Use OLE as necessary to place copies of the illustrations in the catalog, or print the catalog directly from the database.

Part 12 Project

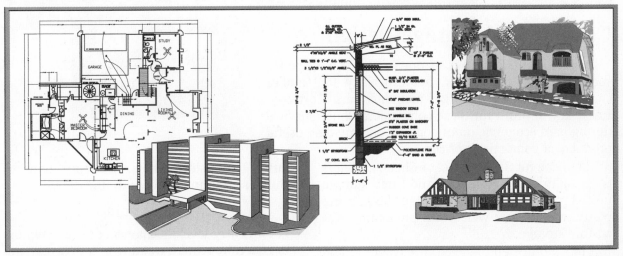

Fig. P12-1

Hints and Suggestions

- Remember that AutoCAD can import many types of electronic art.

- Collect all of the drawings before you begin working on the database. This will allow you to determine the fields you need to include. For example, one of the fields should tell the user in what form the illustration is available (EPS file, AutoCAD file, etc.).

- Ask the advice of database experts to help you set up the database. You may also wish to consult someone who works in the same field as the company for which you are creating this database.

Summary Questions/Self-Evaluation

1. Of what practical use is an illustration database such as the one you have created in this project? Give specific examples related to the database you have created.

2. What "outside help" did you require to complete this project? Explain your answer in terms of the concept of teamwork in the workplace.

3. In most jobs, a combination of skills and knowledge is required to perform the job well. The exact combination depends on the specific job. What knowledge and skills, in addition to those related to AutoCAD, were required to complete this project? Explain why it is helpful to have an understanding of a variety of processes and skills, even if they are not specifically required for your job.

Careers
Using
AutoCAD

Custom Tool Storage Design

Find a niche and keep costs down—this is a recipe for success for all companies, but especially for smaller companies that supply larger manufacturers. Cost control is key to profit for these companies, whose survival during hard economic times depends on providing useful products at competitive prices

The niche for Seibert, Inc. of Chenoa, Illinois is specialty gages, adapters, and fitted storage units for machine tools. Machine tools, being costly to make, are well worth protecting with custom storage.

A tool storage unit begins with a metal frame holding a panel of Formica. Holes are cut in the Formica and bushings are fit in to lightly hold a particular machine tool. Customers order the sizes and arrangement of slots they want from a selection of AutoCAD drawings.

A hand drawing of the customer's order comes to Jeff Shoff, a mechanical drafter at Seibert. Jeff creates an AutoCAD drawing at full size, which is used in the shop as an overlay for cutting the holes in the tool storage unit. In some cases, Jeff may copy or modify an existing AutoCAD file to make the drawing. Like machine tools, mechanical drawings take time and money to produce, and re-using them helps control costs.

Career Focus—Mechanical Drafter

A mechanical drafter prepares technical drawings based on sketches or calculations from a mechanical engineer. The drafter uses a knowledge of manufacturing practices to fill in details. The drawings show multiple views and include specifications such as measurements, materials, and fabrication methods.

A mechanical drafter may work in a shop or office environment, often at a computer workstation alongside others. Drafting requires hours of attention to work in fine detail.

Education and Training

For an entry-level mechanical drafter, employers generally require training at a two-year college or technical institute. Courses cover drafting and mechanical drawing fundamentals, CAD techniques, methods of manufacturing and construction, and mathematics, science, and engineering techniques.

A beginning drafter usually does routine work under close supervision. After gaining experience, the drafter takes on more challenging tasks and may advance to senior drafter or supervisor.

To succeed in drafting, a student should have an aptitude for drawing and for visualizing objects from a two-dimensional plan. An interest in seeing a job done right, down to the last detail, is essential to success as a drafter.

Career Activity

- Think of a niche for which you might start up your own small business. Make a list of five products or services used in your school or office.
- Write a paragraph describing how one of the products/services might be improved at a reasonable cost by your new business.

Additional Problems

Introduction

The following problems provide additional practice with AutoCAD. These problems encompass a variety of disciplines. They will help you expand your knowledge and ability and will offer you new and challenging experiences.

The key to success is to *plan before you begin*. Review the options for setting up a new drawing and consider which commands and features you might use to solve the problem. As you discover new and easier methods of creating drawings, apply these methods to solving the problems. Since there is usually more than one way to complete a drawing, experiment with alternative methods. Discuss these alternatives with other users and create strategies for efficient completion of the problems.

Remember, there is no substitute for practice. The expertise you gain is proportionate to the time you spend on the system. Set aside blocks of time to work with AutoCAD, think through your approach, and enjoy this fascinating technology.

Each problem in this section is preceded by an icon that describes its level of difficulty, as shown below. The level of difficulty assigned to each problem assumes that you have learned the AutoCAD techniques and skills necessary to work the problem successfully. For some of the Level 3 problems, you may be required to expand your experience by combining your knowledge of AutoCAD with knowledge and skills that are not taught in this book.

 Uses basic AutoCAD skills

 Uses intermediate AutoCAD skills

 Uses advanced AutoCAD and problem-solving skills

AdditionalProblems

Problem 1

Create and fully dimension the drawing of a link.

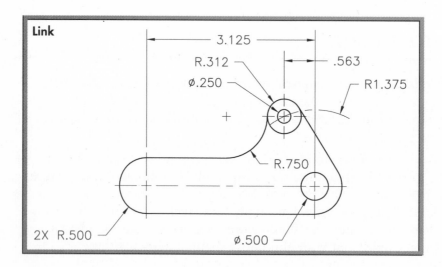

Problem 2

Create and fully dimension the drawing of a filler.

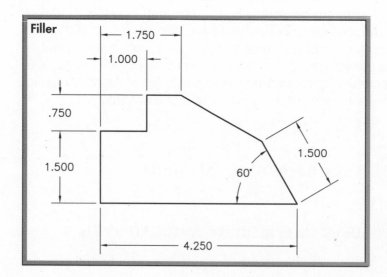

AdditionalProblems

LEVEL 1

Problem 3

Create and fully dimension the drawing of a link.

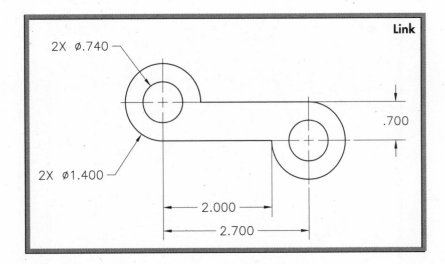

LEVEL 1

Problem 4

Create and fully dimension the drawing of a shaft bracket.

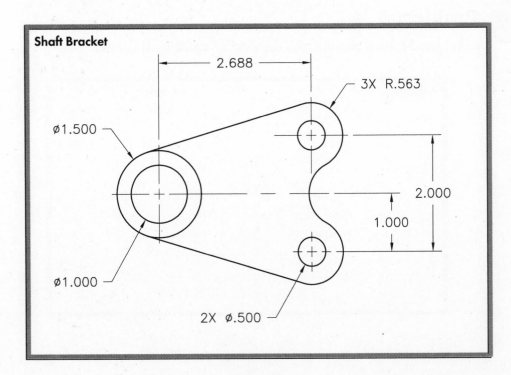

AdditionalProblems

Problem 5

Create and fully dimension the drawing of a gage.

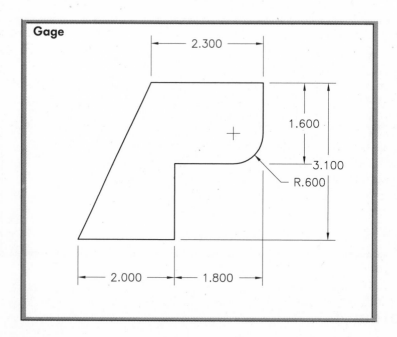

Problem 6

Create and fully dimension the drawing of an adjustable link.

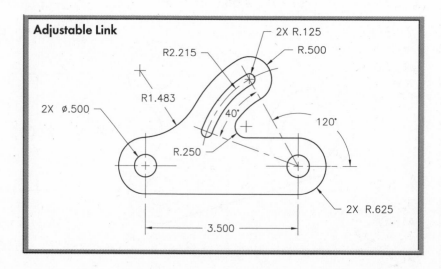

AdditionalProblems

LEVEL
1

Problem 7

Create and dimension the orthographic views of the block.

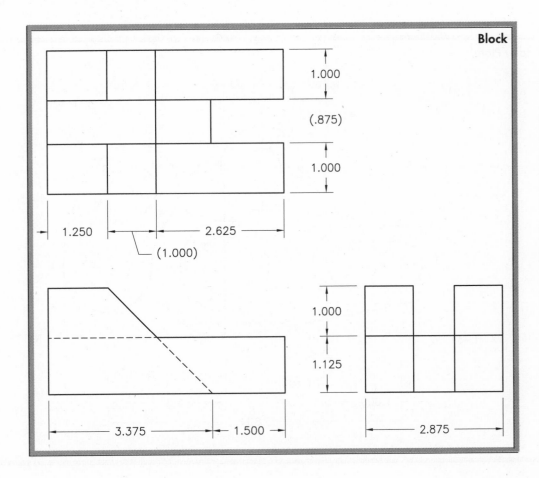

AdditionalProblems

Problem 8

Create and dimension the drawing of an idler plate.

Idler Plate

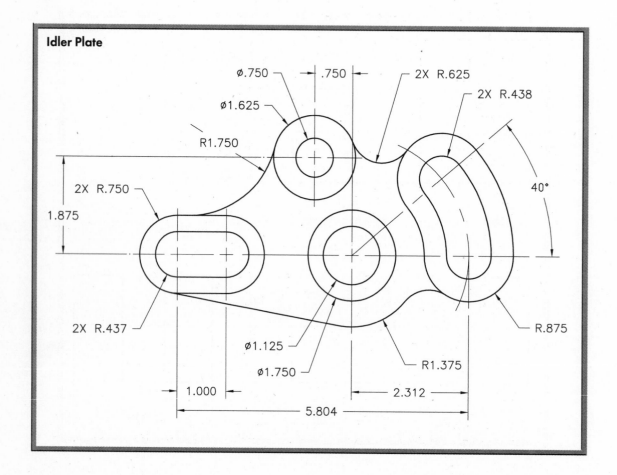

Problem 8 courtesy of Steve Huycke, Lake Michigan College

AdditionalProblems

Problem 9

Create and dimension the drawing of an end block.

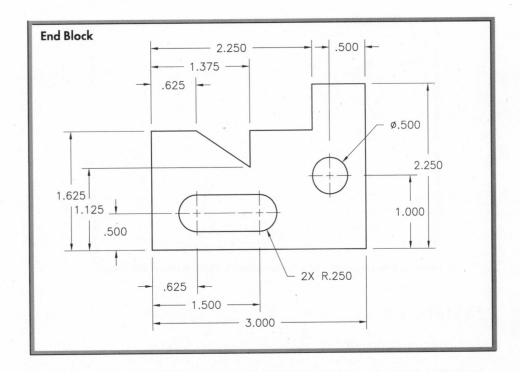

End Block

AdditionalProblems

Problem 10

Shown below is an isometric view of an angle block. Create orthographic views of the block. Use a drawing area of 17×11.

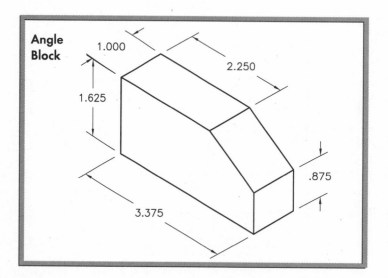

Problem 11

Create and dimension orthographic views of the support.

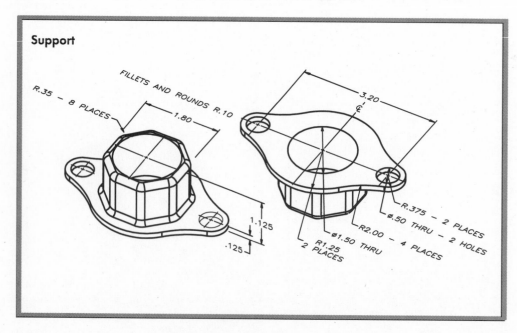

Problem 11 courtesy of Gary J. Hordemann, Gonzaga University

AdditionalProblems

LEVEL 2

Problem 12

Create and dimension the drawing of a cover plate.

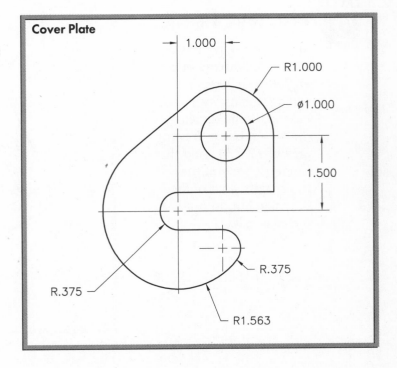

Cover Plate

- 1.000 -
- R1.000
- Ø1.000
- 1.500
- R.375
- R.375
- R1.563

LEVEL 1

Problem 13

Shown at the right is an isometric drawing of a nesting block. Create orthographic views of the block and fully dimension the views. Add the isometric view in the upper right area of the drawing, but do not dimension this view.

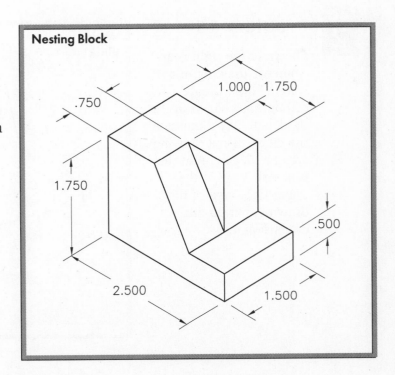

Nesting Block

- .750
- 1.000   1.750
- 1.750
- .500
- 2.500
- 1.500

AdditionalProblems

Problem 14

Shown at the right is an isometric drawing of an angled base. Create orthographic views of the base and fully dimension the views. Use a drawing area of 17 × 11. Add the isometric view in the upper right area of the drawing, but do not dimension it.

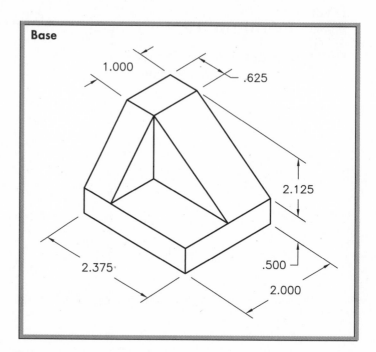

Base

1.000
.625
2.125
2.375
.500
2.000

Problem 15

Shown at the right is an isometric drawing of a wedge block. Create orthographic views of the block and fully dimension the views. Use a drawing area of 17 × 11. Add the isometric view in the upper right area of the drawing, but do not dimension it.

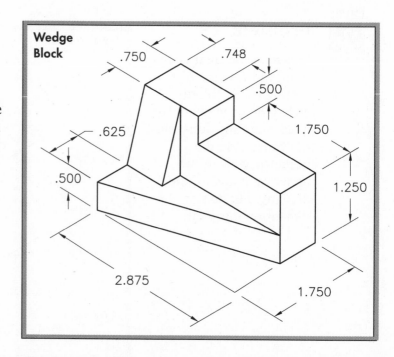

Wedge Block

.750
.748
.500
.625
1.750
.500
1.250
2.875
1.750

AdditionalProblems

LEVEL 2
Problem 16

Shown at the right is an isometric view of an angle block. Create and fully dimension the orthographic views of the angle block. Add the isometric view in the upper right area of the drawing, but do not dimension it.

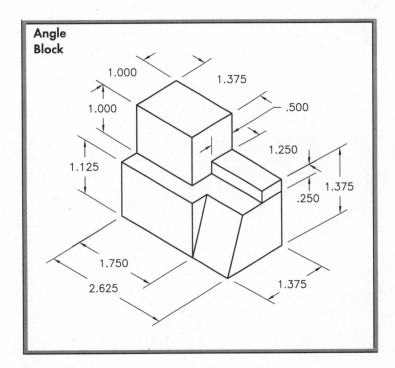

Angle Block

1.000

1.375

1.000

.500

1.125

1.250

1.375

.250

1.750

2.625

1.375

LEVEL 2
Problem 17

Create the isometric view of the 90° link shown at the right.

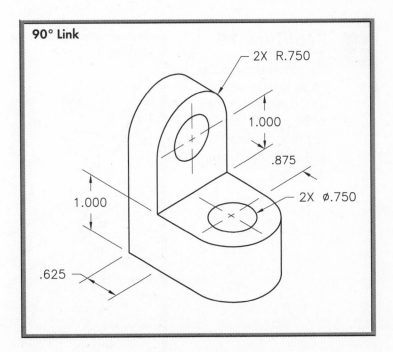

90° Link

2X R.750

1.000

.875

1.000

2X ⌀.750

.625

AdditionalProblems

Problem 18

Shown at the right is an isometric view of a locator. Create orthographic views of the locator and fully dimension the views.

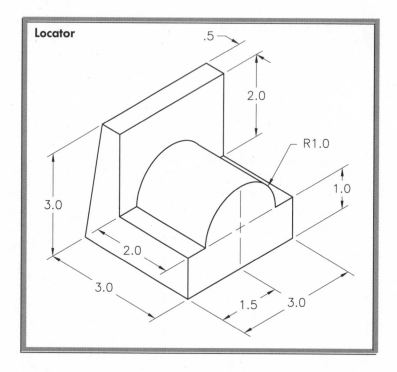

Locator

.5

2.0

R1.0

3.0

1.0

2.0

3.0

1.5   3.0

Problem 19

Shown at the right is an isometric view of a step block. Create orthographic views of the step block and fully dimension the views.

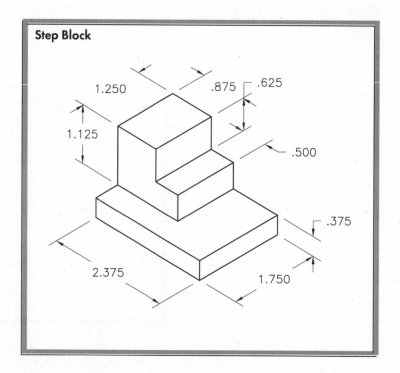

Step Block

1.250   .875   .625

1.125

.500

.375

2.375   1.750

AdditionalProblems

Problem 20

Shown at the right is an isometric view of a cradle. Create orthographic views of the cradle and fully dimension the views. Add the isometric view in the upper right area of the drawing, but do not dimension it.

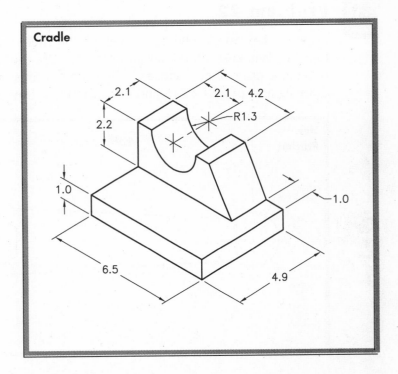

Cradle

Problem 21

Create and fully dimension the drawing of a block.

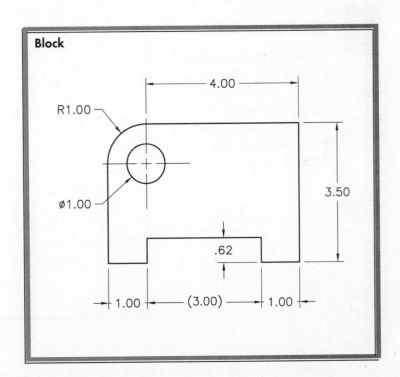

Block

AdditionalProblems

Problem 22

Draw the hex ratchet. Begin by creating the circles for the ends of the ratchet. Then create the lines and polygon. Use the BREAK command to remove parts of both circles as indicated to finish the drawing. Use a drawing area of 280 × 215 and dimension the drawing in millimeters.

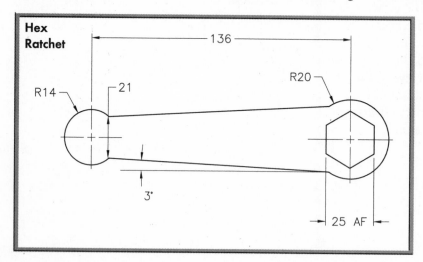

Problem 23

Draw the front and right-side views of the gear as shown. Use the detail drawing to create the teeth accurately. Hatch as indicated using the ANSI31 pattern.

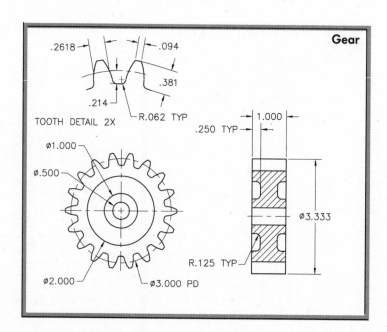

Problem 22 adapted from the textbook *Drafting Fundamentals* by Scott, Foy, and Schwendau
Problem 23 courtesy of Steve Huycke, Lake Michigan College

AdditionalProblems

LEVEL 3

Problem 24

The mechanism shown below consists of three parts: a driver, a follower, and a link. Use the dimensions shown in the illustration and the CIRCLE, OFFSET, and FILLET commands to draw the front view.

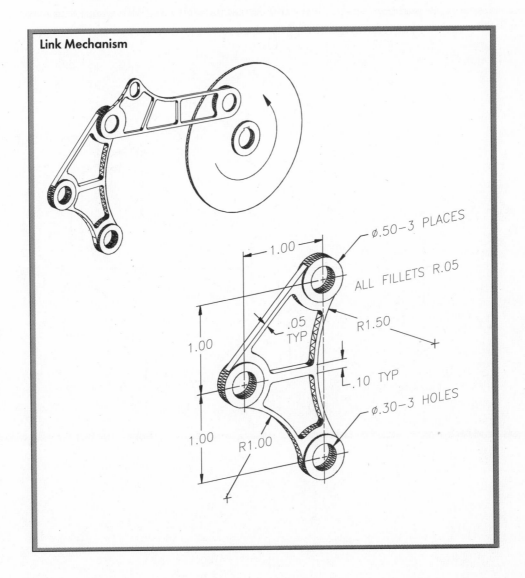

Link Mechanism

ø.50–3 PLACES

ALL FILLETS R.05

R1.50

.05 TYP

.10 TYP

ø.30–3 HOLES

R1.00

1.00

1.00

1.00

Problem 24 courtesy of Gary J. Hordemann, Gonzaga University

AdditionalProblems

Problem 25

Create and fully dimension the isometric view of the strip block shown below.
Use a drawing area of 17 × 11.

Strip Block

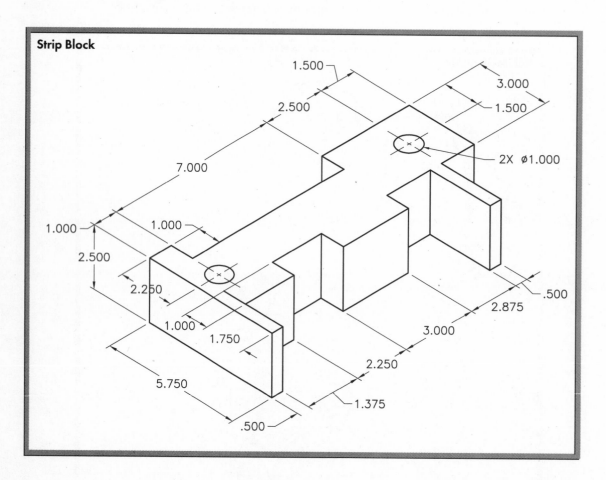

AdditionalProblems

Problem 26

Create and fully dimension the drawing of a master template. Use 24 × 18 for the drawing area.

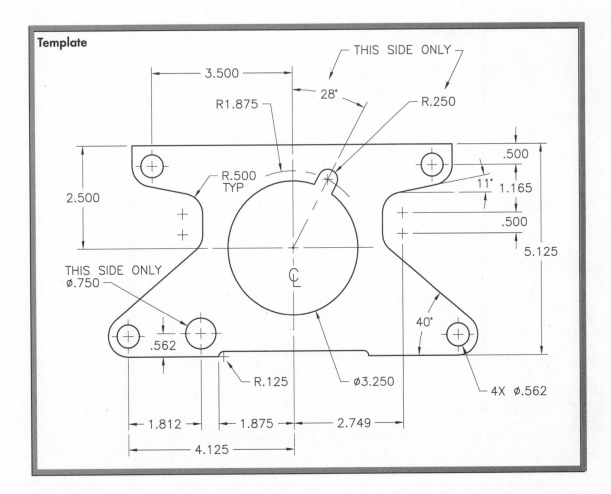

Problem 26 courtesy of Steve Huycke, Lake Michigan College

AdditionalProblems

LEVEL
1

Problem 27

Create the drawing of an angle bracket. Use the dimensions shown, but dimension the drawing using decimal units.

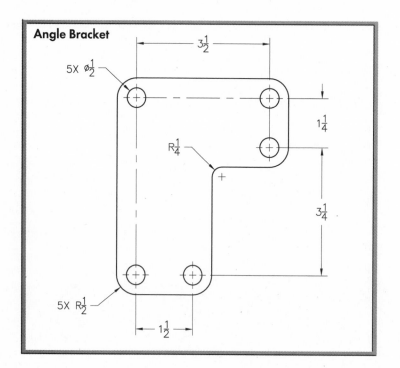

LEVEL
2

Problem 28

Create and dimension the drawing of a rod support.

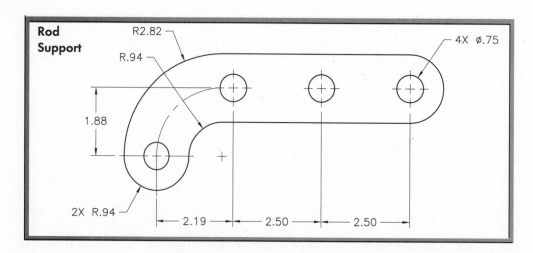

AdditionalProblems

Problem 29

Create and dimension the drawing of a steel plate.

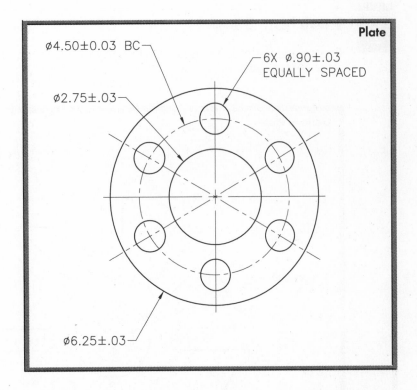

Plate

ø4.50±0.03 BC

6X ø.90±.03
EQUALLY SPACED

ø2.75±.03

ø6.25±.03

Problem 30

Create and dimension the drawing of a rocker arm.

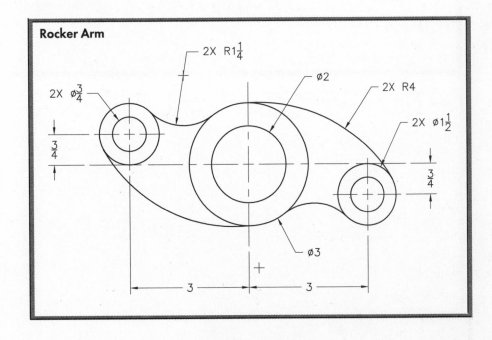

Rocker Arm

2X R1$\frac{1}{4}$

ø2

2X R4

2X ø$\frac{3}{4}$

2X ø1$\frac{1}{2}$

$\frac{3}{4}$

$\frac{3}{4}$

ø3

3

3

AdditionalProblems

Problem 31

Create and fully dimension the drawing of a Geneva plate. Use a drawing area of 24 × 18.

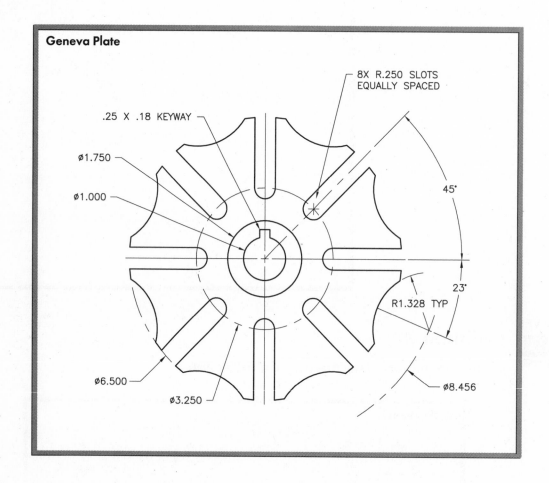

Geneva Plate

AdditionalProblems

LEVEL 2 Problem 32

Create and dimension the drawing of a cassette reel. Use a drawing area of 280 × 215.

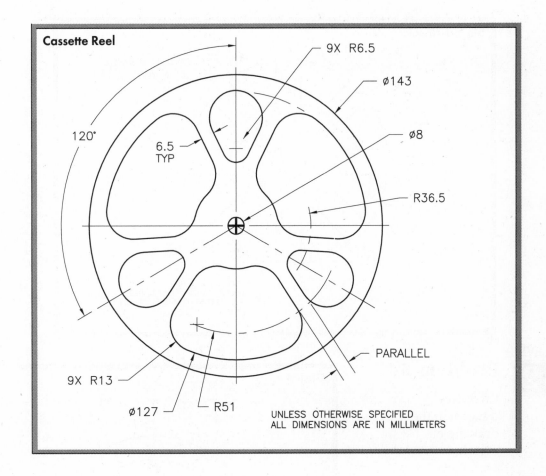

Cassette Reel

9X R6.5

Ø143

Ø8

R36.5

120°

6.5 TYP

PARALLEL

9X R13

Ø127

R51

UNLESS OTHERWISE SPECIFIED
ALL DIMENSIONS ARE IN MILLIMETERS

AdditionalProblems

Problem 33

Create and dimension the drawing of a slotted wheel. Use a drawing area of 17×11.

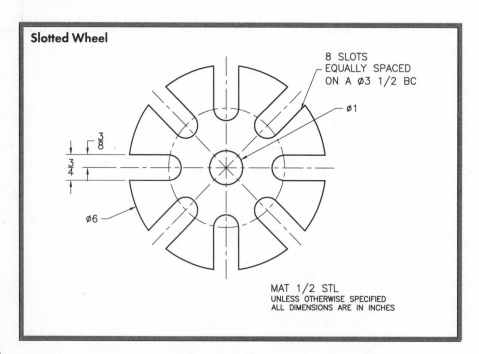

Slotted Wheel

8 SLOTS
EQUALLY SPACED
ON A Ø3 1/2 BC

Ø1

Ø6

MAT 1/2 STL
UNLESS OTHERWISE SPECIFIED
ALL DIMENSIONS ARE IN INCHES

Problem 34

Create and dimension the drawing of a gasket.

Problems 33 and 34 adapted from the textbook *Drafting Fundamentals* by Scott, Foy, and Schwendau

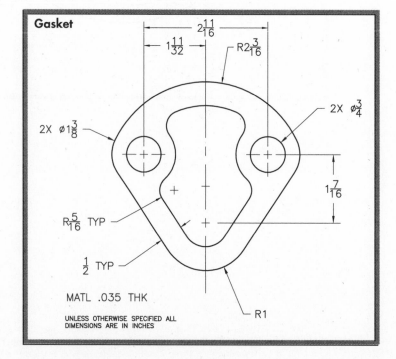

Gasket

$2\frac{11}{16}$

$1\frac{11}{32}$

$R2\frac{3}{16}$

2X Ø$\frac{3}{4}$

2X Ø$1\frac{3}{8}$

$1\frac{7}{16}$

$R\frac{5}{16}$ TYP

$\frac{1}{2}$ TYP

R1

MATL .035 THK

UNLESS OTHERWISE SPECIFIED ALL
DIMENSIONS ARE IN INCHES

AdditionalProblems

LEVEL
2

Problem 35

Create and dimension the drawing of an 18-tooth cutter. Use a drawing area of 17 × 11.

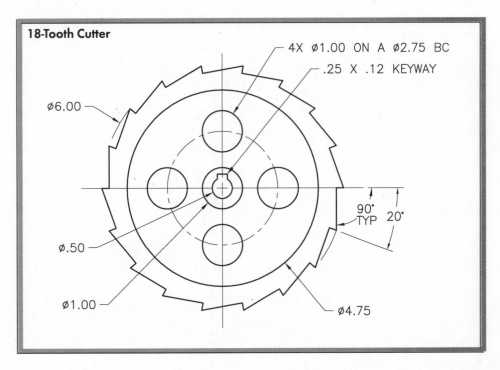

AdditionalProblems

Problem 36

Create and dimension the top and front orthographic views of the slotted shaft.

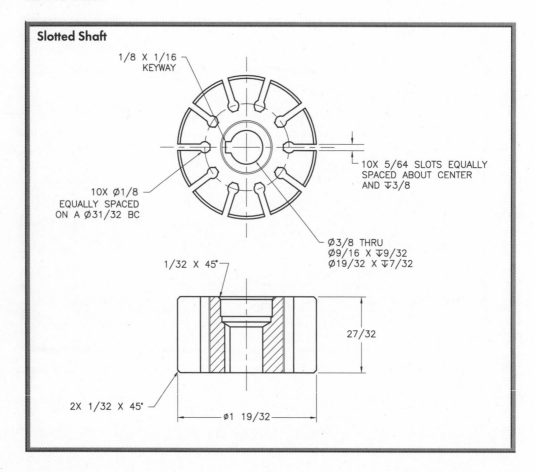

AdditionalProblems

Problem 37

Create and dimension the drawing of an adjustable bracket.

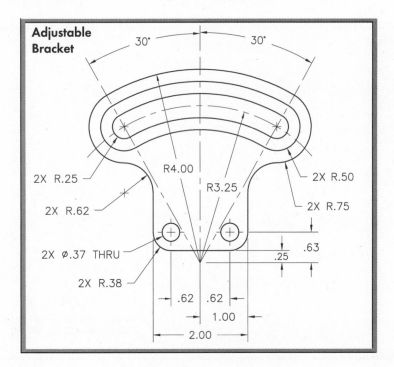

Problem 38

Create and dimension the drawing of a basketball backboard. Use a scale of 1″ = 1′.

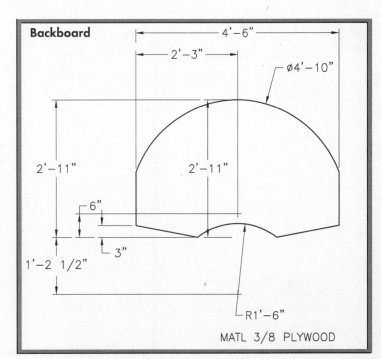

Problems 37 and 38 adapted from the textbook *Drafting Fundamentals* by Scott, Foy, and Schwendau

AdditionalProblems

LEVEL 1 Problem 39

Create and dimension the drawing of a gasket.

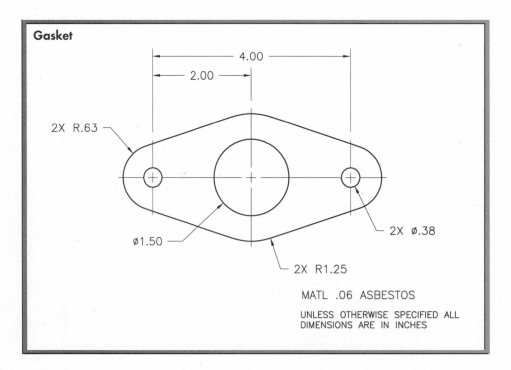

LEVEL 2 Problem 40

Create and dimension
a complete drawing
of the plaque.

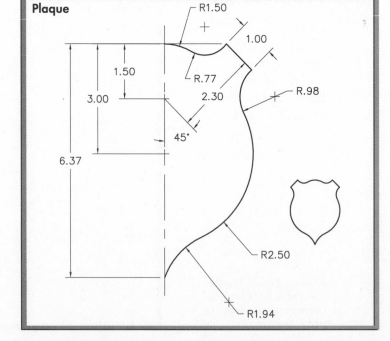

Problem 39 adapted from the
textbook *Drafting Fundamentals*
by Scott, Foy, and Schwendau
Problem 40 courtesy of Mark
Schwendau, Kishwaukee
College

AdditionalProblems

Problem 41

Create and dimension the drawing of a wrench. The units are millimeters. Determine the drawing area on your own, based on the dimensions of the wrench.

Wrench

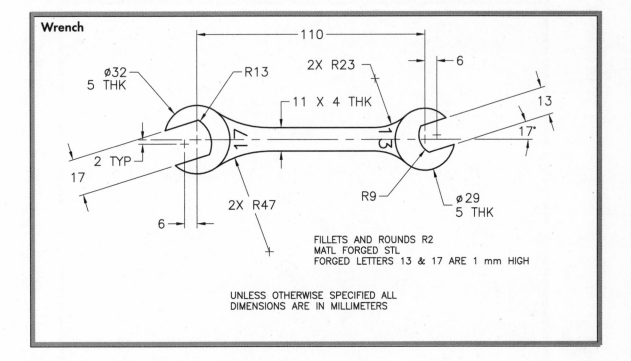

110

2X R23

6

ø32
5 THK

R13

11 X 4 THK

13

17

17°

2 TYP

17

2X R47

R9

ø29
5 THK

6

6

FILLETS AND ROUNDS R2
MATL FORGED STL
FORGED LETTERS 13 & 17 ARE 1 mm HIGH

UNLESS OTHERWISE SPECIFIED ALL
DIMENSIONS ARE IN MILLIMETERS

Problem 41 adapted from the textbook *Drafting Fundamentals* by Scott, Foy, and Schwendau

AdditionalProblems

Problem 42

Create orthographic views of the bracket and fully dimension them.

Bracket

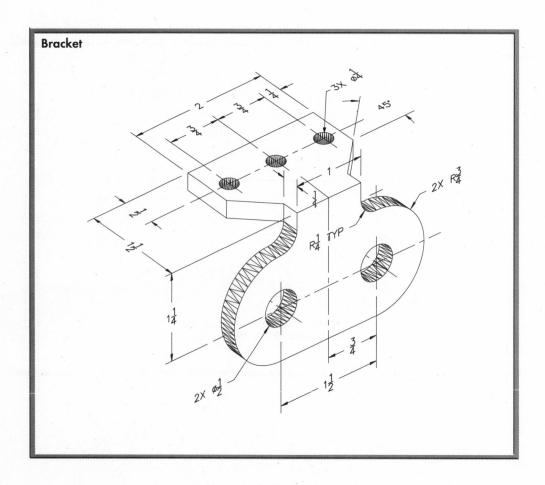

AdditionalProblems

Problem 43

Create and dimension the spacer. Use a drawing area of 17 × 11.

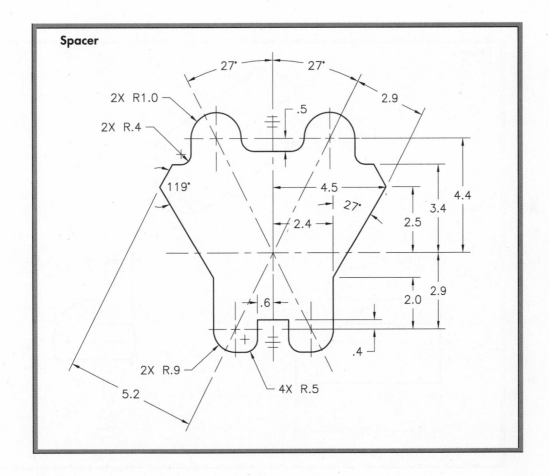

AdditionalProblems

Problem 44

Create the front and right-side views of the fitting and then construct the auxiliary view. Dimension the front and right-side views.

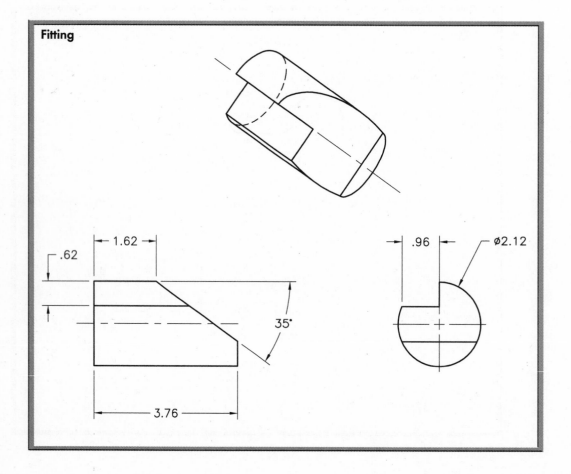

Fitting

AdditionalProblems

LEVEL
3

Problem 45

Create the front and right-side views of the block and then construct the auxiliary view. Dimension the front and right-side views.

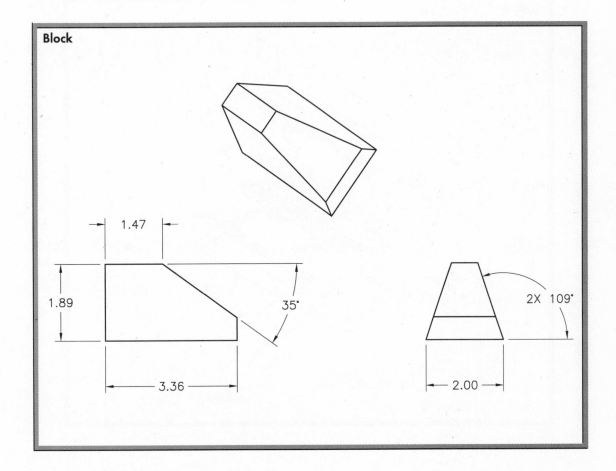

Block

1.47

1.89

35°

3.36

2X 109°

2.00

Problem 45 courtesy of John F. Kirk, Kirk & Associates

AdditionalProblems

Problem 46

Create the front and right-side views of the casting. The right-side view should include a half section. Fully dimension both views.

Casting

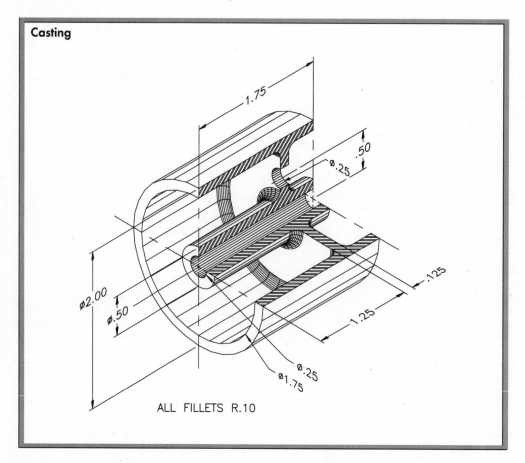

ALL FILLETS R.10

AdditionalProblems

LEVEL
3

Problem 47

Create the top and front views of the alternator bracket. Use a drawing area of 17 × 11.

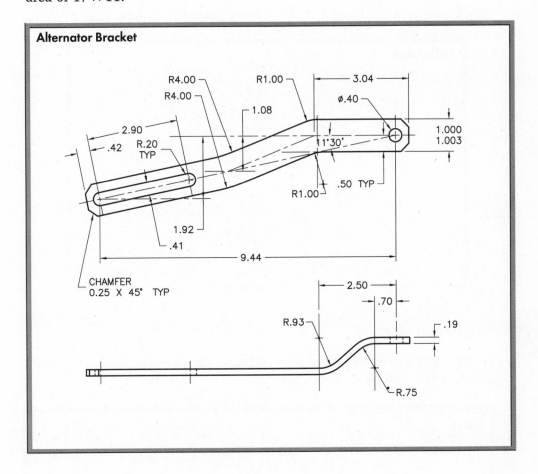

Alternator Bracket

Problem 47 adapted from the textbook *Drafting Fundamentals* by Scott, Foy, and Schwendau

AdditionalProblems

LEVEL 3

Problem 48

Create the necessary orthographic views of the angle block and fully dimension the views. Use a drawing area of 17 × 11. Add an isometric view in the upper right area of the drawing, but do not dimension it.

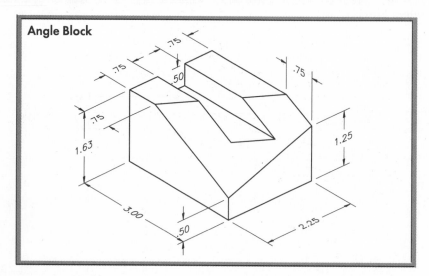

LEVEL 2

Problem 49

Create a single view of the spacer. Dimension the drawing and add a note indicating the thickness of the spacer.

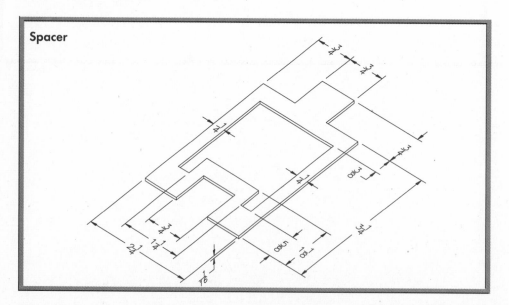

Problems 48 and 49 adapted from the textbook *Drafting Fundamentals* by Scott, Foy, and Schwendau

AdditionalProblems

Problem 50

Using AutoCAD's 3D capabilities, create a model of the impeller. Approximate all dimensions.

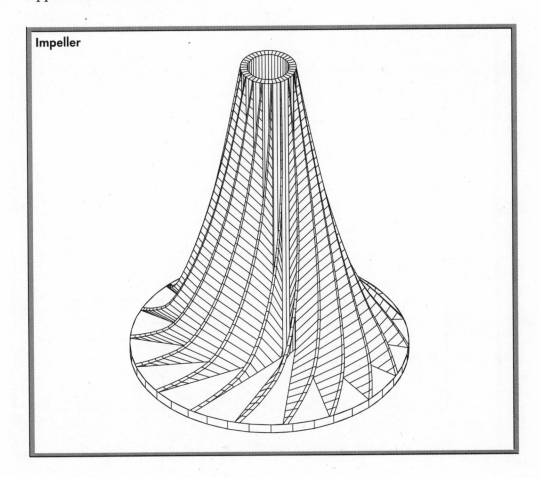

Impeller

AdditionalProblems

Problem 51

Create a solid model of the block.

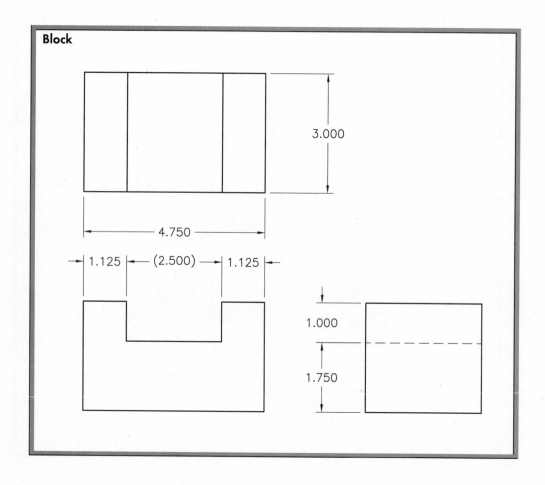

AdditionalProblems

Problem 52

Create a solid model of the block. Then produce orthographic views from the solid model and dimension the views.

Block

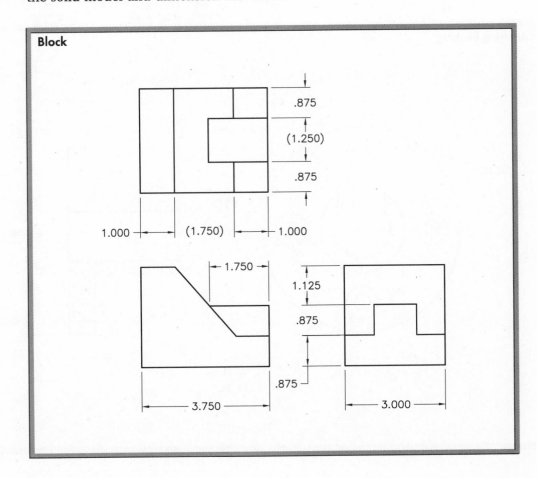

AdditionalProblems

Problem 53

Create the front and right-side views of the centering bushing. Include dimensions.

Centering Bushing

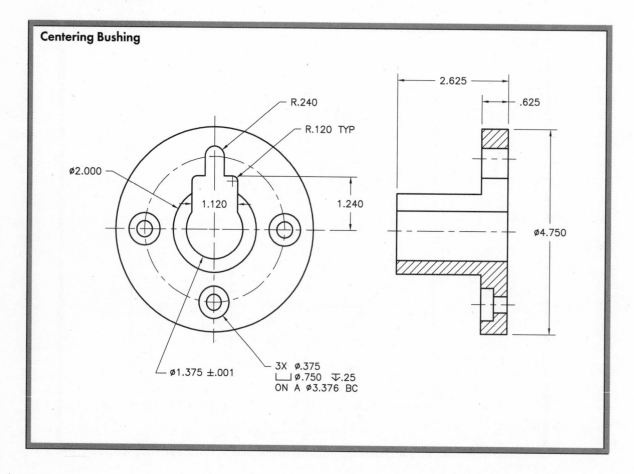

Problem 53 courtesy of Steve Huycke, Lake Michigan College

AdditionalProblems

LEVEL 2 Problem 54

Create the views of the rod support as shown and fully dimension them. Determine the drawing area on your own, based on space requirements.

Rod Support

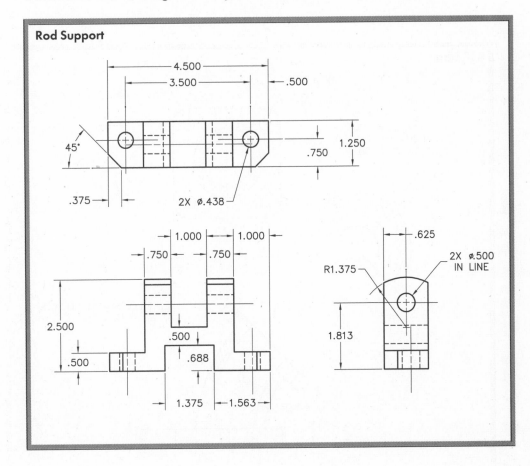

Problem 54 courtesy of Steve Huycke, Lake Michigan College

AdditionalProblems

Problem 55

Create the views of the 45° elbow as shown and fully dimension them. Determine the drawing area on your own, based on space requirements.

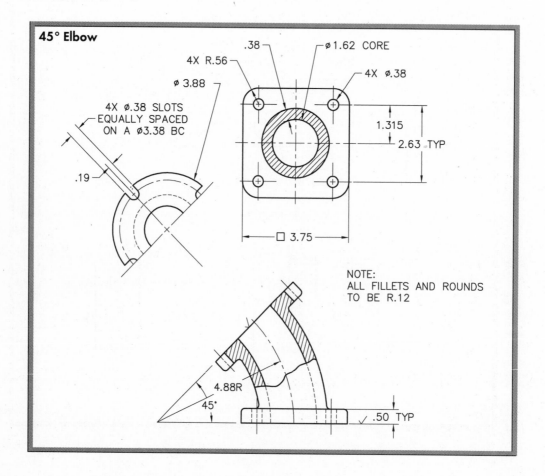

45° Elbow

.38

ø1.62 CORE

4X R.56

4X ø.38

ø 3.88

4X ø.38 SLOTS
EQUALLY SPACED
ON A ø3.38 BC

1.315

2.63 TYP

.19

☐ 3.75

NOTE:
ALL FILLETS AND ROUNDS
TO BE R.12

4.88R

45°

.50 TYP

AdditionalProblems

LEVEL 3

Problem 56

Create the front and full section views of the sprocket and fully dimension them. Units are in millimeters. Determine the drawing area on your own.

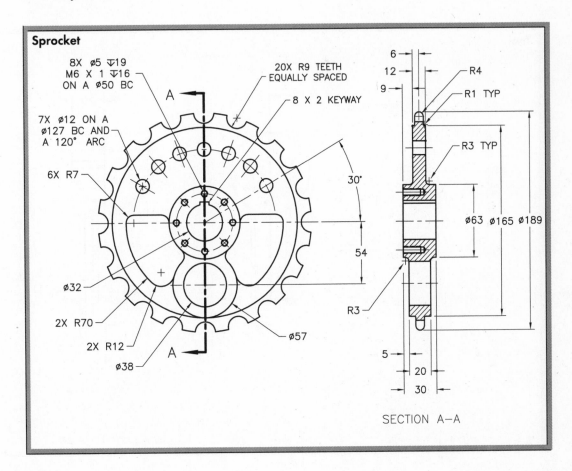

AdditionalProblems

LEVEL
3

Problem 57

Create the front and full section views of the tool and dimension them. Also include the detail as shown. Determine the drawing area, based on space requirements. Units are in millimeters.

Tool

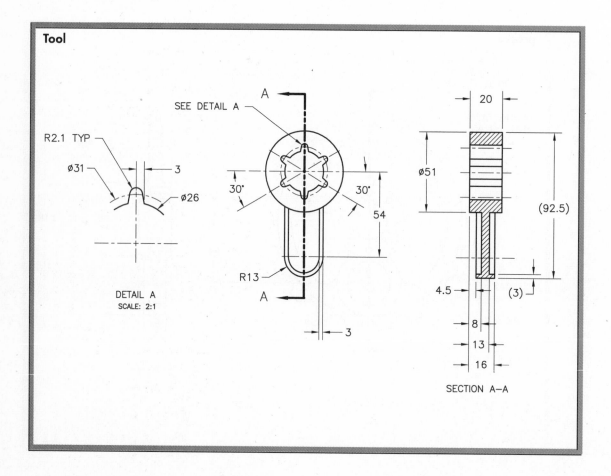

SEE DETAIL A

R2.1 TYP

ø31

3

ø26

30°   30°

54

DETAIL A
SCALE: 2:1

R13

3

20

ø51

(92.5)

4.5   (3)

8

13

16

SECTION A–A

Problem 57 courtesy of Julie H. Wickert, Austin Community College

AdditionalProblems

LEVEL

Problem 58

Create a drawing of the template and dimension it as shown. Determine the drawing area, based on space requirements.

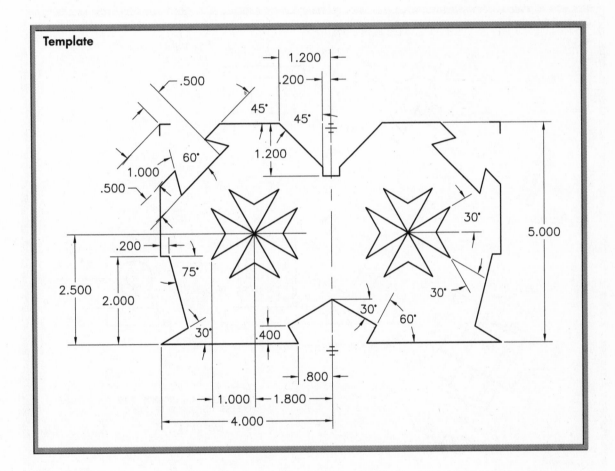

Template

Problem 58 courtesy of Steve Huycke, Lake Michigan College

AdditionalProblems

Problem 59

Create the entire set of drawings of the control bracket. Determine the drawing area, based on space requirements.

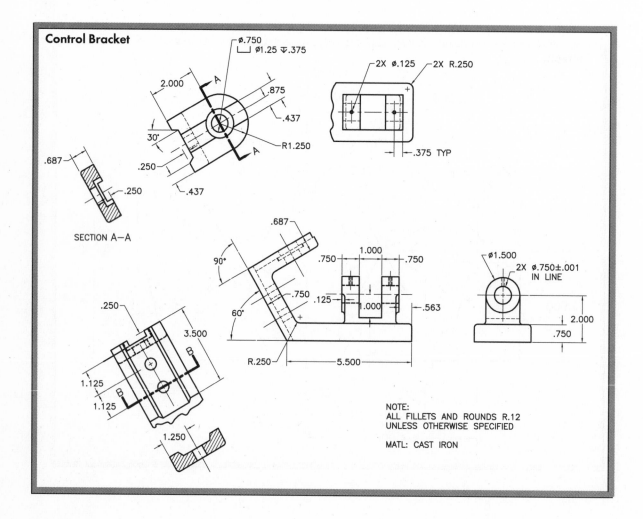

Control Bracket

SECTION A—A

NOTE:
ALL FILLETS AND ROUNDS R.12
UNLESS OTHERWISE SPECIFIED

MATL: CAST IRON

Problem 59 courtesy of Steve Huycke, Lake Michigan College

AdditionalProblems

Problem 60

Create the top and front full section view of the arm. Be sure to include the detail.

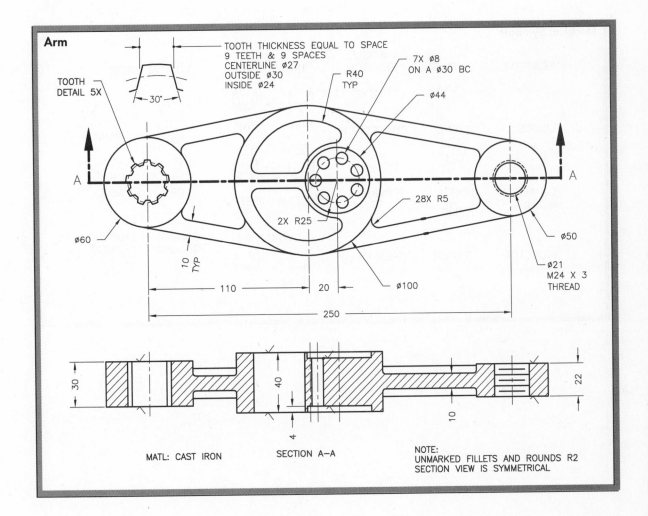

AdditionalProblems

LEVEL

Problem 61

Create the electrical symbols as shown. Estimate their sizes. Block each symbol to create a symbol library.

Electrical Symbols

Problem 61 courtesy of Robert Pruse, Fort Wayne Community Schools

AdditionalProblems

 LEVEL **2**

Problem 62

Create the electrical schematic drawing. Create blocks of each of the components and then insert them as needed.

Electrical Schematic

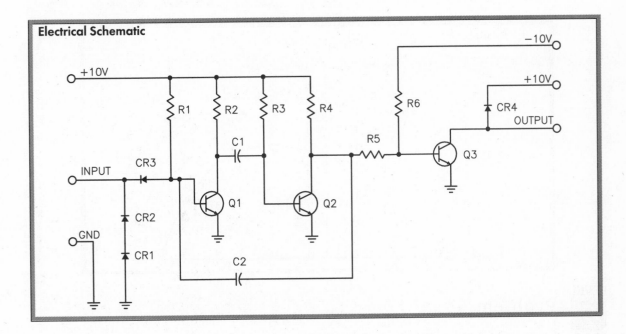

AdditionalProblems

Problem 63

Create and dimension the irregular curve. Determine the drawing area, based on space requirments.

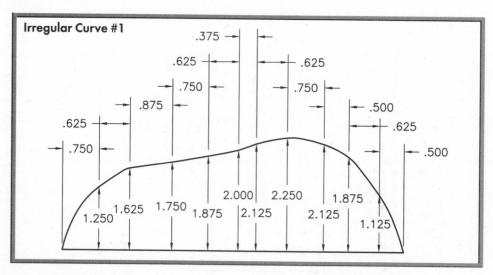

Irregular Curve #1

Problem 64

Create and dimension the irregular curve. Determine the drawing area, based on space requirments.

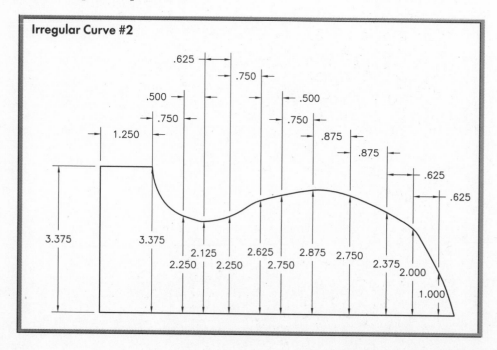

Irregular Curve #2

Additional Problems

LEVEL 3

Problem 65

Create the drawing of the fork lift as well as the bill of materials.

Fork Lift

3pt. HITCH

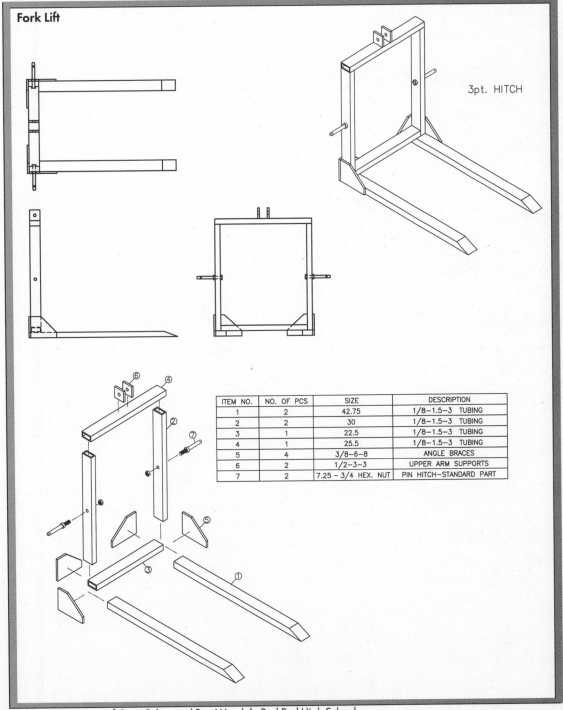

ITEM NO.	NO. OF PCS	SIZE	DESCRIPTION
1	2	42.75	1/8–1.5–3 TUBING
2	2	30	1/8–1.5–3 TUBING
3	1	22.5	1/8–1.5–3 TUBING
4	1	25.5	1/8–1.5–3 TUBING
5	4	3/8–6–8	ANGLE BRACES
6	2	1/2–3–3	UPPER ARM SUPPORTS
7	2	7.25 – 3/4 HEX. NUT	PIN HITCH–STANDARD PART

Problem 65 courtesy of Craig Pelate and Ron Weseloh, Red Bud High School

AdditionalProblems

LEVEL
3

Problem 66

Create the top, front, right-side, and isometric views of the picnic table.

Picnic Table

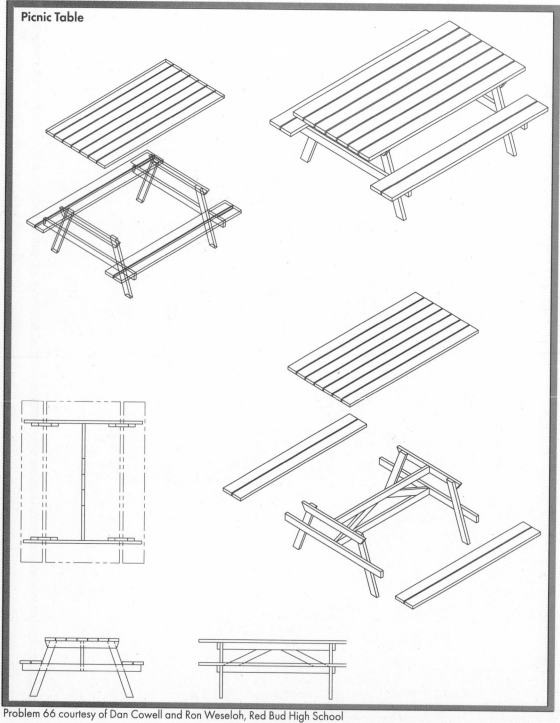

Problem 66 courtesy of Dan Cowell and Ron Weseloh, Red Bud High School

AdditionalProblems

Problem 67

Create and fully dimension the crank handle using decimal inches to three decimal places.

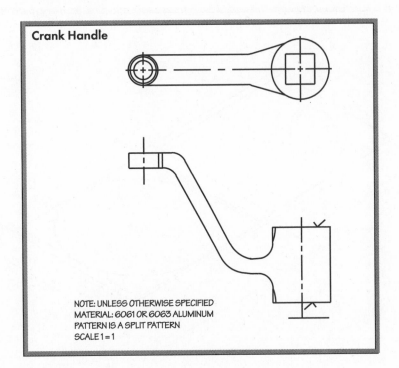

Crank Handle

NOTE: UNLESS OTHERWISE SPECIFIED
MATERIAL: 6061 OR 6063 ALUMINUM
PATTERN IS A SPLIT PATTERN
SCALE 1 = 1

Problem 68

Create the following international symbols.

International Symbols

Problems 67 and 68 courtesy of Joseph K. Yabu, San Jose State University

AdditionalProblems

Problem 69

Create and fully dimension an isometric drawing of the CD jewel case.

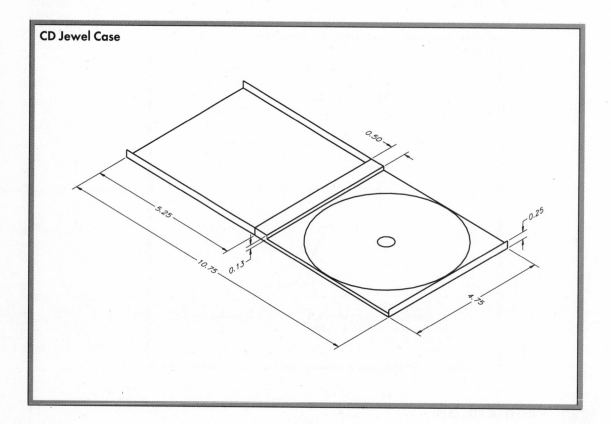

CD Jewel Case

Problems 69 through 71 courtesy of Joseph K. Yabu, San Jose State University

AdditionalProblems

Problem 70

Create a drawing of the the hubcap according to the dimensions given.
Do not dimension.

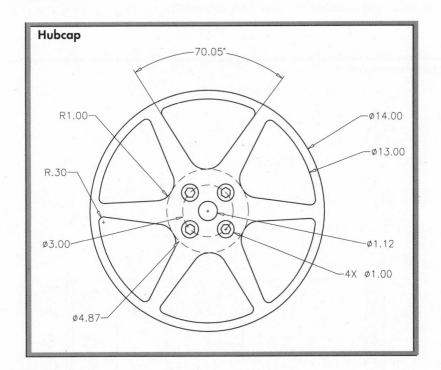

Problem 71

Create drawings of the hubcaps shown below. Estimate all dimensions.

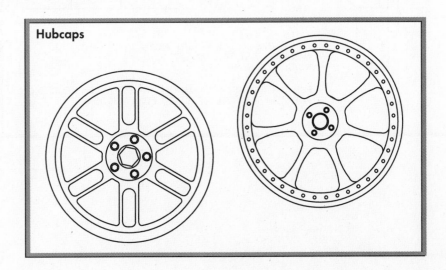

AdditionalProblems

LEVEL
3
Problem 72

Create the front and side views of the structural bracket and dimension them as shown. Also, include a section and isometric views. Determine the drawing area, based on space requirements.

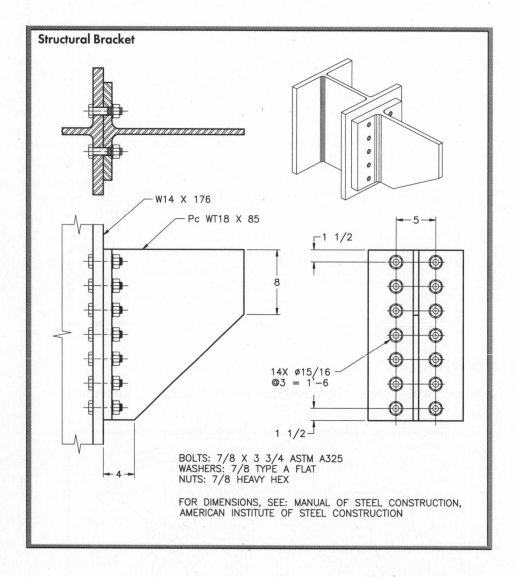

Structural Bracket

W14 X 176

Pc WT18 X 85

8

1 1/2

5

14X Ø15/16
@3 = 1'-6

1 1/2

4

BOLTS: 7/8 X 3 3/4 ASTM A325
WASHERS: 7/8 TYPE A FLAT
NUTS: 7/8 HEAVY HEX

FOR DIMENSIONS, SEE: MANUAL OF STEEL CONSTRUCTION,
AMERICAN INSTITUTE OF STEEL CONSTRUCTION

Problem 72 courtesy of Gary J. Hordemann, Gonzaga University

AdditionalProblems

Problem 73

In a steel frame building, a bracket is to be connected to a column as shown. The front and right-side views of the structural detail are given below. The W 14 × 26 wide flange beam is to be bolted to a 9 × 5 × ³/₈ plate which, in turn, is to be welded to the W 8 × 35 wide flange beam. The dimensions for the two beams are taken from the American Institute of Steel Construction's *Manual of Steel Construction*. Draw the two views.

Structural Assembly

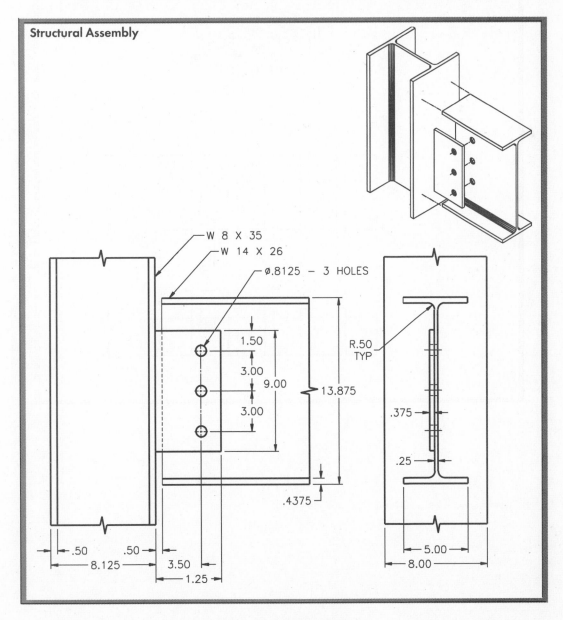

Problem 73 courtesy of Gary J. Hordemann, Gonzaga University

AdditionalProblems

Problem 74

Draw the top and front views of the barrier block shown below. Study the top view and figure out how to minimize drawing and maximize the use of replicating commands such as COPY, MIRROR, and ARRAY.

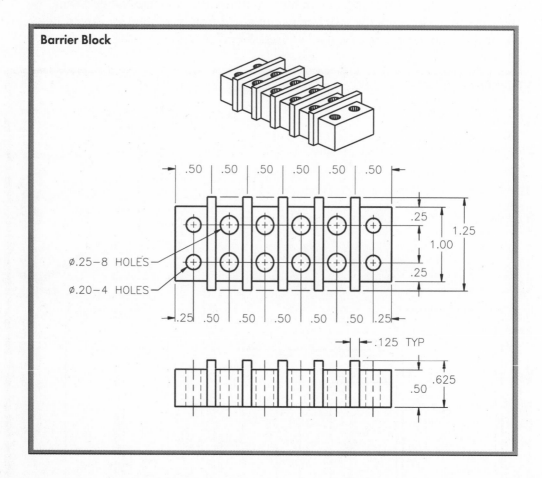

Barrier Block

Ø.25−8 HOLES

Ø.20−4 HOLES

.125 TYP

Problem 74 courtesy of Gary J. Hordemann, Gonzaga University

AdditionalProblems

Problem 75

The front view of the link shown below is amazingly easy to draw if you draw the large circle first; then draw long horizontal and vertical lines snapped to the center of the circle. Offset the lines to locate the centers of the two smaller circles; if necessary, use a zero-distance chamfer or zero-radius fillet to extend the lines until they meet. Insert the two small circles, snapped to the intersections of the lines. Offset again to construct the lines composing the lower left corner and then fillet that corner. Draw a line snapped tangent to the upper small circle and the large circle; use the FILLET command to insert the two arcs joining the small and large circles. Finally, insert the octagon snapped to the center of the large circle.

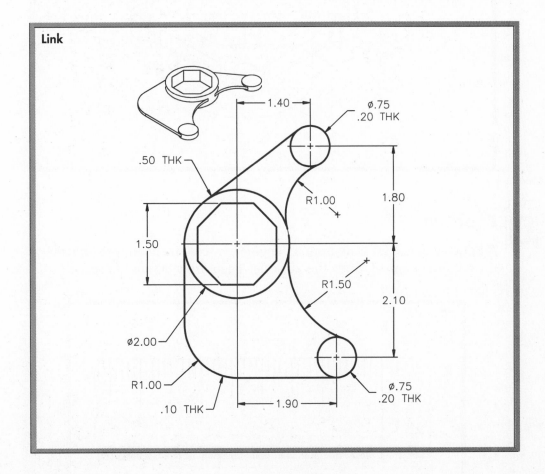

Link

Problem 75 courtesy of Gary J. Hordemann, Gonzaga University

AdditionalProblems

Problem 76

Draw the top view of the stanchion shown below.

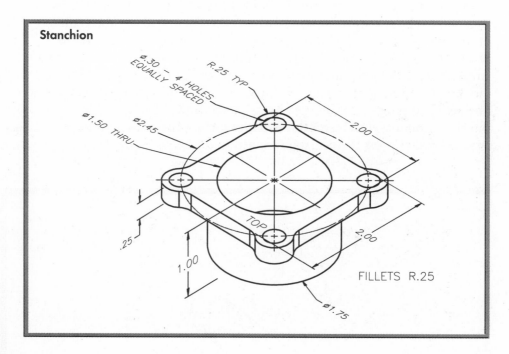

Stanchion

Problem 77

A drawing of the 20 scale from an engineer's scale is shown below. Using the ARRAY command, draw the 20 scale. Then draw 30 and 50 scales.

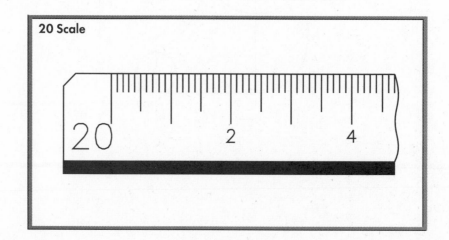

20 Scale

Problems 76 and 77 courtesy of Gary J. Hordemann, Gonzaga University

AdditionalProblems

LEVEL
2

Problem 78

Draw the top view of the spacer shown below. Assume that all of the ribs have a width of .2 and are symmetrical about the center lines. Try this approach: Create the hexagon and offset it. Add one set of circles and array them. Draw lines joining the centers of the circles and polygons. Then offset the lines, fillet them, and array them. *Note:* The fillets are between lines and circles and between lines and the outer polygon, *not* between lines.

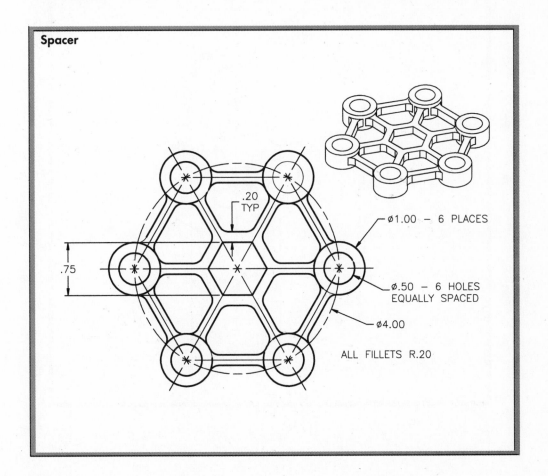

Spacer

.20
TYP

.75

Ø1.00 – 6 PLACES

Ø.50 – 6 HOLES
EQUALLY SPACED

Ø4.00

ALL FILLETS R.20

Problem 78 courtesy of Gary J. Hordemann, Gonzaga University

AdditionalProblems

Problem 79

Draw the front view of the swivel link shown below. Draw half the object using LINE, CIRCLE, OFFSET, and FILLET. Construct the external fillets using circles and lines snapped to tangents. Then use BREAK to trim the circles to arcs. Use MIRROR to complete the other half.

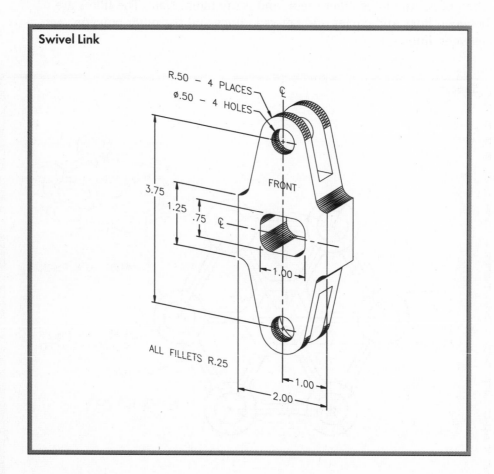

Swivel Link

R.50 – 4 PLACES

Ø.50 – 4 HOLES

FRONT

3.75

1.25

.75

1.00

2.00

1.00

ALL FILLETS R.25

Problem 79 courtesy of Gary J. Hordemann, Gonzaga University

AdditionalProblems

LEVEL 2

Problem 80

The footprints for two cordless mice on a mouse docking (recharging) station are shown below. Draw the two footprints *without* using the FILLET command by drawing circles for all the rounded ends. Use the TTR option of the CIRCLE command for the R3.50 arc. Then use TRIM to create the final objects. Would it be easier to use filleted lines? Try it by drawing the outlines again, this time without using the CIRCLE command.

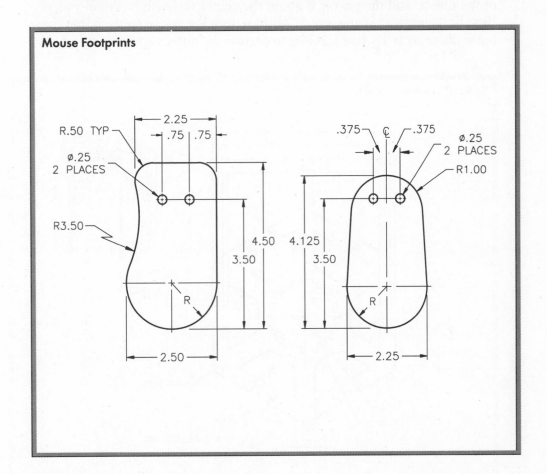

Mouse Footprints

Problem 80 courtesy of Gary J. Hordemann, Gonzaga University

AdditionalProblems

LEVEL
3

Problem 81

Draw the front view of the splined separator block shown below. A suggested approach: Draw two circles bounding the spline teeth. Draw a horizontal line through the center and offset it to form the boundary of one tooth. Trim the circles and lines to form one tooth and array it about the center. Using the OFFSET command and horizontal, vertical, and 30° lines through the center, create 1/12 of the object. Use MIRROR to create 1/6 of the object, and then array it about the center to finish it. When you have finished the drawing, use SCALE to reduce the spline teeth so that the outer diameter is 12. Use STRETCH to narrow the outer gaps from 8 to 6.

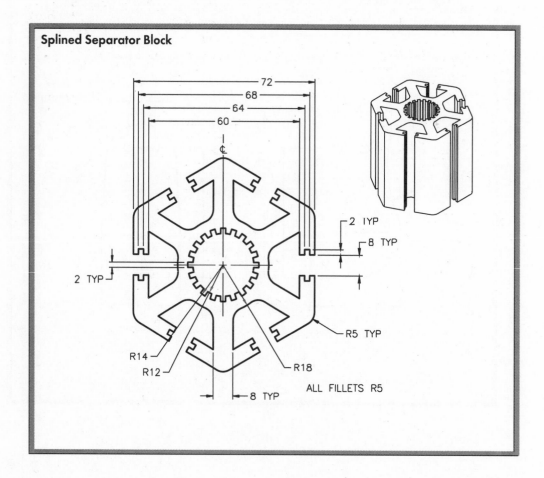

Splined Separator Block

Problem 81 courtesy of Gary J. Hordemann, Gonzaga University

AdditionalProblems

Problem 82

Create and dimension the two views of the cam shown below. Assume the spacing between dots to be .125.

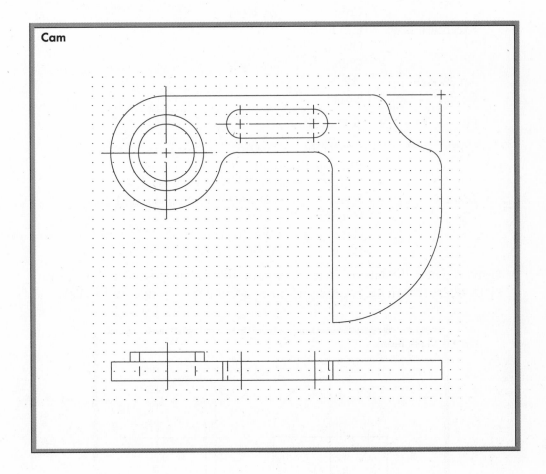

Cam

Problem 82 courtesy of Gary J. Hordemann, Gonzaga University

AdditionalProblems

LEVEL

2

Problem 83

Create a new text style called Inside with the following settings:

Dimension line: Spacing: .30
                 Color: by layer
Extension line: Extension: .06
                 Origin offset: .06
                 Color: by layer
Arrow size: .12
Center: None

Fit: Text only

Units: Precision: 0.00
                 Angular Precision: 0
                 Suppress leading zeros
Text: Style: ROMANS (font romans.shx)
                 Height: .12
                 Gap: .06
                 Color: White

Using the new dimension style, draw and dimension the two views of the clock blank shown below. Assume the spacing between dots to be .125.

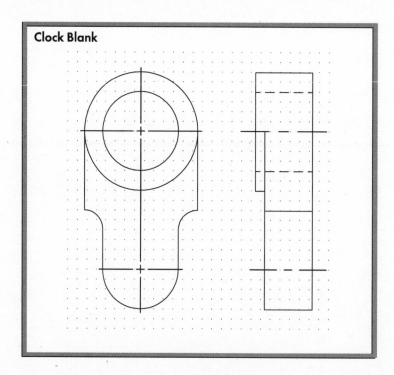

Clock Blank

Problem 83 courtesy of Gary J. Hordemann, Gonzaga University

AdditionalProblems

Problem 84

Create a solid model of the bracket using the dimensions shown in the orthographic views below. Create the orthographic views from the model.

Bracket

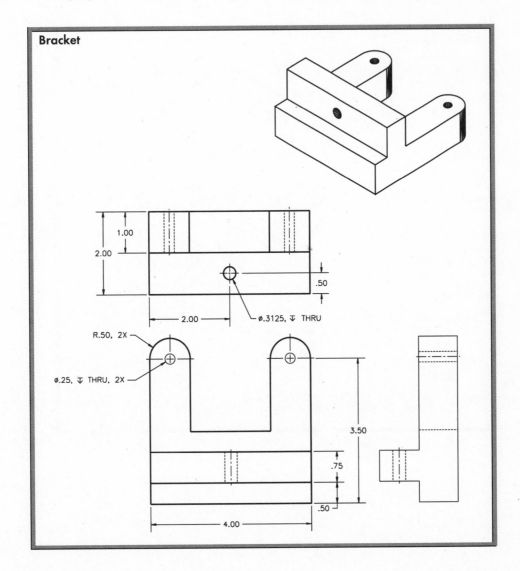

1.00

2.00

.50

2.00

ø.3125, ⊽ THRU

R.50, 2X

ø.25, ⊽ THRU, 2X

3.50

.75

.50

4.00

Problem 84 courtesy of Mark R. Stevenson

AdditionalProblems

Problem 85

Create and dimension a solid model of the locking sleeve according to the dimensions shown below.

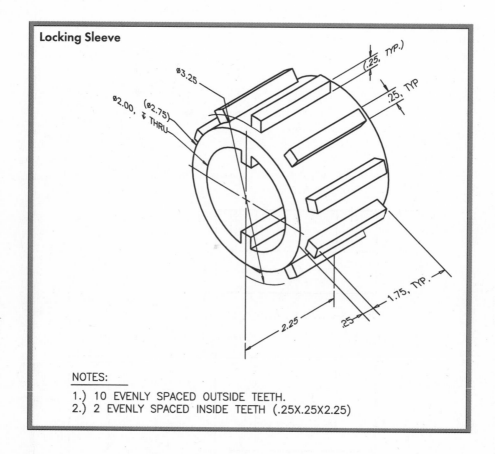

Locking Sleeve

NOTES:

1.) 10 EVENLY SPACED OUTSIDE TEETH.
2.) 2 EVENLY SPACED INSIDE TEETH (.25X.25X2.25)

Problem 85 courtesy of Mark R. Stevenson

AdditionalProblems

LEVEL
3

Problem 86

Using EXTRUDE, create a solid model of the swivel stop shown below. First draw the profile and the circles; turn them into regions; then use SUBTRACT to obtain the final profile for extruding.

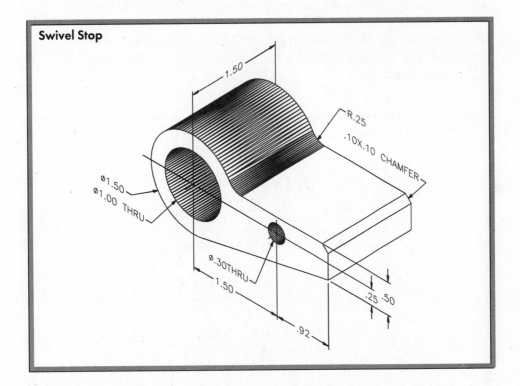

Swivel Stop

Problem 86 courtesy of Gary J. Hordemann, Gonzaga University

AdditionalProblems

Problem 87

Draw each of the three thread representations shown below and write them to files using the WBLOCK command. You will be able to use them in drawings requiring threads by inserting and scaling them. Although you will usually have to do some editing of the threads to make them work in a particular drawing, the blocks will give you a good starting point. The external threads are most easily drawn using the ARRAY command. The distance between dots in the grid is not important, but .125 should work well.

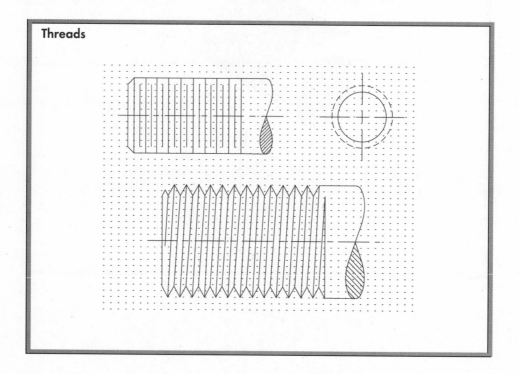

Threads

Problem 87 courtesy of Gary J. Hordemann, Gonzaga University

AdditionalProblems

LEVEL
2

Problem 88

Create a dimensioned isometric view of the roller tube block shown at the right. It is sometimes difficult to draw the line tangent to the two ellipses; if both ellipses are the same size, the Quadrant object snap will yield the same result and is more dependable. Note the complex internal shapes caused by the intersecting holes. Do not attempt to draw them—merely place isometric ellipses on the three surfaces.

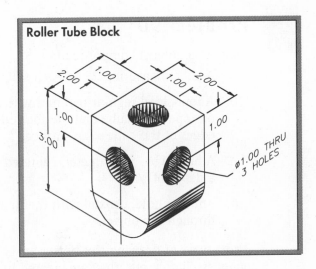

Roller Tube Block

LEVEL
3

Problem 89

Using REVOLVE, create a solid model of the V-belt pulley shown below. Note that by revolving around the axis, you can create a hole without subtracting a cylinder.

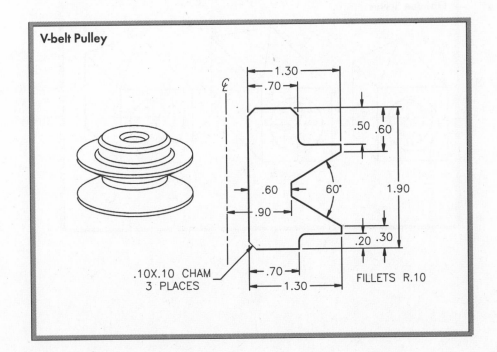

V-belt Pulley

Problems 88 and 89 courtesy of Gary J. Hordemann, Gonzaga University

AdditionalProblems

LEVEL 2

Problem 90

Imagine an object composed of:

a. A cube measuring 2″ on each side.

b. A four-sided regular pyramid with a height of 1″. The pyramid sits on top of the cube and the four edges of its base align with the four edges of the cube's top surface.

c. A hole, 1½″ in diameter, centered in the front cube face and having a depth of ½″.

d. Another hole, coaxial with the first but 1″ in diameter, extending through the object.

Three ways of drawing this object are shown below. They are called *obliques*. Simply put, they are created by drawing one face at its full size and shape with the receding lines at an angle of 45°. The *cabinet* oblique has half-size receding lines; the *cavalier* oblique has full-size receding lines. The receding lines in general obliques vary; those on the one shown here are 0.707-size lines. Draw the three obliques. Then use AutoCAD's isometric grid to draw an isometric view. Which of the four looks most realistic? Why?

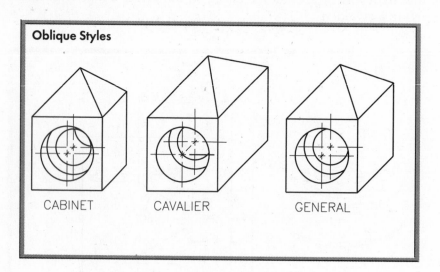

Oblique Styles

CABINET   CAVALIER   GENERAL

Problem 90 courtesy of Gary J. Hordemann, Gonzaga University

AdditionalProblems

LEVEL 2

Problem 91

Create top, front, and isometric views of the fence. Determine the drawing area, based on space requirements. Dimension the top and front views.

Fence

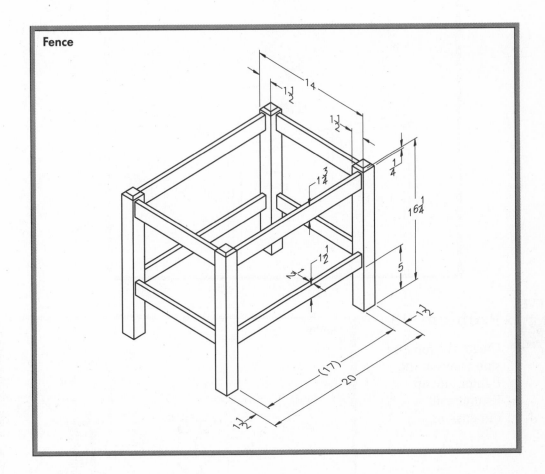

AdditionalProblems

Problem 92

Create solid models of the table and vase. Estimate all dimensions.

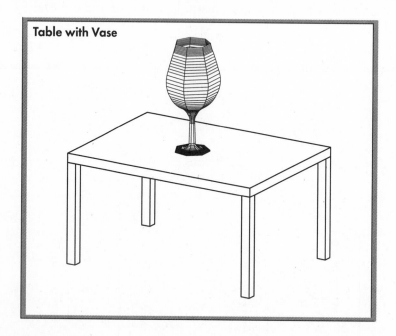

Table with Vase

Problem 93

Create the top and side views of the fighter aircraft. Estimate all dimensions.

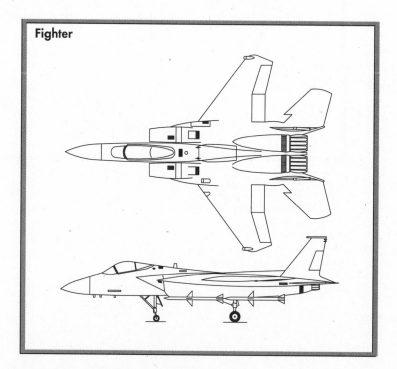

Fighter

Problem 93 courtesy of Matt Melliere and Ronald Weseloh, Red Bud High School

AdditionalProblems

Problem 94

Create the top and side views of the dragster. Estimate all dimensions.

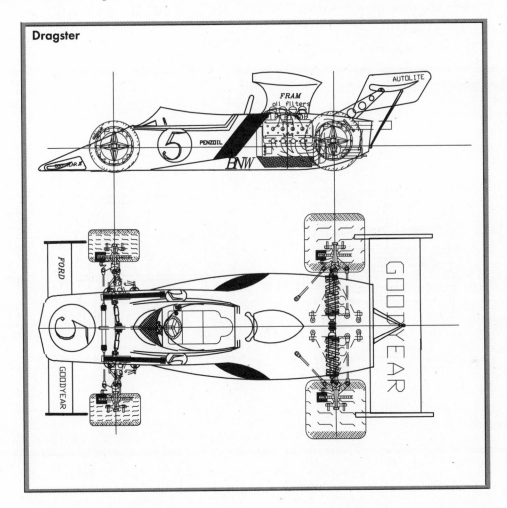

Dragster

Problem 94 courtesy of BNW, Inc.

AdditionalProblems

LEVEL
3

Problem 95

Create the front and back views of the pickup truck. Estimate the
dimensions that are not provided. Determine the drawing area, based
on space requirements.

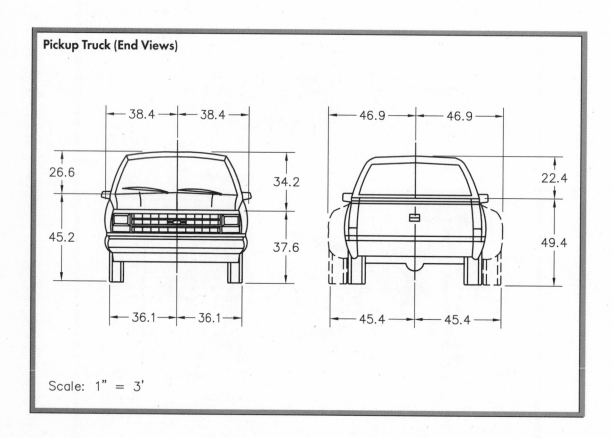

Pickup Truck (End Views)

Scale: 1" = 3'

AdditionalProblems

LEVEL
3
Problem 96

Create the side and top views of the pickup truck. Estimate the dimensions that are not provided. Determine the drawing area based on space requirements.

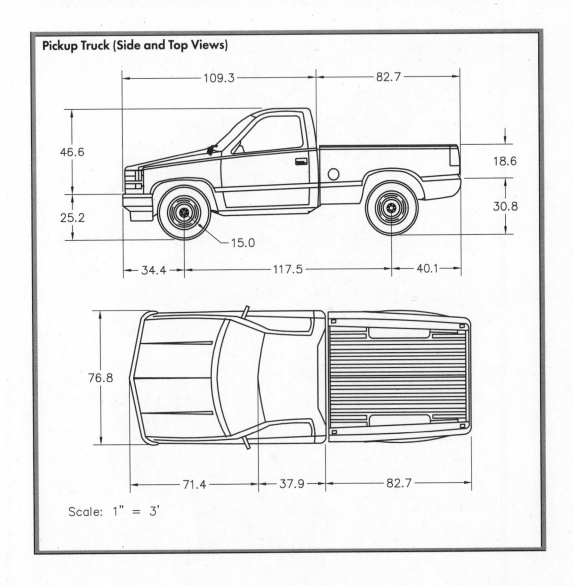

Pickup Truck (Side and Top Views)

Scale: 1" = 3'

AdditionalProblems

Problem 97

Create a wireframe or solid model of the surveyor's transit. Estimate all dimensions.

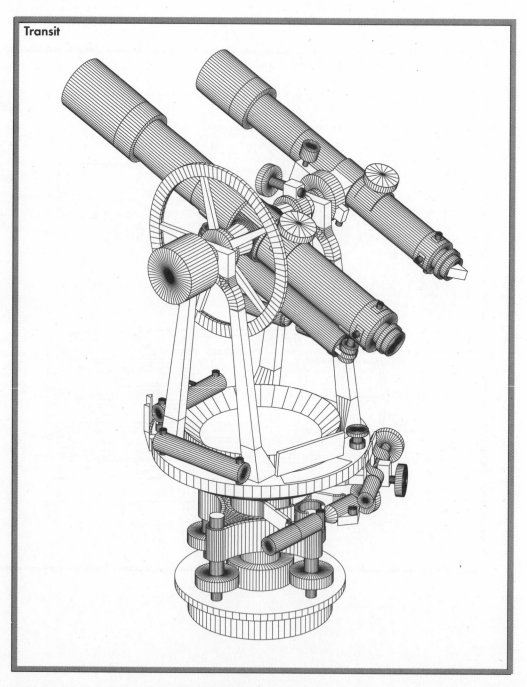

Transit

Problem 97 courtesy of Riley Clark, Hicks & Hartwick, Inc.

AdditionalProblems

LEVEL

Problem 98

Create the landscape drawing as shown. Create blocks of the trees and shrubs and insert them as needed.

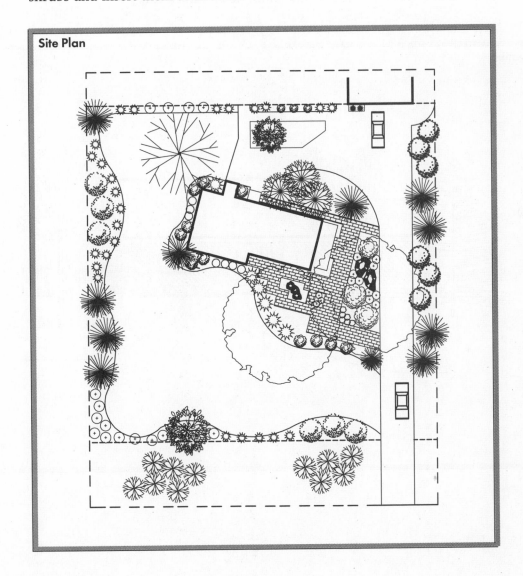

Site Plan

Problem 98 courtesy of Mill Brothers Landscape and Nursery, Inc.

AdditionalProblems

Problem 99

Create the design elevation as shown. Create blocks of the trees and shrubs and insert them as needed. Estimate all dimensions.

Elevation

Problem 99 courtesy of Mill Brothers Landscape and Nursery, Inc.

AdditionalProblems

Problem 100

Create the floor plan drawing of a computer lab. Determine the drawing area according to space requirements. Estimate the dimensions that are not given.

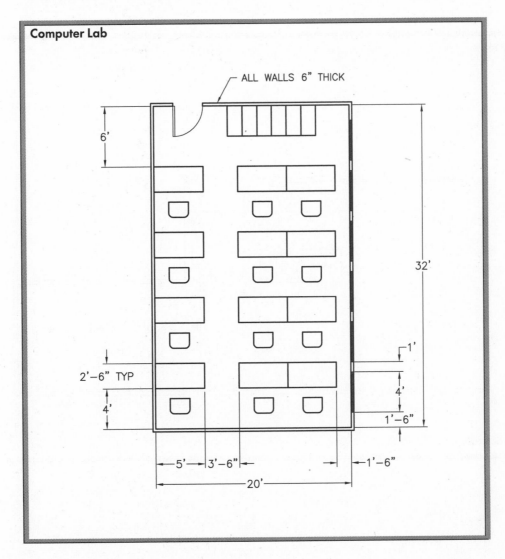

Computer Lab

ALL WALLS 6" THICK

6'

32'

2'-6" TYP

4'

1'

4'

1'-6"

5' 3'-6" 1'-6"

20'

AdditionalProblems

LEVEL 2

Problem 101

Create the floor plan for the first story of a residence. Determine the drawing area according to space requirements. Estimate the dimensions that are not given.

Floor Plan

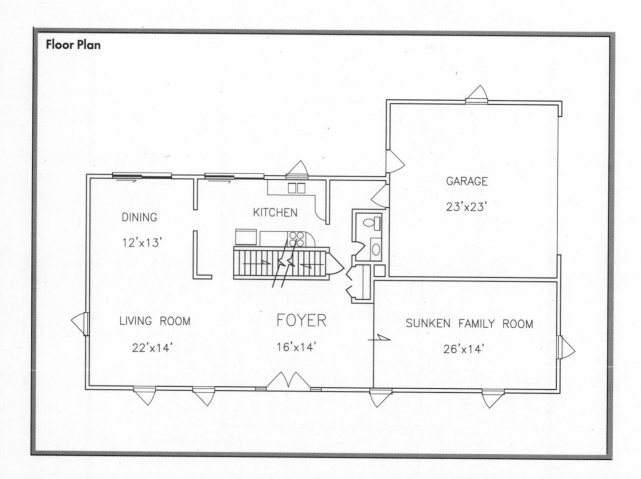

Problem 101 courtesy of Mark Schwendau, Kishwaukee College

AdditionalProblems

LEVEL 3

Problem 102

Create the architectural elevation of the cathedral. Estimate all dimensions.

Musteadt Cathedral

Problem 102 courtesy of David Sala, Forsgren Associates, from *World Atlas of Architecture* (G.K. Hall and Company)

 # AdditionalProblems

Problem 103

Create the wall section as shown below.

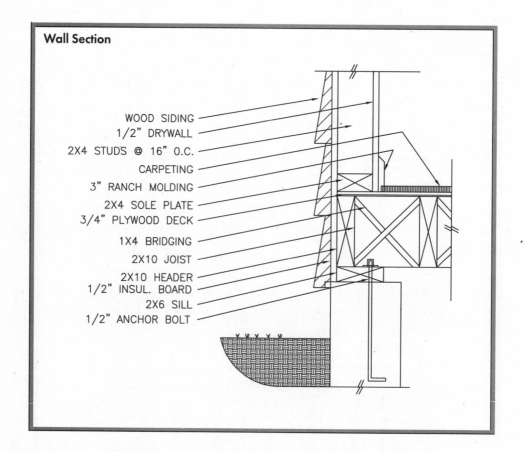

Wall Section

WOOD SIDING
1/2" DRYWALL
2X4 STUDS @ 16" O.C.
CARPETING
3" RANCH MOLDING
2X4 SOLE PLATE
3/4" PLYWOOD DECK
1X4 BRIDGING
2X10 JOIST
2X10 HEADER
1/2" INSUL. BOARD
2X6 SILL
1/2" ANCHOR BOLT

AdditionalProblems

Problem 104

Create a dimension style named Structural that uses feet and inches, oblique arrowheads, and text placed above the dimension lines. Draw and dimension the foundation detail shown below. The hatch patterns are Earth, ANSI37, and AR-Conc.

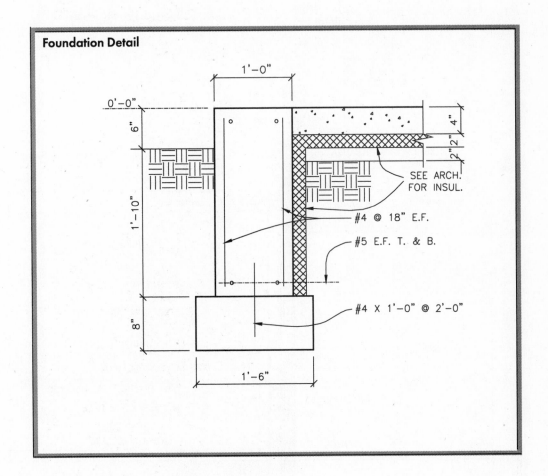

Foundation Detail

Problem 104 courtesy of Gary J. Hordemann, Gonzaga University

AdditionalProblems

Problem 105

Create the wall section as shown below.

Wall Section

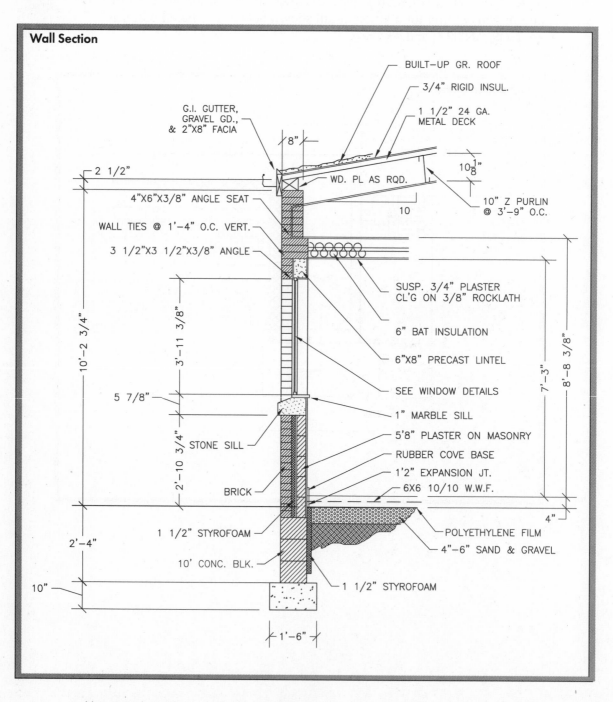

BUILT–UP GR. ROOF

3/4" RIGID INSUL.

1 1/2" 24 GA. METAL DECK

G.I. GUTTER, GRAVEL GD., & 2"X8" FACIA

8"

WD. PL AS RQD.

$10\frac{1}{8}$"

2 1/2"

4"X6"X3/8" ANGLE SEAT

10

10" Z PURLIN @ 3'–9" O.C.

WALL TIES @ 1'–4" O.C. VERT.

3 1/2"X3 1/2"X3/8" ANGLE

SUSP. 3/4" PLASTER CL'G ON 3/8" ROCKLATH

3'–11 3/8"

6" BAT INSULATION

6"X8" PRECAST LINTEL

SEE WINDOW DETAILS

7'–3"

8'–8 3/8"

5 7/8"

1" MARBLE SILL

5'8" PLASTER ON MASONRY

STONE SILL

RUBBER COVE BASE

1'2" EXPANSION JT.

6X6 10/10 W.W.F.

BRICK

4"

10'–2 3/4"

2'–10 3/4"

1 1/2" STYROFOAM

POLYETHYLENE FILM

4"–6" SAND & GRAVEL

2'–4"

10' CONC. BLK.

1 1/2" STYROFOAM

10"

1'–6"

AdditionalProblems

Problem 106

Create the front and left-side architectural elevation drawings. Estimate all dimensions.

Architectural Elevations

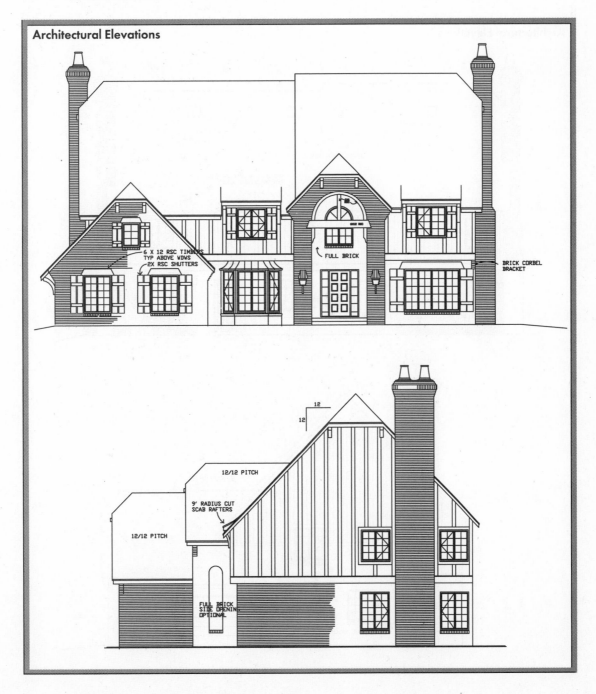

Problem 106 courtesy of Rodger A. Brooks, Architect

AdditionalProblems

Problem 107

Create the back and right-side architectural elevation drawings. Estimate all dimensions.

Architectural Elevations

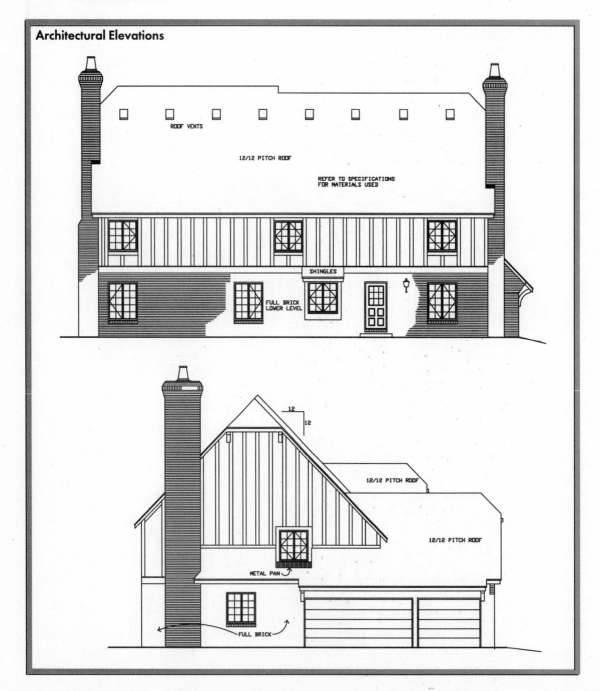

ROOF VENTS

12/12 PITCH ROOF

REFER TO SPECIFICATIONS
FOR MATERIALS USED

SHINGLES

FULL BRICK
LOWER LEVEL

12

12

12/12 PITCH ROOF

12/12 PITCH ROOF

METAL PAN

FULL BRICK

Problem 107 courtesy of Rodger A. Brooks, Architect

AdditionalProblems

LEVEL 3

Problem 108

Create the architectural elevation drawing. Estimate all dimensions.

Architectural Elevation

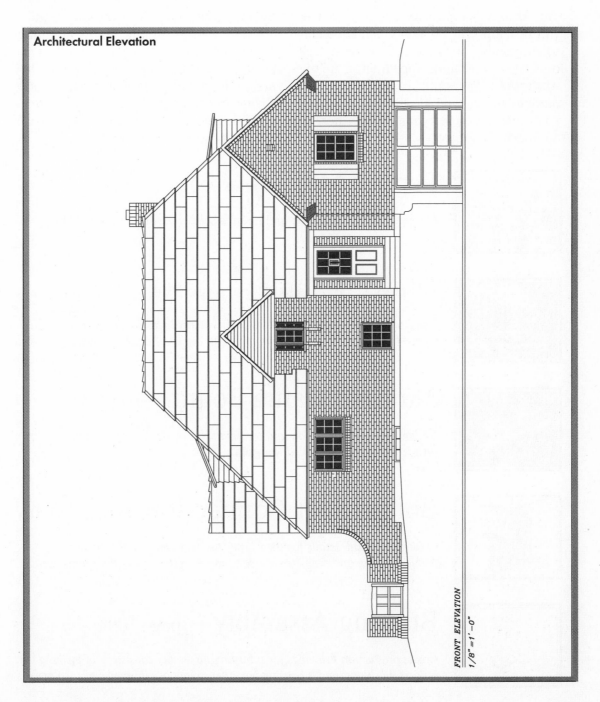

FRONT ELEVATION
1/8" = 1'-0"

Problem 108 courtesy of Rodger A. Brooks, Architect

Advanced

The advanced projects are designed to help you develop your creativity and problem-solving skills. Each project describes a task or job as you might encounter it in the workplace. In some cases, steps are outlined to help you get started. In others, you must decide for yourself how best to approach the project.

Every project in this section requires considerable thought and planning. These projects are suitable for long-term projects for individuals or groups. They are *not* intended to be short-term projects. Give yourself plenty of time, and do not be discouraged if the project does not turn out right on your first attempt.

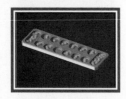
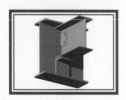
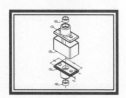

Projects

Architectural Plans – page 1007

Develop a complete set of residential architectural plans according to the client's specifications.

Buffet Tray – page 1008

Design a lap tray for use at a buffet or picnic; document the design, and then create solid models of the pieces and display them as an exploded assembly.

Shackle Assembly – page 1011

Create models of each piece of the shackle assembly and assemble them; slice the assembly and remove half so that the interior is visible.

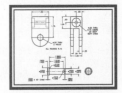

Clothing Rack – page 1014

Create solid models of the pieces of this wall-mounted clothing rack and arrange them into an exploded assembly.

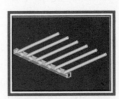

Web-Based School Map – page 1018

Create a clickable Drawing Web Format (DWF) map that allows people to get information about a school or other building by using the Internet.

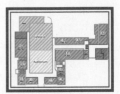

Protective Packaging – page 1020

Design and develop plans for protective packaging for shipping a sensitive electronic device.

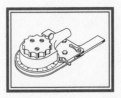

Advanced Project

Project:
Pepper Mill

Shown in Fig. AV1-1 is a disassembled pepper mill. The complete assembly is composed of the following eight parts:

- 1 plastic top
- 1 plastic barrel to hold the peppercorns
- 1 stainless steel stator
- 1 rotor
- 1 threaded shaft
- 2 nuts
- 1 spring

An exploded assembly view of the eight parts is shown in Fig. AV1-2. The shaft and one nut are permanently affixed to the top. When the top is turned, the shaft and rotor rotate, grinding the peppercorns between the rotor and stator.

Your supervisor has asked you to improve the design by creating a more attractive top and barrel. Do not modify the shaft/rotor/stator/spring/nut assembly, except to adjust the length of the shaft to accommodate the new design. (The dimensions of these parts are given in Fig. AV1-4 on page 992.) Concentrate instead on the shape of the pepper mill and the material from which it is made. Consider ergonomic factors: does your new shape make the pepper mill easier to hold and use?

Fig. AV1-1

Advanced**Projects**

■ Part 1

Using the dimensions given, create a new design for the top and barrel. Perform the following tasks:

- Draw the two-dimensional views of all the parts, inserting all hidden and center lines.

- Dimension all of the parts as appropriate.

- Draw the front view of the assembly with the parts exploded. All of the parts should be drawn as full sections, using the hatch patterns appropriate for the materials. Leave out hidden lines, but include the center line.

- Label each part with a circle ("balloon") containing its part number and accompanied by its title and material.

- Add a border and title block similar to that shown on page 282 in Fig. 19-4.

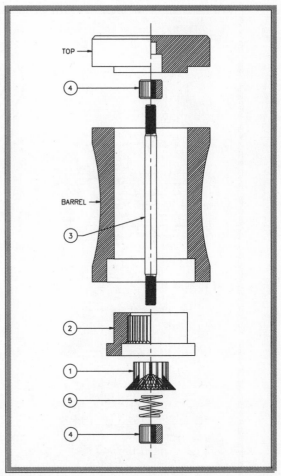

Fig. AV1-2

The complexity of the rotor teeth makes drawing this part especially challenging. A dimensioned single tooth is shown in Fig. AV1-3. While it is a simple object, on the rotor it is rotated about its axis 10° outward from the center of the rotor, rotated 20° clockwise about the midpoint of its lower edge, and leaned in 35° toward the center of the rotor. This twisting and turning results in a drawing complexity that is beyond the scope of this textbook. Consequently, when drawing the rotor, approximate its appearance by following the dimensions given in Fig. AV1-3.

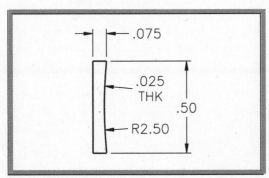

Fig. AV1-3

Advanced**Projects**

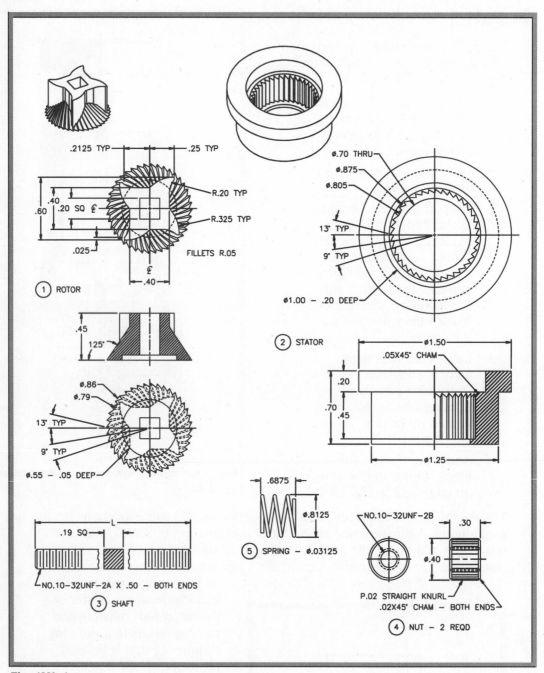

Fig. AV1-4

AdvancedProjects

After creating the top and barrel (a few freehand sketches always help), you should establish how you will arrange the views of the objects in the drawing. There are so many parts to this assembly that it will probably pay you to draw rectangles representing the objects and their dimensions, and then use them to lay out the parts and make decisions about scales and how many pages will be required. Given the smallness of the parts, you should plan to plot them to at least double scale.

Most of the dimensions of the parts are "friendly," that is, they have a common multiple. This makes the use of snaps and grids particularly helpful. Some of the parts, such as the rotor and stator, are especially suitable for the ARRAY and MIRROR commands. (You should always examine the parts for symmetries so you can make use of AutoCAD's powerful replication commands.)

With all your preliminary settings completed and the format of your drawings decided (and assuming that you already have a prototype drawing containing all of the appropriate layers, text style, linetypes, and dimension style) you should outline the steps you will take to create the drawing. The general strategy should be:

1. Draw the needed views of all of the parts. As is often the case with objects that have circular features, it is easiest to begin with the views in which circular features appear as circles or arcs. Then use these views to construct the others.

2. Use copies of the appropriate views to create the exploded assembly.

3. Dimension the orthographic views.

■ Part 2

Make solid models of the pieces of your design. Then perform the following tasks.

- Arrange the solids in an exploded assembly and display them in a view in which you are looking upward at the assembly from below it.

- Determine the volumes and the surface areas of all of the pieces.

- Look up the densities of the materials of your objects (you may have to go to the reference section of a suitable library or search the Internet); use the densities to determine the weights of the parts.

- Render the assembly using suitable materials.

Pepper Mill project courtesy of Gary J. Hordemann, Gonzaga University

Advanced Project

Project:
Drawing Table

Create the solid model of a student drawing table as shown in Fig. AV2-1. Consider making this a group project. Determine which group member will be responsible for each part of the table. When the individual parts have been created, place the files in a central location. Each group member can then use AutoCAD's external referencing capability to assemble the final table.

As a group project, this project works best with a team of 4 people. Assign each person one of the following pieces of the assembly:

- table top
- table drawer
- frame assembly (legs and feet)
- mounting assembly for the top (tubular support and screwheads)

As with any solid model, there are several ways to create a model of the drawing table. Study the parts and choose the method you think will work best. If it doesn't work as well as you'd like, you may want to switch to a different method. The following illustrations and tips may give you some ideas about how to approach each part of the project. For example, one way to create the table frame is to draw a path for the legs and then use the EXTRUDE command. You may want to consider using the SOLREV command to create the feet of the table legs.

Fig. AV2-2 shows a hidden-line view of the table top. Solid primitives of various types may be useful in creating parts such as this.

Fig. AV2-1

Advanced**Projects**

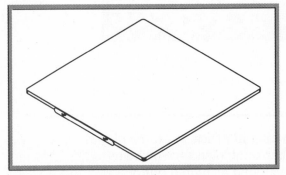

Fig. AV2-2

The drawer handle, hinges, and brackets may also incorporate primitives. Two views of the drawer are provided in Fig. AV2-3. The wireframe view may give you an idea of how the drawer should fit together.

The support assembly, which attaches the table top to the frame, consists of a tubular supporting bracket and holder and Philips-head screws. The tubular bracket is shown in Fig. AV2-4. Fig. AV2-5 shows a closeup of a screw head, and Fig. AV2-6 shows a wireframe view of the assembly in position under the table top.

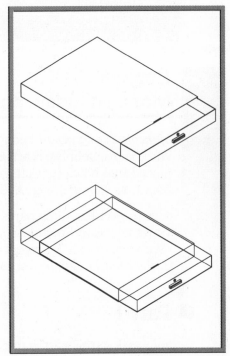

Fig. AV2-3

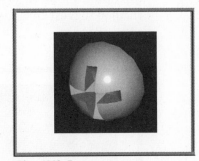

Fig. AV2-5

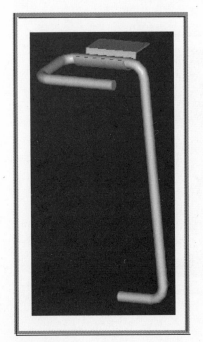

Fig. AV2-4

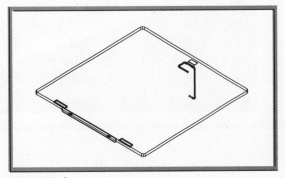

Fig. AV2-6

Drafting Desk project courtesy of Bill Fell, AHST High School

Advanced Project

Project:
Mancala Game Board

Mancala is an ancient bean game originating in Africa. It can be played with beans, pebbles, marbles, or any other small objects, which are placed into holes (pockets) dug into the ground or routed out of a board, as shown in Fig. AV3-1.

This project has three parts. In Part 1, you will create all the orthographic views necessary to build a *Mancala* board. In Part 2, you will design your own version of a *Mancala* board. In Part 3, you will create a solid model of the *Mancala* board from Part 1.

■ Part 1

Your task in this part is to create completely dimensioned views of the board. A number of views can be drawn, but the obvious choice is the top and front views. The fillet and depth dimensions of the pockets further suggest that the front view be drawn as a half or full section.

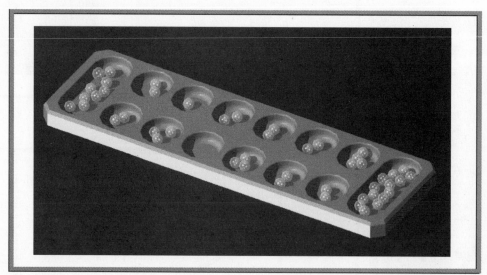

Fig. AV3-1

Advanced**Projects**

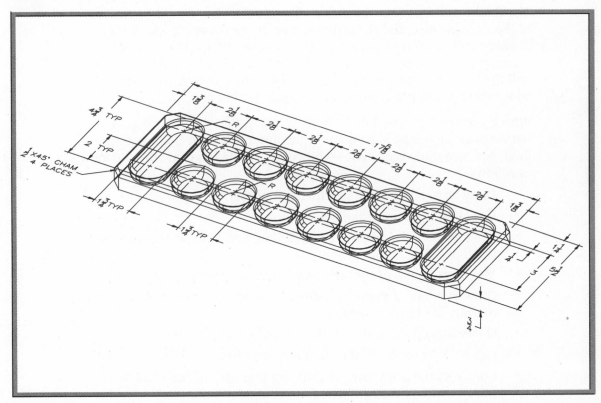

Fig. AV3-2

The board is to have the dimensions shown in Fig. AV3-2. Additional dimensions you will need are:

- Chamfer around the top edge: .125"
- Depth of the pockets: .500"
- Pocket top fillet radius: .062"
- Pocket bottom fillet radius: .250"

The positioning of the pockets on the board is to be symmetrical.

It is probably easiest to begin with the top view and then use it to help construct the front view. You should always begin a drawing by looking for symmetries because AutoCAD is very adept at replication. You can usually save a great deal of time (and boredom!) by taking advantage of commands such as ARRAY, MIRROR, OFFSET, and COPY. The *Mancala* board is particularly suitable for such commands—the entire top view of the pockets could be drawn by replication of just one line and one arc. Another thing you should examine when starting a drawing is the dimensions. Is there a lowest common multiple for at least most of the dimensions? If so, you should use this number for your initial snap and grid settings. What would be good initial snap and grid settings for this project?

AdvancedProjects

The final preliminary considerations for this drawing should be the size of the drawing area and the ultimate size of the paper on which the drawing will be plotted. Assuming size A paper (8.5 × 11), a good drawing limit size for this board is 22 × 17. A final plot scale of 1 = 2 will then give you a nicely sized final plot. Your dimensioning style should then contain a general overall scale setting of 2.

With these settings completed—and assuming that you have begun with a drawing template containing the appropriate layers, text style, linetypes, and dimension style—you could proceed to draw the top view as follows:

- Draw an arc and a line for the large pocket.
- Use MIRROR to finish the pocket.
- Use a copy of the arc, a shortened line, and MIRROR to create a small pocket.
- Make five copies of the small pocket using a rectangular array.
- Draw one-half of the outer edge of the board, using FILLET or snaps to draw the filleted corners.
- Offset these lines to obtain the filleted edges.
- Use MIRROR to obtain the complete top view.
- Construct the front view by drawing one pocket, complete with fillets, and replicating it using a rectangular array.
- Hatch the front view.
- Insert the center lines for all of the circular features, except fillets.
- Dimension the two views.

■ Part 2

Now try creating another, very different *Mancala* board. Incorporate features that you think might be useful to *Mancala* players. See the *Mancala* instructions on the next page for ideas. You may also consider adding the following features:

- handles
- legs
- edges for stacking multiple boards
- a lid
- one or more drawers to store the beans

Create orthographic and isometric views of the board you design.

AdvancedProjects

Part 3

Create a solid model of the board from Part 1. Render the board using different materials such as wood, marble, and brass. Add beans, marbles, or other playing pieces (marbles are shown in Fig. AV3-1) and assign appropriate materials to them also. Be sure to create and position lights so that the board is visible and attractive when you render it. (Remember that you will need to render using either the Photo Real or the Photo Raytrace option to display the materials correctly.)

Rules for *Mancala*

Place four beans in each of the twelve smaller pockets. Each of the two players has six of the small pockets and one of the large pockets. The first person to play picks up the four beans in any one of his or her six pockets. Moving counterclockwise, the player then places one bean in each pocket of the twelve and in his or her own large pocket, but not the opponent's.

If the last bean played falls into the player's large pocket, he or she gets another turn. If it falls into an empty pocket on the player's own side of the board, the bean captures all the beans in the opponent's pocket directly across from that pocket. The capturing bean plus all of the captured beans are placed into the player's large pocket.

The game ends when one player runs out of beans in his or her small pockets. When this happens, the other player places all the beans remaining in his or her small pockets into the large one. The player with the most beans in his or her large pocket wins.

Mancala Game Board Project courtesy of Gary J. Hordemann, Gonzaga University

Advanced Project

Project:
Bolted Seat Connection

When steel frame buildings are designed, the structural engineer prepares *structural details,* which are representations of the connections between beams. The bolted seat connection shown in Fig. AV4-1 is an example of such a connection.

The specifications for structural connections are contained in the *Manual of Steel Construction* published by the American Institute of Steel Construction (AISC). This project uses AISC specifications and is adapted from an example in *Structural Steel Detailing,* also published by AISC.

Bolted seat connections are commonly used to connect grid filler beams to supporting girders (Fig. AV4-2). The filler beam is supported by a seat angle, often merely sitting on the angle without bolts or welds. The other angle is used only for torsional stability.

In its specifications for beams, AISC provides dimensions and physical property data according to beam shape codes. The beams shown here would usually be "W shape," and their identifying codes would be listed in that category.

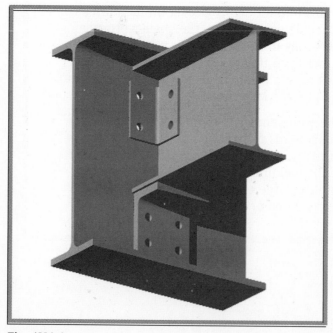

Fig. AV4-1

AdvancedProjects

For the connection in this project, the four members are:

- Girder: W21 × 62 wide flange beam
- Filler: W12 × 26 wide flange beam
- Seat Angle: 6 × 4 × ³/₄ × 6
- Angle: 3¹/₂ × 3¹/₂ × ¹/₄ × 5¹/₂

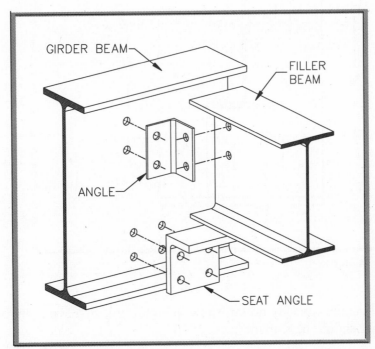

Fig. AV4-2

■ Part 1

Create the front and right-side views of the beam connection. A partially dimensioned front view is shown in Fig. AV4-3 (page 1002).

The remaining dimensions, extracted from the *Manual of Steel Construction,* are given in Fig. AV4-4 (page 1003).

A few other facts you will need:

- Seat angle fillet radius: .5″
- Angle fillet radius: .375″
- Hole diameters: .8125″
- Clearance between the end of the filler beam and the girder: .5″

Advanced**Projects**

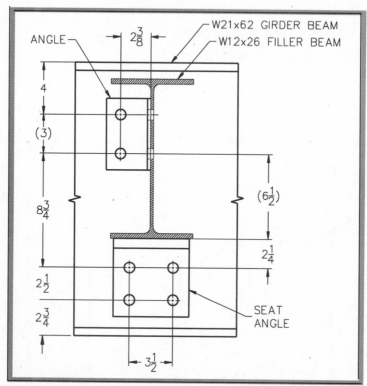

Fig. AV4-3

Structural details typically do not show hidden lines, so leave them out. Center lines should be shown.

This is the type of problem for which snaps and grids are of little use. Furthermore, trying to draw these objects directly, *i.e.,* by using LINE and CIRCLE with coordinates, would be foolhardy. This project is particulary suited to AutoCAD's construction commands OFFSET, EXTEND, and TRIM.

Other preliminary considerations of importance are the size of the drawing area and the size of the paper on which the drawing will ultimately be plotted. Assuming size A paper (8.5 × 11), a good drawing area for this drawing is 44 × 34. A final plot scale of 1 = 4 will then give you a nicely sized final plot.

AdvancedProjects

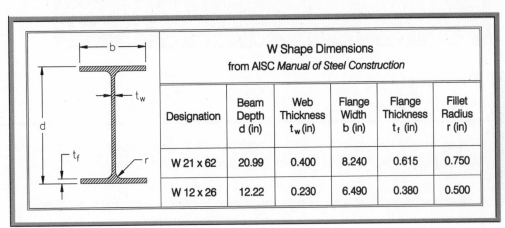

	W Shape Dimensions from AISC *Manual of Steel Construction*					
Designation	Beam Depth d (in)	Web Thickness t_w (in)	Flange Width b (in)	Flange Thickness t_f (in)	Fillet Radius r (in)
W 21 x 62	20.99	0.400	8.240	0.615	0.750
W 12 x 26	12.22	0.230	6.490	0.380	0.500

Fig. AV4-4

Assuming that you have begun with the proper layers, linetypes, and other routine settings, you could create this drawing as follows:

- Start with the front view.
- Draw a vertical line representing the center line of the front view.
- Draw a horizontal line representing the top edge of the girder.
- Use OFFSET to create all the other lines, including center lines for the holes.
- To avoid getting confused by having too many offset construction lines, pause occasionally to trim the lines to their final lengths.
- Use the front view to construct the right-side view, again making extensive use of the OFFSET command.

■ Part 2

If you have access to a copy of the AISC *Manual of Steel Construction,* draw the two views of the bolted column and bracket shown in Problem 72 in the "Additional Problems" section of this book. Leave out the fasteners; they usually are not shown.

■ Part 3

Create solid models of the connection members from Part 1. Render the models using materials that make the members resemble steel.

Bolted Seat Connection project courtesy of Gary J. Hordemann, Gonzaga University

Advanced Project

Project:
Bushing Assembly

The assembly shown in Fig. AV5-1 with nominal dimensions is composed of five parts: two bushing holders, two bushings, and a box.

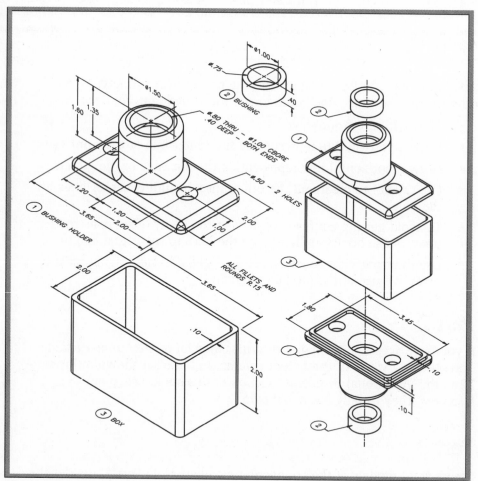

Fig. AV5-1

AdvancedProjects

■ Part 1

Create a full set of drawings to describe the assembly, as follows:

- Draw the necessary two-dimensional views of all the parts, inserting all hidden and center lines.

- Dimension the objects as appropriate, including tolerances, as specified below.

- Draw the front view of the assembly with all of the parts put together. The front view should be drawn as a full section. Assume the holders, box, and bushings to be made of steel, cast iron, and bronze, respectively. Because hidden lines are usually not included in simple sections, leave them out. Include all center lines.

- Add a border and title block similar to that shown earlier in the book in Fig. 19-4.

- Add notes and the properly toleranced dimensions to specify:
  - Overall tolerance of ±.02" on all dimensions unless otherwise specified
  - Bushing outside diameter: 1.000" – 1.0008"
  - Holder inside diameter for bushing: 1.0014" – 1.0019"
  - Bushing inside diameter: .7500 – .7512
  - Perpendicularity tolerance of holder bushing center line to the base: .0020

Assuming that some sort of shaft is to be supported by the bushings, what tolerances would you add to ensure that the two bushings would be properly aligned?

The three different objects are most easily drawn by beginning with their top views. You should first determine how many views of each object will be required to describe it completely. Keep in mind that a view that requires no dimensions is superfluous.

Plan how you will arrange the views and the assembly on the paper. Will you need more than one sheet? What plot scale(s) will you use? Although it would probably be overkill with so few parts, it often helps to draw rectangles representing the outside dimensions of the views and move them around on the drawing to get a preliminary idea of the layout. In planning the layout, be sure to allow plenty of room for dimensions.

AdvancedProjects

When you have finished the dimensioning, ask yourself whether you could redraw it using only your dimensions; this will help you discover any missing dimensions. Since such redrawing usually goes very quickly, you might even take the time to redraw it. (It helps to let a day or so lapse between dimensioning and redrawing). As a practical matter, you should end up with at least as many dimensions as on the original problem statement.

These objects have "nice" dimensions and therefore should be drawn using grids and snaps. Determine the lowest common multiple of the dimensions and use that for your snap setting. Use twice that for your grid setting.

With all of your preliminary settings completed and the format of your drawings decided (and assuming that you have begun with a prototype drawing containing all of the appropriate layers, text style, linetypes, and dimension style) you may wish to proceed as follows:

- Draw the top view of the holder and use it to construct any other view(s) you need.
- Draw the box and bushing views, starting with the top views.
- Use copies of the appropriate views of the three objects to construct the sectioned assembly.
- Dimension the orthographic views of the three pieces.
- Place a circle containing a part number (balloon) with each piece; use a circle four times larger than the number and insert the number using the Middle text option and the Center object snap.
- Use the QLEADER command to add "balloons" to the assembly.

■ Part 2

Create solid models of all the pieces in this assembly. Use the models to create an exploded assembly drawing similar to that shown on the right in Fig. AV5-1.

Bushing Assembly project courtesy of Gary J. Hordemann, Gonzaga University

Advanced Project

Project:
Architectural Plans

You are a senior designer for an architectural firm. A client wants to see design proposals for a new home to be built on property she already owns.

A copy of the property survey is shown in Fig. AV6-1. The client has given you the following list of specifications:

- two-story residence
- 4 bedrooms, 2 baths
- walk-in closets in all bedrooms
- brick construction or brick facing
- attached 2-car garage
- noise level as low as possible in the bedroom area(s)
- total square footage not to exceed 2800 square feet

Design a residence that meets these specifications. Then create a complete set of plans, including (but not necessarily limited to) a site plan, floor plans for each floor, elevations, and a landscaping plan. Include all the information the client needs to make a decision about whether to accept this design.

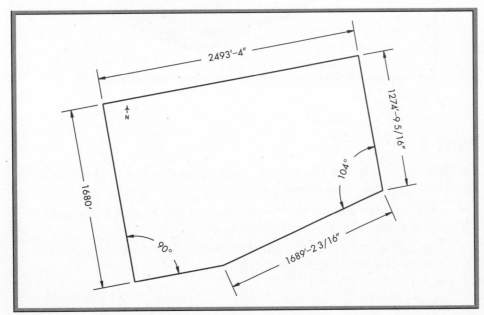

Fig. AV6-1

Advanced Project

Project:
Buffet Tray

Two examples of a lap tray for use at a buffet or picnic are shown in Fig. AV7-1. Note that each example has handles and other adaptations to make it easier for the user to control various dishes and implements.

Fig. AV7-1

■ Part 1

Design your own buffet/picnic tray. In doing so, nearly everything is left up to you. The only constraints are:

- The tray is to contain depressions and/or holders for items such as drink containers, plates, silverware, etc.
- The tray is to have handles that are separate pieces; they may or may not be removable or rotatable.

Advanced**Projects**

Document your creation by drawing completely dimensioned orthographic views. The thinness of the object, plus the need for dimensions on the depressions and their fillets, make section views appropriate. The nature of the tray also suggests the use of offset sections. Thus, you should:

- Draw complete orthographic views of the tray body, the handles, and any other objects that make up your tray.
- Use section views as appropriate.
- Dimension the objects.
- In preparing the orthographic views, show each piece individually. Identify each piece with a name and an associated part number contained in a circle (balloon).
- Draw a section view of the assembly with the connection between the tray and handle magnified.

An example set of dimensioned drawings for the tray shown in the top of Fig. A7-1 is given in Fig. AV7-2 (page 1010). Use this example as a guide in preparing your drawings.

After you have thought out your design (a pencil sketch is a good idea), prepare to draw it by first considering its dimensions. Is there a common multiple of the dimensions that will make it convenient to use grid and snap? What will be the ultimate size of the paper on which the drawing will be plotted? Will you need more than one sheet? Set your limits and overall dimension scale accordingly.

■ Part 2

Prepare solid models of the pieces composing the tray you designed or the tray shown in the sample design drawings.

After creating the solids, perform the following tasks:

- Display the solids as an exploded assembly displayed in a southeast isometric view.
- Determine the volumes and surface areas of the tray and handles.
- Select your material(s) and look up the density of each. You may have to go to the reference section of a suitable library or search the Internet. Use the density to determine the weights of the parts.

The modeling should probably be started in the top view, and EXTRUDE will likely be the "workhorse" command. Tubular handles like those shown in the bottom tray in Fig. AV7-1 can be created by extruding a circle along a polyline.

Try rendering the model with different materials, such as wood or plastic. Be sure to consider the weight of the tray built with each material or combination of materials. Which material(s) will be the most practical to use for the tray you designed?

AdvancedProjects

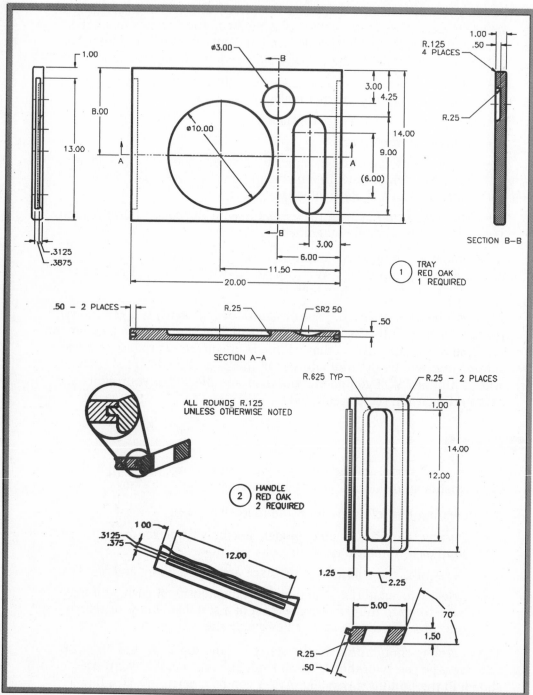

Fig. AV7-2

Buffet Tray project courtesy of Gary J. Hordemann, Gonzaga University

Advanced Project

Project:
Shackle Assembly

The shackle assembly shown here is composed of seven parts. Five of the parts (the chain, two yokes, and two clevis pins that secure the chain to the yokes) are shown in Fig. AV8-1. Not shown are two cotter pins that, when inserted into the holes in the pins, keep the pins from working loose.

The dimensions of the yoke and clevis pin are given in the drawing in Fig. AV8-2. The clevis pin dimensions are those specified for a nominal ½" standard clevis pin by the American National Standards Institute (ANSI) and published as *ANSI B18.8.1-1972, R1983* by the American Society of Mechanical Engineers (ASME).

Each link of the chain has a diameter of .375" and outside measurements of 2.125" × 1.375".

Using these dimensions, perform the following tasks:

- Create solid models of the five components shown and arrange them into an exploded assembly similar to the one shown in Fig. AV8-1.

- Determine the volumes of all pieces, including the chain.

- Assuming the pieces to be steel, with a nominal density of .28 lb/cu in., determine their weights.

- Determine the surface area of a yoke.

- Put the solids together and slice the assembly with a cutting plane oriented along the axes of the pins; remove half so the inside of the assembly can be seen.

- Repeat the slice and remove operations with the cutting plane perpendicular to the axes of the pins.

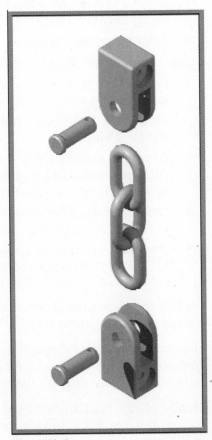

Fig. AV8-1

AdvancedProjects

When creating the solids, it is prudent first to choose the starting view. You should select a view such that the world coordinate system standard orthographic views of top, front, right-side, etc., correspond to those same views of the object. This will minimize the confusion over what you are looking at when you switch views. Assume that this assembly is meant to hang vertically, so it has a natural top view. Make the view facing the heads of the clevis pins the front view. The dimensioned drawing of the yoke shown in Fig. AV8-2 shows its front and right-side views.

After settling on the orientation of the solid, take a little time to consider how you are going to create it. There are usually several ways to build a solid, and some may be very difficult and time-consuming.

One way to create the yoke is to follow this procedure:

- Draw the front view profile, making it a closed polyline.
- Extrude the polyline to the full depth of 1″.
- Insert and subtract a cylinder.
- In the right-side view, insert and subtract cylinders to create the counterbored hole.
- Insert and subtract a block (or extruded rectangle) to create the cavity between the arms.
- Fillet the outside edges.

Some variations on this procedure include:

- Extrude the profile to a depth of .25″ to create one arm; then use a duplicate for the other arm, joining the two arms with a box.
- Make the counterbored hole by drawing and revolving its profile in the front view.
- Draw and extrude the yoke profile in the right-side view. Then use FILLET to obtain the geometry common to both extrusions. (This approach is perhaps the most elegant.)
- Create the space between the arms by making three slices and unions. (This approach, however, is not recommended.)
- You could even create the yoke without using either EXTRUDE or REVOLVE by unioning and subtracting a series of cylinders and boxes. This approach, too, is impractical in reality.

Clearly, there are several practical ways to create this assembly, as well as several very impractical ways.

You should also think through the sequence of operations. The most common error resulting from an incorrect sequence is a hole that does not go all the way through the object.

Finally, always start a drawing at a known and easy-to-remember point such as 0,0,0 or 2,2,2. This is particularly critical when you are working in 3D space.

Advanced**Projects**

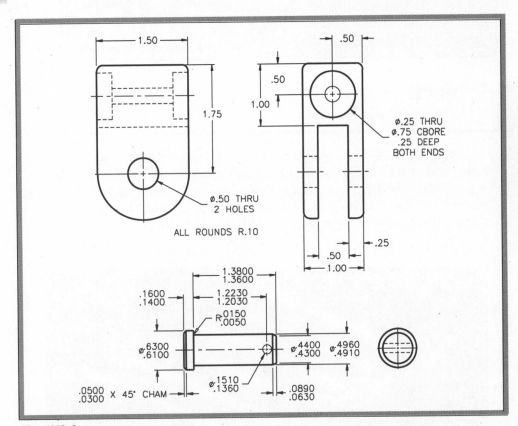

Fig. AV8-2

If you wish to proceed with "elegance," create the yoke as follows:

- Draw the profile of the front view of the yoke and extrude it to 1″.
- Insert a cylinder and subtract it.
- In the front view, draw the half-profile of the counterbored hole and revolve it.
- In the right-side view, draw and extrude the profile, without the holes.
- Fillet the edges of both objects.
- After making sure the two extrusions are coincident, obtain the final object using INTERSECT.

The clevis pins can be created from unioned cylinders or from a revolved profile. The chain link can be made from a circle extruded around a polyline (consider using multiple viewports) or from a sliced torus unioned with two cylinders.

The two cotter pins can probably best be modeled by extruding a half-circle along a polyline. Select the appropriate pin from a table of ANSI standard cotter pins. Such a table can usually be found in the appendix of a graphics textbook.

Shackle Assembly project courtesy of Gary J. Hordemann, Gonzaga University

Advanced Project

Project:
Clothing Rack

Shown in Fig. AV9-1 is a wall-mounted rack for drying clothing. It is composed of six rods inserted into a large cylinder, which is housed in a square tube. The tube has cutouts that allow the rack to be folded downward when the rods and cylinder assembly are slid to the right. In this view, we are looking upward from below the object.

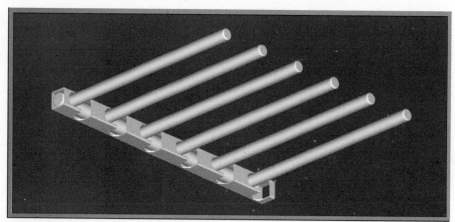

Fig. AV9-1

■ Part 1

The dimensions of the components are shown in Fig. AV9-2. Using these dimensions, perform the following tasks:

- Create solid models of the eight pices and arrange them into an exploded assembly.
- Determine the volumes and surface areas of all eight pieces.
- Assuming the tube to be aluminum with a density of 160 lb/cu ft, determine its weight.
- Assuming the rod assembly to be oak with a density of 45 lb/cu ft, determine its weight.

Advanced**Projects**

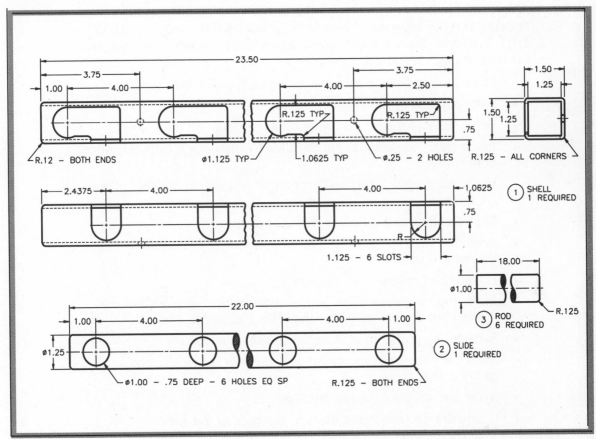

Fig. AV9-2

When creating three-dimensional objects, you will find it least confusing
to select the proper view before you begin. Try to align the object with
the world coordinate system standard orthographic views of top, front,
right-side, etc. The standard views will then make sense when you switch
from view to view. Many objects—such as this one—have a natural top
and front, and you should plan to orient the object accordingly. Following
this orientation, the dimensioned drawing in Fig. AV9-2 shows the front,
bottom, and right-side views.

Before starting the model, consider the different ways it can be carried
out. Always look for symmetries so that you can take advantage of
AutoCAD's replication commands. This object is ideal for use of the
ARRAY command. The modeling of this assembly is straightforward,
being primarily a collection of cylinders. The square tube at first glance
appears complex, but if you realize that cavities usually are best modeled
by creating the positive of the cavity and then subtracting it, the creation
of this object becomes much easier.

AdvancedProjects

After you have laid out your approach to the modeling, begin the drawing at some known and easy-to-remember point, such as 0,0,0 or 1,1,1. The least difficult approach to this model is probably to begin with the rods. Doing so, you could proceed as follows:

- Switch to the front view, change the UCS, and insert a cylinder.
- Fillet one end of the cylinder.
- Use ARRAY to make five copies on 4″ centers.
- In the right-side view, begin the slide by inserting a cylinder; fillet both ends.
- In the front view, insert and array a set of small cylinders for the six holes.
- Subtract the small cylinders from the slide.
- Begin the shell in the right-side view, inserting two squares; fillet the outer one.
- Extrude and subtract the squares.
- In the top view, create the profile of the cavity, joining all of the pieces with the Join option of the PEDIT command.
- Extrude the profile to at least the thickness of the shell.
- Switch to the front view and repeat the process for the profile of the cavity in that view.
- Array and subtract the two profiles.
- Create the two mounting holes in the back of the shell.
- Arrange the objects in an exploded assembly.

■ Part 2

Using AutoCAD's isometric mode, create an isometric exploded assembly of the rack. Using a leader, tag each part with a balloon. A balloon is a circle containing the part number of the piece. The circle (balloon) diameter should be four times the height of the numeral.

Drawing the shell in isometric mode will be a bit of a challenge. Sometimes you can create the isometric view more easily by first drawing an orthographic view such as the front view. Then switch to one of the four standard isometric views to complete the drawing. When switching views, you will want to stay in the world coordinate system. Because different parts of the drawing may end up on different planes, you will discover that TRIM and EXTEND may not work properly. However, TRIM can usually be made to work if you change the coordinate system so that the current view is an XY plane. Once created, the isometric may be transferred into any other view by cutting and pasting. (Do not forget always to change the UCS to the current view before cutting and again before pasting.)

AdvancedProjects

Part 3

The rack as specified in Part 1 has a few shortcomings. First, the mounting holes are not located properly—they should be in line with the cavities so the installer can use a tool on whatever fastener is in the hole. Second, if the mounting fasteners have hex or round heads, the heads will interfere with the movement of the rods/slide assembly. The fasteners should have flat heads, but then the holes should be countersunk. Third, the rectangular tubing might be too expensive.

Address these problems by creating a new design. Use a round tube for the shell. Make the rods and slideout of ¾" PVC pipe. Use ¾" PVC pipe elbows and T's to connect the pipe pieces, and ¾" PVC pipe caps on the ends of the rods. A trip to the local hardware store will provide you with the outside dimensions of these pipe fittings.

Using the drawings on page 1015 as a guide, create a complete set of dimensioned drawings of your design. Remember, when you are drawing orthographic views, draw only as many views as necessary to specify the object completely, and only draw one copy of each part. Tag each part with a balloon specifying the part number, material, and how many are required in the full assembly.

In a drawing with so many parts, you should first plan how you will arrange the parts on the paper. This will help you figure out how many sheets to use and what the plot scales will be. Be sure to allow sufficient room for the dimensions.

Part 4

Using AutoCAD's isometric mode, create an isometric exploded assembly of the rack you designed in Part 3. Using leaders, tag each part with a balloon. Create the isometric drawing as a separate drawing file. Then insert the isometric into the drawing from Part 3.

Part 5

Make solid models of the pieces of your design. Then perform the following tasks:

* Arrange the solids in an exploded assembly and display them in a view in which you are looking upward at the assembly from below.

* Determine the volumes and surface areas of all the pieces.

* Select a material for the shell and look up its density and that of PVC; use the densities to determine the weight of the shell and of the rods/slide assembly.

* Render the rack using suitable materials.

Clothing Rack project courtesy of Gary J. Hordemann, Gonzaga University

Advanced**Projects**

Project:
Web-Based School Map

The goal of this project is to create a Drawing Web Format (DWF) map for use on the Internet. The example shown in Fig. AV10-1 shows a two-story high school. Each room contains links to further information.

When used with the Autodesk *WHIP!* Viewer, anyone can see the map of the school and who teaches in each classroom. (The *WHIP!* Viewer is available on the Instructor's Resource CD-ROM.) The intent of the map in the example is to help students coming from another school or parents who are visiting for a parent-teacher conference.

The basic map in the example is simply a two-dimensional map drawn in AutoCAD. The key to the map is having something to click on. The wall and object lines in the drawing shown in Fig. AV10-1 are so thin that they are hard to select. Hatching was considered the best choice in this case because it offers a large clickable area.

Start by deciding on a building, such as a school, office building, or even a county courthouse, for which having Internet access to a map and associated information would be useful. Then perform the following tasks:

- Create an AutoCAD drawing of the building.
- From the Internet toolbar in AutoCAD, select the Attach URL button. (URL stands for "Universal Resource Locator"; this is the familiar "http://" address that identifies individual Web sites.) With this button, you can select an object in a drawing, such as a circle, line, or hatched region, and link it to an Internet Web site address. If you use a hatched region, be sure to give each separate region its own hatch so that you can attach more than one URL to the drawing.
- Select each area that you want to attach and type in the address.
- After attaching all of the URLs, export the file by selecting Export on the File pull-down menu and saving the drawing in DWF format.

Note: Before you can view the map in a browser such as Netscape or Internet Explorer, you must have access to the *WHIP!* Viewer. If the Instructor's Resource CD-ROM is not available, you can download this free software from Autodesk (www.autodesk.com).

Advanced**Projects**

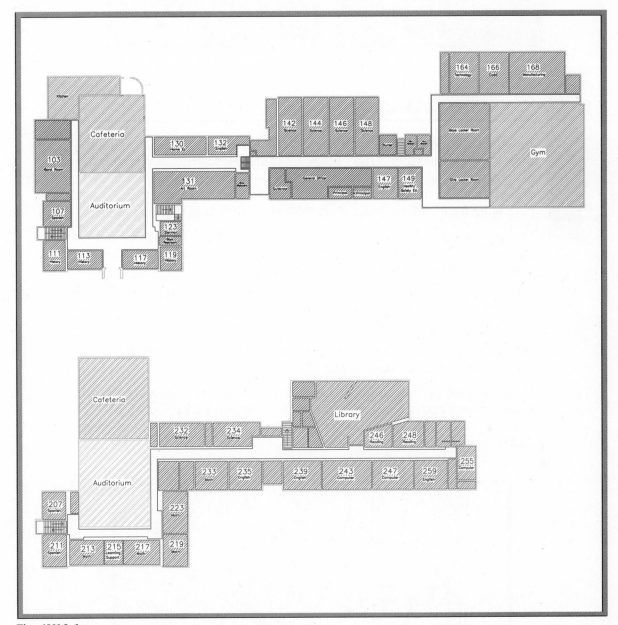

Fig. AV10-1

The example shown in Fig. AV10-1 is an interactive map. Clicking on a room takes the user to a Web page that describes what takes place in that room. The linked pages contain information about who the teacher is, what subject(s) he or she teaches, what classes are held there, and listings of useful Web links related to the teacher or the courses taught. You should determine what information is most useful for the application you are designing and implement it appropriately.

Web-Based School Map project courtesy of Joe Bodenschatz and Dana Driscoll

Advanced Project

Project:
Protective Packaging

Designers for packaging companies typically design custom packaging to protect various products. The packaging company's clients depend on the expertise of these designers to provide safe, effective shipping containers and other packaging. In this project, you will develop protective packaging for a client's new electronic device, shown in Fig. AV11-1.

The level is made of a titanium alloy to reduce weight. The electronics are located in the central housing and are relatively unprotected. The client has specified that the packaging must:

- protect the level from sudden impacts that may result from being dropped or shifting in an airplane cargo hold.

- insulate the level from abrupt temperature changes during shipping.

- provide an attractive appearance that will be appealing to the client's potential customers.

Design custom packaging for this sensitive instrument. Create drawings to show how your packaging will protect the level. Also create assembly drawings as necessary to show the client how to assemble the packaging and how to pack the level properly.

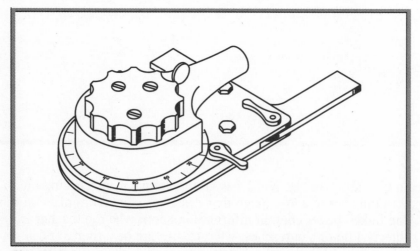

Fig. AV11-1

Appendix A: Options Dialog Box

This appendix focuses on AutoCAD's Options dialog box, which permits you to customize many of AutoCAD's settings. Also, the dialog box enables you to add pointing devices and printers to your AutoCAD system. Select Options... from the Tools pull-down menu to display the Options dialog box, as shown in Fig. A-1.

■ Files

The Files tab specifies the folders in which AutoCAD searches for text fonts, drawings to insert, linetypes, hatch patterns, menus, plug-ins, and other files that are not located in the current folder. It also specifies user-defined settings such as the custom dictionary to use for spell checking.

■ Display

This tab customizes the AutoCAD window. For instance, you can specify the display of scroll bars in the window and indicate the number of lines of text to show in the docked AutoCAD – Command line window. It allows you to set the colors for the model and layout tabs, Command line background and text, and AutoTracking vector. You can also control the size of the crosshairs and the resolution of arcs and circles.

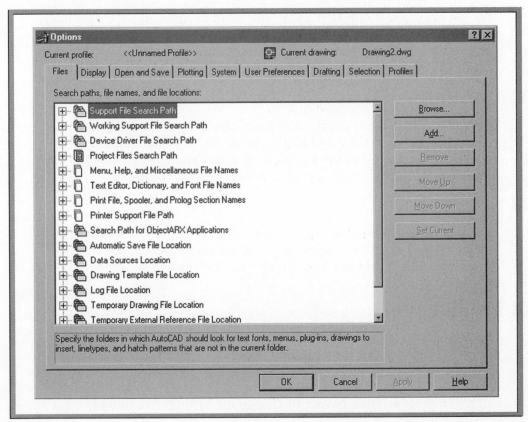

Fig. A-1

Appendix A: Options Dialog Box (Continued)

■ Open and Save

This tab controls the options related to opening and saving files. Examples include the default file type when saving an AutoCAD drawing and whether a thumbnail preview image appears in the Select File dialog box. The settings on this tab also allow you to set the number of minutes between automatic saves.

■ Plotting

The Plotting tab deals with settings and controls related to plotting and plotting devices. You can set the default output device and add and configure a new device. You can also add or edit plot style tables.

■ System

Use this tab to control miscellaneous system settings, such as the display of 3D graphics and whether AutoCAD displays the Startup dialog box. This is also where you change the current pointing device, including the configuration of a Wintab-compatible digitizing tablet.

■ User Preferences

This tab helps you optimize the way you work in AutoCAD. For example, you can customize how the right-click features in AutoCAD behave. Also, you can indicate the source and target drawing units when using AutoCAD DesignCenter.

■ Drafting

The Drafting tab controls many editing options. For instance, you can make adjustments to the AutoSnap features, such as changing the color and size of the AutoSnap marker. Here, you can also change AutoTracking settings, including whether to display polar tracking vectors.

■ Selection

This controls settings related to methods of object selection. Examples include turning noun/verb selection on or off and adjusting the size of the pickbox and grip boxes. You can also set the color of selected and unselected grip boxes.

■ Profiles

This tab, shown in Fig. A-2, controls the use of profiles. In this context, a *profile* is a user-defined configuration that allows each user to set up and save various personal preferences, such as which toolbars are present and where they appear on the screen. Profiles allow several people to share a single computer without having to change the settings individually each time they use it.

Appendix A: Options Dialog Box (Continued)

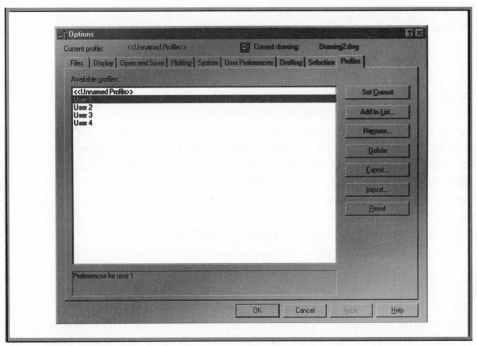

Fig. A-2

If no profiles have been defined for a computer, you can define one by **picking the** Add to List... button. A dialog box appears that allows you to **name the** profile and give it a description. Note that creating a new **profile** does not automatically select it as the current profile. You must **pick the** Set Current button to set the active profile.

When more than one profile exists, any changes you make to AutoCAD's settings are automatically saved to the current profile. To change the current profile, display the Options dialog box, pick the Profiles tab, select the profile you want to make current, and pick the Set Current button.

Appendix B: AutoCAD Management Tips

Organizing an AutoCAD installation requires several considerations. For example, you should store files in specific folders. Name the files and folders using standard naming conventions. If you don't, you may not be able to find the files later. It is also very important that you produce backup files regularly. If you overlook these and other system management tasks, your productivity will decline. Also, you may lose work, causing frustration that you could otherwise avoid.

AutoCAD users and managers alike will find the following information helpful. If you manage an AutoCAD system with care, you should never have to create the same object twice. This comes only with careful planning and cooperation among those who will be using the system.

Bear in mind that the process of implementing, managing, and expanding an AutoCAD system evolves over time. As you and others become more familiar with AutoCAD and the importance of managing files and folders, refer back to this appendix.

■ System Manager

One person (or possibly two, but no more) within the organization should have the responsibility for managing the system and overseeing its use. This person should be the resident CAD authority and should answer questions and provide directions to other users of the system. The manager should oversee the components of the system, including software, documentation, and hardware. The manager should work with the AutoCAD users to establish procedural standards for use with the system.

■ Management Considerations

The system manager should consider questions such as those listed below when installing and organizing AutoCAD. The questions are intended to help guide your thinking, from a management perspective, as you become familiar with the various components of AutoCAD.

- How can I best categorize the files so that each folder does not grow to more than 200 files total?

- If I plan to install three or more AutoCAD stations, should I centralize the storage of user-created files and plotting by using a network and file server?

After AutoCAD is in place and you are familiar with the system, you will create many new files. The following questions address the efficiency with which you create and store these files.

- Are there template files (or existing drawings) on file that may serve as a starting point for new drawings?

- Where should new drawing files be stored, what should they be named, and how can users easily locate them?

Appendix B: AutoCAD Management Tips (Continued)

- Are there predefined libraries of symbols and details that I can use while I develop a drawing? Do any of the symbol libraries that come with AutoCAD DesignCenter meet my needs?

- Is a custom pull-down or tablet menu or toolbar available that lends itself to my drawing application?

- Are AutoLISP routines available that would help me perform certain drafting operations more easily?

If you feel uncomfortable about your answers to these questions, there is probably room for improvement. The following discussion is provided to help you organize and manage your CAD system more effectively.

 NOTE:

Generally, the following discussion applies to almost all AutoCAD users and files. However, there are inevitable differences among users (backgrounds and interests, for example), drawing applications, and the specific hardware and software that make up the system. Take these differences into consideration.

■ Software/Documentation

The AutoCAD system manager spends considerable time organizing and documenting files, establishing rules and guidelines, and tracking new software and hardware developments.

File Management—Know where files are located and the purpose of each. Understand which ones are AutoCAD system files and which are not. Create a system for making backup files, and back up regularly. Emphasize this to all users. Delete "junk" files.

Template Files—Create an easy-to-use system for the development, storage, and retrieval of AutoCAD template files. Allow for ongoing development and improvement of each template. Store the template files in a folder dedicated to templates so that they are accessible by other users.

Document the contents of each template file by printing the drawing status information, layers, text styles, linetype scale, and other relevant information. On the first page of this information, write the name of the template file, its location, sheet size, and plot scale. Keep this information in a three-ring binder for future reference to other users.

User Drawing Files—Store these in separate folders. Place drawing objects on the proper layers. Assign standard colors, linetypes, and lineweights to the standard layers. Make a backup copy of each drawing and store it on a separate disk or tape backup system. Plot the drawings most likely to be used by others and store them in a three-ring binder for future reference.

Appendix B: AutoCAD Management Tips (Continued)

Symbol Libraries—Develop a system for ongoing library development. (See Chapter 32 for details on creating symbol libraries.) Plot each symbol library drawing file, and place the library drawings on the wall near the system(s) or in a binder. Encourage all users to contribute to the libraries.

Menu Files—Develop, set up, and make available custom toolbars, pull-down and tablet menus, and tablet overlays. (See Chapters 53 through 55 for details on creating toolbars, pull-down menus, and tablet menus.) Store the menus in a folder dedicated to menus so that they will be accessible to others.

AutoCAD Upgrades—Handle the acquisition and installation of AutoCAD software upgrades. Inform users of the new features and changes contained in the new software. Coordinate upgrade training.

AutoCAD Third-Party Software—Handle the acquisition and installation of third-party software developed for specific applications and utility purposes. Inform users of its availability and use.

■ Hardware

Oversee the use and maintenance of the hardware components that make up the system. Consider hardware upgrades as user and software requirements change.

■ Procedural Standards

Develop clear, practical standards in the organization to minimize inconsistency and confusion. Each template file should have a standard set of drawing layers, with a specific color and linetype dedicated to each layer. For example, you may reserve a layer called Dimension, with color yellow, a continuous linetype, and a lineweight of .3 mm, for all dimensions. Assign a specific pen to each stall on your pen plotter (if you are using one) and make this information available to others. This will avoid confusion and improve consistency within your organization. Develop similar standards for other AutoCAD-related practices.

In summary, take seriously the management of your AutoCAD system. Encourage users to experiment and to be creative by making software and hardware available to them. Make a team effort out of learning, developing, and managing the AutoCAD system so that everyone can learn and benefit from its power and capability.

Appendix C: Drawing Area Guidelines

	Sheet Size	Approximate Plotting Area	Scale	Area
Architect's Scale	A: 12″ × 9″ B: 18″ × 12″ C: 24″ × 18″ D: 36″ × 24″ E: 48″ × 36″	10″ × 8″ 16″ × 11″ 22″ × 16″ 34″ × 22″ 46″ × 34″	$\frac{1}{8}″ = 1'$ $\frac{1}{2}″ = 1'$ $\frac{1}{4}″ = 1'$ $3' = 1'$ $3' = 1'$	96′ × 72′ 36′ × 24′ 96′ × 72′ 12′ × 8′ 16′ × 12′
Civil Engineer's Scale	A: 12″ × 9″ B: 18″ × 12″ C: 24″ × 18″ D: 36″ × 24″ E: 48″ × 36″	10″ × 8″ 16″ × 11″ 22″ × 16″ 34″ × 22″ 46″ × 34″	1″ = 200′ 1″ = 50′ 1″ = 10′ 1″ = 300′ 1″ = 20′	2400′ × 1800′ 900′ × 600′ 240′ × 180′ 10,800′ × 7200′ 960′ × 720′
Mechanical Engineer's Scale	A: 11″ × 8″ B: 17″ × 11″ C: 22″ × 17″ D: 34″ × 22″ E: 44″ × 34″	9″ × 7″ 15″ × 10″ 20″ × 15″ 32″ × 20″ 42″ × 32″	1″ = 2″ 2″ = 1″ 1″ = 1″ 1″ = 1.5″ 4″ = 1″	22″ × 17″ 8.5″ × 5.5″ 22″ × 17″ 51″ × 33″ 11″ × 8.5″
Metric Scale	A: 279 mm × 216 mm (11″ × 8½″) B: 432 mm × 279 mm (17″ × 11″) C: 55.9 cm × 43.2 cm (22″ × 17″) D: 86.4 cm × 55.9 cm (34″ × 22″) E: 111.8 cm × 86.4 cm (44″ × 34″)	229 mm × 178 mm (9″ × 7″) 381 mm × 254 mm (15″ × 10″) 50.8 cm × 38.1 cm (20″ × 15″) 81.3 cm × 50.8 cm (32″ × 20″) 106.7 cm × 81.3 cm (42″ × 32″)	1 mm = 5 mm 1 mm = 20 mm 1 cm = 10 cm 2 cm = 1 cm 1 cm = 2 cm	1395 × 1080 8640 × 5580 559 × 432 43.2 × 27.95 237.6 × 172.8

Note: 1″ = 25.4 mm

Appendix D: Dimensioning Symbols

Geometric Characteristic Symbols

Type of Tolerance	Symbol	Name
Location	◎	Concentricity
	⊕	Position
	⚌	Symmetry
Orientation	∠	Angularity
	∥	Parallelism
	⊥	Perpendicularity
Form	⌭	Cylindricity
	▱	Flatness
	○	Circularity (roundness)
	—	Straightness
Profile	⌒	Profile of a line
	⌓	Profile of a surface
Runout	↗	Circular runout
	↗↗	Total runout
Supplementary	Ⓜ	Maximum material condition (MMC)
	Ⓛ	Least material condition (LMC)
	Ⓟ	Projected tolerance zone

Dimensioning Symbols

Symbol	Type of Dimension	Symbol	Type of Dimension
⌀	Diameter	∨	Countersink
R	Radius	⌴	Counterbore/Spotface
SR	Spherical radius (ISO name)	⤓	Deep
S⌀	Spherical diameter (ISO name)	X	Places, times, or by
()	Reference		

Appendix E: Standard Hatch Patterns

Shown here are the standard hatch patterns supplied in the file *acad.pat*.

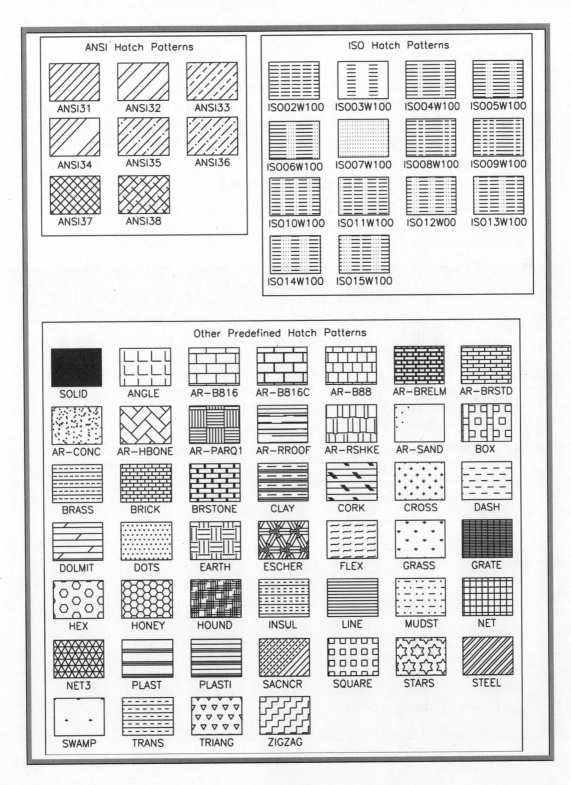

Appendix F: AutoCAD Fonts

AutoCAD provides several standard fonts, which have file extensions of SHX. You can use the STYLE command to apply expansion, compression, or obliquing to any of these fonts, thereby tailoring the characters to your needs. (See Chapter 18 for details on the STYLE command.) You can draw characters of any desired height using any of the fonts.

Examples of some of the fonts supplied with AutoCAD are listed in this appendix, along with samples of their appearance. With the exception of monotxt.shx (not included in the samples below), each font's characters are proportionately spaced. Hence, the space needed for the letter "i," for example, is narrower than that needed for the letter "m."

Each font resides in a separate file. This is the "compiled" form of the font, for direct use by AutoCAD. Examples of standard and TrueType fonts are shown in this appendix.

Standard Fonts

AutoCAD's standard SHX fonts include both text and symbol files that have been created as AutoCAD shapes. You can change the appearance of these fonts by expanding, compressing, or slanting the characters.

Name	Description	Appearance
txt.shx	Basic AutoCAD font; this font is very simple and generates quickly on the screen	ABCDEFGHIJKLMNOPQRSTUVWXYZ abcdefghijklmnopqrstuvwxyz
romans.shx	A "simplex" roman font drawn by means of many short line segments; produces smoother characters than txt.shx	ABCDEFGHIJKLMNOPQRSTUVWXYZ abcdefghijklmnopqrstuvwxyz
romand.shx	Similar to romans.shx, but instead of a single stroke, it uses a double stroke technique to produce darker, thicker lines	ABCDEFGHIJKLMNOPQRSTUVWXYZ abcdefghijklmnopqrstuvwxyz0123456789
itallicc.shx	Complex italic font using double stroke and serifs.	*ABCDEFGHIJKLMNOPQRSTUVWXYZ abcdefghijklmnopqrstuvwxyz0123456789*
scripts.shx	A single-stroke (simplex) script font	*ABCDEFGHIJKLMNOPQRSTUVWXYZ abcdefghijklmnopqrstuvwxyz0123456789*
gothice.shx	Gothic English font	𝔄𝔅ℭ𝔇𝔈𝔉𝔊ℌℑ𝔍𝔎𝔏𝔐𝔑𝔒𝔓𝔔ℜ𝔖𝔗𝔘𝔙𝔚𝔛𝔜ℨ abcdefghijklmnopqrstuvwxyz0123456789
syastro.shx	A symbol font that includes common astronomical symbols	☉♀♁⊕♂♃♄♅♆♇ …
symusic.shx	A symbol font that includes common music symbols	(music symbols)

Appendix F: AutoCAD Fonts (Continued)

TrueType Fonts

AutoCAD uses several TrueType fonts and font "families" (groups of related fonts). In the TrueType fonts, each font in a family has its own font file. The examples below are shown in outline form, as they appear by default in AutoCAD. To display solid characters, set the TEXTFILL system variable to 1. You may also change the print quality by changing the value of the TEXTQLTY system variable.

Name	Description
Arial Narrow	ABCDEFGHIJKLMNOPQRSTUVWXYZ abcdefghijklmnopqrstuvwxyz0123456789
Dutch801 RmBT	*ABCDEFGHIJKLMNOPQRSTUVWXYZ abcdefghijklmnopqrstuvwxyz0123456789*
Lucida Console	ABCDEFGHIJKLMNOPQRSTUVWXYZ abcdefghijklmnopqrstuvwxyz0123456789
Swis721 BdOul BT	ABCDEFGHIJKLMNOPQRSTUVWXYZ abcdefghijklmnopqrstuvwxyz0123456789
UniversalMath1 BT	ΑΒΨΔΕΦΓΗΙΞΚΛΜΝΟΠΘΡΣΤΘΩϬΧΥΖ αβψδεφγηιξκλμνοπϑρστθωφχυζ" + − × ÷ = ± ∓°′
Wingdings	(symbol characters)

Appendix G: Toolbars

AutoCAD comes with a collection of 24 toolbars, each consisting of several buttons that enter commands and options. Some of the toolbars also contain flyouts—sets of additional buttons that you can access by holding down the pick button on the pointing device over certain buttons. Buttons that display flyouts contain a small black triangle in the lower right corner.

This appendix lists the AutoCAD toolbars and the buttons they contain. Some toolbars display additional features when they are displayed horizontally on the screen. These toolbars are presented first.. The remaining toolbars are presented in a vertical orientation, in alphabetical order, for easier reading when you use the pages for reference.

Most of the flyouts also exist as standalone toolbars, and they are shown here in their standalone form. For these flyouts, a note has been placed in parentheses next to the parent button indicating which toolbar the flyout contains. For more information about each button, refer to the index.

3D Orbit Toolbar

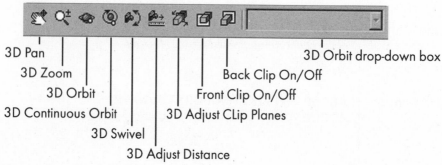

3D Pan
3D Zoom
3D Orbit
3D Continuous Orbit
3D Swivel
3D Adjust Distance
3D Adjust CLip Planes
Front Clip On/Off
Back Clip On/Off
3D Orbit drop-down box

Dimension Toolbar

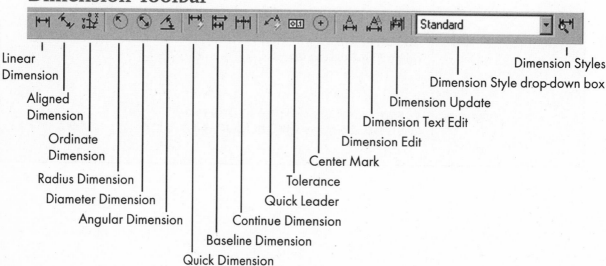

Linear Dimension
Aligned Dimension
Ordinate Dimension
Radius Dimension
Diameter Dimension
Angular Dimension
Quick Dimension
Baseline Dimension
Continue Dimension
Quick Leader
Tolerance
Center Mark
Dimension Edit
Dimension Text Edit
Dimension Update
Dimension Style drop-down box
Dimension Styles

Appendix G: Toolbars (Continued)

Object Properties Toolbar

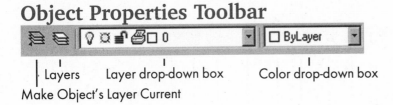

| Layers | Layer drop-down box | Color drop-down box
Make Object's Layer Current

Object Properties Toolbar (continued)

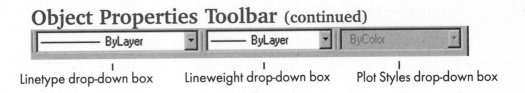

Linetype drop-down box   Lineweight drop-down box   Plot Styles drop-down box

Refedit Toolbar

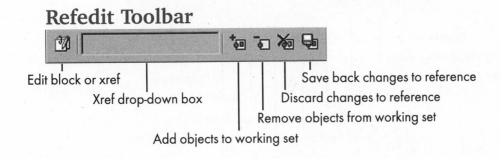

Edit block or xref

Xref drop-down box

Save back changes to reference

Discard changes to reference

Remove objects from working set

Add objects to working set

UCS II Toolbar

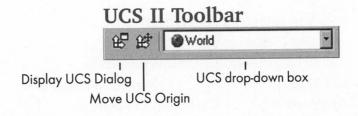

Display UCS Dialog

Move UCS Origin

UCS drop-down box

Appendix G: Toolbars (Continued)

Standard Toolbar

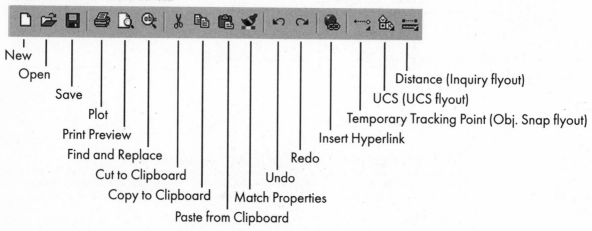

New
Open
Save
Plot
Print Preview
Find and Replace
Cut to Clipboard
Copy to Clipboard
Paste from Clipboard
Match Properties
Undo
Redo
Insert Hyperlink
Temporary Tracking Point (Obj. Snap flyout)
UCS (UCS flyout)
Distance (Inquiry flyout)

Standard Toolbar (continued)

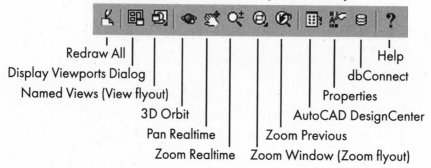

Redraw All
Display Viewports Dialog
Named Views (View flyout)
3D Orbit
Pan Realtime
Zoom Realtime
Zoom Window (Zoom flyout)
Zoom Previous
AutoCAD DesignCenter
Properties
dbConnect
Help

Viewports Toolbar

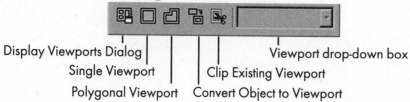

Display Viewports Dialog
Single Viewport
Polygonal Viewport
Convert Object to Viewport
Clip Existing Viewport
Viewport drop-down box

Appendix G: Toolbars (Continued)

Draw Toolbar

- Line
- Construction Line
- Multiline
- Polyline
- Polygon
- Rectangle
- Arc
- Circle
- Spline
- Ellipse
- Insert Block
- Make Block
- Point
- Hatch
- Region
- Multiline Text

Insert Toolbar

- Insert Block
- External Reference
- Image
- Import
- OLE Object

Layouts Toolbar

- New Layout
- Layout from Template
- Page Setup
- Display Viewports Dialog

Modify Toolbar

- Erase
- Copy Object
- Mirror
- Offset
- Array
- Move
- Rotate
- Scale
- Stretch
- Lengthen
- Trim
- Extend
- Break
- Chamfer
- Fillet
- Explode

Modify II Toolbar

- Draworder
- Edit Hatch
- Edit Polyline
- Edit Spline
- Edit Multiline
- Edit Attribute
- Edit Text

Inquiry Toolbar

- Distance
- Area
- Mass Properties
- List
- Locate Point

Appendix G: Toolbars (Continued)

Object Snap Toolbar

- Temporary Tracking Point
- Snap From
- Snap to Endpoint
- Snap to Midpoint
- Snap to Intersection
- Snap to Apparent Intersection
- Snap to Extension
- Snap to Center
- Snap to Quadrant
- Snap to Tangent
- Snap to Perpendicular
- Snap to Parallel
- Snap to Insert
- Snap to Node
- Snap to Nearest
- Snap to None
- Object Snap Settings

Reference Toolbar

- External Reference
- External Reference Attach
- External Reference Clip
- External Reference Bind
- External Reference Clip Frame
- Image
- Image Attach
- Image Clip
- Image Adjust
- Image Quality
- Image Transparency
- Image Frame

Render Toolbar

- Hide
- Render
- Scenes
- Lights
- Materials
- Materials Library
- Mapping
- Background
- Fog
- Landscape New
- Landscape Edit
- Landscape Library
- Render Preferences
- Statistics

Appendix G: Toolbars (Continued)

Shade Toolbar

 2D Wireframe

 3D Wireframe

 Hidden

 Flat Shaded

 Gouraud Shaded

 Flat Shaded, Edges On

 Gouraud Shaded, Edges On

Solids Toolbar

 Box

 Sphere

 Cylinder

 Cone

 Wedge

  Torus

 Extrude

 Revolve

 Slice

 Section

 Interfere

 Setup Drawing

 Setup View

 Setup Profile

Solids Editing Toolbar

 Union

 Subtract

 Intersect

 Extrude Faces

 Move Faces

 Offset Faces

 Delete Faces

 Rotate Faces

 Taper Faces

 copy Faces

 Color Faces

 Copy Edges

 Color Edges

 Imprint

 Clean

 Separate

 Shell

Check

Surfaces Toolbar

 2D Solid

 3D Face

 Box

 Wedge

 Pyramid

 Cone

 Sphere

 Dome

 Dish

 Torus

Edge

 3D Mesh

 Revolved Surface

Tabulated Surface

Ruled Surface

Edge Surface

Appendix G: Toolbars (Continued)

UCS Toolbar

 UCS

 Display UCS Dialog

 UCS Previous

 World UCS

 Object UCS

 Face UCS

 View UCS

 Origin UCS

 Z Axis Vector UCS

 3 Point UCS

 X Axis Rotate UCS

 Y Axis Rotate UCS

 Z Axis Rotate UCS

 Apply UCS

View Toolbar

 Named Views

 Top View

 Bottom View

 Left View

 Right View

 Front View

 Back View

 SW Isometric View

 SE Isometric View

 NE Isometric View

NW Isometric View

Camera

Web Toolbar

 Go Back

 Go Forward

 Stop Navigation

 Browse the Web

Zoom Toolbar

 Zoom Window

 Zoom Dynamic

 Zoom Scale

 Zoom Center

 Zoom In

 Zoom Out

 Zoom All

 Zoom Extents

Appendix H: Command Glossary

This appendix lists brief descriptions of all of the commands included in the basic AutoCAD package. For the locations of more detailed descriptions and instructions on how to apply them, refer to the index.

Some commands can be used transparently (that is, used while another command is in progress) by preceding the command name with an apostrophe. Such commands are listed here with an apostrophe.

3D

Purpose

Creates a three-dimensional polygon mesh object

Options

B	Creates a 3D box polygon mesh
C	Creates a cone-shaped polygon mesh
D	Creates the lower half of a spherical polygon mesh
M	Creates a planar mesh whose *M* and *N* sizes determine the number of lines drawn in each direction along the mesh. The *M* and *N* directions are similar to the X and Y axes of an XY plane.
P	Creates a pyramid or a tetrahedron
S	Creates a spherical polygon mesh
T	Creates a toroidal polygon mesh that is parallel to the XY plane of the current UCS
W	Creates a right-angle wedge-shaped polygon mesh with the sloped face tapering along the X axis

3DARRAY

Purpose

Creates a three-dimensional array

Options

P	Creates a polar array
R	Creates a rectangular array

3DCLIP

Purpose

Invokes the interactive 3D view and opens the Adjust Clipping Planes window

3DCORBIT

Purpose

Invokes the interactive 3D view and enables you to set the objects in the 3D view into continuous motion

3DDISTANCE

Purpose

Invokes the interactive 3D view and makes objects appear closer or farther away

3DFACE

Purpose

Creates a three-dimensional face

3DMESH

Purpose

Creates a free-form polygon mesh

3DORBIT

Purpose

Controls the interactive viewing of objects in 3D

Appendix H: Command Glossary (Continued)

3DPAN

Purpose

Invokes the interactive 3D view and enables you to drag the view horizontally and vertically

3DPOLY

Purpose

Creates a polyline with straight line segments using the CONTINUOUS linetype in 3D space

Options

C   Draws a closing line back to the first point and ends the command

E   Draws a straight line from the previous point to the specified new point

U   Deletes the last line and allows you to continue drawing from the previous point

3DSIN

Purpose

Presents a dialog box to control the import of a 3D Studio (3DS) file

3DSOUT

Purpose

Presents a dialog box to control the export of a 3D Studio (3DS) file

3DSWIVEL

Purpose

Invokes the interactive 3D view and simulates the effect of turning the camera

3DZOOM

Purpose

Invokes the interactive 3D view so you can zoom in and out on the view

'ABOUT

Purpose

Displays the AutoCAD version and serial number, license information, and the contents of the acad.msg file

ACISIN

Purpose

Presents a dialog box from which you can select an ACIS file to import

ACISOUT

Purpose

Presents a dialog box from which you can export an AutoCAD solid, body, or region to an ACIS file

ADCCLOSE

Purpose

Closes AutoCAD DesignCenter

ADCENTER

Purpose

Manages content

ADCNAVIGATE

Purpose

Directs the Desktop in AutoCAD DesignCenter to the file name, folder location, or network path you specify

Appendix H: **Command Glossary** (Continued)

ALIGN

Purpose

Moves and rotates objects in two or three dimensions to align with other objects using one, two, or three sets of points

AMECONVERT

Purpose

Converts AME solid models to AutoCAD solid objects

'APERTURE

Purpose

Controls the size of the object snap target box

'APPLOAD

Purpose

Loads and unloads applications and defines which applications to load at startup

ARC

Purpose

Creates an arc

Options

CE   Specify center point of arc

ENTER (as reply to Start point) sets start point and direction tangent to last line or arc

AREA

Purpose

Calculates the area and perimeter of objects or defined areas

Options

A   Sets Add mode

F   Specifies First Corner Point

O   Calculates area of a selected object

S   Sets Subtract mode

ARRAY

Purpose

Makes multiple copies of selected objects in a specified pattern

Options

P   Polar (circular) array

R   Rectangular array

ARX

Purpose

Loads, unloads, and provides information about ObjectARX™ applications

Options

?   Lists Applications

Commands—Lists the AcEd-registered commands

L   Loads an ObjectARX application

O   Options (for developers of ObjectARX applications)

U   Unloads an ObjectARX application

Appendix H: **Command Glossary** (Continued)

ATTDEF

Purpose

Creates an attribute definition

Options

C	Controls constant/variable mode
I	Controls attribute visibility
P	Controls present mode
V	Controls verify mode

'ATTDISP

Purpose

Controls the visibility of attribute entities on a global basis

Options

N	(Normal) Keeps the current visibility of each attribute (visible attributes are displayed, but invisible attributes are not)
OFF	Makes all attributes invisible
ON	Makes all attributes visible

ATTEDIT

Purpose

Permits editing of attributes

ATTEXT

Purpose

Extracts attribute data from a drawing

Options

C	CDF (comma-delimited) format
D	DXF format
O	Extracts attributes from selected objects
S	SDF (space-delimited) format

ATTREDEF

Purpose

Redefines a block and updates its associated attributes

AUDIT

Purpose

Evaluates the integrity of a drawing

Options

N	Reports, but does not fix, any errors encountered
Y	Fixes errors encountered

BACKGROUND

Purpose

Sets up the background for a scene

'BASE

Purpose

Sets the insertion base point for the current drawing

BHATCH

Purpose

Presents a dialog box from which you can create an associative hatch pattern within an automatically defined boundary; allows you to preview the hatch and make repeated adjustments without starting over each time

Appendix H: **Command Glossary** (Continued)

'BLIPMODE

Purpose

Controls the display of marker blips

Options

OFF   Disables temporary marker blips

ON   Enables temporary marker blips

BLOCK

Purpose

Creates a block definition from a group of selected objects

BLOCKICON

Purpose

Generates preview images for blocks created with Release 14 or earlier

BMPOUT

Purpose

Saves selected objects to a file in device-independent bitmap format

BOUNDARY

Purpose

Presents a dialog box from which you can create a region or polyline from an enclosed area

BOX

Purpose

Creates a three-dimensional solid box

Options

C   Creates a box by using a specified center point

ENTER Defines the corner of the box

BREAK

Purpose

Erases part of an object or splits an object in two

Option

F   Respecifies first point

BROWSER

Purpose

Launches the default Web browser defined in the system registry

'CAL

Purpose

Evaluates mathematical and geometric expressions

CAMERA

Purpose

Sets a different camera and target location

CHAMFER

Purpose

Bevels the edges of objects

Options

A   Sets the chamfer distances using a specified distance and an angle

D   Sets the chamfer distances from the selected edge

First Line   Specifies the first of two edges required to define a 2D chamfer, or the edge of a 3D solid to chamfer

Appendix H: Command Glossary (Continued)

(CHAMFER, *continued*)

M   Controls whether AutoCAD uses two distances or a distance and an angle to create the chamfer

P   Chamfers an entire 2D polyline

T   Controls whether AutoCAD trims the selected edges to the chamfer line endpoints

CHANGE

Purpose

Changes the properties of existing objects

Options

C   Color
E   Elevation
LA   Layer
LT   Linetype
LW   Lineweight
PL   Plot Style
S   Linetype scale
T   Thickness

CHPROP

Purpose

Changes the color, layer, linetype, linetype scale factor, lineweight, thickness, and plot style of an object

Options

C   Color
LA   Layer
LT   Linetype
LW   Lineweight
PL   Plot Style
S   Linetype scale
T   Thickness

CIRCLE

Purpose

Creates a circle

Options

2P   Draws a circle based on two endpoints of the diameter

3P   Draws a circle based on three points on the circumference

C   Draws a circle based on a center point and diameter or radius

TTR   Draws a circle tangent to two objects with a specified radius

CLOSE

Purpose

Closes the current drawing

'COLOR

Purpose

Defines the color for new objects

Options

(name)   Sets entity color to standard color name

(number)   Sets entity color number

BYBLOCK   Sets the floating entity color

BYLAYER   Uses layer's color for entities

COMPILE

Purpose

Presents a dialog box from which you can compile shape files and PostScript font files

Appendix H: Command Glossary (Continued)

CONE

Purpose

Creates a three-dimensional solid cone

Options

C   Center point for circular base of the cone

E   Creates a cone with an elliptical base

CONVERT

Purpose

Optimizes 2D polylines and associative hatches created in AutoCAD Release 13 or earlier

COPY

Purpose

Duplicates selected objects

Option

M   Makes multiple copies of the selected objects

COPYBASE

Purpose

Copies objects with a specified base point

COPYCLIP

Purpose

Copies objects to the Windows Clipboard

COPYHIST

Purpose

Copies text from the command line history window to the Windows Clipboard

COPYLINK

Purpose

Copies the current view to the Windows Clipboard for linking the AutoCAD view to another OLE-capable application

CUTCLIP

Purpose

Copies objects to the Windows Clipboard and erases them from the drawing

CYLINDER

Purpose

Creates a three-dimensional solid cylinder

Options

C   Defines the center of the circular base of the cylinder

E   Creates a cylinder with an elliptical base

DBCCLOSE

Purpose

Closes the dbConnect Manager dialog box

DBCONNECT

Purpose

Provides an AutoCAD interface to external database tables

DBLIST

Purpose

Lists database information for every object in the drawing

Appendix H: Command Glossary (Continued)

DDEDIT

Purpose

Presents a dialog box from which you can edit text and attribute definitions

'DDPTYPE

Purpose

Presents a dialog box from which you can specify the display mode and size of point objects

'DDVPOINT

Purpose

Presents a dialog box that allows you to set the three-dimensional viewing direction

DELAY

Purpose

Provides a timed pause within a script

DIM and DIM1

Purpose

Provides compatibility with previous releases of AutoCAD; enters the Dimensioning mode, in which dimensioning subcommands can be used to dimension objects

DIMALIGNED

Purpose

Creates an aligned linear dimension

DIMANGULAR

Purpose

Creates an angular dimension

Options

A   Specifies the angle at which the dimension text displays

M   Allows you to enter an mtext string in addition to or instead of the default dimension

T   Allows you to enter a single-line text string instead of the default dimension

DIMBASELINE

Purpose

Continues a linear, angular, or ordinate dimension from the baseline of the previous or selected dimension

Options

S   Allows you to select the next dimension in a series

U   Undoes the previous baseline dimension within a series

DIMCENTER

Purpose

Creates the center mark or the center lines of circles and arcs

DIMCONTINUE

Purpose

Continues a linear, angular, or ordinate dimension from the second extension line of the previous or a selected dimension

Appendix H: Command Glossary (Continued)

DIMDIAMETER

Purpose

Creates diameter dimensions for circles and arcs

Options

Dimension Line Location—Uses the point you specify to locate the dimension line.

A   Changes the angle of the dimension text

M   Allows you to enter an mtext string in addition to or instead of the default dimension

T   Allows you to enter a single-line text string instead of the default dimension

DIMEDIT

Purpose

Edits dimensions

Options

H   Moves dimension text back to its default position

N   Changes dimension text

O   Adjusts the obliquing angle of the extension lines for linear dimensions

R   Rotates dimension text

DIMLINEAR

Purpose

Creates linear (vertical and horizontal) dimensions

DIMORDINATE

Purpose

Creates ordinate point dimensions

Options

M   Allows you to customize mtext objects

T   Allows you to customize the text

X   Measures the x coordinate and determines orientation of the leader line and dimension text

Y   Measures the y coordinate and determines orientation of the leader line and dimension text

DIMOVERRIDE

Purpose

Overrides dimension system variables

Option

C   Clears any overrides on selected dimensions

DIMRADIUS

Purpose

Creates radial dimensions for circles and arcs

DIMSTYLE

Purpose

Creates and modifies dimension styles at the Command line

Options

?   Lists the named dimension styles in the current drawing

A   Updates the dimension objects you select so that they use the current settings of the dimensioning system variables

Appendix H: Command Glossary (Continued)

(DIMSTYLE, *continued*)

R   Changes the dimensioning system variable settings by reading new settings from an existing dimension style

S   Saves the current settings of dimensioning system variables to a dimension style

ST   Displays the current values of all dimensioning system variables

V   Lists the dimensioning system variable settings of a dimension style without modifying the current settings

DIMTEDIT

Purpose

Moves and rotates dimension text

Options

A   Changes the angle of the dimension text

C   Centers the dimension text on the dimension line

H   Moves dimension text that has been moved back to its default position

L   Left justifies the dimension text along the dimension line for linear, radial, and diameter dimensions

R   Right justifies the dimension text along the dimension line for linear, radial, and diameter dimensions

'DIST

Purpose

Measures the distance and angle between two points

DIVIDE

Purpose

Places evenly spaced point objects or blocks along the length or perimeter of an object

Option

(number)—Places point objects at equal intervals along the selected objects

B   Places blocks at equal intervals along the selected object

DONUT

Purpose

Draws filled circles and rings

'DRAGMODE

Purpose

Controls the way dragged objects are displayed

Options

A   (Auto) Turns on dragging for every command that supports it and performs drags automatically so that you do not have to enter DRAG each time

OFF   Ignores all dragging requests, including those embedded in menu items

ON   Permits dragging, but you must enter DRAG where appropriate to initiate dragging

Appendix H: **Command Glossary** (Continued)

DRAWORDER

Purpose

Changes the display order of objects and images

Options

A   Moves an object above a specified reference object

B   Object moves to the bottom of the drawing order

F   Object moves to the top of the drawing order

U   Moves an object below a specified reference object

DSETTINGS

Purpose

Specifies settings for snap mode, grid, and polar and object snap tracking

DSVIEWER

Purpose

Opens the Aerial View

DVIEW

Purpose

Defines parallel projection or perspective views

Options

CA   Specifies a new camera position by rotating the camera about the target point

CL   Clips the view, obscuring portions of the drawing that are behind or in front of the front clipping plane

D   Moves the camera in or out along the line of sight relative to target; turns on perspective viewing

H   Performs hidden line suppression on selected objects

O   Turns off perspective viewing

PA   Shifts the image without changing the level of magnification

PO   Locates the camera and target points using x,y,z coordinates

TA   Specifies a new position for the target by rotating it around the camera

TW   Twists or tilts the view around the line of sights

U   Reverses the effects of the last DVIEW operation

X   Ends the DVIEW command

Z   If perspective viewing is off, performs the equivalent of a ZOOM Center; if perspective viewing is on, adjusts the camera lens length, which changes the field of view and causes more or less of the drawing to be visible at a given camera and target distance

DWGPROPS

Purpose

Sets and displays the properties of the current drawing

DXBIN

Purpose

Displays a dialog box from which you can import specially coded binary files

Appendix H: **Command Glossary** (Continued)

EDGE

Purpose

Changes the visibility of three-dimensional face edges

Option

D   Selects invisible edges of 3D faces so that you can redisplay them

EDGESURF

Purpose

Creates a 3D polygon mesh approximating a Coons surface patch (a bicubic surface interpolated between four adjoining edges)

'ELEV

Purpose

Sets elevation and extrusion thickness properties of new objects

ELLIPSE

Purpose

Creates an ellipse or an elliptical arc

Options

A   Creates an elliptical arc

C   Creates the ellipse using a specified center point

I   Creates an isometric circle in the current isometric drawing plane

ERASE

Purpose

Removes objects from a drawing

EXPLODE

Purpose

Breaks a block, polyline, or other compound object into its component parts

EXPORT

Purpose

Saves objects to other file formats

EXPRESSTOOLS

Purpose

Activates the installed AutoCAD Express Tools if currently unavailable

EXTEND

Purpose

Extends a line, arc, elliptical arc, open 2D and 3D polyline, or ray to meet another object

Options

E   Determines whether the object is extended to another object's implied edge or only to an object that actually intersects it in 3D space

P   Specifies the projection mode AutoCAD uses when extending objects (none, UCS, or view)

U   Reverses the most recent change made by EXTEND

Appendix H: **Command Glossary** (Continued)

EXTRUDE

Purpose

Creates unique solid primitives by extruding existing two-dimensional objects

Options

H	Specifies height of extrusion
P	Selects the extrusion path based on a specified object

'FILL

Purpose

Controls the filling of multilines, traces, solids, and wide polylines

Options

OFF	Disables fill mode
ON	Enables fill mode

FILLET

Purpose

Rounds and fillets the edges of objects

Options

P	Inserts fillet arcs at each vertex of a 2D polyline
R	Defines the radius of the fillet arc
T	Controls whether AutoCAD trims the selected edges to the fillet arc endpoints

'FILTER

Purpose

Presents a dialog box from which you can create reusable filters to select objects based on properties

FIND

Purpose

Finds, replaces, selects, or zooms to specified text

FOG

Purpose

Provides visual cues for the apparent distance of objects; allows use of fog and depth cueing, in which white is "fog" and traditional depth cueing is black. With the FOG command, the user can specify any color in between these two extremes to provide visual depth cues to a drawing

'GRAPHSCR

Purpose

Switches from the text screen to the graphics screen; used in command script and menus

'GRID

Purpose

Displays a grid of dots at specified spacing on the screen

Options

A	Sets the grid to a different spacing on the X and Y axes
OFF	Turns off the grid
ON	Turns on the grid at the current spacing
S	Sets the grid spacing to the current snap interval (as set by the SNAP command)

Appendix H: Command Glossary (Continued)

GROUP

Purpose

Presents a dialog box from which you can create a named selection set of objects

HATCH

Purpose

Fills a specified boundary with a pattern

Options

?   Lists and provides a brief description of the hatch patterns defined in the acad.pat file

S   Specifies a solid fill

U   Specifies a pattern of lines using the current linetype

HATCHEDIT

Purpose

Presents a dialog box from which you can modify an existing hatch object

'HELP

Purpose

Displays online help. (The F1 function key also performs this function.)

HIDE

Purpose

Regenerates a three-dimensional model with hidden lines suppressed

HYPERLINK

Purpose

Attaches a hyperlink to a graphical object or modifies an existing hyperlink

Options

I   Inserts (attaches) a hyperlink to a bounded area or selected objects

R   Removes the hyperlink from the selected objects

HYPERLINKOPTIONS

Purpose

Controls the visibility of the hyperlink cursor and the display of hyperlink tooltips

'ID

Purpose

Displays the coordinate values of a location

IMAGE

Purpose

Inserts images in various formats into an AutoCAD drawing file

Options

Note: These options can be entered by picking buttons in the dialog box or by entering them at the keyboard

?   Lists images defined in the drawing database

A   Attaches a new image or a copy of an attached image to the current drawing

Appendix H: Command Glossary (Continued)

D   Detaches the named image from the drawing, marks it for deletion, and erases all occurrences

P   (Path) Allows the user to update the path name associated with an image

R   Reloads the selected images, making that information available for display and plotting

U   Unloads an image from working memory, but image remains in the drawing database

IMAGEADJUST

Purpose

Opens a dialog box from which the user can control brightness, contrast, and fade values of an image in the drawing database

Options

B   (Brightness) Controls brightness of an image on a scale of 0 through 100, with 100 being the brightest

C   (Contrast) Controls the contrast, and indirectly the fading effect, of the image. Values range from 0 through 100. The greater the value, the more each pixel is forced to its primary or secondary color

F   (Fade) Controls the fading effect of the image. Values range from 0 through 100. The greater the value, the more the image blends with the current background color

Note: From the dialog box, you can also preview the image and reset the above options to their default settings.

IMAGEATTACH

Purpose

Opens the Attach Image dialog box directly from the keyboard; allows the user to specify the image name, as well as parameters such as scale factor, rotation angle, and the option to specify parameters on-screen

IMAGECLIP

Purpose

Creates new clipping boundaries for an image object

Options

D   (Delete) Deletes a clipping boundary and displays the entire image

N   (New Boundary) Specifies a new clipping boundary. The boundary can be rectangular or polygonal, and consists only of straight line segments

ON   Turns on image clipping and clips to a previously defined boundary

OFF   Turns off image clipping and shows the entire drawing

IMAGEFRAME

Purpose

Controls whether the image frame is displayed on the screen or hidden from view

Options

OFF   Hides image frames so you cannot select images

ON   Displays image frames so you can select images

Appendix H: Command Glossary (Continued)

IMAGEQUALITY

Purpose

Controls display quality of images

Options

D   (Draft) Produces lower-quality images, but is faster than using the High option

H   (High) Produces high-quality images, but they are displayed and plotted more slowly than draft-quality images

IMPORT

Purpose

Displays a dialog box that allows you to import various file formats into AutoCAD

INSERT

Purpose

Inserts a named block or drawing into the current drawing

Options

?   Lists named blocks in the current drawing

B   Allows you to enter the name of a block or drawing

INSERTOBJ

Purpose

Inserts a linked or embedded object into AutoCAD

INTERFERE

Purpose

Finds the interferences of two or more solids and creates a composite solid from their common volume

INTERSECT

Purpose

Creates composite solids or regions from the intersection of two or more solids or regions

'ISOPLANE

Purpose

Specifies the current isometric plane

Options

ENTER Toggles to the next plane in a clockwise fashion from left to top to right

L   Selects the left isometric plane

R   Selects the right isometric plane

T   Selects the top isometric plane

'LAYER

Purpose

Displays a dialog box that permits you to create named drawing layers and assigns color and linetype properties to those layers

LAYOUT

Purpose

Creates a new layout and renames, copies, saves, or deletes an existing layout

Options

?   Lists all the layouts defined in the drawing

C   Copies a layout

D   Deletes a layout

N   Creates a new layout tab

R   Renames a layout

S   Makes a layout current

SA   Saves a layout. All layouts are stored in the drawing template file

T   Creates a new template based on an existing layout in a template or drawing file

LAYOUTWIZARD

Purpose

Starts the Layout wizard, in which you can designate page and plot settings for a new layout

LEADER

Purpose

Creates a line that connects an annotation to a feature

Options

A   Inserts annotation at the end of the leader line; the annotation can be text, a feature control frame, or a block

F   Controls the way the leader is drawn and whether it has an arrowhead

U   Undoes the last vertex point

LENGTHEN

Purpose

Changes the length of objects and the included angle of arcs

Options

DE   Changes the length of an object or the included angle of an arc by a specified incremental length; a positive value extends the object, and a negative value trims the object

DY   Enters dynamic dragging mode; changes the length of a selected object based on where its endpoint is dragged

P   Sets the length of an object by a specified percentage of its total length; sets the included angle of an arc by a specified percentage of the total angle of the selected arc

T   Sets the length of a selected object by specifying the total absolute length; sets the total angle of a selected arc by a specified total included angle

LIGHT

Purpose

Presents a dialog box from which you can manage lights and lighting effects

'LIMITS

Purpose

Sets and controls the drawing boundaries and grid display

Options

OFF Turns off limits checking but maintains the current values for the next time limits checking is set to ON

ON   Turns on limits checking; causes AutoCAD to reject attempts to enter points outside the drawing limits

Appendix H: Command Glossary (Continued)

LINE

Purpose

Creates straight line segments

Options

C   Draws a line segment from the most recent endpoint of a line to the first point, creating a closed polygon

ENTER Begins the current line at the last endpoint of the most recently drawn line

U   Undoes the most recent line segment.

'LINETYPE

Purpose

Displays a dialog box that enables you to create, load, and set linetypes

LIST

Purpose

Displays database information for selected objects

LOAD

Purpose

Presents a dialog box from which you can make shapes available for use by the SHAPE command

LOGFILEOFF

Purpose

Closes the log file opened by LOGFILEON

LOGFILEON

Purpose

Writes the text window contents to a file; the default log file name is acad.log

LSEDIT

Purpose

Allows you to edit a landscape object

LSLIB

Purpose

Allows you to maintain libraries of landscape objects by modifying or deleting landscape options

LSNEW

Purpose

Adds realistic landscape items such as trees to your drawings

'LTSCALE

Purpose

Sets the linetype scale factor

'LWEIGHT

Purpose

Sets the current lineweight, lineweight display options, and lineweight units

MASSPROP

Purpose

Calculates and displays the mass properties of regions and solids

Appendix H: Command Glossary (Continued)

MATCHPROP

Purpose

Copies the properties from one object to one or more objects

MATLIB

Purpose

Presents a dialog box from which you can import and export materials to and from a library of materials

MEASURE

Purpose

Places point objects or blocks at measured intervals on an object

Option

B   Places blocks at a specified interval along the selected object

MENU

Purpose

Presents a dialog box from which you can load a menu file

MENULOAD

Purpose

Loads partial menu files

MENUUNLOAD

Purpose

Unloads partial menu files

MINSERT

Purpose

Inserts multiple instances of a block in a rectangular array

Options

?   Lists currently defined block definitions in the drawing

~   Displays the Select Drawing File dialog box

MIRROR

Purpose

Creates a mirror-image copy of objects

MIRROR3D

Purpose

Creates a mirror-image copy of objects about a plane

Options

3   Defines the mirroring plane by three points

L   Mirrors the selected objects about the last defined mirrored plane

O   Uses the plane of a selected planar object as the mirroring plane

V   Aligns the mirroring plane with the viewing plane of the current viewport through a point

XY   Aligns the mirroring plane with the standard XY plane through a specified point

YZ   Aligns the mirroring plane with the standard YZ plane through a specified point

Appendix H: **Command Glossary** (Continued)

(MIRROR3D, *continued*)

Z   Defines the mirroring plane by a point on the plane and a point normal to the plane (on the Z axis)

ZX   Aligns the mirroring plane with the standard ZX plane through a specified point

MLEDIT

Purpose

Presents a dialog box from which you can edit multiple parallel lines

MLINE

Purpose

Creates multiple parallel lines

Options

J   Determines how the multiline is drawn between the points you specify; available choices are Top (draws the multiline below the cursor so that the top line aligns with the specified points), Zero (draws the multiline with its origin entered at the cursor so that the multiline is centered around the specified points), and Bottom (draws the multiline above the cursor so that the bottom line aligns with the specified points)

S   Controls the overall width of the multiline; does not affect linetype scale

ST   Specifies a style to use for the multiline

MLSTYLE

Purpose

Presents a dialog box from which you can create, load, or rename multiline styles and control the properties of each style

MODEL

Purpose

Switches from a layout tab to the Model tab and makes it current

MOVE

Purpose

Displaces objects a specified distance in a specified direction

MSLIDE

Purpose

Creates a slide file of the current viewport in model space, or of all viewports in paper space

MSPACE

Purpose

Switches from paper space to a model space viewport

MTEXT

Purpose

Creates paragraphs that fit within a nonprinting text boundary

Options

H   Specifies the height of the font used for mtext characters

Appendix H: **Command Glossary** (Continued)

J Determines both justification and text flow, for new or selected text, in relation to the text boundary

L Specifies line spacing for the mtext object

R Specifies the rotation angle of the text boundary

S Specifies the text style to use for paragraph text

W Specifies the width of the multiline text object

MULTIPLE

Purpose

Repeats the next command until canceled

MVIEW

Purpose

Creates floating viewports and turns on existing floating viewports

Options

2 Divides the specified area horizontally or vertically into two viewports

3 Divides the specified area into three viewports

4 Divides the specified area horizontally and vertically into four viewports of equal size

F Creates one viewport that fills the available display area

H Removes hidden lines from a viewport during plotting from paper space

L Locks the selected viewport

OFF Turns off a viewport

ON Turns on a viewport, making its objects visible

O Specifies a closed polyline, ellipse, spline, region, or circle to convert into a viewport

P Creates an irregularly shaped viewport using specified points

R Translates viewport configurations saved with the VPORTS command into individual viewports in paper space

MVSETUP

Purpose

Sets up the specifications of a drawing using MVSETUP on either the Model tab or the Layout tab

Options for Layout Tab

A (Align) Pans the view in a viewport so that it aligns with a base point in another viewport; the current viewport is the viewport to which the other point moves

C Creates viewports

O Sets MVSETUP preferences; you can set the layer on which to insert the title block, specify whether to reset the limits to the drawing extents after a title block has been inserted, specify paper space units, and specify whether the title block is to be inserted or externally referenced

S Adjusts the scale factor (the ratio between the scale of the border in paper space and the scale of the drawing objects in the viewports)

T Prepares paper space, orients the drawing by setting the origin, and creates a drawing border and title block

U Reverses operations performed in the current MVSETUP session

Appendix H: Command Glossary (Continued)

NEW

Purpose

Presents a dialog box from which you can create a new drawing file

OFFSET

Purpose

Creates concentric circles, parallel lines, and parallel curves

Option

T   Creates an object passing through a specified point

OLELINKS

Purpose

Updates, changes, and cancels existing OLE links

OLESCALE

Purpose

Displays the OLE Properties dialog box

OOPS

Purpose

Restores erased objects

OPEN

Purpose

Presents a dialog box from which you can specify a drawing file to open

OPTIONS

Purpose

Customizes the AutoCAD settings

'ORTHO

Purpose

Constrains cursor movement

Options

OFF   Turns off the ortho mode

ON   Turns on the ortho mode

'OSNAP

Purpose

Displays a dialog box that allows you to set object snap modes

Options

APP   Apparent Intersection includes two separate snap modes: Apparent Intersection and Extended Apparent Intersection. Apparent and Extended Apparent Intersection recognize edges of regions and curves but not edges or corners of 3D solids

CEN   Center of an arc, circle, ellipse, or elliptical arc

END   Closest endpoint of an arc, elliptical arc, line, mline, polyline segment, or ray or to the closest corner of a trace, solid, or 3D face

EXT   Extension point of an object

INS   Insertion point of an attribute, a block, a shape, or text

INT   Intersection of an arc, circle, ellipse, elliptical arc, line, mline, polyline, ray, spline, or xline

MID   Midpoint of an arc, ellipse, elliptical arc, line, mline, polyline segment, solid, spline, or xline

NEA   Nearest point on an arc, circle, ellipse, elliptical arc, line, mline, point, polyline, spline, or xline

Appendix H: Command Glossary (Continued)

NOD   Snaps to a point object

NON   Turns off object snap modes

PAR   Snaps to an extension in parallel with an object

PER   Perpendicular to an arc, circle, ellipse, elliptical arc, line, mline, polyline, ray, solid, spline, or xline

QUA   Snaps to a quadrant point of an arc, circle, ellipse, or elliptical arc

QUI   Snaps to the first snap point found

TAN   Tangent of an arc, circle, ellipse, or elliptical arc

Note: You may get varying results if you have both Intersection and Apparent Intersection on at the same time

PAGESETUP

Purpose

Specifies the layout page, plotting device, paper size, and settings for each new layout

'PAN

Purpose

Moves the drawing display in the current viewport

PARTIALOAD

Purpose

Loads additional geometry into a partially opened drawing

PARTIALOPEN

Purpose

Loads geometry from a selected view or layer into a drawing

PASTEBLOCK

Purpose

Pastes a copied block into a new drawing

PASTECLIP

Purpose

Inserts data from the Windows Clipboard

PASTEORIG

Purpose

Pastes a copied object in a new drawing using the coordinates from the original drawing

PASTESPEC

Purpose

Inserts data from the Windows Clipboard and controls the format of the data; used with OLE

PCINWIZARD

Purpose

Displays a wizard to import PCP and PC2 configuration file plot settings into the Model tab or current layout

Appendix H: Command Glossary (Continued)

PEDIT

Purpose

Edits polylines and three-dimensional polygon meshes

Options for 2D Polylines

C	Closes an open polyline
D	Removes extra vertices inserted by a fit or spline curve and straightens all segments of the polyline
E	Edits vertices
F	Fits curve to polyline
J	Joins to polyline
L	Toggles linetype generation to be either a continuous pattern at vertices, or with dashes generated at the start and end of vertices
O	Opens a closed polyline
S	Uses the polyline vertices as the frame for a spline curve (type set by SPLINETYPE)
U	Reverses operations one at a time as far back as the beginning of the PEDIT session
W	Specifies a new uniform width for the entire polyline

Options for 3D Polylines

C	Closes an open polyline
D	Decurves, or returns a spline curve to its control frame
E	Edits vertices
O	Opens a closed polyline
S	Fits a 3D B-spline curve to its control points
U	Reverses operations one at a time as far back as the beginning of the PEDIT session
X	Exits 3D polyline selection.

Options for 3D Polygon Meshes

D	Restores the original control-point polygon mesh
E	Edits mesh vertices
M	Opens or closes *M*-direction polylines
N	Opens or closes *N*-direction polylines
S	Fits a smooth surface. The SURFTYPE system variable controls the type of surface this option fits
U	Reverses operations one at a time as far back as the beginning of the PEDIT session

PFACE

Purpose

Creates a three-dimensional polyface mesh vertex by vertex

PLAN

Purpose

Displays the plan view of a user coordinate system

Options

C	Establishes a plan view of the current UCS
U	Establishes a plan view of the specified UCS
W	Establishes a plan view of the WCS

PLINE

Purpose

Creates two-dimensional polylines

Options

A	Adds arc segments to the polyline

Appendix H: **Command Glossary** (Continued)

C   Draws a line segment from the current position to the starting point of the polyline, creating a closed polyline

H   Specifies the width from the center of a wide polyline segment to one of its edges

L   Draws a line segment of a specified length at the same angle as the previous segment (or tangent to the arc if the preceding segment was an arc)

U   Removes the most recent line segment added to the polyline

W   Specifies the width of the next line segment

PLOT
Purpose
Presents a dialog box from which you can plot a drawing to a plotter, printer, or file

PLOTSTYLE
Purpose
Sets the current plot style for new objects, or the assigned plot style for selected objects

PLOTTERMANAGER
Purpose
Displays the Plotter Manager dialog box, from which you can launch the Add-a-Plotter wizard and the Plotter Configuration Editor

POINT
Purpose
Creates a point object

POLYGON
Purpose
Creates an equilateral closed polyline

Options

C   Defines the center of the polygon

E   Defines a polygon by specifying the endpoints of the first edge

PREVIEW
Purpose
Shows how a drawing will look when it is printed or plotted

PROPERTIES
Purpose
Controls properties of existing objects

PROPERTIESCLOSE
Purpose
Closes the Properties window

PSDRAG
Purpose
Controls the appearance of a PostScript image as it is dragged into position with PSIN

Options

0   Displays only the image's bounding box as you drag it into place

1   Displays the rendered PostScript image as you drag it into place

Appendix H: Command Glossary (Continued)

PSETUPIN

Purpose

Imports a user-defined page setup into a new drawing layout

PSFILL

Purpose

Fills a 2D polyline outline with a PostScript pattern

PSIN

Purpose

Imports a PostScript file

PSOUT

Purpose

Displays a dialog box that allows you to create an Encapsulated PostScript (EPS) file

PSPACE

Purpose

Switches from a model space viewport to paper space

PURGE

Purpose

Removes unused named objects, such as unused blocks or layers, from the database

QDIM

Purpose

Quickly creates a dimension

Options

B	Creates a series of baseline dimensions
C	Creates a series of continued dimensions
D	Creates a series of diameter dimensions
E	Edits a series of dimensions
O	Creates a series of ordinate dimensions
P	Sets a new datum point for baseline and ordinate dimensions
R	Creates a series of radius dimensions
S	Creates a series of staggered dimensions

QLEADER

Purpose

Quickly creates a leader and leader annotation

QSAVE

Purpose

Quickly saves the current drawing

QSELECT

Purpose

Quickly creates selection sets based on filtering criteria

Appendix H: **Command Glossary** (Continued)

QTEXT

Purpose

Controls the display and plotting of text and attribute objects

Options

OFF   Displays the actual text and attribute objects

ON   Displays all existing text and attribute objects as bounding boxes only

QUIT

Purpose

Quits AutoCAD if there have been no changes since the drawing was last saved; if the drawing has been modified, AutoCAD prompts you to save or discard the changes before quitting

RAY

Purpose

Creates a line that extends infinitely in one direction only

RECOVER

Purpose

Repairs a damaged drawing

RECTANG

Purpose

Draws a rectangular polyline

REDEFINE

Purpose

Restores AutoCAD internal commands overridden by UNDEFINE

REDO

Purpose

Reverses the effects of the previous UNDO or U command

'REDRAW

Purpose

Refreshes the display of the current viewport

'REDRAWALL

Purpose

Refreshes the display in all viewports

REFCLOSE

Purpose

Saves back or discards changes made during in-place editing of a reference (an xref or a block)

REFEDIT

Purpose

Selects a reference for editing

REFSET

Purpose

Adds or removes objects from a working set during in-place editing of a reference (an xref or a block)

REGEN

Purpose

Regenerates the drawing and refreshes the current viewport

Appendix H: Command Glossary (Continued)

REGENALL

Purpose

Regenerates the drawing and refreshes all viewports

'REGENAUTO

Purpose

Controls automatic regeneration of a drawing

Options

ON   Drawing regenerates automatically as needed

OFF   When drawing needs to be regenerated, AutoCAD prompts the user instead of regenerating automatically

REGION

Purpose

Creates a region object from a selection set of existing objects

REINIT

Purpose

Displays a dialog box from which you can reinitialize the digitizer, digitizer input/output port, and program parameters file

RENAME

Purpose

Changes the names of objects

RENDER

Purpose

Creates a photorealistic or realistically shaded image of a three-dimensional wireframe or solid model

RENDSCR

Purpose

Redisplays the last rendering created with the RENDER command

REPLAY

Purpose

Presents a dialog box from which you can specify images to be displayed in the GIF, TGA, or TIFF image formats

'RESUME

Purpose

Continues an interrupted script

REVOLVE

Purpose

Creates a solid by revolving a two-dimensional object about an axis

Options

O   Selects an existing line or single-segment polyline that defines the axis about which to revolve the object

X   Uses the positive X axis of the current UCS as the positive axis direction

Y   Uses the positive Y axis of the current UCS as the positive axis direction

Appendix H: Command Glossary (Continued)

REVSURF

Purpose

Creates a revolved surface about a selected axis

RMAT

Purpose

Presents a dialog box from which you can preview, select, modify, duplicate, create, attach, or detach rendering materials

ROTATE

Purpose

Moves objects about a base point

Option

R   Specifies the absolute current rotation angle and desired new rotation angle

ROTATE3D

Purpose

Moves objects about a three-dimensional axis

Options

2   Uses two points to define the axis of rotation

L   Uses the last axis of rotation

O   Aligns the axis of rotation with an existing object

V   Aligns the axis of rotation with the viewing direction of the current viewport that passes through the selected point

X   Aligns the axis of rotation with the X axis that passes through the selected point

Y   Aligns the axis of rotation with the Y axis that passes through the selected point

Z   Aligns the axis of rotation with the Z axis that passes through the selected point

RPREF

Purpose

Displays a dialog box that allows you to set rendering preferences

RSCRIPT

Purpose

Creates a script that repeats continuously

RULESURF

Purpose

Creates a ruled surface between two curves

SAVE

Purpose

Saves the drawing with the current file name or a specified name

SAVEAS

Purpose

Saves an unnamed drawing with a file name or renames the current drawing

SAVEIMG

Purpose

Presents a dialog box from which you can save a rendered image to a BMP, TIFF, or TGA file

Appendix H: Command Glossary (Continued)

SCALE

Purpose

Enlarges or reduces objects equally in the X, Y, and Z directions

Options

R   Scales the selected objects based on a reference length and a specified new length

S   Sets the scale factor by which the selected objects are multiplied

SCENE

Purpose

Presents a dialog box from which you can manage scenes in model space

'SCRIPT

Purpose

Executes a sequence of commands from a script

SECTION

Purpose

Uses the intersection of a plane and solids to create a region

Options

3   Defines three points on the sectioning plane

O   Aligns the sectioning plane with a circle, ellipse, circular or elliptical arc, 2D spline, or 2D polyline segment

V   Aligns the sectioning plane with the current viewport's viewing plane

XY   Aligns the sectioning plane with the XY plane of the current UCS

YZ   Aligns the sectioning plane with the YZ plane of the current UCS

Z   Defines the sectioning plane by a specified origin point on the Z axis of the plane

ZX   Aligns the sectioning plane with the ZX plane of the current UCS

SELECT

Purpose

Places selected objects in the previous selection set

Options

A   Switches to Add mode

ALL   Selects all objects on thawed layers

AU   Chooses automatic selection

BOX   Selects all objects inside or crossing a rectangle specified by two points

C   Selects objects within and crossing an area defined by two points

CP   Selects objects within and crossing a polygon defined by specifying points around the objects to be selected

F   Selects all objects crossing a selection fence

G   Selects all objects within a specified group

L   Selects the most recently created visible object

M   Allows specification of multiple points without highlighting the objects

P   Selects the most recent selection set

Appendix H: Command Glossary (Continued)

R   Switches to the Remove mode

SI   Places object selection in Single mode and selects the first object or set of objects designated, then ends the selection process

U   Cancels the selection of the object most recently added to the selection set

W   Selects all objects completely inside a rectangle defined by two points, specified from left to right

WP   Selects objects completely inside a polygon defined by points; polygon can be any shape but cannot cross or touch itself

SETUV

Purpose

Lets you map materials onto objects

'SETVAR

Purpose

Lists or changes values of system variables

Option

?   Lists current system variable settings

SHADEMODE

Purpose

Shades the objects in the current viewport

Options

2D   (2D Wireframe) Displays the objects using lines and curves to represent the boundaries

3D   (3D Wireframe) Displays the objects using lines and curves to represent the boundaries; displays a shaded 3D UCS icon

F   Shades the objects between the polygon faces; the objects appear flatter and less smooth than Gouraud-shaded objects

G   Shades the objects and smooths the edges between polygon faces; gives the objects a smooth, realistic appearance

H   Displays the objects using 3D wireframe representation and hides lines representing back faces

L   Combines the Flat Shaded and Wireframe options; objects are flat shaded with the wireframe showing through

O   Combines the Gouraud Shaded and Wireframe options; objects are Gouraud shaded with the wireframe showing through

SHAPE

Purpose

Inserts a shape

Option

?   Lists shapes and the files in which the shapes are defined

SHELL

Purpose

Accesses operating system commands

Option

ENTER   Shells out to DOS until you enter EXIT to return to AutoCAD

Appendix H: Command Glossary (Continued)

SHOWMAT

Purpose

Lists material type and attachment method for a selected object

SKETCH

Purpose

Creates a series of freehand line segments

Options

. (period) Lowers the pen, draws a straight line from the endpoint of the last sketched line to the pen's current location, and returns the pen to the "up" position

C Lowers the pen to continue a sketch sequence from the endpoint of the last sketched line

E Erases any portion of a temporary line and raises the pen if it is down

P Raises and lowers the sketching pen (toggle)

Q Discards all temporary lines sketched since the start of SKETCH or the last use of the Record option, and exits Sketch mode

R Records temporary lines as permanent and does not change the pen's position

X Records and reports the number of temporary lines sketched and exits Sketch mode

SLICE

Purpose

Slices a set of solids with a plane

Options

3 Defines three points on the cutting plane

O Aligns the cutting plane with a circle, ellipse, circular or elliptical arc, 2D spline, or 2D polyline

V Aligns the cutting plane with the current viewport's viewing plane

XY Aligns the cutting plane with the XY plane of the current UCS

YZ Aligns the cutting plane with the YZ plane of the current UCS

Z Defines the cutting plane by a specified origin point on the Z axis (normal) of the plane

ZX Aligns the cutting plane with the ZX plane of the current UCS

'SNAP

Purpose

Restricts cursor movement to specified intervals

Options

A Specifies differing X and Y spacings for the snap grid

OFF Turns off snap mode but retains the current settings

ON Activates snap

R Sets the rotation of the snap grid with respect to the drawing and the display screen

S Specifies the format of the snap grid (standard or isometric)

T Specifies the snap type (polar or grid)

Appendix H: Command Glossary (Continued)

SOLDRAW

Purpose

Generates profiles and sections in viewports created with SOLVIEW

SOLID

Purpose

Creates solid-filled polygons

SOLIDEDIT

Purpose

Edits faces and edges of three-dimensional solid objects

Options

B	Edits the entire solid object
E	Edits 3D solid objects by changing the color of or copying individual edges
F	Edits 3D solid
U	Undoes the editing action
X	Exits the SOLIDEDIT command

SOLPROF

Purpose

Creates profile images of three-dimensional solids

SOLVIEW

Purpose

Creates floating viewports using orthographic projection to lay out multi- and sectional-view drawings of 3D solids

Options

A	Creates an auxiliary view from an existing view (see Glossary)
O	Creates a folded orthographic view from an existing view
U	Creates a profile view relative to a specified UCS
S	Creates a drafting sectional view of solids, complete with cross-hatching

'SPELL

Purpose

Displays a dialog box from which you can check spelling in a drawing

SPHERE

Purpose

Creates a three-dimensional solid sphere

Options

D	Defines the diameter of the sphere
R	Defines the radius of the sphere

SPLINE

Purpose

Creates a quadratic or cubic spline (NURBS) curve

Option

O	Converts 2D or 3D quadratic or cubic spline-fit polylines to equivalent splines and (depending on the setting of the DELOBJ system variable) deletes the polylines

Appendix H: Command Glossary (Continued)

SPLINEDIT

Purpose

Edits a spline object

Options

C	Closes an open spline
E	Reverses the spline's direction
F	Edits fit data
M	Relocates a spline's control vertices and purges the fit points
O	Opens a closed spline
R	Fine-tunes a spline definition
U	Cancels the last editing operation

STATS

Purpose

Presents a dialog box that displays rendering statistics

'STATUS

Purpose

Displays drawing statistics, modes, and extents

STLOUT

Purpose

Stores a solid in an ASCII or binary file

STRETCH

Purpose

Moves or stretches objects

STYLE

Purpose

Creates or modifies named styles and sets the current style for text in your drawing

STYLESMANAGER

Purpose

Displays the Plot Style Manager

SUBTRACT

Purpose

Creates a composite region or solid by subtracting the area of one set of regions from another or subtracting the volume of one set of solids from another

SYSWINDOWS

Purpose

Arrange windows

Options

A	Arranges the window icons
C	Overlaps windows with visible title bars
H	Arranges windows in horizontal, nonoverlapping tiles
V	Arranges windows in vertical, nonoverlapping tiles

TABLET

Purpose

Calibrates, configures, and turns on and off an attached digitizing tablet

Options

CAL	Calibrates tablet for use in the current space
CFG	Configures tablet menus and screen pointing area
OFF	Turns off tablet mode
ON	Turns on tablet mode

Appendix H: **Command Glossary** (Continued)

TABSURF

Purpose

Creates a tabulated surface from a path curve and direction vector

TEXT

Purpose

Displays text on screen as it is entered

Options

J   Controls justification of the text

S   Specifies the text style, which determines the appearance of the text characters

'TEXTSCR

Purpose

Switches from the graphics screen to the text screen

'TIME

Purpose

Displays the date and time statistics of a drawing

Options

D   Repeats the display with updated items

OFF   Stops the user elapsed timer

ON   Starts the user elapsed timer

R   Resets the user elapsed timer

TOLERANCE

Purpose

Creates geometric tolerances

TOOLBAR

Purpose

Displays, hides, and customizes toolbars

TORUS

Purpose

Creates a donut-shaped solid

Options

D   Defines the radius of the torus; allows you to specify the radius or diameter of the tube

R   Defines the diameter of the torus; allows you to specify the radius or diameter of the tube

TRACE

Purpose

Creates solid lines of a specified width

TRANSPARENCY

Purpose

Controls whether background pixels in an image are transparent or opaque

Options

OFF   Turns transparency off so that objects beneath the image are hidden from view

ON   Turns transparency on so that objects beneath the image are visible

'TREESTAT

Purpose

Displays information on the drawing's current spatial index

Appendix H: Command Glossary (Continued)

TRIM

Purpose

Trims objects at a cutting edge defined by other objects

Options

E	Determines whether an object is trimmed at another object's implied edge, or only to an object that intersects it in 3D space
P	Specifies the projection mode for trimming objects (none, UCS, or view)
U	Reverses the most recent trim operation

U

Purpose

Reverses the most recent operation

UCS

Purpose

Manages user coordinate system

Options

?	Lists names of user coordinate systems
A	Applies the current UCS setting to a specified viewport
D	Deletes the specified UCS from the list of saved user coordinate systems
G	Specifies one of the six orthographic UCSs provided with AutoCAD
M	Redefines a UCS by shifting the origin or changing the Z-depth of the current UCS, leaving the orientation of its XY plane unchanged
N	Defines a new coordinate system by one of six methods
P	Restores the previous UCS
R	Restores a saved UCS so that it becomes the current UCS
S	Saves the current UCS to a specified name
W	Sets the current user coordinate system to the world coordinate system

UCSICON

Purpose

Controls the visibility and placement of the UCS icon

Options

All	Applies changes to the icon in all active viewports
N	Displays the icon at the lower left corner of the viewport regardless of the location of the UCS origin
OFF	Disables the coordinate system icon
ON	Enables the coordinate system icon
OR	Forces the icon to appear at the origin of the current coordinate system

UCSMAN

Purpose

Manages defined user coordinate systems

UNDEFINE

Purpose

Allows an application-defined command to override an internal AutoCAD command

Appendix H: **Command Glossary** (Continued)

UNDO

Purpose

Reverses the effect of commands

Options

A   Undoes a menu selection as a single command, reversible by a single U command

B   Undoes all operations in memory until it reaches a mark set by the M option

BE   Begins an UNDO group definition from a sequence of operations

C   Limits or turns off the UNDO command

E   Ends the UNDO group definition begun by the BE option

M   Places a mark in the undo information

(number)—Undoes the specified number of preceding operations

UNION

Purpose

Creates a composite region or solid by addition

'UNITS

Purpose

Controls coordinate and angle display formats and determines precision

VBAIDE

Purpose

Displays the Visual Basic Editor

VBALOAD

Purpose

Loads a global VBA project into the current AutoCAD session

VBAMAN

Purpose

Loads, unloads, saves, creates, embeds, and extracts VBA projects

VBARUN

Purpose

Runs a VBA macro

VBASTMT

Purpose

Executes a VBA statement on the AutoCAD command line

VBAUNLOAD

Purpose

Unloads a global VBA project

'VIEW

Purpose

Saves and restores named views

Options

?   Lists the named views in the drawing

D   Deletes one or more named views

O   Restores the predefined orthographic view you specify to the current viewport

R   Restores the view you specify to the current viewport

Appendix H: Command Glossary (Continued)

(VIEW, continued)

S	Saves the display in the current viewport using the name you supply
UCS	Determines whether the current UCS and elevation settings are saved when a view is saved
W	Saves a portion of the current display as a view

VIEWRES

Purpose

Sets the resolution for object generation in the current viewport

VLISP

Purpose

Displays the Visual LISP interactive development environment (IDE)

VPCLIP

Purpose

Clips viewport objects

VPLAYER

Purpose

Sets layer visibility within viewports

Options

?	Lists frozen layers
F	Freezes specified layers in selected viewports
N	Creates new layers that are frozen in all viewports
R	Sets the visibility of layers in specified viewports to their current default setting

T	Thaws specified layers in selected viewports
V	Determines whether the specified layers are thawed or frozen in subsequently created viewports

VPOINT

Purpose

Sets the viewing direction for a three-dimensional visualization of the drawing

Options

ENTER	Presents an axis tripod from which you can select the new viewing direction
R	Specifies a new direction using two angles
V	Allows you to enter x,y,z coordinates to create a vector that defines a direction from which the drawing can be viewed

VPORTS

Purpose

Divides the graphics area into multiple tiled or floating viewports

Options

?	Lists viewport configurations
2	Divides the current viewport in half
3	Divides current viewport into three viewports
4	Divides current viewport into four viewports of equal size
D	Deletes a named viewport
J	Combines two adjacent viewports into one larger viewport
R	Restores a previously saved viewport configuration

Appendix H: Command Glossary (Continued)

S   Saves the current viewport configuration

SI   Returns the drawing into a single viewport view, using the view from the active viewport

VSLIDE

Purpose

Displays an image slide file in the current viewport

WBLOCK

Purpose

Writes objects or a block to a new drawing file

WEDGE

Purpose

Creates a three-dimensional solid with a tapered, sloping face

Options

C   Defines the first corner of the wedge

CE   Allows you to specify the center point of the wedge

WHOHAS

Purpose

Displays ownership information for opened drawing files

WMFIN

Purpose

Imports a Windows metafile (WMF format)

WMFOPTS

Purpose

Presents a dialog box that allows you to set options for WMFIN

WMFOUT

Purpose

Saves objects in a Windows Metafile (WMF format)

XATTACH

Purpose

Presents a dialog box from which you can attach an external reference to the current drawing

XBIND

Purpose

Binds dependent symbols of an xref to a drawing; dependent symbols may include blocks, dimension styles, layers, linetypes, and text styles

XCLIP

Purpose

Defines an xref or block clipping boundary and sets the front or back clipping planes

Options

C   (Clipdepth) Sets front and back clipping planes on an external reference or block

D   (Delete) Removes a clipping boundary for the selected xref or block (this is a permanent deletion; to remove the clipping boundary temporarily from the screen, use OFF instead)

Appendix H: Command Glossary (Continued)

(XCLIP, *continued*)

N   (New boundary) Creates a new clipping boundary

OFF   Displays all of the geometry of the xref or block, ignoring clipping boundaries

ON   Displays the clipped portion of the xref or block only

P   (Generate Polyline) Automatically draws a polyline (in the current layer) that coincides with the clipping boundary; this allows the clipping boundary to be edited with the PEDIT command

XLINE

Purpose

Creates a line that extends infinitely in both directions

Options

A   Creates an xline at a specified angle

B   Creates an xline that passes through the selected angle's vertex and then bisects the angle

H   Creates a horizontal xline

O   Creates an xline parallel to another object at a specified distance

P   Specifies the location of the infinite line using two points through which it passes

V   Creates a vertical xline

XPLODE

Purpose

Breaks a compound object into its component objects

XREF

Purpose

Displays a dialog box that allows you to control external references to drawing files

'ZOOM

Purpose

Increases or decreases the apparent size of objects in the current viewport

Options

A   Displays the entire drawing in the current viewport; in a plan view, AutoCAD zooms to the drawing limits (area) or current extents, whichever is greater

C   Displays a window defined by a center point and a magnification value or height

D   Displays the generated portion of the drawing with a view box that represents the current viewport, which you can shrink or enlarge and move around the drawing

E   Displays the drawing as large as possible on the screen

P   Displays the previous view

R   The cursor changes to a magnifying glass with plus (+) and minus (–) signs, allowing you to zoom interactively to a logical extent

S   Displays at a specified scale factor. The value you enter is relative to the limits (area) of the drawing; for example, entering 3 triples for the apparent display size of objects

W   Displays an area specified by two opposite corners of a rectangular window

Appendix I: System Variables

This appendix lists the AutoCAD system variables and describes their purpose and settings. You can change all but read-only system variables at the Command prompt using either the SETVAR command or

AutoLISP's getvar and setvar functions. In addition, you can change the values of many of the system variables by simply entering the name of the variable at the Command prompt.

ACADLSPASDOC

Type: Integer

Saved in: Registry

Description: Controls whether AutoCAD loads the acad.lsp file into every drawing or just the first drawing opened in an AutoCAD session.

0   Loads acad.lsp into just the first drawing opened in an AutoCAD session

1   Loads acad.lsp into every drawing opened

ACADPREFIX

Type: String

Saved in: (Not saved)

Description: (Read-only) Stores the directory path specified by the ACAD environment variable, with path separators appended if necessary

ACADVER

Type: String

Saved in: (Not saved)

Description: (Read-only) Stores the AutoCAD version number; note that this variable differs from the DXF file $ACADVER header variable, which contains the drawing database level number

ACISOUTVER

Type: Integer

Saved in: Drawing

Description: Controls the version of ACIS used to create SAT files using the ACISOUT command

AFLAGS

Type: Integer

Saved in: (Not saved)

Description: Sets attribute flags for the ATTDEF command (sum of the following):

0   No attribute mode selected

1   Invisible

2   Constant

4   Verify

8   Preset

ANGBASE

Type: Real

Saved in: Drawing

Description: Sets the base angle at 0 with respect to the current UCS

ANGDIR

Type: Integer

Saved in: Drawing

Description: Sets the angle from angle 0 with respect to the current UCS

0   Counterclockwise

1   Clockwise

Appendix I: System Variables (Continued)

APBOX

Type: Integer

Saved in: Registry

Description: Turns the AutoSnap aperture box on or off

0   Aperture box is not displayed

1   Aperture box is displayed

APERTURE

Type: Integer

Saved in: Registry

Description: Sets object snap target height, in pixels

AREA

Type: Real

Saved in: (Not saved)

Description: (Read-only) Stores the last area computed by AREA, LIST, or DBLIST

ATTDIA

Type: Integer

Saved in: Drawing

Description: Controls whether INSERT uses a dialog box for attribute value entry

0   Issues prompts on the command line

1   Uses a dialog box

ATTMODE

Type: Integer

Saved in: Drawing

Description: Controls attribute display mode

0   Off

1   Normal

2   On

ATTREQ

Type: Integer

Saved in: Drawing

Description: Determines whether INSERT uses default attribute settings during insertion of blocks

0   Assumes the defaults for the values of all attributes

1   Enables prompts or dialog box for attribute values, as selected by ATTDIA

AUDITCTL

Type: Integer

Saved in: Registry

Description: Controls whether AutoCAD creates an ADT file (audit report)

0   Disables or prevents writing of ADT files

1   Enables the writing of ADT files

AUNITS

Type: Integer

Saved in: Drawing

Description: Sets angular units mode

0   Decimal degrees

1   Degrees/minutes/seconds

2   Gradians

3   Radians

4   Surveyor's units

AUPREC

Type: Integer

Saved in: Drawing

Description: Sets angular units decimal places

Appendix I: System Variables (Continued)

AUTOSNAP

Type: Integer

Saved in: Registry

Description: Controls display of the AutoSnap marker and SnapTips; also turns the AutoSnap magnet on and off

0   Turns off the marker, SnapTips, and the AutoSnap magnet

1   Turns on the marker

2   Turns on SnapTips

4   Turns on the magnet

BACKZ

Type: Real

Saved in: Drawing

Description: Stores the back clipping plane offset from the target plane for the current viewport, in drawing units

BINDTYPE

Type: Integer

Saved in: (Not saved)

Description: Controls how xref names are handled when binding xrefs or editing xrefs in-place

0   Traditional binding behavior (xref1|one becomes xref0one)

1   Insert-like behavior (xref1|one becomes one)

BLIPMODE

Type: Integer

Saved in: Drawing

Description: Controls whether marker blips are visible

0   Turns off marker blips

1   Turns on marker blips

CDATE

Type: Real

Saved in: (Not saved)

Description: Sets calendar date and time

CECOLOR

Type: String

Saved in: Drawing

Description: Sets the color of new objects

CELTSCALE

Type: Real

Saved in: Drawing

Description: Sets the current global linetype scale for objects

CELTYPE

Type: String

Saved in: Drawing

Description: Sets linetype of new objects

CELWEIGHT

Type: Integer

Saved in: Drawing

Description: Sets the lineweight of new objects

−1 Sets the lineweight to ByLayer

−2 Sets the lineweight to ByBlock

−3 Sets the lineweight to Default; Default is controlled by the LWDEFAULT system variable

Other valid values include 0, 5, 9, 13, 15, 18, 20, 25, 30, 35, 40, 50, 53, 60, 70, 80, 90, 100, 106, 120, 140, 158, 200, and 211 millimeters.

All values must be entered in millimeters. (Multiply a value by 2.54 to convert values from inches to millimeters.)

Appendix I: System Variables (Continued)

CHAMFERA

Type: Real

Saved in: Drawing

Description: Sets the first chamfer distance

CHAMFERB

Type: Real

Saved in: Drawing

Description: Sets the second chamfer distance

CHAMFERC

Type: Real

Saved in: Drawing

Description: Sets the chamfer length

CHAMFERD

Type: Real

Saved in: Drawing

Description: Sets the chamfer angle

CHAMMODE

Type: Integer

Saved in: (Not saved)

Description: Sets the input method by which AutoCAD creates chamfers

0   Requires two chamfer distances

1   Requires one chamfer length and an angle

CIRCLERAD

Type: Real

Saved in: (Not saved)

Description: Sets the default circle radius; a zero sets no default

CLAYER

Type: String

Saved in: Drawing

Description: Sets the current layer

CMDACTIVE

Type: Integer

Saved in: (Not saved)

Description: (Read-only) Stores bit-code that indicates whether an ordinary command, transparent command, script, or dialog box is active; the sum of the following:

1   Ordinary command

2   Ordinary command and transparent command

4   Script

8   Dialog box

CMDDIA

Type: Integer

Saved in: Registry

Description: Controls whether dialog boxes are enabled for more than just PLOT and external database commands

0   Disables dialog boxes

1   Enables dialog boxes

CMDECHO

Type: Integer

Saved in: (Not saved)

Description: Controls whether AutoCAD echoes prompts and input during the AutoLISP command function

0   Disables echoing

1   Enables echoing

Appendix I: System Variables (Continued)

CMDNAMES

Type: String

Saved in: (Not saved)

Description: (Read-only) Displays the name of the currently active command and transparent command

CMLJUST

Type: Integer

Saved in: Drawing

Description: Specifies multiline justification

0   Top

1   Middle

2   Bottom

CMLSCALE

Type: Real

Saved in: Drawing

Description: Controls the overall width of a multiline as a function of the style definition

CMLSTYLE

Type: String

Saved in: Drawing

Description: Sets the name of the multiline style that AutoCAD uses to draw the multiline

COMPASS

Type: Integer

Saved in: (Not saved)

Description: Controls whether the 3D compass is on or off in the current viewport

0   Turns off the 3D compass

1   Turns on the 3D compass

COORDS

Type: Integer

Saved in: Registry

Description: Controls when coordinates are updated

0   Coordinate display is updated as you specify points with the pointing device

1   Display of absolute coordinates is continuously updated

2   Distance and angle from the last point are displayed when a distance or angle is requested

CPLOTSTYLE

Type: String

Saved in: Drawing

Description: Controls the current plot style for new objects

ByLayer

ByBlock

Normal

User Defined

CPROFILE

Type: Integer

Saved in: Registry

Description: (Read only) Stores the name of the current profile

For more information on profiles, see the OPTIONS command.

Appendix I: System Variables (Continued)

CTAB

Type: String

Saved in: Drawing

Description: (Read only) Returns the name of the current (Model or Layout) tab in the drawing

CURSORSIZE

Type: Integer

Saved in: Registry

Description: Determines the size of the crosshairs as a percentage of the screen size

CVPORT

Type: Integer

Saved in: Drawing

Description: Sets the identification number of the current viewport

DATE

Type: Real

Saved in: (Not saved)

Description: (Read-only) Stores the current date and time represented as a Julian date and fraction in a real number: <Julian date>.<Fraction>

DBMOD

Type: Integer

Saved in: (Not saved)

Description: Indicates the drawing modification status using bit-code; the sum of the following:

0   Object database modified

1   Symbol table modified

2   Database variable modified

8   Window modified

16 View modified

DCTCUST

Type: String

Saved in: Registry

Description: Displays the current custom spelling dictionary path and file name

DCTMAIN

Type: String

Saved in: Registry

Description: Displays current main spelling dictionary file name

DEFLPLSTYLE

Type: String

Saved in: Registry

Description: Specifies the default plot style for new layers

DEFPLSTYLE

Type: String

Saved in: Registry

Description: Specifies the default plot style for new objects

Appendix I: System Variables (Continued)

DELOBJ

Type: Integer

Saved in: Drawing

Description: Controls whether objects used to create other objects are retained or deleted from the drawing database

0   Objects are retained

1   Objects are deleted

DIASTAT

Type: Integer

Saved in: (Not saved)

Description: Stores the exit method of the most recently used dialog box

0   Cancel

1   OK

DIMADEC

Type: Integer

Saved in: Drawing

Description: Controls the number of places of precision displayed for angular dimension text

DIMALT

Type: Switch

Saved in: Drawing

Description: When turned on, enables alternate units dimensioning

DIMALTD

Type: Integer

Saved in: Drawing

Description: Controls number of decimal places used in alternate measurement

DIMALTF

Type: Real

Saved in: Drawing

Description: Controls scale factor of alternate units

DIMALTRND

Type: Real

Saved in: Drawing

Description: Determines rounding of alternate units

DIMALTTD

Type: Integer

Saved in: Drawing

Description: Sets the number of decimal places for the tolerance values of an alternate units dimension

DIMALTTZ

Type: Integer

Saved in: Drawing

Description: Toggles suppression of zeros for tolerance values in alternate units

DIMALTU

Type: Integer

Saved in: Drawing

Description: Sets the units format for alternate units of all dimension style family members except angular

1   Scientific

2   Decimal

3   Engineering

4   Architectural

5   Fractional

Appendix I: System Variables (Continued)

DIMALTZ

Type: Integer

Saved in: Drawing

Description: Toggles suppression of zeros for alternate unit dimension values

0   Turns off suppression of zeros

1   Turns on suppression of zeros

DIMAPOST

Type: String

Saved in: Drawing

Description: Specifies a text prefix or suffix (or both) to the alternate dimension measurement for all types of dimensions except angular

DIMASO

Type: Switch

Saved in: Drawing

Description: Controls the creation of associative dimension objects

OFF   Dimensions are not associative

ON   Dimensions are associative

DIMASZ

Type: Real

Saved in: Drawing

Description: Controls size of dimension line and leader line arrowheads and the size of hook lines

DIMATFIT

Type: Integer

Saved in: Drawing

Description: Determines how dimension text and arrows are arranged when space is not sufficient to place both within the extension lines

0   Places both text and arrows outside extension lines

1   Moves arrows first, then text

2   Moves text first, then arrows

3   Moves either text or arrows, whichever fits best

AutoCAD adds a leader to moved dimension text when DIMTMOVE is set to 1

DIMAUNIT

Type: Integer

Saved in: Drawing

Description: Sets angle format for angular dimensions

0   Decimal degrees

1   Degrees/minutes/seconds

2   Gradians

3   Radians

4   Surveyor's units

DIMAZIN

Type: Integer

Saved in: Drawing

Description: Suppresses zeros for angular dimensions

0   Displays all leading and trailing zeros

1   Suppresses leading zeros in decimal dimensions (for example, 0.5000 becomes .5000)

Appendix I: System Variables (Continued)

2 Suppresses trailing zeros in decimal dimensions (for example, 12.5000 becomes 12.5)

3 Suppresses leading and trailing zeros (for example, 0.5000 becomes .5)

DIMBLK

Type: String

Saved in: Drawing

Description: Sets the name of a block to be drawn instead of the normal arrowhead at the ends of the dimension line or leader line

DIMBLK1

Type: String

Saved in: Drawing

Description: If DIMSAH is on, DIMBLK1 specifies user-defined arrowhead blocks for the first end of the dimension line

DIMBLK2

Type: String

Saved in: Drawing

Description: If DIMSAH is on, DIMBLK2 specifies user-defined arrowhead blocks for the second end of the dimension line

DIMCEN

Type: Real

Saved in: Drawing

Description: Controls drawing of center marks and centerlines by the DIMCENTER, DIMDIAMETER, and DIMRADIUS dimensioning commands

DIMCLRD

Type: Integer

Saved in: Drawing

Description: Assigns colors to leader lines, arrowheads, and dimension lines

DIMCLRE

Type: Integer

Saved in: Drawing

Description: Assigns colors to dimension extension lines

DIMCLRT

Type: Integer

Saved in: Drawing

Description: Assigns colors to dimension text

DIMDEC

Type: Integer

Saved in: Drawing

Description: Sets the number of decimal places displayed for the tolerance values of the primary units of a dimension

DIMDLE

Type: Real

Saved in: Drawing

Description: Sets the distance the dimension line extends beyond the extension line when oblique strokes are drawn instead of arrowheads.

Appendix I: System Variables (Continued)

DIMDLI

Type: Real

Saved in: Drawing

Description: Controls the dimension line spacing for baseline dimensions; each baseline dimension is offset by this amount, if necessary, to avoid drawing over the previous dimension

DIMDSEP

Type: Single character

Saved in: Drawing

Description: Specifies a single character decimal separator to use when creating dimensions whose unit format is decimal.

When prompted, enter a single character at the command line. If dimension units is set to Decimal, the DIMDSEP character is used instead of the default decimal point. If DIMDSEP is set to NULL (default value, reset by entering a period), AutoCAD uses the decimal point as the dimension separator

DIMEXE

Type: Real

Saved in: Drawing

Description: Determines how far to extend the extension line beyond the dimension line

DIMEXO

Type: Real

Saved in: Drawing

Description: Determines how far extension lines are offset from origin points

DIMFIT

Type: Integer

Saved in: Drawing

Description: Controls the placement of text and arrowheads inside or outside extension lines based on available space between extension lines

0   Places text and arrowheads between extension lines if space is available; otherwise places both text and arrowheads outside extension lines

1   Places text and arrowheads between extension lines when space is available; if not, places text between extension lines and arrowheads outside; if not enough space for text inside extension lines, places text and arrowheads outside extension lines

2   Places text and arrowheads between extension lines when space is available; otherwise places text inside if it fits, or places arrowheads inside if they fit; if neither fits inside extension lines, both are placed outside

3   Places whatever best fits between extension lines

4   Creates leader lines when there is not enough space for text between extension lines

DIMFRAC

Type: Integer

Saved in: Drawing

Description: Sets the fraction format when DIMLUNIT is set to 4 (Architectural) or 5 (Fractional)

0   Horizontal

1   Diagonal

2   Not stacked (for example, 1/2)

Appendix I: System Variables (Continued)

DIMGAP

Type: Real

Saved in: Drawing

Description: Sets the distance around the dimension text when you break the dimension line to accommodate dimension text; sets the gap between annotation and hook line in leaders

DIMJUST

Type: Integer

Saved in: Drawing

Description: Controls horizontal dimension text position

0   Center-justifies text between extension lines

1   Positions text next to first extension line

2   Positions text next to second extension line

3   Positions text above and aligned with first extension line

4   Positions text above and aligned with second extension line

DIMLDRBLK

Type: String

Saved in: Drawing

Description: Specifies the arrow type for leaders. To turn off arrowhead display, enter a single period (.); for a list of arrowhead entries, see DIMBLK

DIMLFAC

Type: Real

Saved in: Drawing

Description: Sets global scale factor for linear dimensioning measurements

DIMLIM

Type: Switch

Saved in: Drawing

Description: When turned on, generates dimension limits as the default text

DIMLUNIT

Type: Integer

Saved in: Drawing

Description: Sets units for all dimension types except Angular

1   Scientific

2   Decimal

3   Engineering

4   Architectural

5   Fractional

6   Windows desktop

DIMLWD

Type: Enum

Saved in: Drawing

Description: Assigns lineweight to dimension lines

Values are standard lineweight (BYLAYER, BYBLOCK, integer representing 100th of a millimeter)

DIMLWE

Type: Enum

Saved in: Drawing

Description: Assigns lineweight to extension lines. Values are standard lineweight (BYLAYER, BYBLOCK, integer representing 100th of a millimeter)

Appendix I: System Variables (Continued)

DIMPOST

Type: String

Saved in: Drawing

Description: Specifies a text prefix or suffix (or both) to the dimension measurement

DIMRND

Type: Real

Saved in: Drawing

Description: Rounds all dimensioning distances to the specified value

DIMSAH

Type: Switch

Saved in: Drawing

Description: Controls the use of user-defined arrowhead blocks at the ends of the dimension line

ON   Normal arrowheads or user-defined blocks set by DIMBLK are used

OFF   User-defined arrowhead blocks are used

DIMSCALE

Type: Real

Saved in: Drawing

Description: Sets the overall scale factor applied to dimensioning variables that specify sizes, distances, or offsets

0.0   AutoCAD computes a reasonable default value based on scaling between current model space viewport and paper space

>0   AutoCAD computes a scale factor that leads text sizes, arrowhead sizes, and other scaled distances to plot at their face values

DIMSD1

Type: Switch

Saved in: Drawing

Description: When turned on, suppresses drawing of the first dimension line

DIMSD2

Type: Switch

Saved in: Drawing

Description: When turned on, suppresses drawing of the second dimension line

DIMSE1

Type: Switch

Saved in: Drawing

Description: When turned on, suppresses drawing of the first extension line

DIMSE2

Type: Switch

Saved in: Drawing

Description: When turned on, suppresses drawing of the second extension line

DIMSHO

Type: Switch

Saved in: Drawing

Description: When turned on, controls redefinition of dimension objects while dragging

DIMSOXD

Type: Switch

Saved in: Drawing

Description: When turned on, suppresses drawing of dimension lines outside the extension lines

Appendix I: **System Variables** (Continued)

DIMSTYLE

Type: String

Saved in: Drawing

Description: (Read-only) Sets the current dimension style by name

DIMTAD

Type: Integer

Saved in: Drawing

Description: Controls vertical position of text in relation to the dimension line

0 Centers text between extension lines

1 Places text above dimension line except when dimension line is not horizontal and text inside extension lines is forced horizontal

2 Places text on the side of the dimension line farthest away from the defining points

3 Places text to conform to a JIS representation

DIMTDEC

Type: Integer

Saved in: Drawing

Description: Sets the number of decimal places for the tolerance values for a primary units dimension

DIMTFAC

Type: Real

Saved in: Drawing

Description: Specifies a scale factor for text height of tolerance values relative to the dimension text height as set by DIMTXT:

$$DIMTFAC = \frac{Tolerance\ Height}{Text\ Height}$$

DIMTIH

Type: Switch

Saved in: Drawing

Description: Controls the position of dimension text inside the extension lines for all dimension types except ordinate dimensions

OFF Aligns text with dimension line

ON Draws text horizontally

DIMTIX

Type: Switch

Saved in: Drawing

Description: Draws text between extension lines

OFF For linear and angular dimensions, places text inside extension lines if there is sufficient room; for radius and diameter dimensions, forces text outside the circle or arc

ON Draws dimension text between the extension lines even if AutoCAD would ordinarily place it outside those lines

DIMTM

Type: Real

Saved in: Drawing

Description: When DIMTOL or DIMLIM is on, sets the minimum (lower) tolerance limit for dimension text

Appendix I: System Variables (Continued)

DIMTMOVE

Type: Integer

Saved in: Drawing

Description: Sets dimension text movement rules

0   Moves the dimension line with dimension text

1   Adds a leader when dimension text is moved

2   Allows text to be moved freely without a leader

DIMTOFL

Type: Switch

Saved in: Drawing

Description: When turned on, draws a dimension line between the extension lines even when the text is placed outside the extension lines; for radius and diameter dimensions (while DIMTIX is off), draws a dimension line and arrowheads inside the circle or arc and places the text and leader outside

DIMTOH

Type: Switch

Saved in: Drawing

Description: When turned on, controls the position of dimension text outside the extension lines

0   Aligns text with the dimension line

1   Draws text horizontally

DIMTOL

Type: Switch

Saved in: Drawing

Description: When turned on, appends dimension tolerances to dimension text

DIMTOLJ

Type: Integer

Saved in: Drawing

Description: Sets vertical justification for tolerance values relative to the nominal dimension text

0   Bottom

1   Middle

2   Top

DIMTP

Type: Real

Saved in: Drawing

Description: When DIMTOL or DIMLIM is on, sets the maximum (upper) tolerance limit for dimension text

DIMTSZ

Type: Real

Saved in: Drawing

Description: Specifies size of oblique strokes drawn instead of arrowheads for linear, radius, and diameter dimensioning

0   Draws arrows

>0   Draws oblique strokes; size of strokes is determined by this value multiplied by the DIMSCALE value

Appendix I: System Variables (Continued)

DIMTVP

Type: Real

Saved in: Drawing

Description: Adjusts vertical position of dimension text above or below the dimension line when DIMTAD is off

DIMTXSTY

Type: String

Saved in: Drawing

Description: Specifies the text style of the dimension

DIMTXT

Type: Real

Saved in: Drawing

Description: Specifies the height of dimension text, unless the current text style has a fixed height

DIMTZIN

Type: Integer

Saved in: Drawing

Description: Toggles suppression of zeros for tolerance values

DIMUNIT

Type: Integer

Saved in: Drawing

Description: Sets the units format for all dimension style family members except angular

1   Scientific

2   Decimal

3   Engineering

4   Architectural

5   Fractional

DIMUPT

Type: Switch

Saved in: Drawing

Description: Controls cursor functionality for user-positioned text

0   Cursor controls only dimension line location

1   Cursor controls text position as well as dimension line location

DIMZIN

Type: Integer

Saved in: Drawing

Description: Controls suppression of the inches portion of a feet-and-inches dimension when the distance is an integral number of feet, or the feet portion when the distance is less than one foot

0   Suppresses zero feet and precisely zero inches

1   Includes zero feet and precisely zero inches

2   Includes zero feet and suppresses zero inches

3   Includes zero inches and suppresses zero feet

DISPSILH

Type: Integer

Saved in: Drawing

Description: Controls the display of silhouette curves of body objects in wireframe mode

0   Off

1   On

Appendix I: System Variables (Continued)

DISTANCE

Type: Real

Saved in: Drawing

Description: (Read-only) Stores the distance computed by the DIST command

DONUTID

Type: Integer

Saved in: (Not saved)

Description: Sets the default for the inside diameter of a donut

DONUTOD

Type: Real

Saved in: (Not saved)

Description: Sets the default for the outside diameter of a donut (must be nonzero)

DRAGMODE

Type: Integer

Saved in: drawing

Description: Sets object drag mode

0   No dragging

1   On (if requested)

2   Auto

DRAGP1

Type: Integer

Saved in: Registry

Description: Sets regen-drag input sampling rate

DRAGP2

Type: Integer

Saved in: Registry

Description: Sets fast-drag input sampling rate

DWGCHECK

Type: Integer

Saved in: Registry

Description: Determines whether a drawing was last edited by a product other than AutoCAD.

0   Suppresses dialog box display

1   Displays the dialog box, if warranted

DWGCODEPAGE

Type: String

Saved in: Drawing

Description: (Read-only) Stores the drawing code page

DWGNAME

Type: String

Saved in: (Not saved)

Description: (Read-only) Stores the drawing name as entered by the user

DWGPREFIX

Type: String

Saved in: (Not saved)

Description: Stores the drive/directory prefix for the drawing

Appendix I: System Variables (Continued)

DWGTITLED

Type: Integer

Saved in: (Not saved)

Description: Indicates whether the current drawing has been named

0   Not named

1   Named

EDGEMODE

Type: Integer

Saved in: (Not saved)

Description: Controls determination of cutting edges for TRIM and EXTEND commands

ELEVATION

Type: Real

Saved in: Drawing

Description: Stores the current 3D elevation relative to the current UCS for the current space

EXPERT

Type: Integer

Saved in: (Not saved)

Description: Controls issuance of certain prompts; when prompts are suppressed, the operation is performed as though you had entered y at the prompt

0   Issues all prompts normally

1   Suppresses About to regen, proceed? and Really want to turn the current layer off?

2   Suppresses preceding prompts, Block already defined, Redefine it?, and A drawing with this name already exists. Overwrite it?

3   Suppresses preceding prompts and those issued by LINETYPE if you try to load a linetype that's already loaded or create a new linetype in a file that already defines it

4   Suppresses preceding prompts and those issued by UCS Save and VPORTS Save if the name you supply already exists

5   Suppresses preceding prompts and those issued by DIMSTYLE Save and DIMOVERRIDE if the dimension style name you supply already exists (the entries are redefined)

EXPLMODE

Type: Integer

Saved in: Drawing

Description: Controls whether the EXPLODE command supports non-uniformly scaled (NUS) blocks

0   Does not explode NUS blocks

1   Explodes NUS blocks

EXTMAX

Type: 3D Point

Saved in: Drawing

Description: Stores the upper right point of drawing extents

EXTMIN

Type: 3D Point

Saved in: Drawing

Description: Stores the lower left point of drawing extents

Appendix I: System Variables (Continued)

EXTNAMES

Type: Integer

Saved in: Drawing

Description: Sets the parameters for named object names (such as linetypes and layers) stored in symbol tables

0   Uses Release 14 parameters, which limit names to 31 characters in length. Names can include the letters A to Z, the numerals 0 to 9, and the special characters, dollar sign ($), underscore (_), and hyphen (-).

1   Uses AutoCAD 2000 parameters. Names can be up to 255 characters in length and can include the letters A to Z, the numerals 0 to 9, spaces, and any special characters not used by Microsoft Windows and AutoCAD for other purposes.

FACETRATIO

Type: Integer

Saved in: (Not saved)

Description: Controls the aspect ratio of faceting for cylindrical and conic ACIS solids. A setting of 1 increases the density of the mesh to improve the quality of rendered and shaded models.

0   Creates an *N* by *M* mesh for cylindrical and conic ACIS solids

1   Creates a high-density *N* by *M* mesh for cylindrical and conic ACIS solids

FACETRES

Type: Real

Saved in: Drawing

Description: Further adjusts the smoothness of shaded and hidden line-removed objects; valid values are from .01 to 10.0

FILEDIA

Type: Integer

Saved in: Registry

Description: Suppresses display of file dialog boxes

0   Disables file dialog boxes

1   Enables file dialog boxes

FILLETRAD

Type: Real

Saved in: Drawing

Description: Stores the current fillet radius

FILLMODE

Type: Integer

Saved in: Drawing

Description: Specifies whether objects created with SOLID are filled in

0   Not filled

1   Filled

FONTALT

Type: String

Saved in: Registry

Description: Specifies the alternate font to be used when the specified font file cannot be located

FONTMAP

Type: String

Saved in: Registry

Description: Specifies the font mapping file to be used when the specified font cannot be located

Appendix I: System Variables (Continued)

FRONTZ

Type: Real

Saved in: Drawing

Description: Stores the front clipping plane offset from the target plane for the current viewport, in drawing units

FULLOPEN

Type: Integer

Saved in: (Not saved)

Description: (Read only) Indicates whether the current drawing is fully open

0   Indicates a partially open drawing

1   Indicates a fully open drawing

GRIDMODE

Type: Integer

Saved in: Drawing

Description: Specifies whether the grid is turned on

0   Turns the grid off

1   Turns the grid on

GRIDUNIT

Type: Real

Saved in: Drawing

Description: Specifies the grid spacing (X and Y) for the current viewport

GRIPBLOCK

Type: Integer

Saved in: Registry

Description: Controls assignment of grips in blocks

0   Assigns grips only to insertion point of block

1   Assigns grips to objects within the block

GRIPCOLOR

Type: Integer

Saved in: Registry

Description: Controls the color of nonselected grips

GRIPHOT

Type: Integer

Saved in: Registry

Description: Controls the color of selected grips

GRIPS

Type: Integer

Saved in: Registry

Description: Allows the use of selection set grips for the Stretch, Move, Rotate, Scale, and Mirror grip modes

0   Disables grips

1   Enables grips

GRIPSIZE

Type: Integer

Saved in: Registry

Description: Sets the size of the box drawn to display the grip in pixels

Appendix I: System Variables (Continued)

HANDLES

Type: Integer

Saved in: Drawing

Description: (Read-only) Reports that object handles are enabled and can be accessed by applications

HIDEPRECISION

Type: Integer

Saved in: (Not saved)

Description: Controls the accuracy of hides and shades. Hides can be calculated in double precision or single precision. Setting HIDEPRECISION to 1 produces more accurate hides by using double precision, but this setting also uses more memory and can affect performance, especially when hiding solids

0   Single precision; uses less memory

1   Double precision; uses more memory

HIGHLIGHT

Type: Integer

Saved in: (Not saved)

Description: Controls object highlighting; does not affect objects selected with grips

0   Disables object selection highlighting

1   Enables object selection highlighting

HPANG

Type: Real

Saved in: (Not saved)

Description: Specifies the hatch pattern angle

HPBOUND

Type: Real

Saved in: (Not saved)

Description: Controls the object type created by the BHATCH and BOUNDARY commands

0   Creates a polyline

1   Creates a region

HPDOUBLE

Type: Integer

Saved in: (Not saved)

Description: Specifies hatch pattern doubling for "U" user-defined patterns

0   Disables hatch pattern doubling

1   Enables hatch pattern doubling

HPNAME

Type: String

Saved in: (Not saved)

Description: Sets default hatch pattern name of up to 34 characters, no spaces allowed

HPSCALE

Type: Real

Saved in: (Not saved)

Description: Specifies the hatch pattern scale factor; must be nonzero

HPSPACE

Type: Real

Saved in: (Not saved)

Description: Specifies the hatch pattern line spacing for "U" user-defined hatch patterns; must be nonzero

Appendix I: System Variables (Continued)

HYPERLINKBASE

Type: String

Saved in: Drawing

Description: Specifies the path used for all relative hyperlinks in the drawing; if no value is specifed, the drawing path is used for all relative hyperlinks

IMAGEHLT

Type: Integer

Saved in: Registry

Description: Controls whether the entire raster image or only the raster image frame is highlighted

0   Highlights only the raster image frame

1   Highlights the entire raster image

INDEXCTL

Type: Integer

Saved in: Drawing

Description: Controls whether layer and spatial indexes are created and saved in drawing files

0   No indexes are created

1   Creates layer index

2   Creates spatial index

3   Creates both layer and spatial indexes

INETLOCATION

Type: Real

Saved in: Registry

Description: Stores the Internet location used by BROWSER

INSBASE

Type: 3D point

Saved in: Drawing

Description: Stores insertion base point set by BASE command, expressed in UCS coordinates for the current space

INSNAME

Type: String

Saved in: (Not saved)

Description: Sets default block name for DDINSERT or INSERT

INSUNITS

Type: Integer

Saved in: Drawing

Description: When you drag a block from AutoCAD DesignCenter, specifies a drawing units value as follows:

0   Unspecified (No units)

1   Inches

2   Feet

3   Miles

4   Millimeters

5   Centimeters

6   Meters

7   Kilometers

8   Microinches

9   Mils

10 Yards

11 Angstroms

12 Nanometers

13 Microns

14 Decimeters

Appendix I: System Variables (Continued)

(INSUNITS, continued)

15 Decameters

16 Hectometers

17 Gigameters

18 Astronomical Units

19 Light Years

20 Parsecs

INSUNITSDEFSOURCE

Type: Integer

Saved in: Registry

Description: Sets source content units value; valid range is 0 to 20

INSUNITSDEFTARGET

Type: Integer

Saved in: Registry

Description: Sets target drawing units value; valid range is 0 to 20

ISAVEBAK

Type: Integer

Saved in: Registry

Description: Improves the speed of incremental saves, especially for large drawings in Windows

0   No BAK file is created (even for a full save)

1   A BAK file is created

ISAVEPERCENT

Type: Integer

Saved in: Registry

Description: Determines the amount of wasted space tolerated in a drawing file; values are integers between 0 and 100; default value is 50. Value is the percent of total file size wasted in a file. Wasted space is eliminated with each full save, so when AutoCAD's estimate reaches the value set in ISAVEPERCENT, it automatically performs a full save

ISOLINES

Type: Integer

Saved in: Drawing

Description: Specifies the number of isolines per surface on objects; valid integer values are from 0 to 2047

LASTANGLE

Type: Real

Saved in: (Not saved)

Description: (Read-only) Stores the end angle of the last arc entered, relative to the XY plane of the current UCS for the current space

LASTPOINT

Type: 3D point

Saved in: Drawing

Description: Stores the last point entered, expressed in UCS coordinates for the current space; referenced by @ during keyboard entry

Appendix I: System Variables (Continued)

LASTPROMPT

Type: String

Saved in: (Not saved)

Description: Stores the last string echoed to the Command line; read-only

LENSLENGTH

Type: Real

Saved in: Drawing

Description: Stores lens length (in millimeters) used in perspective viewing for the current viewport

LIMCHECK

Type: Integer

Saved in: Drawing

Description: Controls object creation outside drawing limits

0   Enables object creation

1   Disables object creation

LIMMAX

Type: 2D point

Saved in: Drawing

Description: Stores upper right drawing limits for the current space expressed in world coordinates

LIMMIN

Type: 2D point

Saved in: Drawing

Description: Stores lower left drawing limits for the current space expressed in world coordinates

LISPINIT

Type: Integer

Saved in: Registry

Description: Specifies whether AutoLISP-defined functions and variables are preserved when you open a new drawing

0   AutoLISP functions and variables are preserved from drawing to drawing

1   AutoLISP functions and variables are valid in current drawing only

LOCALE

Type: String

Saved in: (Not saved)

Description:(Read-only) Displays ISO language code of the current AutoCAD version

LOGFILEMODE

Type: Integer

Saved in: Registry

Description: Specifies whether the contents of the text window are written to a log file

0   Log file is not maintained

1   Log file is maintained

LOGFILENAME

Type: String

Saved in: Registry

Description: Specifies path for the log file

Appendix I: System Variables (Continued)

LOGFILEPATH

Type: String

Saved in: Registry

Description: Specifies the path for the log files for all drawings in a session. You can also specify the path by using the OPTIONS command. The initial value varies depending on where you installed AutoCAD

LOGINNAME

Type: String

Saved in: (Not saved)

Description: Displays the user's name as configured or input when AutoCAD is loaded

LTSCALE

Type: Real

Saved in: Drawing

Description: Sets global linetype scale factor

LUNITS

Type: Integer

Saved in: Drawing

Description: Sets mode for linear units

1   Scientific

2   Decimal

3   Engineering

4   Architectural

5   Fractional

LUPREC

Type: Integer

Saved in: Drawing

Description: Sets linear units decimal places or denominator

LWDEFAULT

Type: Enum

Saved in: Registry

Description: Sets the value for the default lineweight to any valid lineweight value in millimeters

LWDISPLAY

Type: Integer

Saved in: Drawing

Description: Controls whether the lineweight is displayed in the Model or Layout tab

0   Lineweight is not displayed

1   Lineweight is displayed

LWUNITS

Type: Integer

Saved in: Registry

Description: Controls whether lineweight units are displayed in inches or millimeters

0   Inches

1   Millimeters

MAXACTVP

Type: Integer

Saved in: (Not saved)

Description: Sets the maximum number of viewports to regenerate at one time

Appendix I: System Variables (Continued)

MAXOBJMEM

Type: Integer

Saved in: (Not saved)

Description: Controls the object pager (specifies how much virtual memory a drawing uses before it starts paging out to disk into the object pager's swap files)

MAXSORT

Type: Integer

Saved in: Registry

Description: Sets maximum number of symbol names or file names to be sorted by listing commands

MBUTTONPAN

Type: Integer

Saved in: Registry

Description: Controls the behavior of the third button on the pointing device

0   Supports the action defined in the AutoCAD menu (MNU) file

1   Supports panning by holding and dragging the button or wheel

MEASUREMENT

Type: Integer

Saved in: Drawing

Description: Sets drawing units as English or metric

0   English; AutoCAD uses the hatch pattern file and linetype file designated by ANSIHatch and ANSILinetype registry settings

1   Metric; AutoCAD uses the hatch pattern file and linetype file designated by the ISOHatch and ISOLinetype registry settings

MENUCTL

Type: Integer

Saved in: Registry

Description: Controls the page switching of the screen menu

0   Screen menu does not switch pages in response to keyboard command entry

1   Screen menu switches pages in response to keyboard command entry

MENUECHO

Type: Integer

Saved in: (Not saved)

Description: Sets menu echo and prompt control bits; the sum of the following:

1   Suppresses echo of menu items (^P in a menu item toggles echoing)

2   Suppresses display of system prompts during menu

4   Disables ^P toggle of menu echoing

8   Displays input/output strings; debugging aid for DIESEL macros

MENUNAME

Type: String

Saved in: (Not saved)

Description: Stores the name and path of the currently loaded base menu file

MIRRTEXT

Type: Integer

Saved in: Drawing

Description: Controls how MIRROR reflects text

0   Retains text direction

1   Mirrors the text

Appendix I: System Variables (Continued)

MODEMACRO

Type: String

Saved in: (Not saved)

Description: Displays a text string on the status line, such as the name of the current drawing or time/date stamp

MTEXTED

Type: String

Saved in: Registry

Description: Sets the name of the program to use for editing mtext objects

NOMUTT

Type: Short

Saved in: (Not saved)

Description: Suppresses the message display (muttering) when it wouldn't normally be suppressed. Displaying messages is the normal mode; but message display is suppressed during scripts, AutoLISP routines, and so on

0   Resumes normal muttering behavior

1   Suppresses muttering indefinitely

OFFSETDIST

Type: Real

Saved in: (Not saved)

Description: Sets the default offset distance

<0 Changes to Through mode

>0 Sets the default offset distance

OFFSETGAPTYPE

Type: Integer

Saved in: Registry

Description: Controls how to offset polylines when a gap is created as a result of offsetting the individual polyline segments

0   Extends the segments to fill the gap

1   Fills the gaps with a filleted arc segment (the radius of the arc segment is equal to the offset distance)

2   Fills the gaps with a chamfered line segment

OLEHIDE

Type: Integer

Saved in: Registry

Description: Controls the display of OLE objects in AutoCAD both on the screen and the plotted image

0   All OLE objects are visible

1   OLE objects are visible in paper space only

2   OLE objects are visible in model space only

3   No OLE objects are visible

OLEQUALITY

Type: Integer

Saved in: Registry

Description: Controls the default quality level for embedded OLE objects

0   Line art quality, such as an embedded spreadsheet

1   Text quality, such as an embedded Word document

2   Graphics quality, such as an embedded pie chart

3   Photograph quality

4   High quality photograph

Appendix I: System Variables (Continued)

OLESTARTUP

Type: Integer

Saved in: Drawing

Description: Controls whether the source application of an embedded OLE object loads when plotting; may improve the plot quality.

0 Does not load the OLE source application

1 Loads the OLE source application when plotting

ORTHOMODE

Type: Integer

Saved in: Drawing

Description: Controls orthogonal display of lines or polylines

0 Off

1 On

OSMODE

Type: Integer

Saved in: Drawing

Description: Sets running object snap modes using the following bit-codes; to enter more than one object snap, enter the sum of their values

0 NONe

1 ENDpoint

2 MIDpoint

4 CENter

8 NODe

16 QUAdrant

32 INTersection

64 INSertion

128 PERpendicular

256 TANgent

512 NEArest

1024 QUIck

2048 APPint

OSNAPCOORD

Type: Integer

Saved in: Registry

Description: Controls whether coordinates entered on the command line override running object snaps

0 Running object snap settings override keyboard coordinate entry

1 Keyboard entry overrides object snap settings

2 Keyboard entry overrides object snap settings except in scripts

PAPERUPDATE

Type: Integer

Saved in: Registry

Description: Controls display of a warning dialog when attempting to print a layout with a paper size different from that specified by the default for the plotter configuration file

0 Displays a warning dialog box if the paper size specified in the layout is not supported by the plotter

1 Sets paper size to the configured paper size of the plotter configuration file

PDMODE

Type: Integer

Saved in: Drawing

Description: Sets point object display mode

Appendix I: System Variables (Continued)

PDSIZE

Type: Real

Saved in: Drawing

Description: Sets point object display size

0 Creates a point at 5% of graphics area height

>0 Specifies an absolute size

<0 Specifies a percentage of the viewport size

PELLIPSE

Type: Integer

Saved in: Drawing

Description: Controls the ellipse type created with ELLIPSE

0 Creates a true ellipse object

1 Creates a polyline representation of an ellipse

PERIMETER

Type: Real

Saved in: (Not saved)

Description: (Read-only) Stores the last perimeter value computed by AREA, LIST, or DBLIST

PCFACEVMAX

Type: Integer

Saved in: (Not saved)

Description: (Read-only) Sets the maximum number of vertices per face

PICKADD

Type: Integer

Saved in: Registry

Description: Controls additive selection of objects

0 Disables PICKADD

1 Enables PICKADD

PICKAUTO

Type: Integer

Saved in: Registry

Description: Controls automatic windowing when the Select objects prompt appears

0 Disables PICKAUTO

1 Draws a selection window (both window and crossing window) automatically at the Select objects prompt

PICKBOX

Type: Integer

Saved in: Registry

Description: Sets object selection target height, in pixels

PICKDRAG

Type: Integer

Saved in: Registry

Description: Controls the method of drawing a selection window

0 Draws window by clicking mouse or digitizer at opposite corners

1 Draws window by clicking at one corner, holding down mouse or digitizer button, dragging, and releasing button at other corner

Appendix I: System Variables (Continued)

PICKFIRST

Type: Integer

Saved in: Registry

Description: Controls method of object selection so that you select objects first and then use an edit or inquiry command

0 Disables PICKFIRST

1 Enables PICKFIRST

PICKSTYLE

Type: Integer

Saved in: Drawing

Description: Controls group selection and associative hatch selection

0 No group selection or associative hatch selection

1 Group selection

2 Associative hatch selection

3 Group selection and associative hatch selection

PLATFORM

Type: String

Saved in: (Not saved)

Description: Indicates which platform of AutoCAD is in use

PLINEGEN

Type: Integer

Saved in: Drawing

Description: Sets the linetype pattern generation around the vertices of a 2D polyline; does not apply to polylines with tapered segments

0 Polylines are generated to start and end with a dash at each vertex

1 Generates the linetype in a continuous pattern around the vertices of the polyline

PLINETYPE

Type: Integer

Saved in: Registry

Description: Specifies whether AutoCAD uses optimized 2D polylines; also affects the polyline type used with BOUNDARY, DONUT, ELLIPSE (when value is set to 1), PEDIT, POLYGON, and SKETCH (when SKPOLY is set to 1)

0 Polylines in older drawings are not converted on open; PLINE creates old-format polylines

1 Polylines in older drawings are not converted on open; PLINE creates optimized polylines

2 Polylines in older drawings are converted on open; PLINE creates optimized polylines

PLINEWID

Type: Real

Saved in: Drawing

Description: Stores the default polyline width

PLOTID

Type: String

Saved in: Registry

Description: Changes the default plotter and retains the text string of the current plotter description

Appendix I: System Variables (Continued)

PLOTROTMODE

Type: Integer

Saved in: Drawing

Description: Controls the orientation of plots

0   Rotates effective plotting area so that corner with rotation icon aligns with paper at lower left (0), top left (90), top right (180) or lower right (270) corner

1   Aligns lower left corner of effective plotting area with lower left corner of the paper

PLOTTER

Type: Integer

Saved in: Config

Description: Changes the default plotter, based on its assigned integer, and retains an integer number that AutoCAD assigns for each plotter

PLQUIET

Type: Integer

Saved in: Registry

Description: Controls the display of optional dialog boxes and nonfatal errors for batch plotting and scripts

0   Displays plot dialog boxes and nonfatal errors

1   Logs nonfatal errors and doesn't display plot-related dialog boxes

POLARADDANG

Type: String

Saved in: Registry

Description: Contains user-defined polar angles. You can add up to 10 angles. Each angle can be up to 25 characters, separated with semicolons (;). AutoCAD displays angles in the format set in the AUNITS system variable

POLARANG

Type: Real

Saved in: Registry

Description: Sets the polar angle increment; values are 90, 45, 30, 22.5, 18, 15,10, and 5

POLARDIST

Type: Real

Saved in: Registry

Description: Sets the snap increment when the SNAPSTYL system variable is set to 1 (polar snap)

POLARMODE

Type: Integer

Saved in: Registry

Description: Controls settings for polar and object snap tracking; the value is the sum of four bitcodes:

Polar angle measurements

0   Measure polar angles based on current UCS (absolute)

1   Measure polar angles from selected objects (relative)

Object snap tracking

0   Track orthogonally only

2   Use polar tracking settings in object snap tracking

Appendix I: **System Variables** (Continued)

Use additional polar tracking angles

0   No

4   Yes

Acquire object snap tracking points

0   Acquire automatically

8   Press SHIFT to acquire

POLYSIDES

Type: Integer

Saved in: (Not saved)

Description: Sets the default number of sides for POLYGON; the range is 3 to 1024

POPUPS

Type: Integer

Saved in: (Not saved)

Description: (Read-only) Displays the status of the currently configured display driver

0   Does not support dialog boxes, the menu bar, pull-down menus, and image tile menus

1   Supports the above features

PROJECTNAME

Type: String

Saved in: Drawing

Description: Stores the current project name; each project name can contain one or more search paths; used when an xref or image is not found in the original search path

PROJMODE

Type: Integer

Saved in: Registry

Description: Sets the current projection mode for TRIM and EXTEND operations

0   True 3D mode (no projection)

1   Project to the XY plane of the current UCS

2   Project to the current view plane

PROXYGRAPHICS

Type: Integer

Saved in: Drawing

Description: Specifies whether images of proxy objects are saved in the drawing

0   Image is not saved with the drawing; a bounding box is displayed instead

1   Image is saved with the drawing

PROXYNOTICE

Type: Integer

Saved in: Registry

Description: Displays a notice when you open a drawing containing custom objects created by an application that is not present

0   No proxy warning is displayed

1   Proxy warning is displayed

Appendix I: System Variables (Continued)

PROXYSHOW

Type: Integer

Saved in: Registry

Description: Controls the display of proxy objects in a drawing

0   Proxy objects are not displayed

1   Graphic images are displayed for all proxy objects

2   Only the bounding box is displayed for all proxy objects

PLTSCALE

Type: Integer

Saved in: Drawing

Description: Controls paper space linetype scaling

0   No special linetype scaling

1   Viewport scaling governs linetype scaling

PSPROLOG

Type: String

Saved in: Registry

Description: Assigns a name for a prologue section to be read from the acad.psf file when using PSOUT

PSQUALITY

Type: Integer

Saved in: Drawing

Description: Controls the rendering quality of PostScript images and whether they are drawn as filled objects or as outlines

0   Disables PostScript image generation

<0 Sets number of pixels per AutoCAD drawing unit for the PostScript resolution

>0 Sets number of pixels per drawing unit, but uses the absolute value; causes AutoCAD to show the PostScript paths as outlines and does not fill them

PSTYLEMODE

Type: Read only

Saved in: Drawing

Description: Indicates whether the current drawing is in a Color-Dependent or Named Plot Style mode

0   Uses named plot style tables in the current drawing

1   Uses color-dependent plot style tables in the current drawing

PSTYLEPOLICY

Type: Integer

Saved in: Registry

Description: Controls whether an object's color property is associated with its plot style; the new value you assign affects only newly created drawings and pre-AutoCAD 2000 drawings

0   No association is made between color and plot style. The plot style for new objects is set to the default defined in DEFPLSTYLE. The plot style for new layers is set to the default defined in DEFLPLSTYLE

1   An object's plot style is associated with its color

Appendix I: System Variables (Continued)

PSVPSCALE

Type: Real

Saved in: (Not saved)

Description: Sets the view scale factor for all newly created viewports. The view scale factor is defined by comparing the ratio of units in paper space to the units in newly created model space viewports. The view scale factor you set is used with the VPORTS command. A value of 0 means the scale factor is Scaled to Fit. A scale must be a positive real value

PUCSBASE

Type: String

Saved in: Drawing

Description: Stores the name of the UCS that defines the origin and orientation of orthographic UCS settings in paper space only

QTEXTMODE

Type: Integer

Saved in: Drawing

Description: Controls quick text mode

0   Off

1   On

RASTERPREVIEW

Type: Integer

Saved in: Drawing

Description: Controls whether BMP previews are saved with the drawing

0   No preview image created

1   Preview image created

REFEDITNAME

Type: Integer

Saved in: Drawing

Description (Read only) Indicates whether a drawing is in a reference-editing state and stores the reference file name

REGENMODE

Type: Integer

Saved in: Drawing

Description: Controls automatic regeneration of the drawing

0   Turns REGENAUTO off

1   Turns REGENAUTO on

RE-INIT

Type: Integer

Saved in: (Not Saved)

Description: Reinitializes the I/O ports, digitizer display, plotter, and acad.pgp file using bit-codes; to specify more than one, enter the sum of their values

0   No initialization

1   Digitizer port reinitialization

2   Plotter port reinitialization

4   Digitizer reinitialization

8   Display reinitialization

16 PGP file reinitialization

RTDDISPLAY

Type: Integer

Saved in: Registry

Description: Controls the display of raster images during realtime zoom or pan

0   Displays raster image content

1   Displays raster image outline only

Appendix I: System Variables (Continued)

SAVEFILE

Type: String

Saved in: Registry

Description: (Read-only) stores current auto-save file name

SAVEFILEPATH

Type: String

Saved in: Registry

Description: Specifies the path to the directory for all automatic save files for the AutoCAD session; also available on the Files tab in the Options dialog box

SAVENAME

Type: String

Saved in: Drawing

Description: (Read-only) Stores the file name you assign to a drawing

SAVETIME

Type: Integer

Saved in: Registry

Description: Sets automatic save interval, in minutes (or 0 to disable automatic saves)

SCREENBOXES

Type: Integer

Saved in: Registry

Description: (Read-only) Stores the number of boxes in the screen menu area of the graphics area

SCREENMODE

Type: Integer

Saved in: Registry

Description: (Read-only) Stores a bit-code indicating the graphics/text state of the AutoCAD display; the sum of the following values

0   Text screen is displayed

1   Graphics mode is displayed

2   Dual-screen display is configured

SCREENSIZE

Type: 2D point

Saved in: (Not saved)

Description: Stores current viewport size in pixels

SDI

Type: Integer

Saved in: Registry

Description: Controls whether AutoCAD runs in single- or multiple-document mode; helps third-party developers update applications to work smoothly with the AutoCAD multiple-drawing mode

0   Turns on multiple-drawing interface

1   Turns off multiple-drawing interface

2   (Read-only) Multiple-drawing interface is disabled because AutoCAD has loaded an application that does not support multiple drawing; SDI setting 2 is not saved

Appendix I: System Variables (Continued)

3   (Read-only) Multiple-drawing interface is disabled because the user has set SDI to 1 and AutoCAD has loaded an application that does not support multiple drawings; SDI setting 3 is not saved

If SDI is set to 3, AutoCAD switches it back to 1 when the application that doesn't support multiple drawings is unloaded.

SHADEDGE

Type: Integer

Saved in: Drawing

Description: Controls shading of edges in rendering

0   Faces shaded, edges not highlighted

1   Faces shaded, edges drawn in background color

2   Faces not filled, edges in object color

3   Faces in object color, edges in background color

SHADEDIF

Type: Integer

Saved in: Drawing

Description: Sets ratio of diffuse reflective light to ambient light (in percent of diffuse reflective light)

SHORTCUTMENU

Type: Integer

Saved in: Registry

Description: Controls whether Default, Edit, and Command mode shortcut menus are available in the drawing area; uses the following bitcodes:

0   Disables all Default, Edit, and Command mode shortcut menus, restoring R14 legacy behavior

1   Enables Default mode shortcut menus

2   Enables Edit mode shortcut menus

4   Enables Command mode shortcut menus; available whenever a command is active

8   Enables Command mode shortcut menus only when command options are currently available from the command line

To enable more than one type of shortcut menu at one time, enter the sum of their values. For example, entering 3 enables both Default (1) and Edit (2) mode shortcut menus.

SHPNAME

Type: String

Saved in: (Not saved)

Description: Sets default shape name

SKETCHINC

Type: Real

Saved in: Drawing

Description: Sets SKETCH record increments

Appendix I: System Variables (Continued)

SKPOLY

Type: Integer

Saved in: Drawing

Description: Determines whether SKETCH generates lines or polylines

0   Lines

1   Polylines

SNAPANG

Type: Real

Saved in: Drawing

Description: Sets (UCS-relative) snap/grid rotation angle for the current viewport

SNAPBASE

Type: 2D point

Saved in: Drawing

Description: Sets snap/grid origin point for the current viewport

SNAPISOPAIR

Type: Integer

Saved in: Drawing

Description: Controls current isometric plane for the current viewport

0   Left

1   Top

2   Right

SNAPMODE

Type: Integer

Saved in: Drawing

Description: Controls the snap mode

0   Off

1   On (for current viewport)

SNAPSTYL

Type: Integer

Saved in: Drawing

Description: Sets snap style for the current viewport

0   Standard

1   Isometric

SNAPTYPE

Type: Integer

Saved in: Registry

Description: Sets the snap style for the current viewport.

0   Grid, or standard snap.

1   Polar snap. Snaps along polar angle increments. Use polar snap with polar and object snap tracking

SNAPUNIT

Type: 2D point

Saved in: Drawing

Description: Sets snap spacing for current viewport

SOLIDCHECK

Type: Integer

Saved in: Not saved

Description: Turns the solid validation on and off for the current AutoCAD session

0   Turns off solid validation

1   Turns on solid validation

Appendix I: System Variables (Continued)

SORTENTS

Type: Integer

Saved in: Config

Description: Controls the display of object sort order operations using the following codes; to select more than one, enter the sum of their codes

0 Disables SORTENTS

1 Sorts for object selection

2 Sorts for object snap

4 Sorts for redraws

8 Sorts for MSLIDE slide creation

16 Sorts for regenerations

32 Sorts for plotting

64 Sorts for PostScript output

SPLFRAME

Type: Integer

Saved in: Drawing

Description: Controls display of spline-fit polylines

0 Does not display control polygon; displays fit surface of polygon mesh; does not display invisible edges of 3D faces or polyface meshes

1 Displays control polygon; displays defining mesh of surface-fit polygon mesh; displays invisible edges of 3D faces or polyface meshes

SPLINESEGS

Type: Integer

Saved in: Drawing

Description: Sets the number of line segments to be generated for each spline

SPLINETYPE

Type: Integer

Saved in: Drawing

Description: Sets the type of spline curve to be generated by PEDIT Spline

5 Quadratic B-spline

6 Cubic B-spline

SURFTAB1

Type: Integer

Saved in: Drawing

Description: Sets the number of tabulations to be generated for RULESURF and TABSURF; sets the mesh density in the M direction for REVSURF and EDGESURF

SURFTAB2

Type: Integer

Saved in: Drawing

Description: Sets the mesh density in the N direction for REVSURF and EDGESURF

SURFTYPE

Type: Integer

Saved in: Drawing

Description: Controls the type of surface fitting to be performed by PEDIT Smooth

5 Quadratic B-spline surface

6 Cubic B-spline surface

8 Bezier surface

SURFU

Type: Integer

Saved in: Drawing

Description: Sets the surface density in the M direction

Appendix I: System Variables (Continued)

SURFV

Type: Integer

Saved in: Drawing

Description: Sets the surface density in the *N* direction

SYSCODEPAGE

Type: String

Saved in: Drawing

Description: Indicates the system code pages specified in acad.xmf

TABMODE

Type: Integer

Saved in: (Not saved)

Description: Controls use of tablet mode

0   Disables tablet mode

1   Enables tablet mode

TARGET

Type: 3D point

Saved in: Drawing

Description: (Read-only) Stores location of the target point for current viewport (in UCS coordinates)

TDCREATE

Type: Real

Saved in: Drawing

Description: (Read-only) Stores time and date of drawing creation

TDINDWG

Type: Real

Saved in: Drawing

Description: (Read only) Stores total editing time

TDUCREATE

Type: Real

Saved in: Drawing

Description: Stores the universal time and date the drawing was created

TDUPDATE

Type: Real

Saved in: Drawing

Description: (Read-only) Stores time and date of last update/save

TDUSRTIMER

Type: Real

Saved in: Drawing

Description: (Read-only) Stores user-elapsed timer

TDUUPDATE

Type: Real

Saved in: Drawing

Description: Stores the universal time and date of the last update/save

TEMPPREFIX

Type: String

Saved in: (Not saved)

Description: Contains the directory name configured for placement of temporary files

Appendix I: System Variables (Continued)

TEXTEVAL

Type: Integer

Saved in: (Not saved)

Description: Controls method of evaluation of text strings

0   All responses to prompts for text strings and attribute values are taken literally

1   Text starting with "(" or "!" is evaluated as an AutoLISP expression, as for nontextual input

TEXTFILL

Type: Integer

Saved in: Drawing

Description: Controls the filling of Bitstream, TrueType, and Adobe Type 1 fonts

0   Outlines

1   Filled images

TEXTQLTY

Type: Real

Saved in: Drawing

Description: Sets resolution of Bitstream, TrueType, and Adobe Type 1 Fonts

TEXTSIZE

Type: Real

Saved in: Drawing

Description: Sets the default height for new text objects drawn with the current text style

TEXTSTYLE

Type: String

Saved in: Drawing

Description: Contains the name of the current text style

THICKNESS

Type: Real

Saved in: Drawing

Description: Sets the current 3D thickness

TILEMODE

Type: Integer

Saved in: Drawing

Description: Makes the Model tab or the last Layout tab current

0   Enables last active Layout tab (paper space)

1   Enables the Model tab

TOOLTIPS

Type: Integer

Saved in: Registry

Description: Controls the display of tooltips

TRACEWID

Type: Real

Saved in: Drawing

Description: Sets default trace width

Appendix I: System Variables (Continued)

TRACKPATH

Type: Integer

Saved in: Registry

Description: Controls the display of polar and object snap tracking alignment paths

0 Displays full screen object snap tracking path

1 Displays object snap tracking path only between the alignment point and From point to cursor location

2 Does not display polar tracking path

3 Does not display polar or object snap tracking paths

TREEDEPTH

Type: Integer

Saved in: Drawing

Description: Specifies maximum depth (the number of times the tree-structured spatial index may divide into branches)

TREEMAX

Type: Integer

Saved in: Registry

Description: Limits memory consumption during drawing regeneration by limiting the maximum number of nodes in the spatial index (oct-tree)

TRIMMODE

Type: Integer

Saved in: (Not saved)

Description: Controls whether AutoCAD trims selected edges for chamfers and fillets

0 Leaves selected edges intact

1 Trims selected edges to the endpoints of the chamfer lines and fillet arcs

TSPACEFAC

Type: Real

Saved in: (Not saved)

Description: Controls the multiline text line spacing distance measured as a factor of text height. Valid values are .25 to 4.0

TSPACETYPE

Type: Integer

Saved in: (Not saved)

Description: Controls the type of line spacing used in multiline text; At Least adjusts line spacing based on tallest characters in a line; Exactly uses the specified line spacing, regardless of individual character sizes

1 At Least

2 Exactly

TSTACKALIGN

Type: Integer

Saved in: Drawing

Description: Controls the vertical alignment of stacked text

0 Bottom aligned

1 Center aligned

2 Top aligned

TSTACKSIZE

Type: Integer

Saved in: Drawing

Description: Controls the percentage of stacked text fraction height relative to selected text's current height; valid values are from 1 to 127

Appendix I: System Variables (Continued)

UCSAXISANG

Type: Integer

Saved in: Registry

Description: Stores the default angle when rotating the UCS around one of its axes using the X, Y, or Z option of the UCS command; its value must be entered as an angle in degrees (valid values are: 5, 10, 15, 18, 22.5, 30, 45, 90, 180)

UCSBASE

Type: String

Saved in: Drawing

Description: Stores the name of the UCS that defines the origin and orientation of orthographic UCS settings

UCSFOLLOW

Type: Integer

Saved in: Drawing

Description: Generates a plan view when you change from one UCS to another; can be set separately for each viewpoint

0   UCS does not affect the view

1   Any UCS change causes a change to plan view of the new UCS in the current viewport

UCSICON

Type: Integer

Saved in: Drawing

Description: Displays the coordinate system icon using bit-code for the current viewport; the sum of the following:

0   Off (disabled)

1   On; icon display is enabled

2   Origin; if icon display is enabled, the icon floats to the UCS origin if possible

UCSNAME

Type: String

Saved in: Drawing

Description: (Read-only) Stores the name of the current coordinate system for the current space

UCSORG

Type: 3D point

Saved in: Drawing

Description: (Read-only) Stores the origin point of the current coordinate system for the current space (in world coordinates)

UCSORTHO

Type: Integer

Saved in: Registry

Description: Determines whether the related orthographic UCS setting is restored automatically when an orthographic view is restored

0   Specifies that the UCS setting remains unchanged when an orthographic view is restored

1   Specifies that the related orthographic UCS setting is restored automatically when an orthographic view is restored

UCSVIEW

Type: Integer

Saved in: Registry

Description: Determines whether the current UCS is saved with a named view

0   Does not save current UCS with a named view

1   Saves current UCS whenever a named view is created

Appendix I: System Variables (Continued)

UCSVP

Type: Integer

Saved in: Drawing

Description: Determines whether the UCS in active viewports remains fixed or changes to reflect the UCS of the currently active viewport

0   Unlocked; UCS reflects the UCS of the current viewport

1   Locked; UCS is stored in the viewport and is independent of the UCS of the current viewport

UCSXDIR

Type: 3D point

Saved in: Drawing

Description: (Read-only) Stores the X direction of the current UCS for the current space

UCSYDIR

Type: 3D point

Saved in: Drawing

Description: (Read-only) Stores the Y direction of the current UCS for the current space

UNDOCTL

Type: Integer

Saved in: (Not saved)

Description: Stores a bit-code indicating the state of the UNDO feature; the sum of the following values

0   UNDO is disabled

1   UNDO is enabled

2   Only one command can be undone

4   Auto-group mode is enabled

8   A group is currently active

UNDOMARKS

Type: Integer

Saved in: (Not saved)

Description: (Read-only) Stores the number of marks placed in the UNDO control stream by the Mark option

UNITMODE

Type: Integer

Saved in: Drawing

Description: Controls units display format

0   Displays as previously set

1   Displays in input format

USERI1-5

Type: Integer

Saved in: Drawing

Description: USERI1, USERI2, USERI3, USERI4, and USERI5 are used to store and retrieve integer values

USERR1-5

Type: Real

Saved in: Drawing

Description: USERR1, USERR2, USERR3, USERR4, and USERR5 are used to store and retrieve real numbers

USERS1-5

Type: String

Saved in: (Not saved)

Description: USERS1, USERS2, USERS3, USERS4, and USERS5 are used to store and retrieve text string data

Appendix I: System Variables (Continued)

VIEWCTR

Type: 3D point

Saved in: Drawing

Description: (Read-only) Stores the center of view in the current viewport, expressed in UCS coordinates

VIEWDIR

Type: 3D vector

Saved in: Drawing

Description: (Read-only) Stores viewing direction in the current viewport, expressed in UCS coordinates

VIEWMODE

Type: Integer

Saved in: Drawing

Description: Controls viewing mode for the current viewport using bit-code; sum of the following values:

0   Disabled

1   Perspective view active

2   Front clipping on

4   Back clipping on

8   UCS follow mode on

16 Front clip not at eye (if on, FRONTZ determines clipping plane; if off, front clipping plane is set to pass through camera point)

VIEWSIZE

Type: Real

Saved in: Drawing

Description: Stores height of view in current viewport, expressed in drawing units

VIEWTWIST

Type: Real

Saved in: Drawing

Description: Stores view twist angle for the current viewport

VISRETAIN

Type: Integer

Saved in: Drawing

Description: Controls visibility of layers in xref files

0   Xref layer definition in the current drawing takes precedence over these settings: On/Off, Freeze/Thaw, color, and linetype for xref-dependent layers

1   The above settings for xref-dependent layers take precedence over the xref layer definition in the current drawing

VSMAX

Type: 3D point

Saved in: Drawing

Description: (Read-only) Stores the upper right corner of the current viewport's virtual screen, expressed in UCS coordinates

VSMIN

Type: 3D point

Saved in: Drawing

Description: (Read-only) Stores the lower left corner of the current viewport virtual screen, expressed in UCS coordinates

Appendix I: System Variables (Continued)

WHIPARC

Type: Integer

Saved in: Registry

Description: Controls whether the display of circles and arcs is smooth

0   Circles and arcs are not smooth, but rather are displayed as a series of vectors.

1   Circles and arcs are smooth, displayed as true circles and arcs.

WMFBKGND

Type: Integer

Saved in: (Not saved)

Description: Controls whether the background display of AutoCAD objects is transparent in other applications when these objects are:

- Output to a Windows metafile using the WMFOUT command

- Copied to the Clipboard in AutoCAD and pasted as a Windows metafile

- Dragged and dropped from AutoCAD as a Windows metafile

The AutoCAD defined values are:

0   The background is transparent

1   The background color is the same as the AutoCAD current background color

WORLDUCS

Type: Integer

Saved in: (Not saved)

Description: (Read-only) Indicates whether the UCS is the same as the world coordinate system

0   Current UCS is different from WCS

1   Current UCS is the same as WCS

WORLDVIEW

Type: Integer

Saved in: Drawing

Description: Controls whether UCS changes to WCS during DVIEW or VPOINT

0   Current UCS remains unchanged

1   Current UCS is changed to the WCS for the duration of the DVIEW or VPOINT command; DVIEW and VPOINT command input is relative to the current UCS

WRITESTAT

Type: Read only

Saved in: (Not saved)

Description: Indicates whether a drawing file is read-only or can be written to, for developers who need to determine write status through AutoLISP

0   Cannot write to the drawing

1   Can write to the drawing

Appendix I: System Variables (Continued)

XCLIPFRAME

Type: Integer

Saved in: Drawing

Description: Controls visibility of xref clipping boundaries

0   Clipping boundary is not visible

1   Clipping boundary is visible

XEDIT

Type: Integer

Saved in: Drawing

Description: Controls whether the current drawing can be edited in-place when being referenced by another drawing

0   Cannot use in-place reference editing

1   Can use in-place reference editing

XFADECTL

Type: Integer

Saved in: Registry

Description: Controls the fading intensity for references being edited in-place

0   Zero percent fading, minimum value

90 Ninety percent fading, maximum value

XLOADCTL

Type: Integer

Saved in: Registry

Description: Turns demand loading on and off and controls whether it loads the original drawing or a copy

0   Turns off demand loading; entire drawing is loaded

1   Turns on demand loading, reference file is kept open

2   Turns on demand loading; a copy of the reference file is loaded

XLOADPATH

Type: String

Saved in: Registry

Description: Creates a path for storing temporary copies of demand-loaded xref files

XREFCTL

Type: Integer

Saved in: Registry

Description: Controls whether AutoCAD writes XLG files (external reference log files)

0   Does not write XLG files

1   Writes XLG files

ZOOMFACTOR

Type: Integer

Saved in: Registry

Description: Controls the incremental change in zoom with each IntelliMouse wheel action, whether forward or backward. Accepts an integer between 3 and 100 as a valid value. The higher the number, the more incremental the change applied by each mouse-wheel's forward/backward movement

Appendix J: Glossary of Terms

A

absolute points Specific coordinate points entered directly (without using relative positioning).

acad.dwt AutoCAD's default drawing template file. Unless you specify a unique template file, AutoCAD uses all of the settings and values stored in acad.dwt when you begin a new drawing. Any drawing file can be saved as a template file.

ACIS 3D geometry engine that AutoCAD uses to create solid models. CAD software products that use ACIS can output ASCII files known as SAT files, which are readable directly by most other ACIS-based products. ACIS was developed and made commercially available by Spatial Technology, Inc. (Boulder, CO).

ActiveX Programming interface for AutoCAD. The Technology, developed by Microsoft, is based on the component object model (COM) architecture.

alignment grid A non-printing grid that can be set at any desired spacing in Standard or Isometric drawing mode.

alignment paths Temporary, non-printing lines that appear at predefined angles to help you create objects more accurately.

alternate units A second set of distances shown inside brackets on some dimensioned drawings.

ambient In rendering, the general light surrounding an object or scene.

ANSI dimensioning Standardized format for dimensioning and tolerancing of engineering drawings established by the American National Standards Institute (ANSI).

aperture box A small box that defines an area around the center of the crosshairs within which an object or point will be selected when you press the pick button.

apex The top point of a cone or pyramid.

API Applications Programming Interface.

arcball An object that displays when 3D Orbit is active and allows the user to manipulate the drawing view in realtime.

area The number of square units needed to cover an enclosed two-dimensional shape or surface.

array An orderly grouping or arrangement of objects.

ASCII American Standard Code for Information Interchange (ASCII) consists of standard text, numbers, and special characters produced by a computer keyboard.

aspect ratio The ratio of height to width of a rectangular object or region.

associative dimensions Dimensions that update automatically when you change the drawing by stretching, scaling, etc.

associative hatch A hatch object that updates automatically when its boundaries are changed; a hatch created with the BHATCH command.

attribute extraction Gathering information stored in attributes and placing it into an electronic file that can be read by a computer program.

attributes Text information stored in blocks.

attribute tag A variable that identifies each occurrence of an attribute in a drawing.

attribute values The information stored by attribute tags in specific instances of an attribute.

Appendix J: Glossary of Terms (Continued)

audits In AutoCAD, to examine the validity of a drawing file.

AutoLISP AutoCAD's version of the LISP programming language.

auxiliary view A view that is projected onto a plane perpendicular to one of the orthographic views and inclined in the adjacent view.

B

base drawing A drawing into which external references have been inserted.

baseline dimensions A set of dimensions that have a common baseline to reduce measurement error that can accumulate from "stacking" dimensions.

base menu A menu that is loaded when you first start AutoCAD, or when you use the MENU command.

basic dimension A dimension to which allowances and tolerances are added to obtain limits.

Bezier curve Smooth curve used for 3D surface models of free-form shapes. A Bezier curve is made up of four control points that influence the shape of the curve. The curve does not necessarily pass through the control points.

binary Standard two-digit numerical system in which 0 and 1 are the only digits; forms the basis for all arithmetic calculations in computers.

binary file A file that uses the binary numbering system. (See also *binary.*)

bind Convert an external reference to a permanent part of a drawing.

bitmap An image in which bits are referenced to pixels

bitmapped image Digital representation of an image in which bits are referenced to pixels. In color graphics, a different value represents each red, green, and blue component of a pixel.

bit plane A portion in computer memory allocated to deliver graphic information.

block A collection of objects that you can group to form a single object.

BMP file Microsoft Windows bitmapped image file format.

Boolean logic (See *Boolean operation.*)

Boolean mathematics (See *Boolean operation.*)

Boolean operation In CAD, combining two solid objects, subtracting one from another, or determining intersecting areas between overlapping solid objects; based on Boolean algebra, a mathematical system designed to analyze symbolic logic.

browsing Looking quickly through thumbnails of a large number of files to find the one you want to open.

B-spline A type of spline. (See *spline.*)

buttons In AutoCAD, buttons are the small areas within toolbars that contain pictures (icons). Pressing a button enters its associated command or function. (See *toolbar.*)

C

CAD Computer-Aided Design or Drafting. Programs used to create and document designs such as manufactured products and commercial buildings. AutoCAD is an example of a CAD program.

cadr An AutoLISP function that returns the second value of a pair, as in the *y* coordinate of a coordinate pair.

Appendix J: Glossary of Terms (Continued)

CAE Computer-Aided Engineering. CAE programs offer capabilities for engineering design and analysis such as determining a design's structural integrity and its capacity to transfer heat.

CAM Computer-Aided Manufacturing. Typically refers to systems that use CAD surface data to drive computer numerical control (CNC) machines such as mills, lathes, and flame cutters to fabricate parts, molds, and dies.

car An AutoLISP function that returns the first value in a sequence; for example, the *x* coordinate of a coordinate pair.

Cartesian coordinate system A coordinate system used in geometry that assigns a specific coordinate pair (*x,y*) or coordinate triplet (*x,y,z*) to points in space so that each point has a unique identifier.

cascading menus Submenus in a pull-down menu that offer further choices; cascading menus appear when you rest the pointer on pull-down menu items that contain a small arrow pointing to the right.

cell An area on a tablet menu that, when picked, enters a command or function in AutoCAD.

center lines Imaginary lines that mark the exact center of an object or feature; shown by a series of alternating long and short line segments.

centroid Center of gravity.

chamfers Beveled edges.

circular array (See *polar array*.)

circumference The distance around a circle.

client document A document into which another document or part of a document is linked using OLE.

clip Change the boundary of an image; also called *cropping*.

clipping plane Plane that slices through one or more 3D objects, causing the part of the object on one side of the plane to be omitted.

comma delimited file (CDF) A file format that allows a user to write attributes to an ASCII text file. Entries in this file format are separated by commas.

command alias Command abbreviation that consists of the first one to three letters of a command name. Entering command aliases at the Command prompt enters the associated commands.

command function An AutoLISP function that executes standard AutoCAD commands from within AutoLISP.

Command prompt A text line in AutoCAD at which you can key in commands and options and view information.

compass When the VPOINT command is active, the compass is a globe representation that allows the user to position the drawing at a specific location relative to the viewing screen.

compiled language program environment Type of computer programming language that requires you to compile the code before users can execute it. Compiling means to convert the programming code into machine language. AutoCAD's ObjectARX and Visual LISP are compiled language programming environments whereas AutoLISP is an interpreted language, which does not require compilation. Each line of code is interpreted directly, one line at a time.

composite region A region composed of two or more region primitives that have been combined.

Appendix J: Glossary of Terms (Continued)

composite solids Solids composed of two or more solid primitives.

concentric Sharing a common center point.

configuration A process that allows AutoCAD to recognize the commands and functions associated with each portion of the digitizing tablet.

construction lines Line objects in AutoCAD that extend infinitely in both directions.

context-sensitive The AutoCAD software "remembers" or stores the last command used and displays it at the top of right-click shortcut menus.

context-sensitive help Help related to the currently entered command; obtained by picking the Help button on the Standard toolbar while a command is active.

control points Points on or near a curve that exert a "pull," influencing its shape.

Coons surface patch Technique for creating complex surfaces that interpolate boundary curves derived from Coons mathematics. AutoCAD's EDGESURF command creates a Coons surface patch.

coordinate pair A combination of one x value and one y value that specifies a unique point in the coordinate system.

coordinates Sets of two (for 2D drawings) or three (for 3D drawings) numbers used to describe specific locations. In 2D, the coordinates consist of an *x,y* coordinate pair. In 3D, they consist of *x,y,z* coordinate triplets.

coordinate system icon An icon, placed by default in the lower left corner of the drawing area, that shows the orientation of the axes in the current coordinate system.

copy and paste A Windows-standard copy and paste feature that allows you to copy objects from one drawing and insert them into another.

crop (See *clip.*)

crosshairs A special cursor that consists of two intersecting lines and a pick box; used to select objects and pick points in the drawing area.

crosshatch (See *hatch.*)

cutting plane The plane through a part from which a sectional view is taken.

D

database A file or group of files that holds a large amount of tabular data.

datum A surface, edge, or point that is assumed to be exact.

datum dimensions (See *ordinate dimensions.*)

defun An AutoLISP function that allows you to define new functions.

delta In mathematics, a Greek symbol used to signify change.

dependent symbols Named objects in an external reference, such as blocks, layers, and text styles.

deviation tolerancing A method of tolerancing that allows the user to set the upper and lower tolerances separately (to different values).

dialog box A window that provides information and permits you to make selections and enter information.

dialog control language file (DCL) File that defines the appearance of dialog boxes in Windows.

diameter The length of a line that extends from one side of the circle to the other and passes through its center.

Appendix J: Glossary of Terms (Continued)

digital Data storage format using a two-digit binary system.

digitize Convert 2D drawings or 3D objects into a digital format.

digitizing tablet An electronic input device that consists of a board or tablet and a pointing device called a *puck* or a *stylus*.

dimensioning A system or method of describing the size and location of features on a drawing.

dimension line The part of a dimension that typically contains arrowheads at its ends and shows the extent of the area being dimensioned.

dimension styles Saved collections of dimension settings that control the format and appearance of dimensions in a drawing.

direct distance method A method of entering points in which the user aims using the pointing device to enter a number to specify the length of a line.

direction vector A 3D line that is used to provide a direction for a path to follow.

docked toolbar A toolbar that is attached to the top or side of the drawing area; toolbar names do not appear when a toolbar is docked.

documentation In AutoLISP, adding remarks to clarify the code and its purpose. Any text that occurs after a semicolon is ignored by AutoLISP and is used for documentation.

donuts Thick-walled or solid circles created using the DONUT command.

double-clicking Positioning the pointer on an object or button and pressing the pick button twice in rapid succession.

drawing area The main portion of the window in AutoCAD; the part of the screen in which the drawing appears. Also refers to the boundaries for constructing the drawing, which should correspond to both the drawing scale and the sheet size.

drawing database Information associated with each AutoCAD file that defines the objects in the drawing, layers, and the current space (model or paper), as well as specific numerical information about individual objects and their locations.

drawing interchange format file (DXF) A *de facto* standard for translating files from one CAD program, such as AutoCAD, to another, such as CADKEY; when you select a drawing interchange format for attribute extraction, AutoCAD creates a variant of this format with a DXX extension.

drawing template file A drawing used as a template for creating new drawings. Template files typically contain frequently used settings and values, such as linetypes and layers, that are inserted into new drawings automatically when the new drawing is created. AutoCAD's default template file is acad.dwt.

E

ellipse A regular oval shape that has two centers of equal radius.

encapsulated PostScript (EPS) A graphic file format that is compatible with many publishing standards and printers.

entity An individual predefined element in AutoCAD; the smallest element that you can add to or erase from a drawing. Also called an *object*.

extension lines The lines in a dimension that extend from the object to the dimension line.

Appendix J: Glossary of Terms (Continued)

external references Ordinary drawing files that have been attached to the current (base) drawing.

extruding Giving thickness to a two-dimensional object so that it becomes three-dimensional.

extrusion thickness Thickness on the Z axis; extrusion thickness can be set in AutoCAD by setting the THICKNESS variable or by applying the EXTRUDE command.

F

facet Three- or four-sided polygon element that represents a piece of a 3D surface. Polygonal mesh models are sometimes referred to as faceted models.

facet shading (See *flat shading*.)

feature control frames The frames used to hold geometric characteristic symbols and their corresponding tolerances.

fields Categories of information in a database.

file attributes The characteristics of a file, such as its size, type, and the date and time it was last modified.

fillets Rounded inside corners.

filtering Including or excluding items from a selection set using specific criteria.

fixed attributes Attributes whose values you define when you first create a block.

fixed screen pointing area The area of a digitizing tablet in which AutoCAD recognizes digitized points.

flat shading Method of rendering a polygonal model that fills each polygon with a single shade of color to give the model a faceted look. Also called *facet shading*.

floating-point format A mathematical format in which the decimal point can be manipulated.

floating toolbar A toolbar that displays as an independent window in the drawing area.

flyout toolbars Secondary toolbars that appear when you pick and hold a button that contains a small triangle in the lower right corner.

font Distinctive set of characters consisting of letters, numbers, punctuation marks, and symbols.

freeze In AutoCAD, to remove a layer from view without turning it off or deleting it.

frustum A truncation of a cone or pyramid that cuts off the apex and presents a flat surface.

full section A section that extends all the way through a part or object.

function In AutoLISP, a computer routine or subroutine that performs a calculation with variables you provide and gives a single result.

G

geometric characteristic symbols Symbols that are used to specify form and position tolerances on drawings.

geometric dimensioning and tolerancing (GD&T) Standardized format for dimensioning and showing variations (called tolerances) in the manufacturing process. Tolerances specify the largest variation allowable for a given dimension. GD&T uses geometric characteristic symbols and feature control frames that follow industry standard practices. Used by many government contractors, GD&T standards are defined in the American National Standards Institute (ANSI)

Appendix J: Glossary of Terms (Continued)

Y14.5M Dimensioning/Tolerancing handbook.

Gouraud shading Method of smooth rendering of polygonal models by interpolating (averaging and blending) adjacent color intensities. The Gouraud method applies light to each vertex of a polygon face and interpolates the results to produce a realistic-looking model.

gradient A gradual transition from one color to another.

grips Small boxes that appear at key points on an object when you select the object without first entering a command. Grips allow you to perform several basic operations such as moving, copying, or changing the shape or size of an object.

group A named set of objects. A circle with a line through it, for instance, could become a group.

GUI Graphical User Interface. Microsoft Windows, and the Macintosh environment, as well as programs specifically designed for them, are GUI examples.

H

hatch A repetitive pattern of lines or symbols that shows a related area of a drawing.

heap An area of memory set aside for storage of all AutoLISP programs and symbols (also called *nodes*). Elaborate AutoLISP programs require greater amounts of heap and stack space.

hidden line removal The deletion of tessellation lines that would not be visible if you were viewing the 3D model as a real object.

hidden lines Lines that would not be visible if you were looking at an object without "see-through" capability; shown as dashed or broken lines.

hypertext link Text that, when selected or picked with the pointing device, takes the user to related information.

I

icon Graphic symbol typically used in GUI software. Examples include the UCS icon and the paper space icon.

image tiles Selectable squares displayed in some dialog boxes that contain thumbnail images from which the user can choose.

Initial Graphics Exchange Specification (IGES) An industry standard format for exchanging CAD data between systems.

inner loops Closed 2D areas contained within another loop that make up part of a composite region.

integer Any positive or negative whole number or 0.

interference The overlap of two objects that occupy the same area in 3D space.

interference solids Solids created in AutoCAD that show the overlapping portions of two solids that interfere in 3D space.

interpolation A method of averaging edges of a rendered object so that they blend and provide a smooth appearance.

intersection A method of creating a solid from the overlapping portion of two intersecting solid objects.

ISO International Standards Organization.

isometric drawing A type of two-dimensional drawing that gives a three-dimensional appearance.

isometric planes Three imaginary drawing planes spaced at 120° from each other; used to draw isometric views.

Appendix J: Glossary of Terms (Continued)

J

justify Align multiple lines of text at a specific point (left, right, center, etc.)

K

key In a database, an identifier for each record; the key consists of one or more fields that uniquely identify the record.

L

landscape A paper orientation in which the wider edge of the paper is at the top and bottom of the sheet.

landscape object An AutoCAD object with a bitmap image mapped onto it.

layers Similar to transparent overlays in manual drawing; allow AutoCAD users to selectively display and hide information to clarify the drawing.

layout An arrangement of one or more views of an object on a single sheet.

leaders Lines with an arrowhead at one end that point out a particular feature in a drawing.

limits (See *drawing area*)

limits tolerancing A method of tolerancing in which only the upper and lower limits of variation for a dimension are shown.

linear dimensions Vertical and horizontal dimensions, as well as straight-line dimensions that are aligned with an edge of the object being dimensioned.

link templates Files that provide the information AutoCAD needs to link objects in a drawing to unique records in a database table.

LISP Stands for "LISt Processing," a programming language well suited to graphics applications.

list An AutoLISP function that allows you to string together more than one value.

loops Closed 2D shapes made up of any combination of lines, polylines, arcs, circles, elliptical arcs, ellipses, splines, 3D faces, traces, and solids.

M

machine-independent software Software that is not dependent on a specific brand or class of machine.

macro In AutoCAD, menu items that perform specific AutoCAD functions.

mass properties Characteristics of an object or material, such as its center of gravity.

material condition symbols Symbols used in geometric dimensioning and tolerancing to modify the geometric tolerance in relation to the produced size or location of the feature.

Maximum Material Condition (MMC) A notation that specifies that a feature such as a hole or shaft is at its maximum size or contains its maximum amount of material.

menu area 1 The top area of AutoCAD's tablet menu, which is reserved for customization and contains a total of 225 cells.

mline (See *multiline.*)

model Representation of a proposed or real object. A "soft" model is a computer mock-up of a design, while a "hard model is a physical object that was

Appendix J: Glossary of Terms (Continued)

fabricated from wood, clay, plastic, metal, or some other material.

model space The space, or drawing mode, in AutoCAD in which most drawing is done.

mtext A multiple-line text object created in AutoCAD using a text editor.

multiline A type of line object in AutoCAD that consists of up to 16 parallel lines created simultaneously and viewed by AutoCAD as a single object.

multiview drawing A drawing that describes a three-dimensional object completely using two or more two-dimensional views.

N

node (See *heap*.)

notes Text that refers to an entire drawing, rather than one specific feature. (See also *specifications*.)

noun/verb selection An object selection method in which you select the object first, then enter a command (verb) to perform an operation on the object.

NURBS Non-uniform rational B-spline. Smooth curve or surface defined by a series of weighted control points. AutoCAD's SPLINE command creates a NURBS curve.

O

object An individual predefined element in AutoCAD; the smallest element that you can add to or erase from a drawing. Also called an *entity*.

ObjectARX Programming environment that includes dynamic link libraries (DLLs) that run in the same address space as AutoCAD. ObjectARX programs operate directly with core AutoCAD data structures and code.

object linking and embedding (OLE) A Windows-standard method of linking information between a client document and a server document.

object snap A magnet-like feature in AutoCAD that allows the user to "snap" to endpoints, midpoints, and other specific points easily and accurately.

object snap tracking A feature that, when used in conjunction with object snap, provides alignment paths that help the user produce objects at precise positions and angles.

offsetting Creating a new object (such as a line or circle) at a specific distance from an existing line or circle.

operating system (OS) Computer software environment normally provided by the computer manufacturer. Microsoft Windows, IBM OS/2, Macintosh System, and Silicon Graphics IRIX (a UNIX variation) are examples of operating systems.

ordinate dimensions Dimensions that use a datum, or reference dimension, to help avoid confusion that could result from "stacking" dimensions; also called *datum dimensioning*.

origin The point at which the axes cross in the world coordinate system.

ortho A mode in AutoCAD that forces all lines to be perfectly vertical or horizontal.

orthogonal Drawn at right angles.

orthographic projection A projection of views at right angles to each other in a multiview drawing.

Appendix J: Glossary of Terms (Continued)

outer loop The outer or overall boundary in a composite region.

P

palette In AutoCAD, a selection of colors, similar to an artist's palette of colors.

panning A feature that allows the user to move the viewing window around the drawing without changing the zoom magnification.

paper space The space, or drawing mode, in AutoCAD from which multiple views can be arranged and plotted on the same sheet.

parallel projection The standard projection mode in AutoCAD, in which objects are shown in parallel regardless of perspective.

parametrics The process of parameterizing, or specifying, shape and size of an object using AutoLISP or another program.

partial menus Menus loaded using the MENULOAD command that work with, not instead of, the current base menu.

path curve A line or polyline about which the REVSURF command creates a revolved surface around a selected axis.

perimeter The distance around a two-dimensional shape or surface.

perpendicular Lines, polylines, or other objects that meet at a precise 90° angle.

perspective projection A type of projection in which objects that are farther away appear smaller, as they would in a photograph.

phantom lines Lines drawn using a thick lineweight, alternating two short dashes with one long dash; used for cutting planes in sectional views.

Phong shading Method of rendering polygonal models; calculates the light at several points across the model's surface, producing more accurate specular highlights than Gouraud shading.

photorealistic Describes a high-quality rendering that resembles a photograph.

pickbox A cursor that consists of a small box that allows you to pick objects on the screen.

pictorial representation A drawing of an object or part that offers a realistic view, compared to orthographic projections.

pixel Picture element; smallest addressable dot on a computer screen or raster device such as an ink jet or laser printer.

plane An imaginary flat surface that can be defined by any three points in space or by a coplanar object.

plan view The top view of an object; the plan view can be generated by entering VPOINT and 0,0,1 at the keyboard.

plot scale The scale at which a drawing is plotted to fit on a drawing sheet.

plot style A collection of property settings saved in a plot style table.

plotter Originally, an output device that could handle very large sheets of paper; now used interchangeably with *printer*.

point filtering The process of using the pointing device to enter any two of the 3D coordinates that define a point and then entering the third coordinate from the keyboard.

polar array An array in which objects are arranged radially around a center point.

polar method A relative method of entering points in which the user can produce lines at specific angles around a central point.

Appendix J: Glossary of Terms (Continued)

polar tracking method A method of entering points in which AutoCAD displays alignment paths at prespecified angles to help the user place the points at precise positions and angles.

polyline A type of line object in AutoCAD that consists of a connected sequence of line and arc segments and is treated by AutoCAD as a single object.

portrait A paper orientation in which the narrower edge of the paper is at the top and bottom of the sheet.

primary units The units that appear by default when you add dimensions to a drawing in AutoCAD.

primitives Predefined 3D shapes in AutoCAD that can be used as building blocks to create more complex shapes.

printer Originally, a small, tabletop output device; now used interchangeably with *plotter*.

profile An outline or contour, such as a side view, of an object. (See also *path curve*.)

projected tolerance zone Controls the height of the extended portion of a perpendicular part.

prototype tooling Molds used to produce models and prototype parts; sometimes referred to as soft tooling.

puck A pointing device used with a digitizing tablet; a puck usually resembles a mouse, but it has more buttons and features.

pull-down menus Menus whose names appear across the top of the screen in AutoCAD, from which you can perform many drawing, editing, and file manipulation tasks.

purge In AutoCAD, selectively deleting blocks or other elements that are not currently in use in a drawing.

Q

quadrant A quarter of a circle, defined by the points at exactly 0°, 90°, 180°, and 270°.

quadrant points The points at 0°, 90°, 180°, and 270° on a circle.

queries In a database, queries are sets of criteria that allow you to view only those records that meet those specific criteria.

R

radial Describes a feature on which every point is the same distance from an imaginary center point.

radius The length of a line extending from the center to one side of a circle.

rapid prototyping (RP) Refers to a class of machines used for producing physical models and prototype parts from 3D computer model data such as CAD. Unlike CNC machines, which subtract material, RP processes join together liquid, powder, and sheet materials to form parts. Layer by layer, these machines fabricate plastic, wood, ceramic, and metal objects from thin horizontal cross-sections taken from the 3D computer model. Stereolithography and Fused Deposition Modeling are examples of RP processes.

raster images Images in which all lines and curves are represented using a series of dots or pixels. Computer screens and most printers (ink jet and laser) use raster techniques.

raster printer Printer that defines lines, curves, and solid objects using small dots. See *raster*.

ray A type of line object in AutoCAD that extends infinitely in one direction from a specified point.

Appendix J: Glossary of Terms (Continued)

raytracing A calculation-intense rendering process in which lines are drawn from the eye to every point on the screen.

realtime zooming A method of changing the level of magnification by picking the Realtime Zoom button and moving the cursor.

records Groups of related fields in a database.

rectangular array An array in which the objects are arranged in rows and columns.

region A closed 2D area defined using AutoCAD's REGION command.

region primitive A basic 2D region created using AutoCAD's REGION command.

regular polygon A polygon in which all the sides are of equal length.

relative method A method of entering points that allows the user to base the new points on the position of a point that has already been defined.

rendering A model that has been given an appearance that is more realistic than that of basic shading.

resolution The number of pixels per inch on a display system; this determines the amount of detail you can see on the screen.

revolve To turn around an axis.

revolved surface A surface created using the REVSURF command to revolve a path curve around an axis.

roll In 3D Orbit, moving the view around an axis at the center of the arcball.

rotating Changing the angle of an object in the drawing without changing its scale.

rounds Curved or rounded outside corners on an object.

routines In AutoLISP, groups of functions.

rubber-band effect The stretching effect when a drawing command is active that lets you see the line, circle, etc., "grow" as you move the cursor.

ruled surface Surface created by linear interpolation between two curves. AutoCAD's RULESURF command creates a ruled surface.

running object snap modes/running object snaps Object snaps that have been preset to run automatically.

S

SAT files Machine-independent geometry files that contain topological and surface information about models created using the ACIS engine. AutoCAD can import and export SAT files using the ACISIN and ACISOUT commands, respectively.

scale Increasing or decreasing the overall size of an object or assembly without changing the proportions of the parts to each other.

scene In AutoCAD, a combination of a named view and one or more defined lights.

screen regeneration Recalculation of each vector, or line segment, in a drawing.

script file An ASCII text file with an SCR file extension that runs a script of AutoCAD commands.

section drawings Drawings used to show the interior detail of a part.

selectable In AutoCAD groups, *selectable* means that when you select one of the group's members (*i.e.*, one of the circles), AutoCAD selects all members in the group.

Appendix J: Glossary of Terms (Continued)

selection set All the objects that are currently highlighted, or selected, for a command or operation in AutoCAD.

server document A document that is linked to another document using OLE.

setq An AutoLISP function that assigns a value to a variable.

shading Production of a shaded image using a basic method that processes more quickly than rendering, especially when the model is large and complex.

shelling A method in AutoCAD that allows the user to remove material from an object to create a shell or hollow object.

shortcut menu A context-sensitive menu that appears when you right-click; shortcut menus contain several of the commands and functions the user is most likely to need in the current drawing situation.

slide An image of a drawing that can be displayed and controlled within AutoCAD using a script file.

slide library A single file in which you can store slides, similar to filling a carousel tray of 35-mm slides. The individual files from which the slides were created need not be present to use a slide library.

slide show A script file that controls a continuous sequence of AutoCAD slides in a sort of electronic flipchart.

smooth shading (See *Gouraud shading*.)

solid models Three-dimensional models defined using solid modeling techniques, which is somewhat similar to using physical materials such as wood or clay to produce shapes. CAD solid modeling programs that take advantage of Constructive Solid Geometry (CSG) techniques use primitives, such as cylinders, cones, and spheres, and

Boolean operations to construct shapes. CAD programs that use the Boundary Representation (B-Rep) solid modeling technique store geometry directly using a mathematical representation of each surface boundary. Some CAD programs use both CSG and B-Rep modeling techniques.

solid primitives Basic solid shapes, such as cylinders or spheres, usually used to produce more complex shapes.

sorting Rearranging the order of records in a database.

space delimited file (SDF) A file format that allows the user to write attributes to an ASCII text file. Entries in this file format are separated by spaces.

special characters Characters that are not normally available on the keyboard or in a basic text font.

specifications Text that provides information about sizes, shapes, and surface finishes that apply to specific portions of an object or part.

spline leader Similar to a regular leader, except that the leader line consists of a spline curve, offering more flexibility with its shape.

splines Curves that use sampling points to approximate mathematical functions that define a complex curve. NURBS is one example of a spline curve.

stack An area of memory set aside by AutoLISP. Stack holds programming arguments and partial results; the deeper you "nest" these items, the more stack space is used.

status bar The line at the bottom of the screen that displays various types of status information about modes and settings, including snap, grid, the ortho mode, object snap, object tracking,

Appendix J: Glossary of Terms (Continued)

lineweight, and the current space (model or layout).

STEP Standard for the Exchange of Product model data. STEP is an internationally accepted method of transferring CAD data from one system to another. Many believe that STEP will someday replace IGES.

stereolithography Rapid prototyping (RP) process involving the solidification of ultraviolet light-sensitive liquid resin using a laser to harden and produce a plastic part of a 3D computer model.

STL File format for converting CAD models to physical parts using rapid prototyping (RP) such as stereolithography. 3D Systems, Inc. (Valencia, CA) developed and published the STL format, which is available in binary and ASCII forms.

surface A mesh of polygons, such as triangles, or a mathematical description using spline information.

surface controls Feature control frames, not associated with a specific dimension, that refer to a surface specification regardless of feature size.

surface of revolution Surface created by rotating a line or curve around an axis. AutoCAD's REVSURF command creates a surface of revolution.

surface mesh A three-dimensional grid that is defined in terms of $M \times N$ vertices.

surface model Three-dimensional (3D) model defined by surfaces. The surface consists of a mesh of polygons, such as triangles, or a mathematical description such as a Bezier B-spline surface or non-uniform rational B-spline (NURBS) surface.

symbol library Drawing file that contains a series of blocks for use in other drawings.

symmetrical deviation (See *symmetrical tolerancing.*)

symmetrical tolerancing A method of tolerancing in which the upper and lower limits of a plus/minus tolerance are equal.

system registry A group of files on a computer that stores important information about the operating system, hardware and software installations, user preferences, application settings, and other types of information that makes a computer run smoothly.

system variable Similar to a command, except that it holds temporary settings and values instead of performing a specific function.

T

tables Collections of records in a database.

tablet menu overlay A drawing or other template that an AutoCAD user can create and place on a digitizing tablet to match custom tablet menus.

tablet menus Menus created for use with a digitizing tablet.

tablet mode The mode in which AutoCAD recognizes input from a tablet menu.

tabulated surface Type of ruled surface in which AutoCAD calculates, or tabulates, a surface based on a defined path curve and a direction vector. AutoCAD's TABSURF command creates a tabulated surface.

tangent Lines, circles, or other objects that meet at a single point.

tangential edges Transition lines that occur in a solid model when a curved face meets a flat face.

Appendix J: Glossary of Terms (Continued)

template extraction file A file that defines text format so that attributes can be displayed using CDF, SDF, or DXX files.

template file A file that contains drawing settings that can be imported into new drawing files.

tessellation Lines that describe a curved surface or 3D model.

text editor A word processor that allows the user to insert mtext objects in AutoCAD. Notepad and WordPad are two examples.

third-angle projection A projection system used for multiview drawings in which the top view of an object is placed above the front view. This is the normal way of presenting views in the United States.

3D Three-dimensional.

3D face A planar object created using AutoCAD's 3DFACE command.

3D Studio A 3D modeling and animation program produced by an Autodesk company named Kinetix.

thumbnails Small representations of drawing files

TIF file Tagged image file format (often called TIFF).

title block A portion of a drawing that is set aside to give important information about the drawing, the drafter, the company, and so on.

tolerance A designation that specifies the largest variation allowable for a given dimension.

toolbar A strip of related buttons that can be displayed or hidden by the user.

tooltips One or more words that appear when you position the pointer over a button for a second or longer to help you understand the purpose or function of the button.

torus An inner-tube shaped primitive supplied with AutoCAD.

traces Thick lines commonly used on printed circuit boards; also, lines drawn using AutoCAD's TRACE command.

transparent zooms Zooms that the user forces to occur while another command is in progress.

2D Two-dimensional.

two-dimensional (2D) representation A single profile view of an object seen typically from the top, front, or side.

U

user coordinate systems (UCSs) Custom coordinate systems that a user can define for use in creating 3D drawings.

V

variable attributes Attributes whose value you can change as you insert the block.

vector Line segment defined by its endpoints or by a starting point and direction in 3D space.

vector-based object An object that is stored as one or more line segments that are described by the position of their endpoints.

verb/noun selection The traditional method of entering a command in AutoCAD, in which the user enters a command and then selects the objects on which the command should operate.

vertex Any endpoint of an individual line or arc segment in a polyline, polygon, or other segmented object.

Appendix J: Glossary of Terms (Continued)

view box The window that appears in Aerial View to define the current screen magnification and viewing window.

viewports Portions of the drawing area that a user can define to show a specific view of a drawing.

view resolution The accuracy with which curved lines appear on the screen in AutoCAD; inversely related to regeneration speed.

Visual BASIC Microsoft Visual Basic for Applications (VBA) is an object-based programming environment that has been implemented in AutoCAD. VBA first appeared in Microsoft Excel and Microsoft Project in 1994. VBA 5.0 is a core component of Microsoft Office 97 and is integrated into Microsoft Word and Microsoft PowerPoint.

Visual LISP Enhanced version of AutoLISP, which offers a development environment that includes a compiler, debugger, and other tools. Visual LISP extends the language to interact with objects using ActiveX.

W

wireframe A "stick" representation of a 3D model whose shape is defined by a series of lines and curves.

wizard A series of dialog boxes that steps you through a sequence.

world coordinate system AutoCAD's name for the Cartesian coordinate system. (See *Cartesian coordinate system*.)

X

xline A type of construction line created by the XLINE command that extends infinitely in both directions

xrefs (See *external reference*.)

X/Y/Z filtering The process of using the pointing device to enter any two of the 3D coordinates that define a point and then entering the third coordinate from the keyboard; also called *point filtering*.

XYZ octant The area in 3D space where the x, y, and z coordinates are greater than 0.

Z

Z axis The third axis in a three-dimensional coordinate system, on which depth is measured.

Index